TEUBNER-TEXTE zur Mathematik
Band 140

Oliver Caps

Evolution Equations
in Scales of Banach Spaces

TEUBNER-TEXTE zur Mathematik

Herausgegeben von:

Prof. Dr. Jochen Brüning, Berlin
Prof. Dr. Herbert Gajewski, Berlin
Prof. Dr. Herbert Kurke, Berlin
Prof. Dr. Hans Triebel, Jena

Die Reihe soll ein Forum für Beiträge zu aktuellen Problemstellungen der Mathematik sein. Besonderes Anliegen ist die Veröffentlichung von Darstellungen unterschiedlicher methodischer Ansätze, die das Wechselspiel zwischen Theorie und Anwendungen sowie zwischen Lehre und Forschung reflektieren. Thematische Schwerpunkte sind Analysis, Geometrie und Algebra.

In den Texten sollen sich sowohl Lebendigkeit und Originalität von Spezialvorlesungen und Seminaren als auch Diskussionsergebnisse aus Arbeitsgruppen widerspiegeln.

TEUBNER-TEXTE erscheinen in deutscher oder englischer Sprache.

Oliver Caps

Evolution Equations in Scales of Banach Spaces

Teubner

B. G. Teubner Stuttgart · Leipzig · Wiesbaden

Die Deutsche Bibliothek – CIP-Einheitsaufnahme
Ein Titeldatensatz für diese Publikation ist bei
der Deutschen Bibliothek erhältlich.

Dr. rer. nat. Oliver Caps

Born in 1972 in Mainz (Germany). From 1991 to 1996 he studied mathematics, physics and computer sciences at the Johannes Gutenberg-Universität Mainz. After a stay at the Università degli Studi of Bologna (Italy) in spring 1997 he returned to the Johannes Gutenberg-Universität Mainz and received his Dr. rer. nat. in 2000.

From summer 1997 to summer 2001 he held a position as scientific assistant at the University of Mainz. Currently he is working on financial mathematics at a financial institute in Germany.

1. Auflage Juli 2002

Der Verlag Teubner ist ein Unternehmen der Fachverlagsgruppe BertelsmannSpringer.
www.teubner.de

Umschlaggestaltung: Ulrike Weigel, www.CorporateDesignGroup.de

Gedruckt auf säurefreiem und chlorfrei gebleichtem Papier.

ISBN-13:978-3-519-00376-2   e-ISBN-13:978-3-322-80039-8
DOI: 10.1007/978-3-322-80039-8

Preface

The book provides a new functional-analytic approach to evolution equations by considering the abstract Cauchy problem in a scale of Banach spaces. The usual functional analytic methods for studying evolution equations are formulated within the setting of unbounded, closed operators in one Banach space. This setting is not adapted very well to the study of many pseudodifferential and differential equations because these operators are naturally not given as closed, unbounded operators in one Banach space but as continuous operators in a scale of function spaces. Thus, applications within the setting of unbounded, closed operators require a considerable amount of additional work because one has to construct suitable closed realizations of these operators. This choice of closed realizations is technically complicated even for simple applications.

The main feature of the new functional analytic approach of the book is to study the operators in scales of Banach spaces that are constructed by simple reference operators. This is a natural setting for many operators acting in scales of function spaces. The operators are only expected to respect the scale and to satisfy certain inequalities but we can avoid completely the choice of any closed realization of these operators which is of great importance in applications. We use the mapping properties of the reference operators to prove sufficient conditions for well-posedness of linear and quasilinear Cauchy problems. In the linear, time-dependent case these conditions are shown to characterize well-posedness. A similar result in the standard setting (i.e., a time-dependent generalization of the Hille-Yosida/Lumer-Phillips theorem) is still an open problem.

The generality of the new functional analytic approach of the book is demonstrated by many applications to several mathematical and physical fields. One of the most important applications is a simultaneous treatment of some parabolic and hyperbolic equations. In the standard approach this is not possible, and hyperbolic and parabolic equations have to be treated by different and incompatible methods. In particular, the approach of the book can be used for applications to (strongly) degenerate parabolic equations appearing in connection with some physical and probabilistic problems. Classical results on hyperbolic and parabolic equations are special cases of these results. A further important example of equations of that type is the parabolic Navier-Stokes equations, which degenerates hyperbolically to the Euler equation. Hence, in contrast to the standard approach, with the new methods of the book we obtain results on Navier-Stokes equations degenerating to Euler equations in some parts of the space. Further results of this book include conditions on symbols for essential selfadjointness of pseudodifferential operators and well-posedness of Schrödinger equations, linear and quasilinear evolution equations in L^q-Sobolev spaces, and spaces of continuously differentiable functions, degenerate-elliptic boundary value problems, evolutions equations on networks, and a unified approach to both types of Kadomtsev-Petviashvili equations with periodic boundary conditions.

The book contains 5 chapters. Chapter 1 provides some functional analytic methods. The abstract theory of linear evolution equations in scales of Banach spaces is developed in chapter 2 and the abstract theory of quasilinear equations in chapter 3. Applications of the abstract methods to linear equations are given in chapter 4 and to quasilinear equations in chapter 5. The abstract part of the book, i.e. chapter 1-3, is kept completely self-contained. Assuming only basic knowledge on functional analysis of bounded, linear operators in Banach spaces, all functional analytic methods needed for further reading are proved in chapter 1. Readers experienced in functional analysis may skip chapter 1, start reading directly chapter 2, and go back to chapter 1 only occasionally. Whereas all auxiliary results in the abstract part are proved in detail in the book, we cannot continue this way of presentation in chapter 4 and 5 because detailed proofs of analytic results and methods necessary for applications would exceed the limit of this book. Therefore, results of this type are formulated in a self-contained way and for proofs we give references to standard monographs treating these topics.

There are several people to thank. First, I wish to thank B. Gramsch for suggesting to study evolution equations in scales of Banach spaces, many helpful discussions, and valuable support.

I am grateful to G. Schleinkofer for proposing to apply the abstract results to Kadomtsev-Petviashvili equations, W. Arendt, M. Hieber, and J. Voigt for some references, and F. Ali Mehmeti for several remarks. Moreover, I profited from numerous discussions with F. Baldus, R. Lauter, and J. Lutgen.

Finally, I wish to thank J. Weiß from Teubner Verlag for strongly supporting the publication of the present manuscript.

Mainz, April 2002   *Oliver Caps*

Contents

List of Symbols

General Notation

$\hookrightarrow$	continuous embedding	
a.e.	almost everywhere	
$\mathrm{signum}(a)$	sign of a	
linh	linear hull	
$[s] := \min\{l \in \mathbb{Z} : l \geq s\}$		p. 105
$I := [t_0, t_1]$		
$\triangle := \{(t,s) \in I^2 : t \geq s\}$		
Σ_θ		p. 45
$\gamma_{R,\varphi}$		p. 45
$\{e_1, \ldots, e_n\}$	canonical base of $\mathbb{R}^n$	
$\oplus$	direct sum	
$x \vee y := \sup(x,y)$		p. 212
$x \wedge y := \inf(x,y)$		p. 212
$x^+ := x \vee 0$		p. 212
$x^- := (-x) \vee 0$		p. 212
$|x| := x^+ + x^- = x \vee (-x)$		p. 212
$\Sigma := \{z \in \mathbb{C} : 0 < \mathrm{Re}\, z < 1\}$		p. 59
$\overline{\Sigma} := \{z \in \mathbb{C} : 0 \leq \mathrm{Re}\, z \leq 1\}$		p. 59
$X_0 + X_1$		p. 60
$\mathcal{F}(X_0, X_1)$		p. 60

$\mathcal{F}_-(X_0, X_1)$		p. 60
$(X_0, X_1)_{\theta,q}$		p. 63
$[X_0, X_1]_\theta$		p. 61
dist		p. 231
$\mathrm{supp}(f)$	support of f	
$\langle x \rangle$		p. 167
∂^α		p. 167
$(f * g)(x) := \int_{\mathbb{R}^n} f(y) g(x-y) \, dx$		
	convolution	
$\widehat{u}(\xi) := \mathcal{F}[u](\xi) = \mathcal{F}_{x \to \xi}[u(x)](\xi)$		
		p. 167
$\check{u}(\xi) := \mathcal{F}^{-1}[u](\xi)$		p. 167
$d\xi := \frac{d\xi}{2\pi}$		p. 167
div_x divergence with respect to x		
grad_x gradient with respect to x		
$\triangle_x$ Laplacian with respect to x		
$e_{\mu,\nu}(x,y) := \frac{1}{2\pi} e^{i\mu x} e^{i\nu y}$		p. 278
$\widehat{u}(\mu,\nu) := \langle u, e_{\mu,nu} \rangle_{L^2(T^2)}$		p. 278
$\frac{\partial}{\partial \nu}$		p. 235
$\mathcal{A}^{-1}$		p. 96

Spaces

$\mathbb{N}$	natural numbers	
$\mathbb{N}_0 := \mathbb{N} \cup \{0\}$		
$\mathbb{Z}$	integers	
$\mathbb{R}$	real numbers	

$\mathbb{C}$	complex numbers	
$M_m(\mathbb{R})$	$m \times m$ matrices over $\mathbb{R}$	
$X_\mathbb{C} := X \oplus iX$		p. 84
$\mathscr{L}(Y, X)$		p. 27

$\mathscr{L}(X) := \mathscr{L}(X, X)$ — p. 27
$Y \hookrightarrow X$ — p. 27
X^* — p. 27
$\langle \cdot, \cdot \rangle_{X^*, X}$ — p. 27
$(X^k)_k$ — p. 79

X^∞ — p. 79
$(X_Z^k, \|\cdot\|_k)$ — p. 89
E — p. 167
Ψ_ε^k — p. 107
Ψ_ε^∞ — p. 108

Operators

$A \subset B$ — p. 27
$\overline{A}$ — p. 27
$\mathcal{R}(A)$ — p. 27
$\mathcal{N}(A)$ — p. 27
$\rho(A)$ — p. 27
$\sigma(A)$ — p. 27
A^* — p. 27
$\operatorname{Re} A = \frac{1}{2}(A + A^*)$ — p. 186
$\operatorname{Im} A := \frac{1}{2i}(A - A^*)$ — p. 186
$A \leq c\mathrm{Id}$ — p. 186
$e^{tA} = \exp(tA)$ — p. 28

$A^{\complement}$ — p. 84
Z^α — p. 93
Λ^z — p. 55
$[P, Q] := PQ - QP$ — p. 168
$Z = \{Z_1, \dots, Z_M\}$ — p. 89
Z^α — p. 93
J_l — p. 94
$\operatorname{ad}^j(\Lambda)(A)$ — p. 103
δ_Z — p. 99
P_σ — p. 272

Function spaces

$\mathcal{C}(\Omega, E)$ — p. 167
$\mathcal{C}(\Omega) := \mathcal{C}(\Omega, \mathbb{C})$
$\mathcal{C}^k(\Omega)$ — p. 167
$\mathcal{C}_c^k(\Omega) := \{f \in \mathcal{C}^k(\Omega) : \operatorname{supp}(f)$ compact$\}$
$\mathcal{C}_0^k(\mathbb{R}^n, \mathbb{R})$ — p. 176
$\mathcal{C}^k(\overline{\Omega})$ — p. 227
$\mathcal{C}_w(I, X)$ — p. 148
$\mathcal{D}(\Omega)$ — p. 227
$\mathcal{D}'(\Omega)$ — p. 227
$(\cdot, \cdot)_{\mathcal{D}'(\Omega), \mathcal{D}(\Omega)}$ — p. 227
$\mathscr{S}(\mathbb{R}^n, E)$ — p. 167
$\mathscr{S}'(\mathbb{R}^n, E)$ — p. 167
$\mathscr{S}(\mathbb{R}^n) := \mathscr{S}(\mathbb{R}^n, \mathbb{C})$ — p. 167
$\mathscr{S}'(\mathbb{R}^n) := \mathscr{S}'(\mathbb{R}^n, \mathbb{C})$ — p. 167

$\mathcal{B}^k(\mathbb{R}^n)$ — p. 176
$\mathcal{B}^\infty(\mathbb{R}^n)$ — p. 225
$\mathcal{B}^{\infty,1}(\mathbb{R}^n)$ — p. 274
$\mathcal{B}^k(\mathbb{R}^n)$ — p. 176
$(L^p(\Omega, E), \|\cdot\|_{L^p(\Omega, E)})$ — p. 167
$L^p(\Omega) := L^p(\Omega, \mathbb{C})$ — p. 167
$W^k(\Omega), \|\cdot\|_{W^k(\Omega)}$ — p. 227
$W_0^k(\Omega)$ — p. 227
$H_{p,\gamma}^{r,r'}[E]$ — p. 170
$H_p^{r,r'}[E] := H_{p,(\cdot)}^{r,r'}[E]$ — p. 170
$H^{r,r'}[E] := H_2^{r,r'}[E]$ — p. 170
$H_p^r[E] := H_p^{r,0}[E]$ — p. 170
$H^r[E] := H^{r,0}[E]$ — p. 170
$H_{p,\gamma}^{r,r'} := H_{p,\gamma}^{r,r'}[\mathbb{C}]$ — p. 170

Pseudodifferential operators and symbols

Conventions

- Sections are denoted by pairs of numbers like 1.2 and definitions, theorems, etc. by triples of numbers, e.g. theorem 1.2.3 in section 1.2. Equations are denoted by triples of numbers in parentheses like formula (1.2.3) in section 1.2.

- References are denoted by numbers in square brackets like [42] or [42, 1.2], which means 1.2 in reference [42].

- Vector spaces are always assumed to be real or complex if not specified.

- Fréchet spaces are topological vector spaces the topology of which is generated by a separating, countable, complete family of seminorms

- $I = [t_0, t_1]$ denotes a compact interval and $\triangle := \{(t, s) \in I^2 : t \geq s\}$.

- By "for" we always mean "for all".

- We write $A \in \bigcap_{j \in J} \mathscr{L}(X_j, Y_j)$ for Fréchet spaces X_j, Y_j and partially ordered sets $(J, \preceq)$,
  if there are linear continuous operators $A_j \in \mathscr{L}(X_j, Y_j)$ for $j \in J$ such that there are continuous embeddings $X_j \hookrightarrow X_{j'}$ and $Y_j \hookrightarrow Y_{j'}$ with $A_{j'}|_{X_j} = A_j$ for $j, j' \in J$ with $j' \preceq j$.
  For brevity, we always set $Ax := A_j x$ for $x \in X_j$ and $j \in J$.

- Properties for $A \in \bigcap_{j \in J} \mathscr{L}(X_j, Y_j)$ are meant to hold for $A \in \mathscr{L}(X_j, Y_j)$ for any $j \in J$.

- X_Z^k resp., H_Z^k denotes the scale of Banach spaces resp., Hilbert spaces generated by a family of closed operators.

- E will always denote a Hilbert space.

Introduction

The time evolution of a physical system is usually described by the Cauchy problem

$$\frac{du}{dt}(t) = F(t, u(t)), \qquad u(0) = u_0 \tag{1}$$

with a function u taking values in a state space X, a mapping F depending on the time t and the state $u(t)$, and an initial state u_0. As state space one can often choose a locally convex vector space like a Banach space or a projective limit of a scale of Banach spaces. The main question about (1) is to establish conditions implying well-posedness, i.e., existence and uniqueness of solutions which depend continuously on the data of the equation. While there is a huge amount of literature concerning well-posedness of (1) in Banach spaces X, the present book is devoted to the question of well-posedness of (1) in scales of Banach spaces for linear and quasilinear evolution equations. This does not only include a study of regularity and C^∞ properties of solutions but also provides a functional analytic setting adapted to pseudodifferential evolution equations. Before we describe the motivation and several advantages of this approach, in particular concerning applications to pseudodifferential evolution equations, we will briefly review some results on evolution operators, abstract Cauchy problems, and pseudodifferential evolution equations.

Well-posedness and evolution operators

In the linear homogeneous case $F(t, v) = A(t)v$ with linear operators $A(t)$ well-posedness of the Cauchy problem is equivalent to the existence of solution operators. Recall that in this case one can obtain the state $u(t)$ at time t from the state $u(s)$ at time $s \leq t$ by application of a linear operator $U(t, s)$, i.e., $u(t) = U(t, s)u(s)$. This propagator or evolution operator $U(t, s)$ satisfies certain continuity properties and formally

$$U(s, s) = \mathrm{Id}, \quad U(t, q)U(q, s) = U(t, s),$$

$$\frac{\partial}{\partial t}U(t, s) = A(t)U(t, s), \quad \frac{\partial}{\partial s}U(t, s) = -U(t, s)A(s)$$

for $s \leq q \leq t$. Moreover, the solution of the inhomogeneous, linear Cauchy problem

$$\frac{du}{dt}(t) = A(t)u(t) + f(t), \qquad u(0) = u_0$$

is given by the variation of constants formula (or Duhamel's principle)

$$u(t) = U(t,0)u_0 + \int_0^t U(t,s)f(s)ds \,.$$

If a well-posed linear Cauchy problem is time-independent (i.e., $A(t) \equiv A$), then $U(t,s) = T(t-s)$ with linear operators $T(t), t \geq 0$, satisfying certain continuity properties and

$$T(0) = \mathrm{Id}, \quad T(t)T(s) = T(t+s), \quad \frac{d}{dt}T(t) = AT(t) = T(t)A$$

for $t, s \geq 0$. Families $(T(t))_{t \geq 0}$ with these properties are called semigroups and one says that A generates $T(t)$.

For building a solution theory for (1) the fundamental question is to understand the connections between the operators $A(t)$ and well-posedness of the corresponding Cauchy problem, i.e., the existence of evolution operators resp., semigroups, and to characterize well-posedness by properties of the operators $A(t)$.

The Cauchy problem for unbounded operators in Banach spaces

The Picard-Lindelöf theory for ordinary differential equations can be used to solve (1) for bounded, linear operators $A(t)$ in Banach spaces. However, for most applications this is not useful because differential operators cannot be realized as bounded, linear operators in interesting Banach function spaces. They only can be realized as unbounded, not everywhere defined operators in Banach spaces or as everywhere defined, continuous operators in Fréchet spaces. Linear differential equations in Fréchet spaces behave badly, cf. e.g. Herzog/Lemmert [61], so it is more promising to use the first approach. In the autonomous case (i.e., $A(t) = A$ is independent of t) and in the case of Banach spaces X a characterization of well-posedness of the Cauchy problem is given by the Hille-Yosida theorem. This theorem states that a linear, densely defined, closed operator A in a Banach space X generates a C_0-semigroup (i.e., a semigroup $T(t) \in \mathscr{L}(X), t \geq 0$, with $[0,\infty) \ni t \mapsto T(t)x \in X$ continuously for $x \in X$) if and only if A satisfies certain spectral conditions and resolvent estimates. For quasi-contractive C_0-semigroups (i.e., $\|T(t)\|_{\mathscr{L}(X)} \leq e^{\beta t}$ for $t \geq 0$ and a suitable $\beta \geq 0$) a characterization useful for applications is given by the Lumer-Phillips theorem: A densely defined, closed, linear operator A

with domain $D(A)$ generates a quasi-contractive C_0-semigroup if and only if A is quasi-dissipative (i.e., $\|\lambda x - Ax\| \geq (\lambda - \beta)\|x\|$ for $x \in D(A), \lambda > \beta$) and satisfies the range condition $(\lambda \mathrm{Id} - A)(D(A)) = X$ for a $\lambda > \beta$. The theory of C_0-semigroups is well-developed by now and can be found e.g. in the monographs of Goldstein [52] or Pazy [122]. There are many applications of C_0-semigroup theory not only to mathematical physics but also to other fields, cf. Goldstein [52], Pazy [122], or Cazenave/Haraux [25] for a first glimpse. Concerning time-dependent linear Cauchy problems Kato has extended these results (cf. [78], [79], [83], [87], [77]) and has given sufficient conditions for well-posedness assuming that the operators $A(t)$ generate C_0-semigroups and satisfy additional properties. Further developments of theses results have been given e.g. by Kisyński [91], Dorroh [40], Da Prato-Sinestrari [34], [35], Okazawa-Unai [115], and many others. An overview of several sufficient conditions for well-posedness of the time-dependent linear Cauchy problem can be found in the monographs of Pazy [122] or Tanabe [144]. The time-dependent case is much more complicated than the time-independent case. The difficulties are caused, among other things, by the time-dependence of the domains (which may vary dramatically even for well-posed problems, cf. Goldstein [51]) and the non-commutativity of the operators $A(t)$ for different times. Moreover, there is no complement to the Hille-Yosida theorem in the time-dependent case. In fact, as noticed by Goldstein [51] a characterization of well-posedness for the time-dependent Cauchy problem in Banach spaces would be impossible without additional assumptions. Some necessary and sufficient conditions for well-posedness under strong assumptions were given by Komura [95] and Herod/McKelvey [60]. Recently, Nickel [112] connected well-posedness of the time-dependent Cauchy problem with the existence of certain evolution semigroups. Moreover, together with Schnaubelt, he showed that Kato's conditions are not necessary for well-posedness (cf. [113]).

Besides applications to partial differential evolution equations some of the most important consequences of these results are applications to quasilinear evolution equations

$$\frac{du}{dt}(t) = A(u(t))u(t) + f(u(t)), \qquad u(t_0) = u_0. \qquad (2)$$

Here $A(u(t))$ is an unbounded, linear operator in X depending nonlinearly on $u(t)$. A solution of (2) can be obtained inductively by solving the linear, inhomogeneous, time-dependent Cauchy problem

$$\frac{du_{k+1}}{dt}(t) = A(u_k(t))u_{k+1}(t) + f(u_k(t)), \qquad u_{k+1}(t_0) = u_0 \qquad (3)$$

and by constructing the solution of (2) as the limit of the sequence $(u_k)_k$.

This has been done first by Kato [82], [83], [85], [87] and developed further e.g. by Kobaysasi/Sanekata [93], Okazawa/Unai [114], Ascoli [11], and Tanaka [145]. Though the semilinear Cauchy problem (i.e., $A(u(t)) \equiv A$ is independent of $u(t)$) is a special case of the quasilinear one, it is worth considering the semilinear Cauchy problem in its own right due to its simplicity. A solution of the semilinear Cauchy problem can be obtained as the solution of the integral equation (which is the variation of constants formula)

$$u(t) = U(t,0)u_0 + \int_0^t U(t,s)f(u(s))ds$$

with the Banach fixed point theorem. Thus, more detailed information on the solutions can be obtained, cf. for instance Segal [139], Goldstein [52], Cazenave/Haraux [25]. There are many applications of these results to linear and quasilinear partial differential evolution equations, cf. Kato [78], [83], [85], [82], [87], [86], [84],[81], Kato/Ponce [88], Massey [106], Yamazaki [159], Wong-Dzung [158], Ali-Mehmeti [6], [5], Isaza/Mejia/Stallbohm [70], [71], Schleinkofer [136], Constantin/Escher [30].

Pseudodifferential evolution equations

Some important types of evolution equations are pseudodifferential evolution equations

$$\frac{du}{dt}(t) = p(t, X, D_x)u(t), \qquad\qquad u(0) = u_0 \qquad\qquad (4)$$

with time-dependent pseudodifferential operators in the spatial variables

$$p(t, X, D_x)v(x) = \frac{1}{(2\pi)^n} \int_{\mathbf{R}^n} \int_{\mathbf{R}^n} e^{i(x-y)\xi} p(t, x, \xi)v(y)dyd\xi \,.$$

Here $p(t, \cdot, \cdot) : \mathbf{R}^n \times \mathbf{R}^n \to \mathbf{C}$, the so-called symbol of the operator $p(t, X, D_x)$, is a function satisfying certain growth estimates. More precisely, a symbol of order m is a smooth function with

$$|\partial_\xi^\alpha \partial_x^\beta q(x, \xi)| \leq c_{\alpha\beta}(1 + |\xi|)^{m-|\alpha|} \,.$$

There are several reasons to study pseudodifferential evolution equations or systems. First, pseudodifferential evolution equations generalize differential ones because the differential operator $\sum_{|\alpha| \leq m} a_\alpha(t, x)\partial^\alpha$ is the pseudodifferential operator with symbol $p(t, x, \xi) = \sum_{|\alpha| \leq m} a_\alpha(t, x)(i\xi)^\alpha$. Furthermore, the transition from classical mechanics to quantum mechanics is described by replacing the classical Hamilton function by the Hamilton operator acting on square-integrable functions. This transition is called quantization and is merely a transition from a symbol to a pseudodifferential operator. The transition is described more precisely by the Weyl-quantization associating to a Hamilton function $h(t, x, \xi)$ the pseudodifferential operator

$$h^w(t, X, D_x)v(x) := \frac{1}{(2\pi)^n} \int_{\mathbf{R}^n} \int_{\mathbf{R}^n} e^{i(x-y)\xi} h\left(t, \frac{x+y}{2}, \xi\right) v(y) dy d\xi.$$

The corresponding quantum mechanical equation of motion, the time-dependent Schrödinger equation, is now the pseudodifferential evolution equation (cf. Nagase/Umeda [109] for a non-differential equation)

$$\frac{du}{dt}(t) = ih^w(t, X, D_x)u(t), \qquad u(t_0) = u_0.$$

Next, pseudodifferential evolution equations arise in connection with probability theory. There is a 1-1-correspondence between special Markov processes and positivity-preserving, contractive semigroups on spaces of continuous functions, so-called Feller semigroups (cf. e.g. Goldstein [52, 9.11]). By a theorem of Courrège [33] generators of regular Feller semigroups on the space of continuous functions on $\mathbf{R}^n$ vanishing at infinity are second-order pseudodifferential operators with non-regular symbols. This motivates the interest in pseudodifferential evolution equations with second-order $p(t, X, D_x)$. As a last occurrence of pseudodifferential (nonlinear) evolution equations we mention some equations of hydrodynamics like the Benjamin-Ono equation or the Smith equation. These equations contain nonlocal terms like $\sqrt{\mathrm{Id} - \triangle}$ or the Hilbert transform, cf. e.g. Benjamin [14], Ono [118], Kenig/Ponce/Vega [90], Iorio [69], Abdelouhab [1].

The basic question in the study of pseudodifferential evolution equations is the connection between well-posedness of the corresponding Cauchy problem and the symbol $p(t, x, \xi)$. Hörmander [67, 23.1.2] has shown well-posedness in the first-order hyperbolic case, i.e., for first-order symbols with real parts semibounded from above. Moreover, Iwasaki [72] (cf. also Kumano-go [97, 7.4]) has shown well-posedness of the pseudodifferential Cauchy problem in the weakly degenerate parabolic case for many hypoelliptic symbols including elliptic ones. Pseudodifferential operators are defined on a Fréchet space or between several function spaces like Sobolev spaces. Therefore, for applications of the abstract results mentioned above one has to choose suitable functional analytic realizations of these operators. The choice of a closed extension of the operators and the verification of the assumptions of the abstract results are non-trivial questions. Although neither Hörmander nor Iwasaki used the theory of abstract evolution equations for their proofs, it is possible to recover their results with this theory. For verifications of the assumptions one can use the same techniques as in their direct proofs. To be more precise, the sharp Gårding inequality, mollifiers, and commutator techniques have to be used in the hyperbolic case considered by Hörmander, whereas in the weakly degenerate parabolic case considered by Iwasaki parametrices and complex powers of pseudodifferential operators are necessary among other things.

Purpose of the book

The purpose of the present book is to develop an abstract approach to the Cauchy problem that is more adapted for pseudodifferential evolution equations than the abstract theory described above. Instead of considering densely defined, closed operators in one Banach space, we consider operators $A(t)$ of order m in a scale of Banach spaces X^k, $k \in \mathbb{N}_0$, i.e., there are continuous and dense embeddings $X^k \hookrightarrow X^l$ for $k \geq l$ and $A(t) \in \mathscr{L}(X^{k+m}, X^k)$ for any $k \in \mathbb{N}_0$. This is a natural setting for many types of pseudodifferential operators. The most important question is a characterization of a notion of well-posedness adapted to this setting. Well-posedness of the Cauchy problem in a scale of Banach spaces in particular gives precise regularity results for the solutions. For special settings in scales of Banach spaces (that are still general enough to cover several interesting situations) we will answer this question by characterizing well-posedness of the Cauchy problem for $A(t)$ by quasi-dissipativity of the operators $A(t)$ everywhere in the scale (cf. section 2.4). In contrast to the abstract situation described above, we in fact obtain a characterization of the time-dependent Cauchy problem in our setting. Even in the time-independent situation one should note that, unlike the Lumer-Phillips theorem, the range condition has disappeared and (as we will show) is a consequence of the quasi-dissipativity everywhere in the scale. The main advantage of this approach in applications is that it is easier to check inequalities than range conditions because one already has to solve partial differential equations to do the later. Morcover, in this setting the consideration of time-dependent equations involves no additional difficulties compared to time-independent equations. As sums of dissipative operators are again dissipative, our results show that the Cauchy problem for $A_1(t) + A_2(t)$ in the considered scales is well-posed with exponential growth provided that the Cauchy problem for $A_1(t)$ and $A_2(t)$ is well-posed. In particular, it is not necessary to develop a perturbation theory as in semigroup theory to treat terms separately. Furthermore, some of our characterizations (cf. 2.4.6 and also 2.4.1) no longer assume commutator estimates like Kato's results [78], [79], cf. also 1.6.8. This will make possible several applications which appear to be non-accessible to Kato's theory, cf. e.g. 4.2.17, 4.4.14, 4.5.5, 5.2.1, 5.3.10.

In applications one can easily recover some known results on hyperbolic and parabolic pseudodifferential evolution equations and generalize them slightly.

However, the most important application and one of the most interesting results in the present book might be the proof of well-posedness of the pseudodifferential Cauchy problem for second-order symbols with semi-bounded real
parts and first-order imaginary parts (cf. 4.2.17). This result includes strongly
degenerate second-order parabolic operators and contains Hörmander's result for hyperbolic equations and parts of Iwasaki's result for second-order
parabolic equations. Besides applications to conditions on symbols for essential selfadjointness of pseudodifferential operators and well-posedness of
Schrödinger equations (section 4.3), linear evolution equations in L^q-Sobolev
spaces (section 4.5) and spaces of continuously differentiable functions (sections
4.4), degenerate-elliptic boundary value problems (section 4.5), and evolutions
equations on networks (section 4.7) the next important application concerns
quasilinear equations. Similarly to Kato's construction (but technically more
complicated) it is possible to use the time-dependent linear theory to construct
solutions of abstract quasilinear equations in scales of Banach space, as well
(cf. 3.3.9). In this way we can construct local solutions and, for reflexive
scales, we can show that a breakdown of a global solution must involve a blow-
up of the solution in a suitable norm. Together with integrals of motion this
yields the possibility of proving existence of global solutions. However, this
question will be beyond of the scope of the present book. Nevertheless, we
use the blow-up result to prove, using inequalities of Nash-Moser type, a regularity result for quasilinear evolution equations (cf. 3.4.4). In particular, for
initial values in the projective limit $X^\infty := \bigcap_{k=0}^\infty X^k$ we can construct a local
solution in the Fréchet space X^∞. As in the linear case the most interesting
applications might concern strongly-degenerate parabolic equations of second
order (cf. section 5.2) like Navier-Stokes equations that degenerate to Euler
equations in some parts of the space (cf. section 5.3). Now we will give a more
precise description of the book.

Organization of the text

Chapter 1
Chapter 1 provides several functional analytic tools for the following chapters. Readers experienced in functional analysis may skip this chapter and go
back only occasionally. Section 1.1 gives a brief introduction into the theory
of C_0-semigroups. Basic definitions are given and some important properties, in particular concerning a-priori estimates and their connections to quasi-
dissipativity, are proved. Then, in section 1.2 we give important properties
of selfadjoint operators and describe their connections to C_0-semigroups and
quadratic forms. In section 1.3 a special type of C_0-semigroups, so-called analytic ones are introduced. In particular, some properties of the powers of

their generators are proved. Fractional powers of operators of positive type are introduced in section 1.4 and a brief introduction into the theory of interpolation spaces is given in section 1.5. Finally, in section 1.6 we collect important properties and results on time-dependent linear evolution equations in Banach spaces, which generalize several results on C_0-semigroups.

Chapter 2
The second chapter of the present book is devoted to the study of the linear, time-dependent Cauchy problem in special scales of Banach spaces. In section 2.1 we introduce the notion of well-posedness with exponential growth in a scale of Banach spaces $(X^k)_k$. To be more precise, we consider an interval $I = [t_0, t_1]$ and a strongly continuous family of operators

$$I \ni t \mapsto A(t) \in \bigcap_{k \in \mathbb{N}_0} \mathscr{L}(X^{k+m}, X^k),$$

and we call the Cauchy problem for $A(t)$ in the scale $(X^k)_k$ well-posed with exponential growth, if for $s \in I, k \in \mathbb{N}_0$, and $u_0 \in X^{k+m}$, there is a unique $u = u(\cdot, s, u_0) \in \mathcal{C}^1([s, t_1], X^k) \cap \mathcal{C}([s, t_1], X^{k+m})$ with

$$\frac{du}{dt}(t) = A(t)u(t), t \in [s, t_1], \qquad u(s) = u_0$$

and $\|u(t, s, u_0)\|_k \le e^{\beta_k(t-s)} \|u_0\|_k$ for $t \in [s, t_1]$ and suitable constants $\beta_k \ge 0$. As remarked above, well-posedness is equivalent to the existence of propagators. This is made precise in proposition 2.1.4, where we give a rigorous formulation of the variation of constants formula for solving inhomogeneous Cauchy problems. Besides some simple properties of well-posed Cauchy problems we further show in the time-independent case that the Cauchy problem for $A(t) = A$ is well-posed with exponential growth if and only if the closure of $A|_{X^{k+m}}$ in X^k generates a quasi-contractive C_0-semigroup for any k. It is important to note that well-posedness with exponential growth of the Cauchy problem for $A(t)$ in the scale of Banach spaces X^k implies quasi-dissipativity of $A(t)$ in any space X^k for $k \in \mathbb{N}_0$. For special settings we prove converses of this statement in later sections and we obtain characterizations of well-posedness with exponential growth in scales of Banach spaces. Our methods do not work for arbitrary scales of Banach spaces and have to restrict ourselves to special scales.

In section 2.2 we introduce scales of Banach spaces that are generated by a finite family of infinitesimal generators of bounded semigroups, cf. also Gramsch/Kalb [55], Gramsch/Ueberberg/Wagner [56], Lauter [101], Triebel [151], Amann [9], Nagel [111], and the references given therein. However, these scales are still general enough to cover several interesting examples, cf. Goodman [53] and 2.2.2. If the family is commuting modulo operators of order zero in the scale (cf. 2.2.4), we construct a mollifier in this scale (cf. 2.2.9). This is a family of regularizing operators approximating the identity in a suitable way.

In section 2.3 we study commutator estimates in connection with scales of Banach spaces. First, we show in 2.3.1 and 2.3.2 how quasi-dissipativity estimates can be extended from one part of the scale to others using commutator estimates. Then we give several propositions showing how estimates, in particular commutator estimates, for fractional powers of operators can be obtained from multi-commutator estimates for the operator itself, cf. 2.3.8, 2.3.9, 2.3.10. This is important for applications because we will often need estimates for fractional powers of differential operators, cf. section 4.5 and 4.7. These estimates usually are difficult to prove, whereas multi-commutator estimates for the differential operator itself can be obtained rather easily with the Leibniz rule. We conclude the section with some remarks concerning C^∞-elements of positive selfadjoint operators Λ in Hilbert spaces H, i.e., those $A \in \mathcal{L}(H)$ such that $\mathbb{R} \ni t \mapsto e^{it\Lambda} A e^{-it\Lambda} \in \mathcal{L}(H)$ is smooth, cf. e.g. Gramsch [54], Cordes [32, ch. 8]. We show in 2.3.12 that the set of C^∞-elements of Λ^ε grows, if ε decreases.

In section 2.4 we can finally characterize well-posedness with exponential growth of the Cauchy problem for operators $A(t)$ in a scale X^k in four situations under additional assumptions. First, in 2.4.1 we achieve this for scales generated by the generator of a quasi-contractive C_0-semigroup and relatively bounded operators $A(t)$ with bound 0 with respect to this generator. Then, in 2.4.3 we can state a characterization, if the scale is generated by the generator of an analytic semigroup and if $A(t)$ satisfies certain commutator estimates with this generator. Here the allowed order of the operators $A(t)$ is related to the angle of analyticity of the semigroup. A special case of this result is described in 2.4.5. There we consider generators of analytic semigroups that are sums of squares of commuting generators of isometric C_0-groups in Banach spaces. This models the Laplacian as the sum of squares of first partial derivatives. Operators of this type are known to generate analytic semigroups with angle $\frac{\pi}{2}$ and arbitrary order of $A(t)$ is allowed. The last characterization 2.4.6 is the most important one for applications. It only works for scales of Hilbert spaces that are generated by commuting families of (skew)-selfadjoint operators, but in contrast to the former characterizations no additional assumptions are needed. The proofs of these results are given in section 2.5. It is evident that solvability of the Cauchy problem in a fixed space of the scale only depends on mapping properties of the operators in finitely many spaces of the scale. This is made precise in section 2.5 for the four characterizations in 2.5.2, 2.5.3, 2.5.7, and 2.5.14. In any case we have to assume better regularity properties of the operators $A(t)$ (that is mapping properties in spaces of higher index) than these of the solutions we can construct.

Chapter 3

The third chapter deals with semilinear and quasilinear evolution equations. In section 3.1 we consider semilinear evolution equations. The semilinear Cauchy problem can be solved with the variation of constants formula as soon as the linear one is well-behaved, cf. Segal [139], Goldstein [52], Cazenave/Haraux [25]. We apply this method to our situation in 3.1.2. Moreover, using estimates of Nash-Moser type we give a regularity result, cf. 3.1.6. The section is completed by a simple condition for existence of global solutions of semilinear evolution equations (cf. 3.1.10).

In section 3.2 we establish (similarly to section 2.3) several propositions extending quasi-dissipativity estimates from one part of the scale to others. This will be needed in the following sections and in applications. In section 3.3 we prove, motivated by (3), a local existence and uniqueness result for quasilinear evolution equations. To this end, in 3.3.9 we describe a general method for solving quasilinear evolution equations in scales of Banach spaces, if the linear, time-dependent Cauchy problem can be solved. Concerning the solvability of the linear, time-dependent Cauchy problem in a space with index k we also permit $A(t)$ to operate in spaces of order higher than k. This is necessary to make 3.3.9 accessible to the results in chapter 1. As an additional difficulty we have to use therefore a supplementary regularization procedure in (3) to avoid a loss of regularity. However, we will show that the regularization does not disturb the limiting procedure, and we can construct local and unique solutions of the quasilinear Cauchy problem. Finally, in section 3.4 we give a regularity result for quasilinear evolution equations using inequalities of Nash-Moser type.

Chapter 4

In the forth chapter we will give several applications to linear, time-dependent differential and pseudodifferential evolution equations. First, in section 4.1 we recall and collect several results on pseudodifferential calculi in scales of weighted, vector-valued Sobolev spaces.

Section 4.2 is devoted to a systematic study of well-posedness of the pseudodifferential Cauchy problem in the usual scale of vector-valued Sobolev spaces by means of conditions on symbols. As remarked above we recover the well-known well-posedness results in the first-order hyperbolic case and in the weakly degenerate parabolic case. Whereas one uses the sharp Gårding inequality and commutator arguments in the hyperbolic case, one has to use parametrices and complex powers of pseudodifferential operators in the weakly degenerate parabolic case. Now, there is a strong generalization of the sharp Gårding inequality, the so-called Fefferman-Phong inequality (cf. 4.1.13). Under its much more general assumptions neither the commutator methods from the proofs in the hyperbolic case nor Kato's theory can be applied any longer. Nevertheless, using a double commutator argument and the Fefferman-Phong inequality we can apply the results of chapter 1 to prove well-posedness for pseudodifferential evolution equations with second-order symbols whose imaginary parts are symbols of order 1 and real parts are semi-bounded from above (cf. 4.2.17). In particular, this class of symbols contains scalar symbols in the hyperbolic cases and second-order symbols in the weakly degenerate parabolic cases mentioned above. The proof of well-posedness does not only work in the usual scale of Sobolev spaces but also in weighted scales of Sobolev spaces, and we obtain well-posedness in the space of rapidly decreasing functions in this way.

After proving well-posedness of the Cauchy problem for pseudodifferential wave equations in 4.2.19, we focus on the question of essential self-adjointness of pseudodifferential operators and well-posedness of Schrödinger type equations in section 4.3. As mentioned above, it is an important question to establish essential selfadjointness of Weyl-quantized pseudodifferential operators. Similarly to the question of well-posedness of the pseudodifferential Cauchy problem it is well-known that Weyl-quantized pseudodifferential operators with real-valued symbols are essentially selfadjoint provided that the symbol has order one or is a special hypoelliptic symbol admitting a parametrix (cf. Kumano-go [97, 3.5.8]). The Fefferman-Phong inequality and our methods from former sections can also be used for this question and we prove in 4.3.2 and 4.3.4 that Weyl-quantized pseudodifferential operators with semibounded second-order symbols are essentially selfadjoint. Moreover, using a method of Chernoff [26] we prove a result on essential selfadjointness of all powers of special symmetric pseudodifferential operators (cf. 4.3.11). As a consequence we can easily characterize the scale generated by these operators with distributional conditions (cf. 4.3.10). We conclude the section with a result on pseudodifferential Schrödinger equations in weighted Sobolev spaces (cf. 4.3.19). Thus, we recover in part a result of Yamazaki [159] with a simplified proof.

In section 4.4 we turn towards evolution equations in spaces of continuous functions. As mentioned above it is an important question motivated by probability theory to examine generators of Feller semigroups on $\mathbb{R}^n$, i.e., contractive, positivity preserving C_0-semigroups in the space of continuous functions vanishing at infinity. Due to a theorem of Courrège (mentioned above) generators of regular Feller semigroups on $\mathbb{R}^n$ are always second-order pseudodifferential operators with non-regular symbols the additive inverses of which are negative-definite. The opposite question to determine all pseudodifferential operators generating Feller semigroups on $\mathbb{R}^n$ is not completely answered and only given for regular and, in part, also for irregular symbols that satisfy certain ellipticity assumptions, cf. Jacob [74], Hoh [64], Baldus [12] and the references given therein. In this section we prove a result of that kind for regular symbols without any ellipticity assumption, cf. 4.4.15. This is also done in a time-dependent situation, cf. 4.4.14. At the end of the section we give a result on well-posedness of the Cauchy problem for first-order, symmetric hyperbolic evolution equations in the scale of k-times continuously differentiable functions vanishing at infinity. In section 4.5 we prove a result on well-posedness of a special pseudodifferential Cauchy problem in the scale of L^q-Sobolev spaces. This result includes L^q-well-posedness of the Cauchy problem for second-order degenerate-elliptic differential operators $A(t)$ (cf. 4.5.6) and proves a conjecture in Wong-Dzung [158].

Then, in section 4.5 we consider a degenerate-elliptic second-order differential operator A which is non-characteristic near the boundary. There is an elliptic, second-order differential operator Λ which coincides with A near the boundary. We can use the well-understood theory on elliptic boundary value problems (cf. Agmon [4]) to generate a scale of Hilbert spaces with Λ. Then A operates in this scale and the results of section 2.4 can be applied to prove well-posedness of the Cauchy problem for A in 4.6.8.

The last section 4.7 of this chapter gives some applications to evolution equations on networks. Following Ali Mehmeti [6], [5] a one-dimensional network is defined to be a set of unit intervals the ends of which are glued together by so-called transmission conditions, cf. 4.7.1. Whereas in the previous section we started with an operator A and constructed a suitable scale in which A operates, we use a different method for the application of 2.4.6 in this section by starting with a scale of Hilbert spaces and looking for operators in this scale afterwards. More precisely, using the Friedrichs extension a natural Laplacian with appropriate Sobolev spaces can be defined. After identifying differential operators operating in this scale, we give applications to special Schrödinger type equations (cf. 4.7.7) and equations of degenerate diffusion type (cf. 4.7.8) on networks.

Chapter 5

In section 5.1 we collect several estimates of Gagliardo-Moser-Nirenberg type in L^q-spaces that we will need in the chapter. The estimates seem to be well-known but due to a lack of suitable references we give their proofs. Then, in section 5.2 we extend the results of section 4.2 to special nonlinear evolution equations. We admit nonlinear differential operators of order 1 in (degenerate) second-order pseudodifferential evolution equations and prove solvability of the corresponding Cauchy problem (cf. 5.2.1). This includes e.g. the degenerate Burger equations, the Korteweg-de Vries equation, the Benjamin-Ono-equation, and the Smith equation, cf. 5.2.3. Moreover, second-order equations are included in this result that degenerate to first order equations in some parts of the space.

A particular example of equations of that type is the degenerate Navier-Stokes equation

$$u_t = \nu(x)\triangle u - (u \cdot \operatorname{grad}_x)u - \operatorname{grad}_x \pi, \qquad \operatorname{div}_x u = 0. \qquad (5)$$

Usually, this equation is treated only for $\nu(x) \equiv \nu = \text{const.}$ in the literature, cf. e.g. Fujita/Kato [50], Ladyzhenskaya [99], Temam [150], von Wahl [154], Taylor [148]. For $\nu > 0$ (in the viscous case) this equation is parabolic and called the Navier-Stokes equation and for $\nu = 0$ (in the non-viscous case) (5) is hyperbolic and called the Euler equation. In section 5.3 we apply the results of chapter 3 to (5), and we treat equations with space-varying $\nu(x) \geq 0$. In particular, fluid motions can be described that satisfy the Euler equation in some parts of the space (i.e., without viscosity) and the Navier-Stokes equation in other parts (i.e., with viscosity).

Then, in section 5.3 a unified approach to both types of Kadomtsev-Petviashvili equations with periodic boundary values is given. Proceeding similarly to Isaza/Mejía/Stallbohm (cf. [70], [71], cf. also Schleinkofer [136]) who applied Kato's theory to Kadomtsev-Petiashvili equations (and could only treat equations of type I with this method), we show how the results of chapter 3 can be applied to existence and regularity of solutions in both cases simultaneously. In particular, equations of mixed type can also be considered.

Finally, in section 5.5 and section 5.6 we recover some known results on first-order, hyperbolic evolution equations in L^q-Sobolev spaces resp., spaces of k-times continuously differentiable functions vanishing at infinity. Through the application of the results in chapter 3 we obtain simple proofs for these statements.

Chapter 1

Tools from functional analysis

1.1 A brief introduction into the theory of semi-groups

The purpose of this section is to provide briefly some results on C_0-semigroup theory that we will need in later sections. We assume the reader to be familiar with elementary functional analysis of bounded, linear operators in Banach spaces.

For real or complex Fréchet spaces X, Y we denote the continuous, linear mappings of Y into X by $\mathscr{L}(Y, X)$, and we write $\mathscr{L}(X)$ instead of $\mathscr{L}(X, X)$. We write $Y \hookrightarrow X$, if Y is continuously embedded into X and $A \in \mathscr{L}(X) \cap \mathscr{L}(Y)$, if $A \in \mathscr{L}(X)$ and $A|_Y \in \mathscr{L}(Y)$, where $A|_Y$ denotes the restriction of A to Y. For a subset $\Omega \subset \mathbb{R}^n$ a function $\Omega \ni t \mapsto A(t) \in \mathscr{L}(Y, X)$ is called strongly continuous (resp., strongly continuously differentiable), if $\Omega \ni t \mapsto A(t)y \in X$ is continuous (resp., continuously differentiable) for every $y \in Y$. For a real or complex Banach spaces X we denote its norm by $\|\cdot\|_X$ or simply by $\|\cdot\|$, if no confusion can occur, and the operator norm of $A \in \mathscr{L}(Y, X)$ by $\|A\|_{\mathscr{L}(Y,X)}$. If not specified a Banach space is always considered to be real or complex. For a complex (resp., real) Banach space X we write $X^* := \mathscr{L}(X, \mathbb{C})$ (resp., $X^* = \mathscr{L}(X, \mathbb{R})$) for its dual and $\langle \cdot, \cdot \rangle : X^* \times X \to \mathbb{C}$ (resp., $\mathbb{R}$) for its duality section. A sequence $(x_k)_k \subset X$ is called weakly convergent to $x \in X$, if $\langle x^*, x_k \rangle \xrightarrow[k \to \infty]{} \langle x^*, x \rangle$ for any $x^* \in X^*$.

If X is a Banach space, $D(A) \subset X$ is a subspace, and $A : D(A) \to X$ is linear, then A is called a linear operator with domain $D(A)$ in X. For a second linear operator $B : D(B) \to X$ in X with domain $D(B)$ we call B an extension of A and write $A \subset B$, if $D(A) \subset D(B)$ and $Ax = Bx$ for $x \in D(A)$. We write $A = B$, if $A \subset B$ and $B \subset A$. A linear operator $A : D(A) \to X$ with domain

$D(A) \subset X$ in a Banach space X is called densely-defined, if its domain is a dense subspace $D(A) \subset X$, and closed if for every sequence $(x_n)_{n \in \mathbb{N}} \subset D(A)$ with $x_n \xrightarrow[n \to \infty]{} x \in X$ and $A x_n \xrightarrow[n \to \infty]{} y \in X$ we have $x \in D(A)$ and $Ax = y$. The operator A is called closable, if for every sequence $(x_n)_{n \in \mathbb{N}} \subset D(A)$ and $x \in X$ with $x_n \xrightarrow[n \to \infty]{} x \in X$ and $A x_n \xrightarrow[n \to \infty]{} 0$ we have $x = 0$. In this case there is a minimal closed extension $\overline{A}$ of A, called the closure of A. The range of A is denoted by $\mathcal{R}(A)$ and the kernel by $\mathcal{N}(A)$. The resolvent set of A in a complex Banach space is defined by

$$\rho(A) := \{\lambda \in \mathbb{C} : \lambda \mathrm{Id} - A : D(A) \to X \text{ is bijective and } (\lambda \mathrm{Id} - A)^{-1} \in \mathscr{L}(X)\},$$

where Id denotes the identity mapping, and the spectrum by $\sigma(A) := \mathbb{C} \setminus \rho(A)$. For operators in real Banach spaces the resolvent set and the spectrum are defined as the resolvent set and the spectrum of their complexifications.

1.1.1 Definition. Let X be a Banach space.

(a) A C_0-semigroup or strongly continuous semigroup $(T(t))_{t \geq 0} \subset \mathscr{L}(X)$ is a family of linear operators satisfying the following properties:

- $T(0) = \mathrm{Id}$ and $T(t)T(s) = T(t + s)$ for $t, s \geq 0$.
- $[0, \infty) \ni t \mapsto T(t) \in \mathscr{L}(X)$ is strongly continuous.

To a C_0-semigroup we associate its infinitesimal generator $A : D(A) \to X$ with domain $D(A)$ by

$$Ax := y_x \quad \text{for } x \in D(A) := \left\{ x \in X : \lim_{h \to 0+} \frac{T(h)x - x}{h} =: y_x \text{ exists} \right\}.$$

We call $(T(t))_{t \geq 0}$ the semigroup generated by A and write symbolically $e^{tA} := \exp(tA) := T(t)$ for $t \geq 0$.
A semigroup $(T(t))_{t \geq 0} \subset \mathscr{L}(X)$ is called quasi-contractive with constant β, if there is a $\beta \geq 0$ with $\|T(t)\|_{\mathscr{L}(X)} \leq e^{\beta t}$ for $t \geq 0$.
The semigroup is called contractive, if $\beta = 0$.

(b) A C_0-group or strongly continuous group $(T(t))_{t \in \mathbb{R}} \subset \mathscr{L}(X)$ is a family of linear operators satisfying the following properties:

- $T(0) = \mathrm{Id}$ and $T(t)T(s) = T(t + s)$ for $t, s \in \mathbb{R}$.
- $\mathbb{R} \ni t \mapsto T(t) \in \mathscr{L}(X)$ is strongly continuous.

Its infinitesimal generator is defined as in (a).

1.1.2 Lemma. Let $(T(t))_{t \geq 0} \subset \mathcal{L}(X)$ be a C_0-semigroup in a Banach space X. Then there are $M \geq 1$ and $\beta \geq 0$ with $\|T(t)\|_{\mathcal{L}(X)} \leq M e^{\beta t}$ for $t \geq 0$.

PROOF: The uniform boundedness principle implies the existence of a constant $M \geq 1$ with $\|T(t)\|_{\mathcal{L}(X)} \leq M$ for $0 \leq t \leq 1$. Let $\beta := \log M, t > 0$, and take an $n \in \mathbb{N}$ with $t \leq n < t + 1$. Then

$$\|T(t)\|_{\mathcal{L}(X)} = \left\| T \left(\frac{t}{n} \right)^n \right\|_{\mathcal{L}(X)} \leq M^n \leq M^{t+1} = M e^{\beta t} .$$

$\square$

We will often use the following lemma in connection with density arguments.

1.1.3 Lemma. Let X, Y be Banach spaces, $D \subset X$ be a dense subset, $(A_n)_n \subset \mathcal{L}(X, Y)$, and suppose that there is an $M \geq 0$ with $\|A_n\|_{\mathcal{L}(X,Y)} \leq M$ for $n \in \mathbb{N}$. If $(A_n x)_n \subset Y$ converges for any $x \in D$, then $(A_n x)_n \subset Y$ converges for any $x \in X$ to an element $Ax \in Y$. Moreover, we have $A \in \mathcal{L}(X, Y)$.

PROOF: Let $\varepsilon > 0, x \in X$, and $y \in D$ with $\|x - y\|_X \leq \frac{\varepsilon}{3M}$. Let $N \in \mathbb{N}$ with $\|A_n y - A_m x\|_Y \leq \frac{\varepsilon}{3}$ for $n, m \geq N$. Then

$$\|A_n x - A_m y\|_Y$$
$$\leq \|A_n\|_{\mathcal{L}(X,Y)} \|x - y\|_X + \|A_n y - A_m y\|_Y + \|A_m\|_{\mathcal{L}(X,Y)} \|x - y\|_X \leq \varepsilon$$

for $n, m \geq N$. This proves the convergence of $(A_n x)_n$. Clearly, A is a linear operator. Finally, $\|A_n x\|_Y \leq M \|x\|_X$ for $x \in X$ implies $\|Ax\|_Y \leq M \|x\|_X$ for $x \in X$, which proves the assertion. $\square$

It is natural to ask which linear operators occur as generators of C_0-semigroups. This question is answered by the fundamental Hille-Yosida theorem 1.1.4.

1.1.4 Theorem. For a linear operator $A : D(A) \to X$ in a Banach space X we have equivalently:

(a) A is the generator of a strongly continuous semigroup $(T(t))_{t \geq 0} \subset \mathcal{L}(X)$.

(b) A is densely defined, closed, and there are $M \geq 1, \beta \geq 0$ such that for λ with $\operatorname{Re} \lambda > \beta$ we have $\lambda \in \rho(A)$ and $\left\| (\lambda \operatorname{Id} - A)^{-n} \right\|_{\mathcal{L}(X)} \leq \dfrac{M}{(\operatorname{Re} \lambda - \beta)^n}$ for $n \in \mathbb{N}$.

(c) A is a densely defined, closed operator, and there are $M \geq 1, \beta \geq 0$ with $(\beta, \infty) \subset \rho(A)$ and $\left\| (\lambda \operatorname{Id} - A)^{-n} \right\|_{\mathcal{L}(X)} \leq \dfrac{M}{(\lambda - \beta)^n}$ for $n \in \mathbb{N}, \lambda > \beta$.

Moreover, in this case $(\lambda \operatorname{Id} - A)^{-1} = \displaystyle\int_0^\infty e^{-\lambda t} T(t) dt$ for λ with $\operatorname{Re} \lambda > \beta$.

PROOF: First assume (a). Let $M \geq 1, \beta > 0$ with $\|T(t)\|_{\mathscr{L}(X)} \leq Me^{\beta t}$ for $t \geq 0$. For $x \in X$ and $\varepsilon > 0$ let $x_\varepsilon := \frac{1}{\varepsilon} \int_0^\varepsilon T(s)xds$. Then

$$\frac{T(h)x_\varepsilon - x_\varepsilon}{h} = \frac{1}{\varepsilon}\left(\frac{1}{h}\int_\varepsilon^{h+\varepsilon} T(s)xds - \frac{1}{h}\int_0^h T(s)xds\right) \xrightarrow{h\to 0} \frac{1}{\varepsilon}(T(\varepsilon)x - x) .$$

This shows $x_\varepsilon \in D(A)$ and proves $D(A) \subset X$ densely because $x_\varepsilon \xrightarrow{\varepsilon\to 0} x$.

Since $\|e^{-\lambda t}T(t)\|_{\mathscr{L}(X)} \leq Me^{-(\operatorname{Re}\lambda - \beta)t}$, $R(\lambda) := \int_0^\infty e^{-\lambda t}T(t)dt \in \mathscr{L}(X)$ is absolutely convergent for $\operatorname{Re}\lambda > \beta$. We have for $x \in X$

$$\frac{T(h) - \operatorname{Id}}{h}R(\lambda)x = \frac{1}{h}\int_h^\infty e^{-\lambda(t-h)}T(t)xdt - \frac{1}{h}\int_0^\infty e^{-\lambda t}T(t)xdt$$

$$= \frac{e^{\lambda h} - 1}{h}\int_0^\infty e^{-\lambda t}T(t)xdt - \frac{e^{\lambda h}}{h}\int_0^h e^{-\lambda t}T(t)xdt \xrightarrow{h\to 0} \lambda R(\lambda)x - x .$$

Thus $R(\lambda)x \in D(A)$ with $AR(\lambda)x = \lambda R(\lambda)x - x$ and $(\lambda\operatorname{Id} - A)R(\lambda)x = x$ for $x \in X$. Moreover, similarly one can show $R(\lambda)Ax = \lambda R(\lambda)x - x$ for $x \in D(A)$. This proves $R(\lambda)(\lambda\operatorname{Id} - A)x = x$ for $x \in D(A)$. Therefore we have shown $\lambda \in \rho(A)$ for $\operatorname{Re}\lambda > \beta$ and $(\lambda\operatorname{Id} - A)^{-1} = R(\lambda)$.

Now take a sequence $(x_k)_k \subset D(A)$ with $x_k \xrightarrow{k\to\infty} x \in X$ and $Ax_k \xrightarrow{k\to\infty} y \in X$. This implies $x_k = (\lambda\operatorname{Id} - A)^{-1}(\lambda x_k - Ax_k) \xrightarrow{k\to\infty} (\lambda\operatorname{Id} - A)^{-1}(\lambda x - y)$ and therefore $x = (\lambda\operatorname{Id} - A)^{-1}(\lambda x - y) \in D(A)$ and $Ax = y$. Thus A is closed. Finally, for $\lambda, \lambda' \in \rho(A)$ we have

$$\frac{(\lambda'\operatorname{Id} - A)^{-1} - (\lambda\operatorname{Id} - A)^{-1}}{\lambda' - \lambda} = -(\lambda'\operatorname{Id} - A)^{-1}(\lambda\operatorname{Id} - A)^{-1},$$

hence $\rho(A) \ni \lambda \mapsto (\lambda\operatorname{Id} - A)^{-1} \in \mathscr{L}(X)$ is smooth with n-th derivatives $\frac{d^n}{d\lambda^n}(\lambda\operatorname{Id} - A)^{-1} = (-1)^n n!(\lambda\operatorname{Id} - A)^{-(n+1)}$ for $n \in \mathbb{N}_0$. Therefore, for $n \in \mathbb{N}$ and $\operatorname{Re}\lambda > \beta$ we have

$$(-1)^{n-1}(n-1)!(\lambda\operatorname{Id} - A)^{-n} = \frac{d^{n-1}}{d\lambda^{n-1}}(\lambda\operatorname{Id} - A)^{-1} = \frac{d^{n-1}}{d\lambda^{n-1}}\int_0^\infty e^{-\lambda t}T(t)dt$$

$$= (-1)^{n-1}\int_0^\infty e^{-\lambda t}t^{n-1}T(t)dt$$

Using $\int_0^\infty e^{-\alpha t}t^{n-1}dt = \frac{(n-1)!}{\alpha^n}$ for $\alpha > 0$ this proves

$$\|(\lambda\operatorname{Id} - A)^{-n}\|_{\mathscr{L}(X)} \leq \frac{1}{(n-1)!}\int_0^\infty e^{-t\operatorname{Re}\lambda}t^{n-1}Me^{\beta t}dt = \frac{M}{(\operatorname{Re}\lambda - \beta)^n} .$$

Hence we have proved (b). Trivially, (b) implies (c).

Finally suppose (c). Let $\|x\|_\mu := \sup_{n \in \mathbb{N}_0} \|(\mu - \beta)^n (\mu \mathrm{Id} - A)^{-n} x\|$ for $\mu > \beta$
and $x \in X$. Then $\|x\| \leq \|x\|_\mu \leq M \|x\|$ and $\left\|(\mu - \beta)(\mu \mathrm{Id} - A)^{-1} x\right\|_\mu \leq \|x\|_\mu$
for $x \in X$. Moreover, for $\beta < \lambda < \mu$

$$\left\|(\lambda \mathrm{Id} - A)^{-1} x\right\|_\mu = \left\|(\mu \mathrm{Id} - A)^{-1} x + (\mu - \lambda)(\mu \mathrm{Id} - A)^{-1}(\lambda \mathrm{Id} - A)^{-1} x\right\|_\mu$$

$$\leq \frac{1}{\mu - \beta} \|x\|_\mu + \frac{\mu - \lambda}{\mu - \beta} \left\|(\lambda \mathrm{Id} - A)^{-1} x\right\|_\mu$$

which shows $(\lambda - \beta) \left\|(\lambda \mathrm{Id} - A)^{-1} x\right\|_\mu \leq \|x\|_\mu$. This implies for any $n \in \mathbb{N}$
$\|(\lambda - \beta)^n (\lambda \mathrm{Id} - A)^{-n} x\| \leq \|(\lambda - \beta)^n (\lambda \mathrm{Id} - A)^{-n} x\|_\mu \leq \|x\|_\mu$ and $\|x\|_\mu \leq \|x\|_\lambda$
for $\beta < \lambda < \mu$. Hence $\|x\|^\sim := \lim_{\mu \to \infty} \|x\|_\mu$ for $x \in X$ defines a norm on X
with $\|x\| \leq \|x\|^\sim \leq M \|x\|$ and $\left\|(\lambda - \beta)(\lambda \mathrm{Id} - A)^{-1} x\right\|^\sim \leq \|x\|^\sim$. Therefore
we can assume $M = 1$ without restriction.
Let $A_\lambda := \lambda A (\lambda \mathrm{Id} - A)^{-1} = \lambda^2 (\lambda \mathrm{Id} - A)^{-1} - \lambda \mathrm{Id} \in \mathscr{L}(X)$ for $\lambda > \beta$. Then

$$\left\|e^{tA_\lambda}\right\|_{\mathscr{L}(X)} \leq e^{-t\lambda} e^{t\lambda^2 \left\|(\lambda \mathrm{Id} - A)^{-1}\right\|_{\mathscr{L}(X)}} \leq e^{-t\lambda} e^{\frac{t\lambda^2}{\lambda - \beta}} = e^{\frac{\lambda \beta t}{\lambda - \beta}} .$$

Moreover, for $x \in D(A)$ and $\lambda, \mu \geq \beta + 1$ we have

$$e^{tA_\lambda} x - e^{tA_\mu} x = \int_0^1 \frac{d}{ds} \left(e^{tsA_\lambda} e^{t(1-s)A_\mu} x\right) dx = \int_0^1 t e^{tsA_\lambda} (A_\lambda - A_\mu) e^{t(1-s)A_\mu} x \, ds$$

This shows $\left\|e^{tA_\lambda} x - e^{tA_\mu} x\right\| \leq e^{\beta(\beta+1)t} t \left\|A_\lambda x - A_\mu x\right\|$ for $\lambda, \mu \geq \beta + 1$ and
therefore $\lim_{\lambda \to \infty} e^{tA_\lambda} x \in X$ exists for $x \in D(A)$. Since $(e^{tA_\lambda})_{\lambda > \beta} \subset \mathscr{L}(X)$ is
uniformly bounded,

$$T(t)x := \lim_{\lambda \to \infty} e^{tA_\lambda} x$$

exists for $x \in X$ due to 1.1.3 and defines a continuous operator $T(t) \in \mathscr{L}(X)$ for
$t \geq 0$. Clearly, we have $T(0) = \mathrm{Id}, T(s + t) = T(s)T(t)$ and $\|T(t)\|_{\mathscr{L}(X)} \leq e^{\beta t}$
for $t, s \geq 0$. Moreover, for $x \in D(A), t \geq 0$, and small $h \in \mathbb{R}$ we have

$$\|T(t + h)x - T(t)x\| = \lim_{\lambda \to \infty} \left\| \int_t^{t+h} e^{sA_\lambda} A_\lambda x \, ds \right\| \leq |h| C \|Ax\|$$

with a suitable $C \geq 0$. This shows the continuity of $[0, \infty) \ni t \mapsto T(t)x \in X$
for $x \in D(A)$. Since $(T(t))_{t \geq 0} \subset \mathscr{L}(X)$ is locally bounded, 1.1.3 implies the
continuity of $[0, \infty) \ni t \mapsto T(t)x \in X$ for all $x \in X$ and $(T(t))_{t \geq 0}$ defines a
C_0-semigroup. Let B be the infinitesimal generator of this C_0-semigroup. We
still have to show $A = B$. First, for $x \in D(A)$ we have

$$T(t)x - x = \lim_{\lambda \to \infty} (e^{tA_\lambda} x - x) = \lim_{\lambda \to \infty} \int_0^t e^{sA_\lambda} A_\lambda x \, ds = \int_0^t T(s) Ax \, ds .$$

This shows $A \subset B$. Since $(\beta, \infty) \subset \rho(A) \cap \rho(B)$, this implies $A = B$. Hence
we have proved (a). $\qquad \square$

The connection between generators of C_0-semigroups and C_0-groups is given by the next proposition. In particular, together with the Hille-Yosida theorem we obtain a characterization of generators of C_0-groups.

1.1.5 Proposition. A linear operator $A : D(A) \to X$ in a Banach space X with domain $D(A) \subset X$ generates a C_0-group if and only if A and $-A$ generate C_0-semigroups.

PROOF: We only have to show that A generates a C_0-group, if $\pm A$ generate C_0-semigroups. Let $(T_\pm(t))_{t\geq 0}$ be the C_0-semigroups generated by $\pm A$. Then, for $x \in D(A)$ we have

$$\frac{d}{dt}T_+(t)T_-(t)x = AT_+(t)T_-(t)x + T_+(t)(-A)T_-(t)x = 0 .$$

Since $D(A) \subset X$ densely, this shows $T_+(t)T_-(t) = \mathrm{Id}$ and similarly $T_-(t)T_+(t) = \mathrm{Id}$. Hence $T_+(t)$ is invertible with $(T_+(t))^{-1} = T_-(t)$ for $t \geq 0$. Now it is straightforward to check that A generates the C_0-group

$$T(t) := \begin{cases} T_+(t) & \text{for } t \geq 0 \\ T_-(-t) & \text{for } t < 0 \end{cases}$$

$\square$

Next, we introduce the notion of dissipativity and quasi-dissipativity of linear operators in a Banach space. As we will show, quasi-dissipativity of linear operators is closely connected with apriori-estimates for solutions of related differential equations.

1.1.6 Proposition. Let $A : D(A) \to X$ be a linear operator with dense domain $D(A) \subset X$ in a Banach space X. Then the following statements are equivalent:

(a) $\|\lambda x - Ax\| \geq \mathrm{Re}\,\lambda \,\|x\|$ for λ with $\mathrm{Re}\,\lambda > 0, x \in D(A)$.

(b) $\|\lambda x - Ax\| \geq \lambda \,\|x\|$ for $\lambda > 0, x \in D(A)$.

(c) For $x \in D(A)$ and $x^* \in X^*$ with $\langle x^*, x \rangle = \|x^*\|^2 = \|x\|^2$ we have $\mathrm{Re}\,\langle x^*, Ax \rangle \leq 0$.

(d) For $x \in D(A)$ there is a $x^* \in X^*$ with $\langle x^*, x \rangle = \|x^*\|^2 = \|x\|^2$ and $\mathrm{Re}\,\langle x^*, Ax \rangle \leq 0$.

In this case $(A, D(A))$ is called dissipative. A is called quasi-dissipative with constant $\beta \geq 0$, if $A - \beta\mathrm{Id}$ is dissipative.

PROOF: We clearly have (a) $\Rightarrow$ (b) and (c) $\Rightarrow$ (d). To prove (b) $\Rightarrow$ (c) let $x^* \in X^*$ with $\langle x^*, x \rangle = \|x^*\|^2 = \|x\|^2$ and choose $y_k \in D(A)$ with $y_k \underset{k \to \infty}{\longrightarrow} Ax$, then for $t > 0$ and $k \in \mathbb{N}$

$$\text{Re } \langle x^*, Ax \rangle = \frac{1}{t} \left(\text{Re } \langle x^*, x + tAx \rangle - \text{Re } \langle x^*, x \rangle \right)$$

$$\leq \frac{1}{t} \|x\| \left(\|x + tAx\| - \|x\| \right) = \frac{1}{t} \|x\| \left(\|x + ty_k + tAx - ty_k\| - \|x\| \right)$$

$$\leq \frac{1}{t} \|x\| \left(\|x + ty_k\| + t \|Ax - y_k\| - \|x\| \right)$$

$$\leq \frac{1}{t} \|x\| \left(t \left\| \left(\frac{1}{t}\text{Id} - A \right) (x + ty_k) \right\| + t \|Ax - y_k\| - \|x\| \right)$$

$$\leq \frac{1}{t} \|x\| \left(\|x\| + t \|y_k - Ax\| + t^2 \|Ay_k\| + t \|Ax - y_k\| - \|x\| \right)$$

$$\leq \|x\| \left(2 \|y_k - Ax\| + t \|Ay_k\| \right)$$

Now let first $t \to 0$ and then $k \to \infty$. Then this implies (b) $\Rightarrow$ (c). Finally, (d) $\Rightarrow$ (a) is a consequence of

$$\|x\| \|\lambda x - Ax\| \geq |\langle x^*, \lambda x - Ax \rangle| \geq \text{Re } \lambda \langle x^*, x \rangle - \text{Re } \langle x^*, Ax \rangle \geq \text{Re } \lambda \|x\|^2$$

$\square$

This proposition has two important corollaries.

1.1.7 Corollary.

(a) Let $A_j : D \to X, j = 1, 2$, be two quasi-dissipative operators with constants $\beta_j \geq 0$ and a common and dense domain D in a Banach space X. Then $A_1 + A_2 : D \to X$ is quasi-dissipative with constant $\beta_1 + \beta_2$.

(b) Let $A : D(A) \to X$ be a densely defined, quasi-dissipative operator in a Hilbert space $(H, \langle \cdot, \cdot \rangle)$. Then A is quasi-dissipative with constant $\beta \geq 0$ if and only if Re $\langle Ax, x \rangle \leq \beta \|x\|^2$ for $x \in D(A)$.

Quasi-dissipativity implies a-priori estimates for solutions of evolution equations as it is shown by the following proposition and its corollary. In the proof we will have to make use of the Dini-derivative of a continuous function.

1.1.8 Lemma. Let $I \subset \mathbb{R}$ be an interval and $f : I \to \mathbb{R}$ be continuous. Then the left upper Dini-derivative of f is defined by

$$D^- f(t) := \limsup_{h \to 0+} \frac{f(t - h) - f(t)}{-h}.$$

If $D^- f(t) \leq 0$ for all $t \in I$ then f is decreasing.
Similarly, right upper Dini-derivatives and lower Dini-derivatives are defined. Then the same statement holds also for these derivatives.

PROOF: Let $\varepsilon > 0$ and consider $f_\varepsilon(t) := f(t) - \varepsilon t$. Let $a, b \in I$ with $a < b$ and suppose $f_\varepsilon(a) < f_\varepsilon(b)$. There exists an $\alpha \in \mathbb{R}$ with $f_\varepsilon(a) < \alpha < f_\varepsilon(b)$ and $c_\alpha := \inf\{x \in [a, b] : f_\varepsilon(x) \geq \alpha\}$. Now $f_\varepsilon(c_\alpha) = \alpha$ and $f_\varepsilon(c_\alpha - h) < f_\varepsilon(c_\alpha)$ for $h > 0$ sufficiently small. Therefore, for small $h > 0$ we obtain the contradiction

$$0 = \frac{\alpha - \alpha}{-h} < \frac{f_\varepsilon(c_\alpha - h) - f_\varepsilon(c_\alpha)}{-h} = \frac{f(c_\alpha - h) - f(c_\alpha)}{-h} - \varepsilon.$$

Hence $f(a) - \varepsilon a = f_\varepsilon(a) \geq f_\varepsilon(b) = f(b) - \varepsilon b$ and with $\varepsilon \to 0$ we obtain the assertion. $\qquad \square$

1.1.9 Proposition. Assume that

(a) $(X, \|\cdot\|)$ is a Banach space, $I = [t_0, t_1]$ is a compact interval, $\beta \geq 0$.

(b) $A(t) : D(A(t)) \to X$ for $t \in I$ is a family of linear operators in X with domains $D(A(t))$.

(c) $u \in C^1(I, X)$ with $u(t) \in D(A(t))$ and $u'(t) = A(t)u(t)$ for $t \in I$.

(d) $\Phi : X \to [0, \infty)$ is a function with $\Phi(\lambda x) = \lambda \Phi(x), \Phi(x+y) \leq \Phi(x) + \Phi(y)$, and $\Phi(x) \leq \|x\|$ for $x, y \in X, \lambda \in [0, \infty)$.

(e) $\Phi(\lambda x - A(t)x) \geq (\lambda - \beta)\Phi(x)$ for $\lambda > \beta, x \in D(A(t)), t \in I$.

Then we have $\Phi(u(t)) \leq e^{\beta(t-t_0)}\Phi(u(t_0))$ for $t \in I$.

PROOF: Let $v(t) := e^{-\beta t}u(t)$ and $f(t) := \Phi(v(t)) = e^{-\beta t}\Phi(u(t))$ for $t \in I$. Then $v \in C^1(I, X)$ with

$$v'(t) = -\beta e^{-\beta t}u(t) + e^{-\beta t}A(t)u(t) = (A(t) - \beta)v(t)$$

and with $r_t(h) := \Phi(v(t-h)) - \Phi(v(t) + v'(t)(-h))$ for $t \in (t_0, t_1], h \in (0, t - t_0)$ we have

$$
\begin{aligned}
f(t - h) &= \Phi(v(t) + v'(t)(-h)) + r_t(h) = \Phi(v(t) - h(A(t) - \beta)v(t)) + r_t(h) \\
&= h\Phi((\beta + \tfrac{1}{h})v(t) - A(t)v(t)) + r_t(h) \geq f(t) + r_t(h).
\end{aligned}
$$

Hence for the left upper Dini-derivative of f we obtain

$$D^- \leq \limsup_{h \to 0+} \frac{\Phi(v(t) + v'(t)(-h)) - \Phi(v(t-h))}{h}$$

$$\leq \limsup_{h \to 0+} \frac{\Phi(-v(t-h) + v(t) + v'(t)(-h)) + \Phi(v(t-h)) - \Phi(v(t-h))}{h}$$

$$\leq \limsup_{h \to 0} \frac{\|-v(t-h) + v(t) + v'(t)(-h)\|}{h} = 0$$

Hence f is decreasing, which proves the assertion. $\qquad\square$

As a corollary we obtain the following apriori estimate.

1.1.10 Corollary. Let $(X, \|\cdot\|)$ be a Banach space, $A(t) : D(A(t)) \to X$ be a family of linear operators in X for $t \in I := [t_0, t_1]$, and $u \in C^1([t_0, t_1], X)$ with $u(t) \in D(A(t))$ and

$$\frac{du}{dt}(t) = A(t)u(t) \quad \text{for } t \in I.$$

Moreover, suppose there is a $\beta \geq 0$ with $\|\lambda x - A(t)x\| \geq (\lambda - \beta)\|x\|$ for $x \in D(A(t)), \lambda > \beta, t \in I$. Then $\|u(t)\| \leq e^{\beta(t-t_0)}\|u(t_0)\|$ for $t \in I$.

This a-priori estimate has the following converse.

1.1.11 Lemma. Let $(X, \|\cdot\|)$ be a Banach space with a dense subspace D, $I = [t_0, t_1] \subset \mathbb{R}$ be an interval, $A(t) : D \to D$ for $t \in I$ be a family of linear operators with $I \ni t \mapsto A(t)x \in X$ continuously for $x \in D$, and $U(t, s) \in \mathcal{L}(X)$ be linear operators for $t_0 \leq s \leq t \leq t_1$ with $U(s, s) = \mathrm{Id}, U(t, s)(D) \subset D$ and $\|U(t, s)\|_{\mathcal{L}(X)} \leq e^{\beta(t-s)}$ for $t_0 \leq s \leq t \leq t_1$ and a constant $\beta \geq 0$.
If $[s, t_1] \ni t \mapsto U(t, s)x \in X$ is differentiable for any $x \in D$ with derivative $A(t)U(t, s)x$, then $\|\lambda x - A(t)x\| \geq (\lambda - \beta)\|x\|$ for $x \in D, t \in I, \lambda > \beta$.

PROOF: By continuity we can assume that t is not the right boundary point of I. Choose an x^* in the dual X^* of X with $\langle x^*, x \rangle = \|x\|^2 = \|x^*\|^2$, then

$$\mathrm{Re}\,\langle x^*, A(t)x \rangle \xleftarrow[+0 \leftarrow h]{} \mathrm{Re}\,\left\langle x^*, \frac{U(t+h, t)x - U(t, t)x}{h} \right\rangle$$

$$= \frac{\mathrm{Re}\,\langle x^*, U(t+h, t)x \rangle - \langle x^*, x \rangle}{h} \leq \frac{e^{\beta h} - 1}{h}\|x\|^2 \xrightarrow[h \to 0+]{} \beta\|x\|^2.$$

Thus

$$\begin{aligned}
\|x\|\,\|\lambda x - A(t)x\| &\geq |\mathrm{Re}\,\langle x^*, \lambda x - A(t)x \rangle| \geq \lambda\langle x^*, x \rangle - \mathrm{Re}\,\langle x^*, A(t)x \rangle \\
&\geq (\lambda - \beta)\|x\|^2.
\end{aligned}$$

$\qquad\square$

As a consequence, we have the Lumer-Phillips theorem.

1.1.12 Theorem. For a closed linear operator $A : D(A) \to X$ with dense domain $D(A) \subset X$ and $\beta > 0$ the following statements are equivalent:

(a) A generates a C_0-semigroup $(T(t))_{t \geq 0}$ with $\|T(t)\|_{\mathscr{L}(X)} \leq e^{\beta t}$ for $t \geq 0$.

(b) A is quasi-dissipative with constant β and $\mathcal{R}(\lambda \mathrm{Id} - A) = X$ for all λ with $\mathrm{Re}\, \lambda > \beta$.

(c) A is quasi-dissipative with constant β and there is a λ with $\mathrm{Re}\, \lambda > \beta$ and $\mathcal{R}(\lambda \mathrm{Id} - A) = X$.

PROOF: We only have to prove that (c) implies (b). It is sufficient to show that $\Lambda := \{\lambda : \mathrm{Re}\, \lambda \in (\beta, \infty), \mathcal{R}(\lambda \mathrm{Id} - A) = X\} \subset \{\lambda : \mathrm{Re}\, \lambda \in (\beta, \infty)\}$ is open and closed. If $\lambda \in \Lambda$, then $R(\lambda') := \sum_{k=0}^{\infty} (\lambda - \lambda')^k (\lambda \mathrm{Id} - A)^{-(k+1)} \in \mathscr{L}(X)$ exists for $|\lambda' - \lambda| < \|(\lambda \mathrm{Id} - A)^{-1}\|^{-1}$. It is easy to show $(\lambda' \mathrm{Id} - A) R(\lambda') = \mathrm{Id}$. This proves that Λ is open. Moreover, take $(\lambda_k)_k \subset \Lambda$ with $\lambda_k \xrightarrow{k \to \infty} \lambda$ and $\mathrm{Re}\, \lambda \in (\beta, \infty)$. We have $\|(\lambda_k \mathrm{Id} - A)^{-1}\| \leq \frac{1}{\lambda_k - \beta}$. Hence there is a $\varepsilon > 0$ such that $R(\lambda')$ exists for $|\lambda' - \lambda_k| < \varepsilon$ and $k \in \mathbb{N}_0$. For k sufficiently large we have $|\lambda - \lambda_k| < \varepsilon$. This proves that Λ is closed. $\qquad\square$

For closable operators, we have the following corollary.

1.1.13 Corollary. Let $A : D(A) \to X$ be a densely-defined, closable linear operator with domain $D(A)$ in a Banach space X.
Then the closure of A generates a quasi-contractive C_0-group with constant β if and only if A is quasi-dissipative with constant β and $\mathcal{R}(\lambda \mathrm{Id} - A) \subset X$ densely for one (resp. for all) $\lambda > \beta$.

PROOF: We only have to show $\overline{\mathcal{R}(\lambda \mathrm{Id} - A)} = \mathcal{R}(\lambda \mathrm{Id} - \overline{A})$ for the closure $\overline{A}$ of A and $\lambda > 0$. But this can be proved easily using the quasi-dissipativity of A. $\square$

Generators of semigroups can be reconstructed if they are known on dense invariant subsets of their semigroup. We will use this fact to show that closures of several operators generate C_0-semigroups.

1.1.14 Proposition. Let $(T(t))_{t \geq 0}$ be a C_0-semigroup of linear operators on a Banach space X with infinitesimal generator $A : D(A) \to X$. Moreover, let $D \subset D(A)$ be a dense subspace and suppose $T(t)(D) \subset D$ for all $t \geq 0$. Then A is the closure of its restriction to D, i.e., D is a core of A.

PROOF: There are $M \geq 1, \beta > 0$ with $\|T(t)\|_{\mathscr{L}(X)} \leq Me^{\beta t}$ for $t \geq 0$. Let $\lambda > \beta$ and assume that $(\lambda\text{Id} - A)(D) \subset X$ is not dense. Then there is an $x^* \in X^*$ with $x^* \neq 0$ and $x^*|_{(\lambda\text{Id}-A)(D)} = 0$. Pick an $x \in D$ with $\langle x^*, x \rangle \neq 0$. Then

$$\frac{d}{dt}\langle x^*, T(t)x \rangle = \langle x^*, AT(t)x \rangle = \lambda \langle x^*, T(t)x \rangle$$

because $T(t)x \in D$. Therefore $\langle x^*, T(t)x \rangle = e^{\lambda t}\langle x^*, x \rangle$. But this is impossible because $e^{\lambda t}\langle x^*, x \rangle = \langle x^*, T(t)x \rangle \leq Me^{\beta t} \|x\|_X \|x^*\|_{X^*}$. Since $\lambda\text{Id} - A$ is invertible, $(\lambda\text{Id} - A)(D)$ densely implies the assertion.

□

At the end of this section we prove a perturbation result. Perturbation results are often used to verify that a sum of operators generates a C_0-semigroup provided that further information on its terms are known. These terms are usually simpler to treat than the whole sum.

1.1.15 Proposition. Let $S : D(S) \to X$ be the generator of a C_0-semigroup $(e^{tS})_{t \geq 0}$ in a Banach space X with $\|e^{tS}\|_{\mathscr{L}(X)} \leq e^{\gamma t}$ for $t \geq 0$ and a suitable $\gamma \geq 0$, and let $A : D(S) \to X$ be quasi-dissipative with constant $\beta \geq 0$. Moreover, assume that there are $0 < \delta < 1, b_\delta > 0$ with $\|Ax\| \leq \delta \|Sx\| + b_\delta \|x\|$ for $x \in D(S)$. Then $S + A : D(S) \to X$ generates a C_0-semigroup with $\|e^{t(S+A)}\|_{\mathscr{L}(X)} \leq e^{(\gamma+\beta)t}$ for $t \geq 0$.

PROOF: If S generates the C_0-semigroup $(T(t))_{t \geq 0}$ and if $\lambda > 0$, then $S - \lambda\text{Id}$ generates the C_0-semigroup $(e^{-\lambda t}T(t))_{t \geq 0}$. Therefore by considering $S - \gamma\text{Id}$ and $A - \beta\text{Id}$ at the place of S and A we can assume without restriction that S generates a C_0-semigroup of contractions and that A is dissipative. First we assume $\delta < \frac{1}{2}$. Since $S + A$ is dissipative due to 1.1.7 we only have to show $\mathcal{R}(\lambda\text{Id} - S - A) = X$ for large $\lambda > 0$ using the Lumer-Phillips theorem 1.1.12. For $x \in X$ we have

$$\|A(\lambda\text{Id} - S)^{-1}x\| \leq \delta \|S(\lambda\text{Id} - S)^{-1}x\| + b_\delta \|(\lambda\text{Id} - S)^{-1}x\|$$
$$= \delta \|\lambda(\lambda\text{Id} - S)^{-1}x - x\| + b_\delta \|(\lambda\text{Id} - S)^{-1}x\| \leq \left(2\delta + \frac{b_\delta}{\lambda}\right)\|x\|$$

This shows that $\text{Id} - A(\lambda\text{Id} - S)^{-1}$ is invertible for large $\lambda > 0$ and therefore

$$\mathcal{R}(\lambda\text{Id} - (S+A)) \supset \mathcal{R}((\lambda\text{Id} - (S+A))(\lambda\text{Id} - S)^{-1}) = \mathcal{R}(\text{Id} - A(\lambda\text{Id} - S)^{-1}) = X.$$

This implies the assertion for $\delta < \frac{1}{2}$.
Now assume $\delta < 1$ and let $0 \leq t \leq 1$. For $x \in D(A)$ we have

$$\|(S + tA)x\| \geq \|Sx\| - t\|Ax\| \geq (1 - \delta)\|Sx\| - b_\delta \|x\|$$

Fix an integer n with $\frac{\delta}{n} < \frac{1-\delta}{4}$. Then

$$\left\| \frac{1}{n} Ax \right\| \le \frac{\delta}{n} \|Sx\| + \frac{b_\delta}{n} \|x\| \le \frac{1}{4} \|(S + tA)x\| + \left(\frac{b_\delta}{n} + \frac{b_\delta}{4} \right) \|x\| .$$

Therefore the assertion for $\delta < \frac{1}{2}$ implies that $S + tA + \frac{1}{n}A$ generates a contractive C_0-semigroup provided that $S + tA$ does. Together with the assertion for $\delta < 1$ this completes the proof.   □

This result contains an important and simple perturbation result as a special case: If we perturb the generator of a C_0-semigroup by a bounded operator, the perturbed operator still generates a C_0-semigroup:

1.1.16 Corollary. Let $S : D(S) \to X$ be the infinitesimal generator of a C_0-semigroup in a Banach space X and let $B \in \mathscr{L}(X)$.
Then $S + B : D(S) \to X$ generates a C_0-semigroup.

PROOF: Let $(T(t))_{t\ge 0}$ be the C_0-semigroup generated by S. Then there are $M \ge 1, \beta \ge 0$ with $\|T(t)\|_{\mathscr{L}(X)} \le Me^{\beta t}$ for $t \ge 0$. By considering $S - \beta \mathrm{Id}$ at the place of S we can assume without restriction that $\beta = 0$. Moreover, $\|x\|^{\sim} := \sup_{t>0} \|T(t)x\|$ defines a norm on X with $\|x\| \le \|x\|^{\sim} \le M \|x\|$ and $\|T(t)x\|^{\sim} \le \|x\|^{\sim}$ for $t \ge 0$ and $x \in X$. Using the norm $\|\cdot\|^{\sim}$ at the place of $\|\cdot\|$ we can therefore assume without restriction that $M = 1$. Since B is dissipative with constant $\|B\|_{\mathscr{L}(X)}$, the assertion is a consequence of 1.1.15.
                                          □

1.1.17 Bibliographical remarks. Many books on functional analysis deal with unbounded, linear operators and C_0-semigroups. Let us mention here for example Hille-Phillips [63], Fattorini [44], Goldstein [52], Pazy [122], Davies [36], Reed-Simon [127], Clément et al. [28], or Engel-Nagel [43]. These books provide thorough introductions into the theory of C_0-semigroups and several results in this section are discussed in detail in these books. Other results in this section are rarely discussed in textbooks on C_0-semigroups. The characterization of dissipativity 1.1.6 has been given by Arendt-Chernoff-Kato [10], the apriori estimate 1.1.10 by Kisyński [91].

1.2   Selfadjoint operators

In this section we treat a special type of generators of C_0-semigroups in Hilbert spaces, so-called self-adjoint operators. Throughout this section $(H, \langle \cdot, \cdot \rangle, \|\cdot\|)$ will denote a complex Hilbert space.

1.2.1 Definition. For a densely-defined linear operator A in H with domain $D(A) \subset H$ we define the adjoint operator A^* with domain

$$D(A^*) := \{x \in H : \text{there is an } \tilde{x} \in H \text{ with } \langle Ay, x \rangle = \langle y, \tilde{x} \rangle \text{ for } y \in D(A)\}$$

by $A^*x := \tilde{x}$ for $x \in D(A^*)$.

The operator A is called symmetric if $A \subset A^*$, i.e., $\langle Ax, y \rangle = \langle x, Ay \rangle$ for $x, y \in D(A)$, selfadjoint if $A^* = A$, and skew-selfadjoint if $A^* = -A$. A closable linear operator $A : D(A) \to H$ is called essentially selfadjoint, if its closure is selfadjoint.

The next lemma collects some basic properties of linear operators in Hilbert spaces and their adjoints.

1.2.2 Lemma. Let $A : D(A) \to H$ be a densely-defined linear operator with domain $D(A) \subset H$ in H. Then

(a) A^* is a closed operator.

(b) A is closable if and only if $D(A^*) \subset H$ densely. In this case we have $\overline{A} = A^{**}$.

(c) $\mathcal{N}(A^*) = \mathcal{R}(A)^\perp$ and $\overline{\mathcal{R}(A)} = \mathcal{N}(A^*)^\perp$.

(d) If A is symmetric, then A is closable and $\|\lambda x - Ax\| \geq |\text{Im } \lambda| \|x\|$ for $\lambda \in \mathbb{C}$ and $x \in D(A)$.

PROOF: We prove the closedness of A^* by proving the closedness of the graph $\mathcal{G}(A^*) := \{(x, A^*x) : x \in D(A^*)\}$ of A^* in the Hilbert space $H \times H$. Note that $U : H \times H \ni (x, y) \mapsto (-y, x) \in H \times H$ defines a unitary operator. Take $y \in D(A^*)$ and $x \in D(A)$, then $\langle (y, A^*y), U(x, Ax) \rangle = -\langle y, Ax \rangle + \langle A^*y, x \rangle = 0$, which proves $\mathcal{G}(A^*) \subset (U(\mathcal{G}(A)))^\perp$. Conversely, for $(y, z) \in (U(\mathcal{G}(A)))^\perp$ we have $0 = \langle (y, z), U(x, Ax) \rangle = -\langle y, Ax \rangle + \langle z, x \rangle \quad$ for $x \in D(A)$, which proves $y \in D(A^*)$ with $A^*y = z$ showing $(U(\mathcal{G}(A)))^\perp \subset \mathcal{G}(A^*)$. This proves (a) because $\mathcal{G}(A^*) = (U(\mathcal{G}(A)))^\perp$. Moreover, we have

$$\overline{\mathcal{G}(A)} = (\mathcal{G}(A)^\perp)^\perp = U^*U(\mathcal{G}(A)^\perp)^\perp = U^*(\mathcal{G}(A^*))^\perp$$

Hence $\overline{\mathcal{G}(A)} = \mathcal{G}(A^{**})$, if A^* is densely-defined. If $D(A^*) \subset H$ is not a dense subspace, then there is a $0 \neq y \in D(A^*)^\perp$ and $(0, y) \in U^*(\mathcal{G}(A^*))^\perp = \overline{\mathcal{G}(A)}$, hence A is not closable. Therefore, we have shown (b).

The first part of (c) follows from

$$x \in \mathcal{N}(A^*) \iff \langle Ay, x \rangle = \langle y, A^*x \rangle = 0 \text{ for } y \in D(A) \iff x \in \mathcal{R}(A)^\perp$$

and the second part from $\overline{\mathcal{R}(A)} = (\mathcal{R}(A)^\perp)^\perp = \mathcal{N}(A^*)^\perp$.

If A is symmetric and if $(x_k)_k \subset D(A)$ with $x_k \xrightarrow[k\to\infty]{} 0$ and $Ax_k \xrightarrow[k\to\infty]{} y \in X$, then

$$\langle y, x \rangle \xleftarrow[\infty\leftarrow k]{} \langle Ax_k, x \rangle = \langle x_k, Ax \rangle \xrightarrow[k\to\infty]{} 0 \quad \text{for } x \in D(A),$$

which implies $y = 0$ and A is closable. Finally, the second assertion in (d) follows from

$$\|\lambda x - Ax\|^2 = \langle (\operatorname{Re} \lambda x - Ax) + i\operatorname{Im} \lambda x), (\operatorname{Re} \lambda x - Ax) + i\operatorname{Im} \lambda x)\rangle$$
$$= \|\operatorname{Re} \lambda x - Ax\|^2 + |\operatorname{Im} \lambda|^2 \|x\|^2 \geq |\operatorname{Im} \lambda|^2 \|x\|^2$$

$\square$

Now, we give a fundamental characterization of selfadjointness. On the one hand, the next theorem shows that a linear operator is selfadjoint if and only if it is symmetric and satisfies a range condition. On the other hand, the theorem shows that A is self-adjoint if and only if iA generates a C_0-group of unitary operators. This second result is also called Stone's theorem.

1.2.3 Theorem. For a closed linear operator $A : D(A) \to H$ with dense domain $D(A) \subset H$ in H the following statements are equivalent:

(a) A is selfadjoint.

(b) A is symmetric and $\mathcal{R}(\lambda\operatorname{Id} - A) = H$ for $\lambda \in \mathbb{C} \setminus \mathbb{R}$.

(c) A is symmetric and there are $\lambda_+, \lambda_- \in \mathbb{C}$ with $\operatorname{Im} \lambda_+ > 0, \operatorname{Im} \lambda_- < 0$, and $\mathcal{R}(\lambda_\pm\operatorname{Id} - A) = H$

(d) iA generates a C_0-group of unitary operators , i.e., iA generates a C_0-group $(T(t))_{t\in\mathbb{R}} \subset \mathcal{L}(H)$ with $T(t)^* = T(-t)$ for $t \in \mathbb{R}$.

PROOF: First, assume (a). Let $\lambda \in \mathbb{C} \setminus \mathbb{R}$ and take $x \in \mathcal{N}(\overline{\lambda}\operatorname{Id} - A)$, then we have $\overline{\lambda}\langle x, x \rangle = \langle Ax, x \rangle = \langle x, Ax \rangle = \lambda\langle x, x \rangle$. This shows $x = 0$ for $\lambda \in \mathbb{C} \setminus \mathbb{R}$. Therefore $\mathcal{R}(\lambda\operatorname{Id} - A)^\perp = \mathcal{N}(\lambda\operatorname{Id} - A^*) = \mathcal{N}(\overline{\lambda}\operatorname{Id} - A) = \{0\}$ due to 1.2.2(c) and $\mathcal{R}(\lambda\operatorname{Id} - A) = \overline{\mathcal{R}(\lambda\operatorname{Id} - A)} = H$ because $\mathcal{R}(\lambda\operatorname{Id} - A)$ is closed due to 1.2.2(d). This proves (b). Clearly, (b) implies (c). Now assume (c), then

$$2\operatorname{Re} \langle iAx, x \rangle = \langle iAx, x \rangle + \langle x, iAx \rangle = i\langle Ax, x \rangle - i\langle Ax, x \rangle = 0 \quad \text{for } x \in D(A).$$

Hence $\pm iA$ are dissipative and the Lumer-Phillips theorem and 1.1.5 show that iA generates a C_0-group $(T(t))_{t\in\mathbb{R}}$. Moreover, for $x, y \in D(A)$ we have

$$\frac{d}{dt}\langle T(t)x, T(t)y \rangle = \langle iAT(t)x, T(t)y \rangle + \langle T(t)x, iAT(t)y \rangle = 0.$$

Hence $\langle T(t)x, T(t)y \rangle = \langle x, y \rangle$ for $x, y \in H$ because $D(A) \subset H$ densely. This proves (d).

If iA generates a C_0-group of unitary operators $(T(t))_{t\in\mathbb{R}}$, then

$$\langle Ax, y\rangle = \lim_{t\to 0}\frac{1}{i}\left\langle \frac{T(t)x - x}{t}, y\right\rangle = \lim_{t\to 0}\frac{1}{i}\left\langle x, \frac{T(-t)y - y}{t}\right\rangle = \langle x, Ay\rangle$$

for $x, y \in D(A)$. Therefore, the Lumer-Phillips theorem implies (d) $\Rightarrow$ (b). Finally, assume (b). Since A is symmetric we have to show $D(A^*) \subset D(A)$ for the proof of (a). Note that $\mathbb{C} \setminus \mathbb{R} \subset \rho(A)$ due to 1.2.2(d) and take $x \in D(A^*)$. Then $x_0 := (i\mathrm{Id} - A)^{-1}(i\mathrm{Id} - A^*)x \in D(A) \subset D(A^*)$ with $Ax_0 = A^*x_0$. Then

$$(i\mathrm{Id} - A^*)(x - x_0) = (i\mathrm{Id} - A^*)x - (i\mathrm{Id} - A)(i\mathrm{Id} - A)^{-1}(i\mathrm{Id} - A^*)x = 0.$$

This shows $x - x_0 \in \mathcal{N}(i\mathrm{Id} - A^*) = \mathcal{R}(-i\mathrm{Id} - A)^{\perp} = H^{\perp} = \{0\}$ due to 1.2.2(c) and therefore $x = x_0 \in D(A)$. $\qquad\square$

This result implies immediately the next corollary.

1.2.4 Corollary. For a densely-defined operator $A : D(A) \to H$ with domain $D(A) \subset H$ in H the following statements are equivalent:

(a) A is essentially selfadjoint.

(b) A is symmetric and $\mathcal{R}(\lambda\mathrm{Id} - A) \subset H$ densely for $\lambda \in \mathbb{C}\setminus\mathbb{R}$.

(c) A is symmetric and there are $\lambda_+, \lambda_- \in \mathbb{C}$ with $\mathrm{Im}\,\lambda_+ > 0, \mathrm{Im}\,\lambda_- < 0$, and $\mathcal{R}(\lambda_{\pm}\mathrm{Id} - A) \subset H$ densely.

(d) iA is closable and its closure generates a C_0-group of unitary operators.

PROOF: We only have to note $\overline{\mathcal{R}(\lambda\mathrm{Id} - A)} = \mathcal{R}(\lambda\mathrm{Id} - \overline{A})$ for $\lambda \in \mathbb{C}\setminus\mathbb{R}$ and symmetric A due to 1.2.2(d). $\qquad\square$

For semibounded, symmetric operators the situation is especially nice.

1.2.5 Proposition. Let $A : D(A) \to H$ be a symmetric, densely defined, linear operator in H with $\langle Ax, x\rangle \leq \beta\,\|x\|^2$ for $x \in D(A)$ and a $\beta > 0$. If $\mathcal{R}(\lambda\mathrm{Id} - A) \subset H$ densely for a $\lambda > \beta$, then A is essentially selfadjoint on $D(A)$ in H.

PROOF: We only have to show $D(A^*) \subset D(\overline{A})$ due to the symmetry of A. 1.1.7 implies $\mathcal{R}(\lambda\mathrm{Id} - \overline{A}) = \overline{\mathcal{R}(\lambda\mathrm{Id} - A)} = H$. Moreover, take $x \in D(A^*)$ and let $y := (\lambda\mathrm{Id} - A^*)x$. Then there is an $y_0 \in D(\overline{A}) \subset D(A^*)$ satisfying $(\lambda\mathrm{Id} - A^*)y_0 = (\lambda\mathrm{Id} - \overline{A})(y_0) = y = (\lambda\mathrm{Id} - A^*)x \in H$. This implies that $x - y_0 \in \mathcal{N}(\lambda\mathrm{Id} - A^*) = \mathcal{R}(\lambda\mathrm{Id} - A)^{\perp} = \{0\}$, which proves $x = y_0 \in D(\overline{A})$. $\qquad\square$

Chernoff has given a condition related to 1.1.14 that implies essential selfadjointness of all powers of symmetric operators.

1.2.6 Proposition. Let $A : D(A) \to H$ be a symmetric operator in H, and
$(T(t))_{t \in \mathbb{R}}$ be a C_0-group of unitary operators in H. Moreover, let $D \subset D(A)$
be a dense subspace of H with $A(D) \subset D$ and $T(t)D \subset D$ for all $t \in \mathbb{R}$ such
that $\mathbb{R} \ni t \mapsto T(t)x \in H$ is differentiable with $\frac{d}{dt}T(t)x = iAT(t)x = iT(t)Ax$
for $x \in D$. Then every power of A is essentially selfadjoint on D.

PROOF: $S := A^n$ for an $n \in \mathbb{N}_0$ defines a symmetric operator with domain
D. Let $x \in D(S^*)$ with $S^*x = ix$ and consider $f(t) := \langle T(t)y, x \rangle$ for a fixed
$y \in D$. Then

$$f^{(n)}(t) = \langle i^n A^n T(t)y, x \rangle = \langle i^n ST(t)y, x \rangle = \langle i^n T(t)y, S^*x \rangle = -i^{n+1} f(t) .$$

Hence $f(t)$ is a linear combination of exponentials $\exp(\alpha t)$, where α runs
through the solutions of $\alpha^n = -i^{n+1}$. None of these α is purely imaginary.
But f is bounded because $T(t)$ is unitary, hence $f(t) = 0$ for $t \in \mathbb{R}$. In particular, $\langle y, x \rangle = 0$ for any $y \in D$, which implies $x = 0$. Hence $\mathcal{N}(-i\mathrm{Id} - S^*) = \{0\}$
and similarly $\mathcal{N}(i\mathrm{Id} - S^*) = \{0\}$. Therefore, 1.2.2(c) implies the assertion
because $\mathcal{R}(\pm i\mathrm{Id} - A^n)^{\perp} = \mathcal{N}(\mp i\mathrm{Id} - S^*) = \{0\}$.   □

Finally, we will prove some properties of quadratic forms. Quadratic forms and
their closures are strongly connected with symmetric operators and selfadjoint
extensions.

1.2.7 Definition. A quadratic form $q : D(q) \times D(q) \to \mathbb{C}$ in H with domain
$D(q)$ is a mapping such that

- $D(q)$ is a dense subset of H.

- $q(\lambda x_1 + \mu x_2, y) = \lambda q(x_1, y) + \mu q(x_2, y)$ for $\lambda, \mu \in \mathbb{C}$ and $x_1, x_2, y \in D(q)$.

- $q(x, \lambda y_1 + \mu y_2) = \overline{\lambda} q(x, y_1) + \overline{\mu} q(x, y_2)$ for $\lambda, \mu \in \mathbb{C}$ and $x, y_1, y_2 \in D(q)$.

q is called symmetric, if $q(x, y) = \overline{q(y, x)}$ for $x, y \in D(q)$ and called semi-
bounded by $M \in \mathbb{R}$, if q is symmetric and $q(x, x) \geq M \|x\|^2$ for $x \in D(q)$. A
semibounded quadratic form $q : D(q) \times D(q) \to \mathbb{C}$ is called closed, if for any se-
quence $(x_n)_n \subset D(q)$ and $x \in H$ with $x_n \xrightarrow[n \to \infty]{} x$ and $q(x_n - x_m, x_n - x_m) \xrightarrow[n,m \to \infty]{} 0$
we have $x \in D(q)$ and $q(x_n - x, x_n - x) \xrightarrow[n \to \infty]{} 0$. q is called closable, if for any
sequence $(x_n)_n \subset D(q)$ with $x_n \xrightarrow[n \to \infty]{} 0$ and $q(x_n - x_m, x_n - x_m) \xrightarrow[n,m \to \infty]{} 0$ we have
$q(x_n, x_n) \xrightarrow[n \to \infty]{} 0$.

1.2.8 Lemma. Let $q : D(q) \times D(q) \to \mathbb{C}$ be a quadratic form in H that is semibounded by $M \in \mathbb{R}$ and let $\langle x, y \rangle_q := q(x, y) + (1 - M)\langle x, y \rangle$ for $x, y \in D(q)$. Then the following holds:

(a) q is closed if and only if $(D(q), \langle \cdot, \cdot \rangle_q)$ is a Hilbert space.

(b) If q is closable, then there is a closed extension $\bar{q} : D(\bar{q}) \times D(\bar{q}) \to \mathbb{C}$ of q to a quadratic form $\bar{q}$, called the closure $\bar{q}$ of q.

PROOF: It is straightforward to verify (a). Note that $\|x\|_q := \sqrt{\langle x, x \rangle_q} \geq \|x\|$ for $x \in D(q)$. Let $D(\bar{q})$ be the closure of $D(q)$ with respect to $\langle \cdot, \cdot \rangle_q$ and let $\iota : D(q) \hookrightarrow X$ denote the embedding. Since ι is continuous there is a continuous linear extension $\bar{\iota} : D(\bar{q}) \to X$. We have to show that $\bar{\iota}$ is injective, i.e., $D(\bar{q}) \hookrightarrow H$. To this end let $x \in D(\bar{q})$ with $\bar{\iota}(x) = 0$. Then there is a sequence $(x_n)_n \subset D(q)$ with $x_n \xrightarrow[n \to \infty]{\|\cdot\|_q} x$. Hence $x_n = \iota(x_n) \xrightarrow{\|\cdot\|} \bar{\iota}(x) = 0$. Since $q(x_n - x_m, x_n - x_m) = \|x_n - x_m\|_q^2 - (1 - M)\|x_n - x_m\|^2 \xrightarrow[n,m \to \infty]{} 0$ the closability of q implies $q(x_n, x_n) \xrightarrow[n \to \infty]{} 0$. Hence $\|x_n\|_q^2 \xrightarrow[n \to \infty]{} 0$, which shows $x = 0$. Let $\langle \cdot, \cdot \rangle_{\bar{q}}$ be the scalar product on $D(\bar{q})$ and $\bar{q}(x, y) := \langle x, y \rangle_{\bar{q}} - (1 - M)\langle x, y \rangle$ for $x, y \in D(\bar{q})$. Then $\bar{q} : D(\bar{q}) \times D(\bar{q}) \to \mathbb{C}$ satisfies the assertion. $\square$

1.2.9 Proposition. Let $q : D(q) \times D(q) \to \mathbb{C}$ be a closed quadratic form in H that is semibounded by $M \in \mathbb{R}$.
Then there is a unique selfadjoint linear operator $Q : D(Q) \to H$ in H with $D(Q) \subset D(q)$ and $q(x, y) = \langle Qx, y \rangle$ for $x \in D(Q), y \in D(q)$.
Moreover, $D(Q) = \{x \in D(q) : \exists\, z_x \in H \text{ with } q(x, y) = \langle z_x, y \rangle \text{ for } y \in D(q)\}$, $Qx = z_x$ for $x \in D(Q)$ and $\langle Qx, x \rangle \geq M \|x\|^2$ for $x \in D(Q)$. Furthermore $D(Q) \subset (D(q), \langle \cdot, \cdot \rangle_q)$ is a dense subspace.
Q is called the selfadjoint operator induced by q.

PROOF: $H_q := D(q)$ is a Hilbert space with respect to $\langle \cdot, \cdot \rangle_q$ satisfying $H_q \hookrightarrow H$. Let $D(P) := \{x \in H_q : \exists\, z_x \in H \text{ with } \langle x, y \rangle_q = \langle z_x, y \rangle \text{ for } y \in H_q\}$ and $Px := z_x$. Moreover, let $P_0 x := Px$ for $x \in D(P_0) := D(P) \subset H_q$ and $Rx := x$ for $x \in D(R) := H_q$. Then P_0 is a linear operator with domain $D(P_0)$ in H_q to H and R is a linear operator with domain $D(R)$ in H to H_q. Clearly, $R^* = P_0$. Moreover, R is closed. Hence $P_0 = R^*$ is densely defined (due to 1.2.2) with $P_0^* = R^{**} = R$. Therefore $D(P) = D(P_0) \subset H_q$ densely, which shows that P is a densely defined, linear operator in H.
Moreover, for $x, y \in D(P) = D(P_0) = D(R^*) \subset H_q = D(R) \subset H$

$$\langle Px, y \rangle = \langle x, y \rangle_q = \langle Rx, y \rangle_q = \langle x, R^* y \rangle = \langle x, P_0 y \rangle = \langle x, Py \rangle$$

Finally, let $z \in H$ and $f_z(y) := \langle y, z \rangle$ for $y \in H_q$. Since $f_z \in H_q^*$ there is an $x \in H_q$ with $\langle y, z \rangle = f_z(y) = \langle y, x \rangle_q$ for $y \in H_q$ and $x \in D(P)$ with $Px = z$. Hence $\mathcal{R}(P) = H$ and P and $Q = P + (M - 1)\mathrm{Id}$ are selfadjoint due to 1.2.5. Finally, suppose that there is a selfadjoint linear operator $S : D(S) \to H$ with $D(S) \subset D(q)$ and $q(x, y) = \langle Sx, y \rangle$ for $x \in D(S), y \in D(q)$. Hence $S \subset Q$. This implies $Q = Q^* \subset S^* = S$ and $S = Q$, which shows the uniqueness of Q. $\square$

With these results on quadratic forms we can easily show that any semibounded symmetric operator has a natural, semibounded, selfadjoint extension, the so-called Friedrichs extension.

1.2.10 Proposition. Let $Z : D(Z) \to H$ be a densely defined, symmetric operator in H with $\langle Zx, x \rangle \geq M \|x\|^2$ for $x \in D(Z)$ and a suitable $M \in \mathbb{R}$. Then $q(x, y) := \langle Zx, y \rangle$ for $x, y \in D(q) := D(Z)$ defines a closable, semibounded quadratic form. The selfadjoint operator $Q : D(Q) \to H$ induced by the closure $\bar{q}$ of q extends Z and satisfies $\langle Qx, x \rangle \geq M \|x\|^2$ for $x \in D(Q)$. It is called the Friedrichs extension of Z.

PROOF: First we have to show that q is closable. Let $(x_n)_n \subset D(q)$ with $x_n \xrightarrow[n \to \infty]{} 0$ and $q(x_n - x_m, x_n - x_m) \xrightarrow[n,m \to \infty]{} 0$. Let $\langle \cdot, \cdot \rangle_q$ and $\|\cdot\|_q$ be as in the previous proofs. Then

$$\|x_n\|_q^2 \leq \|x_n\|_q \|x_n - x_m\|_q + \|(Z + (1 - M)\mathrm{Id})x_n\| \, \|x_m\| \ .$$

Let $\varepsilon > 0$, then there is an $N \in \mathbb{N}$ such that for any $n \geq N$ there is an $m = m_n \geq N$ with $\|x_n - x_m\|_q < \varepsilon$ and $\|(Z + (1 - M)\mathrm{Id})x_n\| \, \|x_m\| < \varepsilon \|x_n\|_q$. Hence $\|x_n\|_q^2 \leq \|x_n\|_q \varepsilon + \varepsilon \|x_n\|_q$, which implies $\|x_n\|_q < 2\varepsilon$. This implies the closability of q.

To complete the proof we still have to show that $Z \subset Q$, if Q is the selfadjoint operator induced by the closure $\bar{q}$ of q. Let $x \in D(Z) = D(q)$ and choose an $y \in D(\bar{q})$. Then there is a sequence $(y_n)_n \subset D(q)$ with $y_n \xrightarrow[n \to \infty]{\|\cdot\|_{\bar{q}}} y$. Hence

$$\langle Zx, y \rangle \xleftarrow[\infty \leftarrow n]{} \langle Zx, y_n \rangle = q(x, y_n) = \bar{q}(x, y_n) = \langle x, y_n \rangle_{\bar{q}} - (1 - M)\langle x, y_n \rangle$$
$$\xrightarrow[n \to \infty]{} \langle x, y \rangle_{\bar{q}} - (1 - M)\langle x, y \rangle = \bar{q}(x, y) \ .$$

Thus $\langle Zx, y \rangle = \bar{q}(x, y)$ for $y \in D(\bar{q})$, which implies $x \in D(Q)$ with $Qx = Zx$. This completes the proof. $\square$

1.2.11 Bibliographical remarks. Properties of selfadjoint operators and quadratic forms can be found in many textbooks on functional analysis, for example Reed/Simon [128], [127], Weidmann [155], Achiezer/Glasmann [2], or Rudin [134]. Proposition 1.2.6 is due to Chernoff [26].

1.3 Generators of analytic semigroups and their powers

It is very difficult to use the Hille-Yosida in the case $M > 1$ for verifying whether a given linear operator generates a C_0-semigroup because one has to satisfy estimates for all powers of the resolvent. It is remarkable that an apparently slightly modified estimate for the resolvent only suffices for proving that a given operator generates a semigroup. Moreover, the semigroup has much better properties than the C_0-semigroup constructed in the proof of the Hille-Yosida theorem. These semigroups are called analytic semigroups.

1.3.1 Definition.

(a) Let $\Sigma_\theta := \{re^{i\varphi} : r > 0, |\varphi| < \theta\}$ for $\theta \in (0, \pi]$.

(b) A bounded analytic semigroup $T(z) \in \mathscr{L}(X), z \in \Sigma_\theta \cup \{0\}$, with angle $0 < \theta \le \frac{\pi}{2}$, where X is a complex Banach space, is a family of linear operators satisfying the following properties:

- $T(0) = \mathrm{Id}$ and $T(z_1 + z_2) = T(z_1)T(z_2)$ for $z_1, z_2 \in \Sigma_\theta$.
- $\Sigma_\theta \ni z \mapsto T(z) \in \mathscr{L}(X)$ is analytic.
- For $0 < \theta' < \theta$ we have $\lim_{\Sigma_{\theta'} \ni z \to 0} T(z)x = x$ for $x \in X$.
- For $0 < \theta' < \theta$ there is an $M_{\theta'} \ge 1$ with $\|T(z)\|_{\mathscr{L}(X)} \le M_{\theta'}$ for $z \in \Sigma_{\theta'}$.

In particular, $(T(t))_{t \ge 0}$ is a C_0-semigroup. Its generator is also called generator for $(T(z))_z$.

(c) A densely defined, linear operator $S : D(S) \to X$ in a complex Banach space X is called sectorial with angle $0 < \theta \le \pi$, if the following conditions are satisfied:

- $\Sigma_\theta \subset \rho(S)$.

- for $0 < \theta' < \theta$ there is an $M_{\theta'} \geq 1$ with

$$\left\| (\lambda \mathrm{Id} - S)^{-1} \right\|_{\mathscr{L}(X)} \leq \frac{M_{\theta'}}{|\lambda|} \quad \text{for } \lambda \in \Sigma_{\theta'}.$$

(d) Let $\gamma_{R,\varphi} = \gamma_{R,\varphi}^{(3)} \circ \gamma_{R,\varphi}^{(2)} \circ \gamma_{R,\varphi}^{(1)}$, where $R > 0, 0 < \varphi < \frac{\pi}{2}$ be the integration path formed by the sub-paths

$$\gamma_{R,\varphi}^{(3)} : (-\infty, -R] \ni t \mapsto (-t)e^{-(\varphi + \frac{\pi}{2})} \in \mathbb{C}$$
$$\gamma_{R,\varphi}^{(3)} : [-(\varphi + \frac{\pi}{2}), \varphi + \frac{\pi}{2}] \ni t \mapsto Re^{it} \in \mathbb{C}$$
$$\gamma_{R,\varphi}^{(3)} : [r, \infty) \ni t \mapsto te^{i(\varphi + \frac{\pi}{2})} \in \mathbb{C}$$

Now be prove a result for analytic semigroups which corresponds to the Hille-Yosida theorem for C_0-semigroups. Note, that only estimates for the resolvent itself are used in the theorem.

1.3.2 Theorem. Let $S : D(S) \to X$ be a linear operator in a complex Banach space X and $0 < \theta \leq \frac{\pi}{2}$.
Then the following statements are equivalent:

(a) S is sectorial with angle $\theta + \frac{\pi}{2}$.

(b) S generates a bounded, analytic semigroup $T(z) \in \mathscr{L}(X), z \in \Sigma_\theta \cup \{0\}$, with angle θ.

In this case, we have

$$T(z) = \frac{1}{2\pi i} \int_{\gamma_{R,\theta''}} e^{z\lambda}(\lambda \mathrm{Id} - S)^{-1} d\lambda$$

for $z \in \Sigma_{\theta'}, 0 < \theta' < \theta'' < \theta$, and $R > 0$.

PROOF: First assume (b) and let $(T(z))_{z \in \Sigma_\theta \cup \{0\}} \subset \mathscr{L}(X)$ be the bounded analytic semigroup generated by S. Define $T_\varphi(t) := T(e^{i\varphi}t)$ for $t \geq 0$ and $|\varphi| < \theta$. Then $(T_\varphi(t))_{t \geq 0}$ is a C_0-semigroup with $\|T_\varphi(t)\|_{\mathscr{L}(X)} \leq M_{\theta'}$ for $t \geq 0$ and $|\varphi| < \theta' < \theta$. Denote its generator by $S_\varphi : D(S_\varphi) \to X$. Let $\gamma : [0, \infty) \ni r \mapsto re^{i\varphi} \in \mathbb{C}$ for $|\varphi| < \theta'$. Then, by analyticity, we have due to the Hille-Yosida theorem 1.1.4

$$(\mathrm{Id} - S)^{-1} = \int_0^\infty e^{-t}T(t)dt = \int_\gamma e^{-z}T(z)dz = \int_0^\infty e^{-re^{i\varphi}}T_\varphi(r)e^{i\varphi}dr$$
$$= e^{i\varphi}(e^{i\varphi}\mathrm{Id} - S_\varphi)^{-1} = (\mathrm{Id} - e^{-i\varphi}S_\varphi)^{-1}.$$

This shows $S_\varphi = e^{i\varphi}S$. Now, let $\lambda = re^{i\varphi} \in \Sigma_{\theta'}$ with $r > 0$, $|\varphi| < \theta' + \frac{\pi}{2}$, and $0 < \theta' < \theta'' < \theta$. We will prove the assertion due to its symmetry only for $\operatorname{Im} \lambda > 0$, i.e. $\varphi > 0$. If $\operatorname{Re} \lambda > 0$, then let $\lambda = a + ib$ with $a, b > 0$ and $e^{-i\theta'} := u - iv$ with $u, v > 0$. Since $(\lambda \operatorname{Id} - S) = e^{i\theta'}(\lambda e^{-i\theta'} - S_{-\theta'})$ and $\operatorname{Re}(\lambda e^{-i\theta'}) = au + bv > 0$, the Hille-Yosida theorem shows $\lambda \in \rho(S)$ with

$$\left\|(\lambda \operatorname{Id} - S)^{-1}\right\|_{\mathscr{L}(X)} \leq \frac{M_{\theta''}}{\operatorname{Re}(\lambda e^{-i\theta'})} \leq \frac{\left(\frac{M_{\theta''}}{\sin(\theta')}\right)}{\operatorname{Im} \lambda}. \tag{1.3.1}$$

The Hille-Yosida theorem shows further $\left\|(\lambda \operatorname{Id} - S)^{-1}\right\|_{\mathscr{L}(X)} \leq \frac{M_{\theta''}}{\operatorname{Re} \lambda}$, hence

$$\left\|(\lambda \operatorname{Id} - S)^{-1}\right\|_{\mathscr{L}(X)} \leq \frac{M_{\theta''}\sqrt{1 + \frac{1}{(\sin(\theta'))^2}}}{|\lambda|}$$

for $\lambda \in \Sigma_{\theta'}$ with $\operatorname{Re} \lambda > 0, \operatorname{Im} \lambda > 0$. Moreover, if $\operatorname{Re} \lambda \leq 0$, then $\frac{|\operatorname{Re} \lambda|}{|\operatorname{Im} \lambda|} \leq \tan \theta'$, hence (1.3.1) shows

$$\left\|(\lambda \operatorname{Id} - S)^{-1}\right\|_{\mathscr{L}(X)} \leq \frac{\frac{M_{\theta''}}{\sin(\theta')}}{\frac{|\operatorname{Re} \lambda|}{\tan(\theta')}} = \frac{\frac{M_{\theta''}}{\tan(\theta')\sin(\theta')}}{|\operatorname{Re} \lambda|}.$$

Applying (1.3.1) once more we obtain

$$\left\|(\lambda \operatorname{Id} - S)^{-1}\right\|_{\mathscr{L}(X)} \leq \frac{\frac{M_{\theta''}}{\sin(\theta')}\sqrt{1 + \frac{1}{(\tan(\theta'))^2}}}{|\lambda|}.$$

This proves (a).

Now assume conversely (a). Define $T(0) := \operatorname{Id}$ and $T(z)$ for $z \in \Sigma_\theta$ as in the assertion of the theorem. For well-definedness note that for $z = re^{i\varphi} \in \Sigma_{\theta'}$ and $\lambda = te^{\pm i(\theta'' + \frac{\pi}{2})}$ with $r > 0, |\varphi| < \theta', t > R > 0$, and $0 < \theta' < \theta'' < \tilde{\theta} < \theta$

$$\left\|e^{z\lambda}(\lambda \operatorname{Id} - S)^{-1}\right\|_{\mathscr{L}(X)} \leq e^{\operatorname{Re}(z\lambda)}\frac{M_{\tilde{\theta}}}{|\lambda|} = e^{rt\cos(\varphi \pm (\theta'' + \frac{\pi}{2}))}\frac{M_{\tilde{\theta}}}{t}$$

and $\cos(\varphi \pm (\theta'' + \frac{\pi}{2})) < 0$. This proves the existence of $T(z)$. Moreover, the analyticity of $\Sigma_\theta \ni \lambda \mapsto (\lambda \operatorname{Id} - S)^{-1}$ and a standard argument using Cauchy's theorem show the independence of the choice of the integration path $\gamma_{R,\theta''}$. (For different paths connect them to closed paths by 2 horizontal lines of increasing distance to $\operatorname{Im} z = 0$). The existence proof for $T(z)$ shows further that for $\theta' < \theta$ there is an $M_{\theta'}$ with $\|T(z)\| \leq M_{\theta'}$ for $z \in \Sigma_{\theta'}$. The analyticity of $\Sigma_\theta \ni z \mapsto T(z) \in \mathscr{L}(X)$ is a consequence of Morrera's theorem. Hence it only remains to prove the semigroup property, the strong continuity in 0, and that A generates the C_0-semigroup $(T(t))_{t \geq 0}$.

To prove the semigroup property let $z_1, z_2 \in \Sigma_{\theta'}$ with $z_1 + z_2 \in \Sigma_{\theta'}$ for suitable $0 < \theta' < \theta$ and choose $0 < R_1 < R_2$ and $0 < \theta' < \theta_2 < \theta_1 < \theta$. Then using

$$(\mu - \lambda)(\lambda\mathrm{Id} - S)^{-1}(\mu\mathrm{Id} - S)^{-1} = ((\lambda\mathrm{Id} - S)^{-1} - (\mu\mathrm{Id} - S)^{-1})$$

for $\lambda, \mu \in \rho(S)$ we have

$$
\begin{aligned}
T(z_1)T(z_2) &= \frac{1}{(2\pi i)^2} \int_{\gamma_{R_1,\theta_1}} \int_{\gamma_{R_2,\theta_2}} e^{z_1\lambda} e^{z_2\mu} (\lambda\mathrm{Id} - S)^{-1}(\mu\mathrm{Id} - S)^{-1} d\mu d\lambda \\
&= \frac{1}{(2\pi i)^2} \int_{\gamma_{R_1,\theta_1}} \underbrace{\int_{\gamma_{R_2,\theta_2}} e^{z_2\mu}(\mu - \lambda)^{-1} d\mu}_{=2\pi i e^{z_2\lambda}} (\lambda\mathrm{Id} - S)^{-1} e^{z_1\lambda} d\lambda \\
&\quad - \frac{1}{(2\pi i)^2} \int_{\gamma_{R_2,\theta_2}} \underbrace{\int_{\gamma_{R_1,\theta_1}} e^{z_1\lambda}(\mu - \lambda)^{-1} d\lambda}_{=0} e^{z_2\mu}(\mu\mathrm{Id} - S)^{-1} d\mu \\
&= \frac{1}{2\pi i} \int_{\gamma_{R_1,\theta_1}} e^{z_1\lambda} e^{z_2\lambda}(\lambda\mathrm{Id} - S)^{-1} d\lambda = T(z_1 + z_2).
\end{aligned}
$$

Here, we have used

$$\int_{\gamma_{R,\theta''}} \frac{e^{z\lambda}}{\lambda - \mu} d\lambda = \begin{cases} 2\pi i e^{z\mu} & , \mu \in \mathbb{C} \setminus \overline{\Sigma_{\theta''}} \\ 0 & , \mu \in \Sigma_{\theta''} \end{cases}$$

for $0 < \theta' < \theta'' < \theta$ and $z \in \Sigma_{\theta'}$, which can be proved with Cauchy's theorem by closing the integration path by a vertical line of increasing length tending to the left. Next, for $x \in D(S), 0 < \theta' < \theta$, and $0 < t < 1$ we have similarly

$$T(t)x - x = \frac{1}{2\pi i} \int_{\gamma_{\theta',1}} e^{\lambda t} \left((\lambda\mathrm{Id} - S)^{-1} - \lambda^{-1}\mathrm{Id}\right) x d\lambda$$

$$= \frac{1}{2\pi i} \int_{\gamma_{\theta',1}} e^{\lambda t}(\lambda\mathrm{Id} - S)^{-1} Sx \frac{d\lambda}{\lambda} \xrightarrow[t\to 0]{} \frac{1}{2\pi i} \int_{\gamma_{\theta',1}} (\lambda\mathrm{Id} - S)^{-1} Sx \frac{d\lambda}{\lambda} = 0$$

by the dominated convergence theorem and by Cauchy's theorem. (For the last equality close the integration path by circles on the right of increasing diameter.) Thus, 1.1.3 implies $T(t)x \xrightarrow[t\to 0]{} x$ for $x \in X$. Let $A : D(A) \to X$ be the infinitesimal generator of the C_0-semigroup $(T(t))_{t\geq 0}$. We have to show $A = S$ to complete the proof. For $x \in D(S)$ and $0 < t < R, 0 < \theta' < \theta$ we have due to Cauchy's theorem (applied similarly as before)

$$\frac{d}{dt}T(t)x \;=\; \frac{1}{2\pi i}\int_{\gamma_{\theta'},R}\lambda e^{\lambda t}(\lambda\mathrm{Id}-S)^{-1}x\,d\lambda$$

$$=\; \underbrace{\frac{1}{2\pi i}\int_{\gamma_{\theta'},R}e^{\lambda t}x\,d\lambda}_{=0}+\frac{1}{2\pi i}\int_{\gamma_{\theta'},R}e^{\lambda t}(\lambda\mathrm{Id}-S)^{-1}Sx\,d\lambda = T(t)Sx\,.$$

Hence $T(t)x \in D(A)$ for $t > 0$ with $AT(t)x = T(t)Sx$. Since $T(t)x \xrightarrow{t\to0} x$ and $AT(t)x = T(t)Sx \xrightarrow{t\to0} Sx$, the closedness of A shows $x \in D(A)$ with $Ax = Sx$. Thus, $S \subset A$ and $\lambda\mathrm{Id}-A$ extends $\lambda\mathrm{Id}-S$ for $\lambda > 0$. But now $A = S$ is implied by $(0,\infty) \subset \rho(A) \cap \rho(S)$.   □

In the sequel we will show that powers of sectorial operators are sectorial again. First, we give a precise definition of powers of unbounded operators.

1.3.3 Definition. For a linear operator $A : D(A) \to X$ in a Banach space X with domain $D(A) \subset X$ let inductively for $j \in \mathbb{N}$:

- $D(A^0) := X, D(A^1) := D(A), D(A^{j+1}) := \{x \in D(A^j) : Ax \in D(A^j)\}$

- $A^0 := \mathrm{Id}, A^1 := A$, and $A^{j+1}x := A^j(Ax)$ for $x \in D(A^{j+1})$.

- $D(A^\infty) := \bigcap_{j\in\mathbb{N}_0} D(A^j)$

Then $A^j : D(A^j) \to X$ is a well-defined linear operator for $j \in \mathbb{N}_0$.

1.3.4 Lemma. Let $A : D(A) \to X$ be a linear operator with dense domain $D(A) \subset X$ in a Banach space X and suppose $\rho(A) \neq 0$. Then A is closed and $D(A^\infty) \subset X$ is a dense subspace. Moreover, $A^j : D(A^j) \to X$ is densely-defined and closed for $j \in \mathbb{N}_0$.

PROOF: Let $\mu \in \rho(A)$. By considering $A - \mu\mathrm{Id}$ at the place of A we can assume $\mu = 0 \in \rho(A)$ for the proof of the density of $D(A^\infty) \subset X$. Let $(x_k)_k \subset D(A)$ with $x_k \xrightarrow{k\to\infty} x \in X$ and $Ax_k \xrightarrow{k\to\infty} y \in X$. Then we have $x_k = A^{-1}Ax_k \xrightarrow{k\to\infty} A^{-1}y$. Hence $x = A^{-1}y \in D(A)$ and $Ax = y$. Next, let $x_0 \in X$ and $\varepsilon > 0$. We construct inductively a sequence $(x_k)_k \subset X$ with $x_k \in D(A^k)$ by choosing suitable $y_{k+1} \in D(A)$ (using $D(A) \subset H$ densely) with $\|y_{k+1} - A^k x_k\| < \frac{\varepsilon}{2^{k+1}\|A^{-1}\|^k_{\mathscr{L}(X)}}$ and setting $x_{k+1} := A^{-k}y_{k+1} \in D(A^{k+1})$.

Then $x_k \in D(A^k)$ and $\left\| A^k x_{k+1} - A^k x_k \right\| < \frac{\varepsilon}{2^{k+1} \|A^{-1}\|_{\mathscr{L}(X)}^k}$. Moreover,

$$\left\| A^k x_{k+j} - A^k x_k \right\| \leq \sum_{l=0}^{j-1} \left\| A^k x_{k+l+1} - A^k x_{k+l} \right\|$$

$$\leq \sum_{l=0}^{j-1} \left\| A^{-1} \right\|_{\mathscr{L}(X)}^l \left\| A^{k+l} x_{k+l+1} - A^{k+l} x_{k+l} \right\| \leq \frac{\varepsilon}{(2 \|A^{-1}\|_{\mathscr{L}(X)})^k}. \quad (1.3.2)$$

for $j, k \in \mathbb{N}_0$. This shows

$$\left\| A^k x_{j+l} - A^k x_j \right\| \leq \left\| A^{-1} \right\|_{\mathscr{L}(X)}^{j-k} \left\| A^j x_{j+l} - A^j x_j \right\| \leq \frac{\varepsilon}{2^j \|A^{-1}\|_{\mathscr{L}(X)}^k}$$

for $j \geq k$ and $l \in \mathbb{N}_0$. Hence $(A^k x_j)_{j \geq k} \subset X$ is a Cauchy sequence for any $k \in \mathbb{N}_0$. With a simple induction the closedness of A implies the existence of an $x \in D(A^\infty)$ with $A^k x_j \xrightarrow[j \to \infty]{} A^k x$ for all $k \in \mathbb{N}_0$. With $k = 0$ and $j \to \infty$ in (1.3.2) we obtain $\|x - x_0\| \leq \varepsilon$. This proves that $D(A^\infty) \subset X$ densely.

To show the closedness of $A^j : D(A^j) \to X$ for any $j \in \mathbb{N}$ let $\lambda \in \rho(A)$ and $p(z) := \sum_{l=0}^{j} \binom{j}{l} \lambda^l \mu^l$, then $(z - \lambda)^j p((z - \lambda)^{-1}) = \sum_{l=0}^{j} \binom{j}{l} \lambda^l (z - \lambda)^{j-l} = z^j$. Hence $A^j x = (A - \lambda \mathrm{Id})^j p((A - \lambda \mathrm{Id})^{-1}) x = p((A - \lambda \mathrm{Id})^{-1})(A - \lambda \mathrm{Id})^j x$ for $x \in D(A^j)$. Let $(x_k)_k \subset D(A^j)$ with $x_k \xrightarrow[k \to \infty]{} x$ and $A^j x_k \xrightarrow[k \to \infty]{} y$. Since $(A - \lambda \mathrm{Id})^{-j}$ is bounded and closed, $(A - \lambda \mathrm{Id})^j : D(A^j) \to X$ is a closed operator. Thus $p((A - \lambda \mathrm{Id})^{-1}) x \in D(A^j)$ because $p((A - \lambda \mathrm{Id})^{-1}) x_k \xrightarrow[k \to \infty]{} p((A - \lambda \mathrm{Id})^{-1}) x$ and $(A - \lambda \mathrm{Id})^j p((A - \lambda \mathrm{Id})^{-1}) x_k = A^j x_k \xrightarrow[k \to \infty]{} y$. Therefore, we have shown $x = p((A - \lambda \mathrm{Id})^{-1}) x - \sum_{l=1}^{j} \binom{j}{l} \lambda^l (A - \lambda \mathrm{Id})^{-l} x \in D(A)$. Now assume inductively $x \in D(A^m)$ for $1 \leq m < j$. Then

$$(A - \lambda \mathrm{Id})^m x = (A - \lambda \mathrm{Id})^m p((A - \lambda \mathrm{Id})^{-1}) x - \sum_{l=1}^{j} \binom{j}{l} \lambda^l (A - \lambda \mathrm{Id})^{m-l} x \in D(A)$$

and hence $x \in D(A^{m+1})$. This completes the proof. $\square$

1.3.5 Lemma. Let $\lambda \in \Sigma_\theta, 0 < \theta \leq \pi$ and $q \in \mathbb{N}$.
Then there exist $\lambda_1, \ldots, \lambda_q \in \Sigma_{(1-\frac{1}{q})\pi + \frac{\theta}{q}}$ with $|\lambda_j|^q = |\lambda|$ and

$$\lambda + (-x)^q = \prod_{j=1}^{q} (\lambda_j - x) \quad \text{for } x \in \mathbb{C}.$$

PROOF: Let $\lambda = re^{i\varphi} \in \Sigma_\theta$. Then there are $\varphi_1^{(q)}, \ldots, \varphi_q^{(q)} \in (-\pi, \pi]$ with

$$e^{i\varphi} + (-x)^q = \prod_{j=1}^q (e^{i\varphi_j^{(q)}} - x) \quad \text{for } x \in \mathbb{C}.$$

Suppose that there is a $j \in \{1, \ldots, q\}$ with $\varphi_j^{(q)} = \pm\pi + \varepsilon$ and $|\varepsilon| \leq \frac{\pi - \theta}{q}$. Then

$$0 = e^{i\varphi} + (-1)^q \left(e^{i\varphi_j^{(q)}}\right)^q = e^{i\varphi} + (-1)^q(-1)^q e^{i\varepsilon q} = e^{i\varphi} + e^{i\varepsilon q}$$

Hence

$$
\begin{aligned}
0 &= \operatorname{Re}\left(e^{i\varphi} + e^{i\varepsilon q}\right) = \cos(\varphi) + \cos(\varepsilon q) > \cos(\theta) + \cos(\pi - \theta) \\
&= \cos(\theta) - \cos(\theta) = 0,
\end{aligned}
$$

which is a contradiction. Thus $|\varphi_j^{(q)}| < \pi - \frac{\pi - \theta}{q} = \left(1 - \frac{1}{q}\right)\pi + \frac{\theta}{q}$ for $j = 1, \ldots, q$ and

$$\lambda + (-x)^q = r\left(e^{i\varphi} + \left(-\frac{x}{\sqrt[q]{r}}\right)^q\right) = r \prod_{j=1}^q \left(e^{i\varphi_j^{(q)}} - \frac{x}{\sqrt[q]{r}}\right) = \prod_{j=1}^q \left(\underbrace{\sqrt[q]{r}e^{i\varphi_j^{(q)}}}_{=:\lambda_j} - x\right).$$

$\square$

Now we can show that powers of sectorial operators are sectorial again provided that their angle satisfies certain conditions.

1.3.6 Proposition. Let $S : D(S) \to X$ be a sectorial operator in X with angle $\theta > \left(1 - \frac{1}{q}\right)\pi$, where $q \in \mathbb{N}$.
Then $-(-S)^q : D(S^q) \to X$ is sectorial with angle $\theta^{(q)} := \pi - q(\pi - \theta) > 0$, and for $0 < \theta' < \theta^{(q)}$ there are $M_{\theta',q} \geq 0$ with

$$\left\|S^j\left(\lambda\operatorname{Id} + (-S)^q\right)^{-1}\right\|_{\mathscr{L}(X)} \leq \frac{M_{\theta',q}}{|\lambda|^{\frac{q-j}{q}}} \quad \text{for } \lambda \in \Sigma_{\theta'}, j = 0, \ldots, q.$$

PROOF: Let $0 < \theta' < \theta^{(q)}, \lambda \in \Sigma_{\theta'}$. There are $\lambda_1, \ldots, \lambda_q \in \Sigma_{\left(1 - \frac{1}{q}\right)\pi + \frac{\theta'}{q}} \subset \Sigma_\theta$, (due to 1.3.5) because $\theta'' := \left(1 - \frac{1}{q}\right)\pi + \frac{\theta'}{q} < \left(1 - \frac{1}{q}\right)\pi + \frac{\pi}{q} - \pi + \theta = \theta$. Then

$$(\lambda\operatorname{Id} + (-S)^q)x = \prod_{j=1}^q (\lambda_j\operatorname{Id} - S)x \quad \text{for } x \in D(S^q).$$

Since $\lambda_j \mathrm{Id} - S : D(S^{k+1}) \to D(S^k)$ is invertible, $\lambda \in \rho(-(-S)^q)$ and

$$(\lambda \mathrm{Id} + (-S)^q)^{-1} = \prod_{j=1}^{q} (\lambda_j \mathrm{Id} - S)^{-1}.$$

Moreover

$$\|S(\lambda_j \mathrm{Id} - S)^{-1} x\| = \|(\lambda_j \mathrm{Id} - (\lambda_j \mathrm{Id} - S))(\lambda_j \mathrm{Id} - S)^{-1} x\| \le (M'_{\theta''} + 1)\|x\|$$

for $x \in X$. Thus

$$\|S^j(\lambda \mathrm{Id} + (-S)^q)^{-1}\| = \left\| \prod_{l=1}^{j} S(\lambda_l \mathrm{Id} - S)^{-1} \prod_{l=j+1}^{q} (\lambda_l \mathrm{Id} - S)^{-1} \right\| \le \frac{M_{\theta',q}}{|\lambda|^{\frac{q-j}{q}}}.$$

$\square$

1.3.7 Remark. Many generators of bounded analytic semigroups generate semigroups with an angle of holomorphy of $\frac{\pi}{2}$ (cf. [62], [119]) and hence arbitrary values of m are allowed in theorem 1.3.6. Moreover, symmetric Markov semigroups in L^2 can always be extended to bounded analytic semigroups in L^p and an estimate for the sector of holomorphy is available (cf. [105], [153]).

Finally, we will show that squares of generators of isometric C_0-groups generate bounded analytic semigroup with angle $\frac{\pi}{2}$. An isometric C_0-group is a C_0-group $(T(t))_{t \in \mathbb{R}}$ of linear operators satisfying $\|T(t)\|_{\mathscr{L}(X)} = 1$ for $t \in \mathbb{R}$.

1.3.8 Proposition. Let $A : D(A) \to X$ be the generator of an isometric C_0-group with domain $D(A) \subset X$ in a Banach space X.
Then $A^2 : D(A^2) \to X$ is sectorial with angle π.

PROOF: Let $0 < \theta < \pi$ and $\lambda = re^{i\varphi} \in \Sigma_\theta$. Note that $\mu \in \rho(\pm A)$ for $\mu \in \mathbb{C}$ with $\mathrm{Re}\,\mu > 0$. Then $(\lambda - A^2) = (\sqrt{r}e^{i\varphi/2} + A)(\sqrt{r}e^{i\varphi/2} - A) : D(A^2) \to X$ is invertible with

$$\|(\lambda - A^2)^{-1}\|_{\mathscr{L}(X)} \le \frac{1}{|\mathrm{Re}\,(\sqrt{r}e^{i\varphi/2})|} \frac{1}{|\mathrm{Re}\,(\sqrt{r}e^{i\varphi/2})|} \le \frac{1}{|\cos(\theta/2)|^2} \frac{1}{|\lambda|}.$$

$\square$

1.3.9 Bibliographical remarks. The characterization of generators of analytic semigroups 1.3.2 is a standard result in semigroup theory and goes back to Hille. It can be found in many textbooks on semigroup theory. 1.3.4 is due to [102] and [42, VII.9], and 1.3.6 is due to de Laubenfels [38]. We took the proof of 1.3.6 in [132, 4.1] and obtained a resolvent estimate as well. Proposition 1.3.8 is taken from [110, A-II,1.13].

1.4 Fractional Powers of operators of positive type

In section 1.3 we already introduced natural powers of densely-defined linear operators with non-void resolvent set. In this section we will extend this definition to fractional powers of densely-defined operators, if we assume in addition the resolvent set to include the negative real axis and certain resolvent estimates. These operators are called operators of positive type.

1.4.1 Definition. A linear operator $\Lambda : D(\Lambda) \to X$ with domain $D(\Lambda)$ is said to be of positive type $M \geq 1$, if it is densely-defined, $[0, \infty) \subset \rho(-\Lambda)$, and

$$\left\|(\lambda \mathrm{Id} + \Lambda)^{-1}\right\|_{\mathscr{L}(X)} \leq \frac{M}{1 + \lambda} \qquad \text{for } \lambda \in [0, \infty) \, .$$

One of the most important example for operators of positive type are strictly positive operators.

1.4.2 Example. Let $\Lambda : D(\Lambda) \to H$ be a strictly positive , selfadjoint linear operator in a complex Hilbert space H, i.e. there is a $\delta > 0$ with $\langle \Lambda x, x \rangle \geq \delta \|x\|^2$ for $x \in D(\Lambda)$.
Then Λ is of positive type.

PROOF: The Lumer-Phillips theorem 1.1.12 and 1.2.3 imply $(-\delta, \infty) \subset \rho(-\Lambda)$. Moreover, $(\lambda + \delta) \|x\|^2 \leq \langle (\lambda \mathrm{Id} + \Lambda)x, x \rangle \leq \|(\lambda \mathrm{Id} + \Lambda)x\| \, \|x\|$ for $x \in D(\Lambda)$ and $\lambda > 0$ implies

$$\left\|(\lambda \mathrm{Id} + \Lambda)^{-1}\right\|_{\mathscr{L}(X)} \leq \frac{1}{\lambda + \delta} \leq \frac{M}{1 + \lambda}$$

with $M := \sup\{\frac{1+\lambda}{\delta+\lambda} : \lambda \geq 0\} \geq 1$. $\qquad\square$

1.4.3 Lemma. Let $0 < \sigma < m$, then $\displaystyle\int_0^\infty \frac{\lambda^{\sigma-1}}{(1+\lambda)^m} d\lambda \leq \frac{m}{\sigma(m-\sigma)}.$

PROOF:

$$\int_0^\infty \frac{\lambda^{\sigma-1}}{(1+\lambda)^m} d\lambda \leq \int_0^1 \lambda^{\sigma-1} d\lambda + \int_1^\infty \lambda^{\sigma-m-1} d\lambda = \frac{1}{\sigma} + \frac{1}{m-\sigma} = \frac{m}{\sigma(m-\sigma)}$$

$\square$

1.4.4 Remark.

(a) The Gamma function $\Gamma : \mathbb{C} \setminus \{-n : n \in \mathbb{N}_0\} \to \mathbb{C}$ is defined by

$$\Gamma(x) := \lim_{n \to \infty} \frac{n! n^z}{z(z+1) \cdots (z+n)} \qquad \text{for } z \in \mathbb{C} \setminus \{-n : n \in \mathbb{N}_0\} .$$

We have $\Gamma(z+1) = z\Gamma(z)$ for $z \in \mathbb{C} \setminus \{-n : n \in \mathbb{N}_0\}$, $\Gamma(n+1) = n!$ for $n \in \mathbb{N}_0$, and $\Gamma(z)\Gamma(1-z) = \frac{\pi}{\sin(\pi z)}$ for $z \in \mathbb{C} \setminus \mathbb{Z}$. Proofs and more detailed information on the Gamma function can be found e.g. in [31, VII, 7].

(b) On can show with the residue theorem (cf. e.g. [31, V, 2.12]) that

$$\int_0^\infty \frac{t^{-\alpha}}{1+t} dt = \frac{\pi}{\sin(\pi\alpha)} \qquad \text{for } 0 < \alpha < 1 .$$

As a consequence we obtain $\dfrac{\sin(\pi z)}{\pi} \displaystyle\int_0^\infty \frac{\lambda^{z-1}}{\lambda + a} a d\lambda = a^z$ for $0 < z < 1$ and $a > 0$. This formula motivates the definition of fractional powers for operators of positive type.

1.4.5 Lemma.
Let $\Lambda : D(\Lambda) \to X$ be an operator of positive type, $n \in \mathbb{N}_0$, $m \in \mathbb{N}$, and $z \in \mathbb{C}$ with $-n < \operatorname{Re} z < m - n$. Then

(a) $\Lambda_0^z x := \dfrac{\Gamma(m)}{\Gamma(z+n)\Gamma(m-n-z)} \displaystyle\int_0^\infty \lambda^{z+n-1} \Lambda^{m-n} (\lambda \mathrm{Id} + \Lambda)^{-m} x d\lambda \in X$ for $x \in D(\Lambda^m)$ is well-defined and independent of the choice of n and m.

(b) $\Lambda_0^z : D(\Lambda^m) \to X$ is a closable operator.

PROOF: Let $0 < \sigma < m$ with $z + n < \sigma$. Then

$$\left\| \lambda^{z+n-1} \Lambda^{m-n} (\lambda \mathrm{Id} + \Lambda)^{-m} x \right\| \leq M^m \frac{\lambda^{\sigma-1}}{(1+\lambda)^m} \left\| \Lambda^{m-n} x \right\| \tag{1.4.1}$$

for $x \in D(\Lambda^m)$ and $\Lambda_0^z x$ exits due to lemma 1.4.3. Moreover, for $\lambda > 0$ we have $\frac{d}{d\lambda}[\lambda^m \Lambda^m (\lambda \mathrm{Id} + \Lambda)^{-m}] = m\lambda^{m-1} \Lambda^{m+1} (\lambda \mathrm{Id} + \Lambda)^{-(m+1)}$. Hence partial integration shows

$$\begin{aligned}
\Lambda_0^z x &= \frac{\Gamma(m)}{\Gamma(z+n)\Gamma(m-n-z)} \int_0^\infty \frac{\left[\frac{d}{d\lambda} \lambda^{-m+n+z} \right]}{-m+n+z} \lambda^m \Lambda^m (\lambda \mathrm{Id} + \Lambda)^{-m} \Lambda^{-n} x d\lambda \\
&= \frac{\Gamma(m+1)}{\Gamma(z+n)\Gamma(m+1-n-z)} \int_0^\infty \lambda^{z+n-1} \Lambda^{(m+1)-n} (\lambda \mathrm{Id} + \Lambda)^{-(m+1)} x d\lambda,
\end{aligned}$$

which shows that $\Lambda_0^z x$ is independent of m. The independence of n is a consequence of the independence of m and

$$
\begin{aligned}
\Lambda_0^z x &= \frac{\Gamma(m)}{\Gamma(z+n)\Gamma(m-n-z)} \int_0^\infty \frac{1}{z+n} \left[\frac{d}{d\lambda}\lambda^{z+n}\right] \Lambda^{m-n}(\lambda \mathrm{Id} + \Lambda)^{-m} x d\lambda \\
&= \frac{\Gamma(m)m}{\Gamma(z+n)(z+n)\Gamma(m-n-z)} \int_0^\infty \lambda^{z+n}\Lambda^{m-n}(\lambda \mathrm{Id} + \Lambda)^{-(m+1)} x d\lambda .
\end{aligned}
$$

Moreover, (1.4.1) and 1.4.3 show $\|\Lambda_0^z x\| \le c_z \|\Lambda^m x\|$ for $x \in D(\Lambda^m)$ and a suitable constant $c_z \ge 0$. Hence $\Lambda^{-m}\Lambda_\sigma^z$ can be extended to a bounded linear operator. Let $(x_k)_k \subset D(\Lambda^m)$ with $x_k \xrightarrow[k\to\infty]{} 0$ and $\Lambda_0^z x_k \xrightarrow[k\to\infty]{} x \in X$. Then $\Lambda^{-m}x = 0$ and $x = 0$. Hence Λ_0^z is closable.   □

1.4.6 Definition. If Λ is an operator of positive type in a complex Banach space X and $z \in \mathbb{C}$, then the fractional power Λ^z of Λ is defined to be the closure of Λ_0^z.

1.4.7 Theorem.   Let $\Lambda : D(\Lambda) \to X$ be an operator of positive type in a complex Banach space X. Then

(a) $\Lambda^z : D(\Lambda^z) \to X$ is the "usual power" of Λ, if $z \in \mathbb{Z}$.

(b) $\Lambda^z \Lambda^w x = \Lambda^{z+w} x$ for $x \in D(\Lambda^{2m}), m \in \mathbb{N}$ with $\mathrm{Re}\, z < m, \mathrm{Re}\, w < m$.

(c) $\Lambda^z \in \mathscr{L}(X)$, if $\mathrm{Re}\, z < 0$.

(d) $\Lambda^{-z} = \dfrac{\sin(\pi z)}{\pi} \int_0^\infty \lambda^{-z}(\lambda \mathrm{Id} + \Lambda)^{-1} d\lambda$ for $0 < \mathrm{Re}\, z < 1$.

(e) $\Lambda^z x = \dfrac{\sin(\pi z)}{\pi} \int_0^\infty \lambda^{z-1}(\lambda \mathrm{Id} + \Lambda)^{-1}\Lambda x d\lambda$ for $0 < \mathrm{Re}\, z < 1, x \in D(\Lambda)$.

(f) $\Lambda^z x = \dfrac{\sin(\pi z)}{\pi z} \int_0^\infty \lambda^z(\lambda \mathrm{Id} + \Lambda)^{-2}\Lambda x d\lambda$ for $-1 < \mathrm{Re}\, z < 1, x \in D(\Lambda)$.

PROOF: First, for $x \in X$ we have $\lambda(\lambda\mathrm{Id} + \Lambda)^{-1} \xrightarrow[\lambda\to\infty]{} \mathrm{Id}$ strongly due to lemma 1.1.3, because $\left\|\lambda(\lambda\mathrm{Id} + \Lambda)^{-1}x - x\right\| = \left\|(\lambda\mathrm{Id} + \Lambda)^{-1}(\Lambda x)\right\| \leq \frac{M}{1+\lambda}\|\Lambda x\| \xrightarrow[\lambda\to\infty]{} 0$ for $x \in D(\Lambda)$. Moreover, $\frac{d}{d\lambda}[\lambda^k\Lambda^k(\lambda\mathrm{Id} + \Lambda)^{-k}] = k\lambda^{k-1}\Lambda^{k+1}(\lambda\mathrm{Id} + \Lambda)^{-(k+1)}$ for $\lambda > 0$ (as noted in the proof of 1.4.5). Hence, by iterated use of this formula and partial integration

$$\frac{k\cdots(2k-1)}{(k-1)!}\int_0^\infty \lambda^{k-1}\Lambda^{2k}(\lambda\mathrm{Id} + \Lambda)^{-2k}\Lambda^{-k}x\,d\lambda$$

$$= \frac{k\cdots(2k-2)}{(k-1)!}\lim_{R\to\infty}\int_0^R \lambda^{1-k}\frac{d}{d\lambda}\left(\lambda^{2k-1}\Lambda^{2k-1}(\lambda\mathrm{Id} + \Lambda)^{-(2k-1)}\right)\Lambda^{-k}x\,d\lambda$$

$$= \frac{k\cdots(2k-2)}{(k-2)!}\lim_{R\to\infty}\int_0^R \lambda^{k-1}\Lambda^{2k-1}(\lambda\mathrm{Id} + \Lambda)^{-(2k-1)}\Lambda^{-k}x\,d\lambda$$

$$= \cdots = \lim_{R\to\infty} k\int_0^R \lambda^{k-1}\Lambda^{k+1}(\lambda\mathrm{Id} + \Lambda)^{-(k+1)}\Lambda^{-k}x\,d\lambda$$

$$= \lim_{R\to\infty}\int_0^R \frac{d}{d\lambda}\left[\lambda^k\Lambda^k(\lambda\mathrm{Id} + \Lambda)^{-k}\right]\Lambda^{-k}x = \lim_{R\to\infty} R^k(R\mathrm{Id} + \Lambda)^{-k}x\,d\lambda = x$$

This shows

$$\Lambda^k x = \frac{\Gamma(2k)}{\Gamma(k)\Gamma(k)}\int_0^\infty \lambda^{k-1}\Lambda^{2k}(\lambda\mathrm{Id} + \Lambda)^{-2k}x\,d\lambda \quad \text{for } x \in D(\Lambda^k) \qquad (1.4.2)$$

With $m = 2k, n = 0, z = k$ this shows $\Lambda^k x = \Lambda_0^k x$ for $x \in D(\Lambda^{2k})$. Therefore 1.3.4 shows that $\overline{\Lambda_0^k}$ is the usual power for $k \in \mathbb{N}$. Moreover, with $m = 2$, $n = 1, z = 0$ we have

$$\Lambda_0^0 x = \int_0^\infty \Lambda(\lambda\mathrm{Id} + \Lambda)^{-2}x\,d\lambda = -\int_0^\infty \frac{d}{d\lambda}\left[\Lambda(\lambda\mathrm{Id} + \Lambda)^{-1}x\right]d\lambda = x$$

for $x \in D(\Lambda^2)$, which proves that $\overline{\Lambda_0^0} = \mathrm{Id}$ is the usual power. Finally, for $k \in \mathbb{N}_0$ we have with $m = 2k, n = 2k, z = -k$ using (1.4.2) $\Lambda_0^{-k}x = \Lambda^{-k}x$ for $x \in D(\Lambda^{2k})$. Since $\Lambda^{-k} \in \mathscr{L}(X)$ this completes the proof of (a). For $x \in D(\Lambda^{2m})$ the closedness of Λ^m shows $\Lambda^z x = \Lambda_0^z x \in D(\Lambda^m)$ for $\mathrm{Re}\, z < m$. Hence $\Lambda_0^z\Lambda_0^w x = \Lambda^z\Lambda^w x$ is well-defined for $x \in D(\Lambda^{2m})$. Since the mapping $\{(z,w) \in \mathbb{C}^2 : \mathrm{Re}\, z < m, \mathrm{Re}\, w < m\} \ni (z,w) \mapsto \Lambda_0^z\Lambda_0^w x, \Lambda_0^{z+w}x \in X$ is analytic, we only have to show $\Lambda_0^z\Lambda_0^w x = \Lambda_0^{z+w}x$ for $0 < z < 1, 0 < w < 1$ for the proof of (b) due to the uniqueness of analytic functions. Now, using 1.4.4 and $\mu(\mu\mathrm{Id} + \Lambda)^{-1} - \lambda(\lambda\mathrm{Id} + \Lambda)^{-1} = (\mu - \lambda)\Lambda(\lambda\mathrm{Id} + \Lambda)^{-1}(\mu\mathrm{Id} + \Lambda)^{-1}$ for $\lambda, \mu \in \rho(\Lambda)$ with $s(z,w) := \frac{\sin(\pi z)}{\pi}\frac{\sin(\pi w)}{\pi}$

$$\Lambda_0^z \Lambda_0^w x = s(z,w) \int_0^\infty \int_0^\lambda \mu^{z-1}(\mu\mathrm{Id}+\Lambda)^{-1}\lambda^{w-1}(\lambda\mathrm{Id}+\Lambda)^{-1}\Lambda^2 x\,d\mu\,d\lambda$$

$$+s(z,w)\int_0^\infty \int_0^\lambda \mu^{w-1}(\mu\mathrm{Id}+\Lambda)^{-1}\lambda^{z-1}(\lambda\mathrm{Id}+\Lambda)^{-1}\Lambda^2 x\,d\mu\,d\lambda$$

$$= s(z,w)\int_0^\infty \int_0^1 (\sigma^{z-1}+\sigma^{w-1})\lambda^{z+w-1}(\lambda\sigma\mathrm{Id}+\Lambda)^{-1}(\lambda\mathrm{Id}+\Lambda)^{-1}\Lambda^2 x\,d\sigma\,d\lambda$$

$$= s(z,w)\int_0^\infty \int_0^1 \frac{\sigma^{z-1}+\sigma^{w-1}}{\sigma-1}\lambda^{z+w-1}(\lambda\mathrm{Id}+\Lambda)^{-1}\Lambda x(\sigma^{-z-w+1}-1)\,d\sigma\,d\lambda$$

$$= s(z,w)\int_0^\infty \int_0^1 \frac{\sigma^{-z}+\sigma^{-w}-\sigma^{z-1}-\sigma^{w-1}}{\sigma-1}\,d\sigma\,\lambda^{z+w-1}(\lambda\mathrm{Id}+\Lambda)^{-1}\Lambda x\,d\lambda$$

$$= c(z,w)\Lambda_0^{z+w}x$$

with a constant $c(z,w)$ depending only on z,w. Since $c(z,w)$ does not depend on the space X and the operator Λ, we can choose special spaces X and operators Λ to determine $c(z,w)$. Choosing $X = \mathbb{C}$ and $\Lambda = a \in (0,\infty)$ we obtain $c(z,w) = 1$ with 1.4.4(b). Hence we have proved (b). Clearly, we have (c) and the further assertions are consequences of 1.4.5. $\qquad\square$

1.4.8 Example. Let H be a separable Hilbert space, (Ω,μ) a finite measure space, $\delta > 0$, $f : M \to [\delta,\infty)$ a measurable function, and $U : H \to L^2(\Omega,\mu)$ be a unitary operator. Moreover, let $D(\Lambda) := \{x \in H : f(\cdot)(Ux)(\cdot) \in L^2(\Omega,\mu)\}$ and $\Lambda x := U^{-1}(f(\cdot)(Ux)(\cdot))$ for $x \in D(\Lambda)$.
Then $\Lambda : D(\Lambda) \to X$ is a strictly-positive, selfadjoint linear operator in H. In particular, it is of positive type and its fractional powers $\Lambda^z : D(\Lambda^z) \to H$ are defined for $z \in \mathbb{C}$. We have $D(\Lambda^z) = \{x \in D(\Lambda) : [f(\cdot)]^z(Ux)(\cdot) \in L^2(\Omega,\mu)\}$ and $\Lambda^z x = U^{-1}([f(\cdot)]^z(Ux)(\cdot))$ for $x \in D(\Lambda^z)$ and $z \in \mathbb{C}$.

PROOF: We neglect no difficulty by assuming $H = L^2(\Omega,\mu)$ and $U = \mathrm{Id}$. Λ is clearly symmetric and satisfies $\langle\Lambda f,f\rangle \geq \delta\|f\|^2$ for $f \in D(\Lambda)$. Moreover, let $g \in D(\Lambda^*)$ and

$$\chi_k(\omega) := \begin{cases} 1 & ,\text{ if } |f(\omega)| \leq k \\ 0 & ,\text{ otherwise} \end{cases}$$

for $\omega \in \Omega$. Then, the monotone convergence theorem shows

$$\|\Lambda^* g\| = \lim_{k\to\infty} \|\chi_k\Lambda^* g\| = \lim_{k\to\infty} \sup_{\|h\|=1} \langle h,\chi_k\Lambda^* g\rangle = \lim_{k\to\infty} \sup_{\|h\|=1} \langle\Lambda(\chi_k h),g\rangle$$

$$= \lim_{k\to\infty} \sup_{\|h\|=1} \langle h,\chi_k fg\rangle = \lim_{k\to\infty} \|\chi_k fg\|$$

This shows $fg \in L^2(\Omega, \mu)$. Hence $g \in D(\Lambda)$ and Λ is selfadjoint. We can therefore define fractional powers $\Lambda^z : D(\Lambda^z) \to H$ for Λ for $z \in \mathbb{C}$. It remains to show $\Lambda^z = \widetilde{\Lambda}^z$ with $D(\widetilde{\Lambda}^z) := \{g \in L^2(\Omega, \mu) : f^z g \in L^2(\Omega, \mu)\}$ and $\widetilde{\Lambda}^z g = f^z g$ for $g \in D(\widetilde{\Lambda}^z)$. Since $\chi_k g \in D(\Lambda^\infty)$ for $k \in \mathbb{N}$ and $g \in D(\Lambda^z) \cup D(\widetilde{\Lambda}^z)$, we know that $D(\Lambda^\infty)$ is a core for both Λ^z and $\widetilde{\Lambda}^z$ and we only have to show $\Lambda^z g = \widetilde{\Lambda}^z g$ for $g \in D(\Lambda^\infty)$ and $z \in \mathbb{C}$. Since $\mathbb{C} \ni z \mapsto \Lambda^z g \in H$ and $\mathbb{C} \ni z \mapsto \widetilde{\Lambda}^z g \in H$ are analytic it is therefore sufficient to show $\Lambda^z g = \widetilde{\Lambda}^z g$ for $g \in D(\Lambda^\infty)$ and $0 < z < 1$. But this is a consequence of 1.4.4(b). □

1.4.9 Remark. The spectral theorem for unbounded, selfadjoint linear operators (cf. e.g. [128, VIII.4]) shows that any strictly positive operator $\Lambda : D(\Lambda) \to H$ in a separable, complex Hilbert space can be written as in example 1.4.8 with a suitable finite measure space (Ω, μ) and a measurable function $f : \Omega \to (0, \infty)$. In particular, example 1.4.8 shows that fractional powers defined with the spectral theorem (cf. e.g. [128, VIII.5]) coincide with the definitions in this section.

1.4.10 Remark. Let q be a closed, positive quadratic form (i.e., q is semi-bounded by a constant $M > 0$) in a complex Hilbert space H with domain $D(q)$ and let $Q : D(Q) \to H$ be the selfadjoint operator induced by q. Since Q is strictly positive, $Q^{1/2} : D(Q^{1/2}) \to H$ is defined. Then $D(Q^{1/2}) = D(q)$ and $\langle Q^{1/2}x, Q^{1/2}y \rangle = q(x, y)$ for $x, y \in D(Q^{1/2}) = D(q)$.

PROOF: Let $\|\cdot\|_q, \langle \cdot, \cdot \rangle_q$ be as in 1.2.8. Then $q(x, y) = \langle Qx, y \rangle = \langle Q^{1/2}x, Q^{1/2}y \rangle$ for $x, y \in D(Q)$. Let $\|x\|_{D(Q^{1/2})} := \|x\| + \|Q^{1/2}x\|$ for $x \in D(Q^{1/2})$. Then $\|\cdot\|_q$ and $\| \|_{D(Q^{1/2})}$ define equivalent norms on $D(Q)$. Since $Q^{1/2} : D(Q^{1/2}) \to H$ is the closure of $Q^{1/2}|_{D(Q)} : D(Q) \to H$ and $D(Q) \subset (D(q), \|\cdot\|_q)$ densely, we have $D(Q^{1/2}) = D(q)$ and $\langle Q^{1/2}x, Q^{1/2}y \rangle = q(x, y)$ for $x, y \in D(Q^{1/2}) = D(q)$. □

1.4.11 Bibliographical remarks. There are many and slightly different approaches to fractional powers. For example, Komatsu [94] has given a systematic treatment in a series of papers, cf. also the work of Kato [76]. For more detailed information on literature and results of fractional powers we refer to the monographs of Triebel [151], Yosida [160], or Amann [9], which also served as a guideline for the treatment discussed in this section.

1.5 Complex interpolation spaces

The theory of interpolation spaces provides a powerful tool for proving estimates for linear operators. Once a linear operator is known to be a continuous
map from a Banach space X_0 to a Banach space Y_0 and from a Banach space
X_1 to a Banach space Y_1, it is automatically known to be a continuous map
between certain intermediate spaces $[X_0, X_1]_\theta$ and $[Y_0, Y_1]_\theta$, so-called interpolation spaces. For example, the theory will show that a linear map which is an
endomorphism of $L^2(\mathbf{R}^n)$ and of $L^\infty(\mathbf{R}^n)$ is automatically an endomorphism of
$L^p(\mathbf{R}^n)$ for $2 < p < \infty$ as well. This is the Riesz-Thorin theorem. There are
several, different methods to construct interpolation spaces. We will focus on
the so-called complex one.
Throughout this section we will use the following notation.

1.5.1 Definition. Let $\Sigma := \{z \in \mathbf{C} : 0 < \operatorname{Re} z < 1\} \subset \mathbf{C}$ and denotes its
closure by $\overline{\Sigma} := \{z \in \mathbf{C} : 0 \le \operatorname{Re} z \le 1\}$.

The theory of complex interpolation spaces is based on Hadamard's three lines
theorem.

1.5.2 Theorem. Let $\varphi : \overline{\Sigma} \to X$ be a continuous and bounded function
with values in a complex Banach space $(X, \|\cdot\|)$ such that $\varphi|_\Sigma$ is analytic and
such that there are constants $M_0, M_1 \ge 0$ with $\|\varphi(z)\| \le M_0$ for $\operatorname{Re} z = 0$ and
$\|\varphi(z)\| \le M_1$ for $\operatorname{Re} z = 1$. Then $\|\varphi(z)\| \le M_0^{1-\operatorname{Re} z} M_1^{\operatorname{Re} z}$ for $z \in \overline{\Sigma}$.

PROOF: By considering $\widetilde{\varphi}(z) := M_0^{z-1} M_1^{-z} \varphi(z)$ at the place of φ we can
assume without restriction that $M_0 = 1 = M_1$. We have to show $\|\varphi(z)\| \le 1$
for $z \in \overline{\Sigma}$. The assertion implies the existence of a $c \ge 0$ with $\|\varphi(z)\| \le c$ for
$z \in \overline{\Sigma}$. Consider $\varphi_\varepsilon(z) := \frac{\varphi(z)}{1+\varepsilon z}$ for $\varepsilon > 0, z \in \overline{\Sigma}$. Then

$$\|\varphi_\varepsilon(z)\| \le \frac{\|\varphi(z)\|}{1 + \varepsilon \operatorname{Re} z} \le \|\varphi(z)\| \le 1 \quad \text{for } |z| = 0 \text{ or } |z| = 1$$

and

$$\|\varphi_\varepsilon(z)\| \le \frac{\|\varphi(z)\|}{\varepsilon |\operatorname{Im} z|} \le \frac{c}{\varepsilon |\operatorname{Im} z|} \quad \text{for } z \in \overline{\Sigma}. \tag{1.5.1}$$

Therefore, the maximum principle for analytic functions applied with the rec-
tangle R_ε with corners $\pm i \frac{c}{\varepsilon}, 1 \pm i \frac{c}{\varepsilon}$ shows $\|\varphi_\varepsilon(z)\| \le 1$ for $z \in R_\varepsilon$. Moreover,
(1.5.1) implies $\|\varphi_\varepsilon(z)\| \le 1$ for $z \in \overline{\Sigma} \setminus R_\varepsilon$. Hence we have shown $\|\varphi_\varepsilon(z)\| \le 1$
for $z \in \overline{\Sigma}$ and obtain the assertion with $\varepsilon \to 0$. $\qquad \square$

1.5.3 Definition. Two complex Banach spaces $(X_0, \|\cdot\|_0)$ and $(X_1, \|\cdot\|_1)$ are called an interpolation couple provided that there is a complex Hausdorff vector space $\mathcal{X}$ with $X_j \hookrightarrow \mathcal{X}$ continuously and with $X_0 \cap X_1 \subset X_j$ densely for $j = 0, 1$. Let

- $\|x\|_{X_0 \cap X_1} := \max(\|x\|_0, \|x\|_1)$ for $x \in X_0 \cap X_1$,

- $X_0 + X_1 := \{x \in \mathcal{X} : \text{ for } j = 0, 1 \, \exists \, x_j \in X_j \text{ with } x = x_0 + x_1\}$,

- $\|x\|_{X_0 + X_1} := \inf\{\|x_0\|_0 + \|x_1\|_1 : \exists \, x_j \in X_j, j = 0, 1, \text{ with } x = x_0 + x_1\}$.

It is easy to show that $(X_0 \cap X_1, \|\cdot\|_{X_0 \cap X_1})$ and $(X_0 + X_1, \|\cdot\|_{X_0 + X_1})$ are Banach spaces with $X_0 \cap X_1 \hookrightarrow X_j \hookrightarrow X_0 + X_1$ for $j = 0, 1$.

1.5.4 Proposition. Let (X_0, X_1) be an interpolation couple and let $\mathcal{F}(X_0, X_1)$ be the set of all functions $f \in C(\overline{\Sigma}, X_0 + X_1)$ such that

- $f|_\Sigma : \Sigma \to X_0 + X_1$ is analytic
- $\sup\limits_{z \in \overline{\Sigma}} \|f(z)\|_{X_0 + X_1} < \infty$
- $\mathbb{R} \ni t \mapsto f(j + it) \in X_j$ continuously for $j = 0, 1$
- $\|f\|_{\mathcal{F}(X_0, X_1)} := \max\limits_{j=0,1} \left(\sup\limits_{t \in \mathbb{R}} \|f(j + it)\|_{X_j} \right) < \infty \Big\}$

Moreover, let $\mathcal{F}_-(X_0, X_1)$ denote the set of those $f \in \mathcal{F}(X_0, X_1)$ with $\lim_{|t| \to \infty} \|f(j + it)\|_j = 0$ for $j = 0, 1$. Then $(\mathcal{F}(X_0, X_1), \|\cdot\|_{\mathcal{F}(X_0, X_1)})$ is a complex Banach space, $\mathcal{F}_-(X_0, X_1)$ is a closed subspace of $\mathcal{F}(X_0, X_1)$ and

$$\mathrm{linh}\{e^{\delta z^2 + \lambda z} x : \delta > 0, \lambda \in \mathbb{R}, x \in X_0 \cap X_1\} \subset \mathcal{F}_-(X_0, X_1) \quad \text{densely.}$$

PROOF: 1.5.2 implies $\sup_{z \in \overline{\Sigma}} \|f(z)\|_{X_0 + X_1} \leq \|f\|_{\mathcal{F}(X_0, X_1)}$ for $f \in \mathcal{F}(X_0, X_1)$. This proves that $\mathcal{F}(X_0, X_1)$ is a Banach space and that $\mathcal{F}(X_0, X_1)$ is a closed subspace. We only have to prove the third statement. Let

$$D := \mathrm{linh}\{e^{\delta z^2 + \lambda z} x : \delta > 0, \lambda \in \mathbb{R}, x \in X_0 \cap X_1\}.$$

Since $\left\| e^{\delta z^2} f(z) - f(z) \right\|_{\mathcal{F}(X_0, X_1)} \to 0$ as $\delta \to 0$ for all $f \in \mathcal{F}_-(X_0, X_1)$, it is sufficient to show that all functions $g(z) := e^{\delta z^2} f(z)$ with $f \in \mathcal{F}_-(X_0, X_1)$ can be approximated by functions in D. Let

$$g_n(z) := \sum_{k \in \mathbb{Z}} g(z + 2\pi i k n) \quad \text{for } n \in \mathbb{N}.$$

Then $e^{sz^2}g_n(z) \in \mathcal{F}_-(X_0, X_1)$ for all $s > 0$, and for $\varepsilon > 0$ we can find $s > 0$ and $n \in \mathbb{N}$ with $\left\|e^{sz^2}g_n(z) - g(z)\right\|_{\mathcal{F}(X_0,X_1)} < \varepsilon$. But now g_n is periodic with period $2\pi i n$ and can be represented by a Fourier series

$$g_n(x+iy) = \sum_{k \in \mathbb{Z}} a_{kn}(x)e^{\frac{k(x+iy)}{n}} \quad \text{with} \quad a_{kn}(x) = \frac{1}{2\pi nm} \int_{-\pi nm}^{\pi nm} g_n(x+iy')e^{\frac{-k(x+iy')}{n}}\,dy'.$$

Note that, by periodicity, $a_{kn}(x)$ is independent of m. Moreover, for given $\eta > 0$ and $x, x' \in (0, 1)$ we have $|a_{kn}(x) - a_{kn}(x')| < \eta$ for sufficiently large m. But $a_{kn}(x)$ is independent of m. Thus $a_{kn}(x) =: a_{kn}$ is independent of $x \in [0, 1]$ and it follows that $a_{kn} \in X_0 \cap X_1$ by choosing $x = 0, 1$. Consider the Cesaro-means

$$\sigma_m(g_n)(z) := \sum_{|k| \leq m} \left(1 - \frac{|k|}{m+1}\right) a_{kn} e^{\frac{kz}{n}}$$

Then the theory of Fourier series shows $\|\sigma_m(g_n)(j + it) - g_n(j + it)\|_j \xrightarrow[m \to \infty]{} 0$ uniformly for $n \in \mathbb{N}$. Thus $\left\|e^{sz^2}(\sigma_m(g_n) - g_n)\right\|_{\mathcal{F}(X_0,X_1)} \xrightarrow[m \to \infty]{} 0$ and therefore $\left\|e^{sz^2}\sigma_m(g_n) - g\right\|_{\mathcal{F}(X_0,X_1)} < 2\varepsilon$ for suitable $s > 0$ and $n, m \in \mathbb{N}$. Since $e^{sz^2}\sigma_m(g_n) \in D$, this implies the assertion. $\qquad\square$

1.5.5 Theorem. Let (X_0, X_1) be an interpolation couple, let $0 < \theta < 1$, and let

$$[X_0, X_1]_\theta := \{x \in X_0 + X_1 : \exists\, f \in \mathcal{F}(X_0, X_1) : f(\theta) = x\} \quad \text{and}$$
$$\|x\|_\theta := \inf_{\substack{f \in \mathcal{F}(X_0,X_1) \\ f(\theta)=x}} \|f\|_{\mathcal{F}(X_0,X_1)} \quad \text{for } f \in [X_0, X_1]_\theta$$

$([X_0, X_1]_\theta, \|\cdot\|_\theta)$ is called the complex interpolation space of X_0 and X_1. It has the following properties:

(a) $([X_0, X_1]_\theta, \|\cdot\|_\theta)$ is a Banach space.

(b) $X_0 \cap X_1 \hookrightarrow [X_0, X_1]_\theta$ densely and $[X_0, X_1]_\theta \hookrightarrow X_0 + X_1$.

(c) If $X_0 \hookrightarrow X_1$, then $X_0 \hookrightarrow [X_0, X_1]_\theta \hookrightarrow [X_0, X_1]_{\theta'} \hookrightarrow X_1$ for $0 < \theta < \theta' < 1$.

(d) Let $A : X_0 + X_1 \to Y_0 + Y_1$ be a linear with $A \in \mathcal{L}(X_0, Y_0) \cap \mathcal{L}(X_1, Y_1)$. Then $A : [X_0, X_1]_\theta \to [Y_0, Y_1]_\theta$ continuously with

$$\|A\|_{\mathcal{L}([X_0,X_1]_\theta,[Y_0,Y_1]_\theta)} \leq \|A\|_{\mathcal{L}(X_0,Y_0)}^{1-\theta} \|A\|_{\mathcal{L}(X_1,Y_1)}^{\theta} \,.$$

PROOF: 1.5.2 shows that $\mathcal{N}(\theta) := \{f \in \mathcal{F}(X_0, X_1) : f(\theta) = 0\}$ is a closed subspace of $\mathcal{F}(X_0, X_1)$. Hence $[X_0, X_1]_\theta = \mathcal{F}(X_0, X_1)/\mathcal{N}(\theta)$ is a quotient Banach space and we have proved (a). Moreover, let $([X_0, X_1]_{\theta,-}, \|\cdot\|_{\theta,-})$ be the Banach space, where $\mathcal{F}(X_0, X_1)$ is substituted by $\mathcal{F}_-(X_0, X_1)$. Let $x \in [X_0, X_1]_\theta$ and take $f \in \mathcal{F}(X_0, X_1)$ with $f(\theta) = x$. Let $g(z) := e^{(z-\theta)^2} f(z)$ for $z \in \overline{\Sigma}$. Then $g \in \mathcal{F}_-(X_0, X_1)$ with $g(\theta) = x$, i.e. $x \in [X_0, X_1]_{\theta,-}$. Therefore, the closed graph theorem shows $[X_0, X_1]_\theta = [X_0, X_1]_{\theta,-}$ with equal norms. Since 1.5.4 implies $X_0 \cap X_1 \subset [X_0, X_1]_{\theta,-}$ densely, we have proved (b).
Let $0 < \theta < \theta' < 1$ and $\lambda := \frac{\theta}{\theta'}$, then $(X_0, [X_0, X_1]_{\theta'})$ is an interpolation couple. Let $x \in X_0$ and $f \in D := \mathrm{linh}\{e^{\delta z^2 + \lambda z} y : \delta > 0, \lambda \in \mathbb{R}, y \in X_0\}$ with $f(\theta) = x$. Let $g(z) := f(\theta' z)$ for $z \in \overline{\Sigma}$. Then $g \in \mathcal{F}(X_0, [X_0, X_1]_{\theta'})$ with $g(\lambda) = x$ and $\|g\|_{\mathcal{F}(X_0,[X_0,X_1]_{\theta'})} \leq \|f\|_{\mathcal{F}_-(X_0,X_1)}$. This shows $\|x\|_{[X_0,[X_0,X_1]_{\theta'}]_\lambda} \leq \|f\|_{\mathcal{F}_-(X_0,X_1)}$. Since $D \subset \mathcal{F}_-(X_0, X_1)$ densely, this implies $[X_0, X_1]_\theta \hookrightarrow [X_0, [X_0, X_1]_{\theta'}]_\lambda$. Hence we have proved (c) because $[X_0, [X_0, X_1]_{\theta'}]_\lambda \hookrightarrow [X_0, X_1]_{\theta'}$ due to (c).
Finally, for the proof of (d) let $M_j > 0$ with $\|Ax\|_{Y_j} \leq M_j \|x\|_{X_j}$ for $x \in X_j$ and $j = 1, 2$. Let $x \in [X_0, X_1]_\theta$ and $f \in \mathcal{F}(X_0, X_1)$ with $f(\theta) = x$. Let $g_f(z) := \left(\frac{M_0}{M_1}\right)^{z-\theta} T(f(z))$ for $z \in \overline{\Sigma}$. Then $g_f \in \mathcal{F}(Y_0, Y_1)$ with $g(\theta) = Tx$ and $\|g_f\|_{\mathcal{F}(Y_0,Y_1)} \leq M_0^{1-\theta} M_1^\theta \|f\|_{\mathcal{F}(X_0,X_1)}$. This implies the assertion. $\qquad\square$

1.5.6 Corollary. Let $(X_0, X_1), (Y_0, Y_1)$ be two interpolation couples with $X_1 \hookrightarrow X_0$ and $Y_1 \hookrightarrow Y_0$, and $I \ni t \mapsto A(t) \in \mathcal{L}(X_j, Y_j), j = 0, 1$, be strongly continuous (resp., strongly continuously differentiable), and let $0 < \theta < 1$. Then

$$I \ni t \mapsto A(t) \in \mathcal{L}([X_0, X_1]_\theta, [Y_0, Y_1]_\theta) \tag{1.5.2}$$

is strongly continuous (resp., strongly continuously differentiable).

PROOF: Due to the uniform boundedness principle there is an $a > 0$ with $\|A(t)\|_{\mathcal{L}(X_j,Y_j)} \leq a$, $j = 0, 1$. Hence 1.5.5 implies $\|A(t)\|_{\mathcal{L}([X_0,X_1]_\theta,[Y_0,Y_1]_\theta)} \leq a'$ for $t \in I$ and a suitable $a' \geq 0$. Moreover, $X_1 \subset [X_0, X_1]_\theta$ densely and $I \ni t \mapsto A(t)x \in Y_1 \hookrightarrow [Y_0, Y_1]_\theta$ continuously for $x \in X_1$. This implies the strong continuity of (1.5.2) due to 1.1.3.
Suppose in addition that $I \ni t \mapsto A(t) \in \mathcal{L}(X_j, Y_j), j = 0, 1$, is strongly continuously differentiable with derivative $A'(t)$. Then, with suitable $b \geq 0$

$$\left\| \frac{A(t+h)x - A(t)x}{h} \right\|_{Y_j} \leq \frac{1}{|h|} \left| \int_t^{t+h} \|A'(\tau)x\|_{Y_j} \, d\tau \right| \leq b \|x\|_{X_j} \quad \text{for } x \in X_j .$$

Hence $\left\{ \frac{A(t+h)-A(t)}{h} : t, t+h \in I, h \neq 0 \right\} \subset \mathcal{L}([X_0, X_1]_\theta, [Y_0, Y_1]_\theta)$ boundedly. Moreover, $I \ni t \mapsto A(t)x \in Y_1 \hookrightarrow [Y_0, Y_1]_\theta$ is differentiable for $x \in X_1$ and $X_1 \hookrightarrow [X_0, X_1]_\theta$ densely. This implies the assertion due to 1.1.3. $\qquad\square$

1.5.7 Remark. A different method to construct interpolation spaces for an interpolation couple (X_0, X_1) is given by the K-method: Let $t \in (0, \infty)$ and

$$K(t, x) := \inf_{\substack{x = x_0 + x_1 \\ x_j \in X_j, j = 0, 1}} \left(\|x_0\|_{X_0} + t \|x_1\|_{X_1} \right) \quad \text{for } x \in X_0 + X_1.$$

Then, for $\theta \in (0, 1)$ and $1 \le q < \infty$ let

$$(X_0, X_1)_{\theta, q} := \left\{ x \in X_0 + X_1 : \|x\|_{(X_0, X_1)_{\theta, q}} := \left(\int_0^\infty [t^{-\theta} K(t, x)]^q \frac{dt}{t} \right)^{\frac{1}{q}} < \infty \right\}.$$

$(X_0, X_1)_{\theta, q}$ is called a real interpolation space of X_0 and X_1. Then statements similar to 1.5.5 and 1.5.6 holds.

Now, we will prove the Riesz-Thorin theorem mentioned at the beginning of this section. To this end, we have to identify the complex interpolation space of $L^2(\mathbf{R}^n)$ and $L^\infty(\mathbf{R}^n)$ as $L^p(\mathbf{R}^n)$ for a suitable $2 < p < \infty$.

1.5.8 Theorem.   Let (Ω, μ) be a σ-finite measure space and let $\theta \in (0, 1)$, $p_0, p_1 \in [1, \infty)$, $p \in (1, \infty)$.

(a) If $\frac{1}{p} = \frac{1-\theta}{p_0} + \frac{\theta}{p_1}$, then $[L^{p_0}(\Omega, \mu), L^{p_1}(\Omega, \mu)]_\theta = L^p(\Omega, \mu)$ with equal norms.

(b) If $\frac{1}{p} = \frac{1-\theta}{p_0}$, then $[L^{p_0}(\Omega, \mu), \widetilde{L}^\infty(\Omega, \mu)]_\theta = L^p(\Omega, \mu)$ with equal norms.
Here $\widetilde{L}^\infty(\Omega, \mu)$ denotes the closure of the simple functions in $L^\infty(\Omega, \mu)$.

PROOF: We will only prove (a). The proof of (b) can be established in the same manner. We have to show that $\|\varphi\|_\theta = \|\varphi\|_{L^p}$ for all simple functions φ.

Let φ be simple with $\|\varphi\|_{L^p} = 1$ and let $f(z) := |\varphi(\cdot)|^{p\left(\frac{z}{p_1} + \frac{1-z}{p_0}\right)} e^{i \arg \varphi(\cdot)}$. Then it is straightforward to check that $f(\theta) = \varphi(\cdot)$ and $f \in \mathcal{F}(L^{p_0}(\Omega, \mu), L^{p_1}(\Omega, \mu))$ with $\|f\|_{\mathcal{F}(L^{p_0}(\Omega, \mu), L^{p_1}(\Omega, \mu))} \le 1$. This implies $\|\varphi\|_\theta \le \|\varphi\|_{L^p}$ for all simple functions φ. Conversely, we have to show $\|\varphi\|_{L^p} \le \|f\|_{\mathcal{F}(L^{p_0}(\Omega, \mu), L^{p_1}(\Omega, \mu))}$ for any simple function φ and any $f \in \mathcal{F}(L^{p_0}(\Omega, \mu), L^{p_1}(\Omega, \mu))$ with $f(\theta) = \varphi$. Take a further simple function ψ and let $g(z) := |\psi(\cdot)|^{q\left(\frac{z}{q_1} + \frac{1-z}{q_0}\right)} e^{i \arg \psi(\cdot)}$, where $\frac{1}{p} + \frac{1}{q} = 1, \frac{1}{p_j} + \frac{1}{q_j} = 1, j = 0, 1$. Applying Hadamard's three lines theorem to $H(z) := \int_\Omega f(z)(x) g(z)(x) d\mu(x)$ we have

$$|H(\theta)| \le \sup_{y \in \mathbf{R}} \{ \|f(iy) g(iy)\|_{L^1}, \|f(1 + iy) g(1 + iy)\|_{L^1} \}$$

$$\le \sup_{y \in \mathbf{R}} \{ \|f(iy)\|_{L^{p_0}} \|g(iy)\|_{L^{q_0}}, \|f(1 + iy)\|_{L^{p_1}} \|g(1 + iy)\|_{L^{q_1}} \}$$

$$\le \|f\|_{\mathcal{F}(L^{p_0}(\Omega, \mu), L^{p_1}(\Omega, \mu))} \|g\|_{\mathcal{F}(L^{q_0}(\Omega, \mu), L^{q_1}(\Omega, \mu))} = \|f\|_{\mathcal{F}(L^{p_0}(\Omega, \mu), L^{p_1}(\Omega, \mu))} \|\psi\|_{L^q}$$

Hence

$$\left| \int_\Omega \varphi(x)\psi(x)d\mu(x) \right| = |H(\theta)| \le \|f\|_{\mathcal{F}(L^{p_0}(\Omega,\mu),L^{p_1}(\Omega,\mu))} \|\psi\|_{L^q}$$

for any simple function ψ. This implies the assertion.   □

Finally, we will show that the domains of fractional powers of strictly positive, selfadjoint operators can also be determined with the complex interpolation method. But first, we have to define a Banach space topology on the domain of a closed operator.

1.5.9 Lemma. Let $\Lambda : D(\Lambda) \to X$ be a linear closed operator with domain $D(\Lambda)$ in a Banach space X. Let $\|x\|_{D(\Lambda)} := \|x\| + \|\Lambda x\|$ for $x \in D(\Lambda)$ be the so-called graph norm of Λ . Then $(D(\Lambda), \|\cdot\|_{D(\Lambda)})$ is a Banach space.

PROOF:The assertion is an immediate consequence of the closedness of Λ. □

1.5.10 Theorem. Let H be a separable Hilbert space and $\Lambda : D(\Lambda) \to H$ be a strictly positive, selfadjoint operator in H.
Then $[D(\Lambda^\alpha), D(\Lambda^\beta)]_\theta = D(\Lambda^{(1-\theta)\alpha+\theta\beta})$ with equivalent norms for $0 \le \alpha < \beta$ and $0 < \theta < 1$.

PROOF: Denote the graph norm of $D(\Lambda^s)$ by $\|\cdot\|_s$. Since $D(\Lambda^\infty) \subset D(\Lambda^z)$ densely for any $z \in \mathbb{R}$ we only have to show that the norms $\|\cdot\|_{(1-\theta)\alpha+\theta\beta}$ and $\|\cdot\|_{[D(\Lambda^\alpha),D(\Lambda^\beta)]_\theta}$ are equivalent on $D(\Lambda^\infty)$. Due to 1.4.9 there is a finite measure space (Ω,μ), a $\delta > 0$, a measurable function $f : M \to [\delta,\infty)$, and a unitary operator $U : H \to L^2(\Omega,\mu)$ with $D(\Lambda^z) := \{x \in H : f(\cdot)^z(Ux)(\cdot) \in L^2(\Omega,\mu)\}$ and $\Lambda^z x := U^{-1}(f(\cdot)^z(Ux)(\cdot))$ for $x \in D(\Lambda^z), z \in \mathbb{R}$. Let $x \in D(\Lambda^\infty)$ and let $F(z) := U^{-1}[f(\cdot)^{(z-\theta)(\alpha-\beta)}(Ux)(\cdot)]$ for $z \in \overline{\Sigma}$. Then $F : \overline{\Sigma} \to D(\Lambda^\gamma)$ is continuous and $F|_\Sigma : \Sigma \to D(\Lambda^\gamma)$ is analytic for any $\gamma \ge 0$ and

$$\|\Lambda^\gamma F(z)\|^2 = \int_\Omega f(\omega)^{2\gamma+2(\operatorname{Re} z-\theta)(\alpha-\beta)}|(Ux)(\omega)|^2 d\mu(\omega) \le \left\| \Lambda^{\gamma+(1-\theta)(\alpha-\beta)} x \right\|^2$$

for $z \in \overline{\Sigma}$. Hence $F \in \mathcal{F}(D(\Lambda^\alpha), D(\Lambda^\beta))$ with

$$\|F\|_{\mathcal{F}(D(\Lambda^\alpha),D(\Lambda^\beta))} = \max\{ \sup_{\operatorname{Re} z=0} \|f(z)\|_\alpha , \sup_{\operatorname{Re} z=1} \|f(z)\|_\beta\} \le \|x\|_{(1-\theta)\alpha+\theta\beta} .$$

Since $F(\theta) = x$ this shows

$$\|x\|_{[D(\Lambda^\alpha),D(\Lambda^\beta)]_\theta} \le \|F\|_{\mathcal{F}(D(\Lambda^\alpha),D(\Lambda^\beta))} \le \|x\|_{(1-\theta)\alpha+\theta\beta} .$$

Conversely, let $F \in \mathcal{F}(D(\Lambda^\alpha), D(\Lambda^\beta))$ with $F(\theta) = x$. Then $F \in \mathcal{C}(\overline{\Sigma}, D(\Lambda^\alpha))$ is bounded. Let

$$\chi_k(\omega) := \begin{cases} 1 & , \text{ if } |f(\omega)| \leq k \\ 0 & , \text{ otherwise} \end{cases}$$

and let $\varphi_k(z)(\omega) := f(\omega)^{(1-z)\alpha + z\beta} \chi_k(\omega)[U(F(z))](\omega)$ for $z \in \overline{\Sigma}, \omega \in \Omega$. Then $\varphi_k : \overline{\Sigma} \to L^2(\Omega, \mu)$ is continuous and $\varphi_k|_\Sigma : \Sigma \to L^2(\Omega, \mu)$ is analytic with

$$
\begin{aligned}
\|\varphi_k(z)\|_{L^2(\Omega,\mu)}^2 &= \int_\Omega f(\omega)^{2(1-\operatorname{Re} z)\alpha + 2(\operatorname{Re} z)\beta} |\chi_k(\omega)|^2 |U(F(z))(\omega)|^2 d\mu(\omega) \\
&\leq \int_\Omega f(\omega)^{2\beta} |\chi_k(\omega)|^2 |U(F(z))(\omega)|^2 d\mu(\omega) \\
&\leq c_k \int_\Omega f(\omega)^{2\alpha} |U(F(z))(\omega)|^2 d\mu(\omega) \leq c_k \|F(z)\|_\alpha^2 \leq d_k
\end{aligned}
$$

for suitable constants $c_k \geq 0, d_k \geq 0$ depending on $k \in \mathbb{N}$. Moreover, for $\operatorname{Re} z = 0$ we have

$$\|\varphi_k(z)\|_{L^2(\Omega,\mu)}^2 \leq \int_\Omega f(\omega)^{2\alpha} |U(F(z))(\omega)|^2 d\mu(\omega) \leq \|F\|_{\mathcal{F}(D(\Lambda^\alpha), D(\Lambda^\beta))}^2$$

and for $\operatorname{Re} z = 1$ we have

$$\|\varphi_k(z)\|_{L^2(\Omega,\mu)}^2 \leq \int_\Omega f(\omega)^{2\beta} |U(F(z))(\omega)|^2 d\mu(\omega) \leq \|F\|_{\mathcal{F}(D(\Lambda^\alpha), D(\Lambda^\beta))}^2 .$$

Therefore, the three lines theorem 1.5.2 implies

$$\left\| f(\cdot)^{(1-z)\alpha + z\beta} \chi_k(\cdot)[U(F(z))](\cdot) \right\|_{L^2(\Omega;\mu)} = \|\varphi_k(z)(\cdot)\|_{L^2(\Omega,\mu)} \leq \|F\|_{\mathcal{F}(D(\Lambda^\alpha), D(\Lambda^\beta))}$$

and with $k \to \infty$ we have

$$\left\| \Lambda^{(1-z)\alpha + z\beta} F(z) \right\| = \left\| f(\cdot)^{(1-z)\alpha + z\beta} [U(F(z))](\cdot) \right\|_{L^2(\Omega,\mu)} \leq \|F\|_{\mathcal{F}(D(\Lambda^\alpha), D(\Lambda^\beta))} .$$

Hence $\left\| \Lambda^{(1-\theta)\alpha + \theta\beta} x \right\| = \left\| \Lambda^{(1-z)\alpha + z\beta} F(\theta) \right\| \leq \|F\|_{\mathcal{F}(D(\Lambda^\alpha), D(\Lambda^\beta))}$ with $z = \theta$, which completes the proof of the theorem.

$\square$

1.5.11 Bibliographical remarks. Complex interpolation spaces have been introduced by Calderón [21] and J.L. Lions [104]. The monographs of Bergh/Löfström [16] and Triebel [151] describe many important aspects of different interpolation methods.

1.6   Time-dependent, linear evolution equations

So far we have only looked at an abstract approach to time-independent, linear
evolution equations $u'(t) = Au(t)$. The purpose of this section is to collect some
results on generalizations of the notion of C_0-semigroups to time-dependent,
linear evolution equations $u'(t) = A(t)u(t)$. These generalizations are called
propagators and are families of linear operators $U(t, s)$ for $t \geq s$ satisfying
$U(s, s) = \mathrm{Id}$ and $U(t, s) = U(t, q)U(q, s)$ for $s \leq q \leq t$. Roughly speaking
$t \mapsto U(t, s)x$ is the solution of $u'(t) = A(t)u(t)$ for $t \geq s$ with initial value
$u(s) = x$.

Throughout this book $I := [t_0, t_1]$ will always denote a compact interval and
$\Delta := \{(t, s) \in I^2 : t \geq s\}$.

1.6.1 Definition. A propagator $U : \Delta \rightarrow \mathscr{L}(X)$ on a Banach space X
is a strongly continuous mapping $U : \Delta \rightarrow \mathscr{L}(X)$ with $U(s, s) = \mathrm{Id}$ and
$U(t, s) = U(t, q)U(q, s)$ for $t_0 \leq s \leq q \leq t \leq t_1$.

1.6.2 Definition. A family of C_0-semigroups $\{(T_t(\tau))_{\tau \geq 0} : t \in I\}$ in a Banach space X is called Kato-stable with stability constants (M, β), if there
exist $M \geq 1, \beta \geq 0$ with

$$\left\| \prod_{j=1}^{k} T_{s_j}(\tau_j) \right\|_{\mathscr{L}(X)} := \left\| T_{s_k}(\tau_k) \cdots T_{s_1}(\tau_1) \right\|_{\mathscr{L}(X)} \leq M e^{\beta \sum_{j=1}^{k} \tau_j}$$

for $t_0 \leq s_1 \leq \ldots \leq s_k \leq t_1, \tau_j \geq 0, j = 1, \ldots, k, \ k \in \mathbb{N}$.

The first theorem in this section is a fundamental existence result for propagators. However, propagators constructed in this result are so far not useful
for solving time-dependent evolution equations because differentiability with
respect to t cannot be verified.

1.6.3 Theorem. Suppose that X, Y are Banach spaces with $Y \hookrightarrow X$ densely
such that the following statements hold:

(a) $P(t) : D(P(t)) \rightarrow X$ generates a C_0-semigroup for any $t \in I$ and
$\{(e^{\tau P(t)})_{\tau \geq 0} : t \in I\}$ is Kato-stable with stability constants (M, β). Here
$(e^{\tau P(t)})_{\tau \geq 0}$ denotes the C_0-semigroup generated by $P(t)$.

(b) For any $t \in I$ we have $e^{\tau P(t)}(Y) \subset Y$ and $\{(e^{\tau P(t)}|_Y)_{\tau \geq 0} : t \in I\}$ is a
Kato-stable family of C_0-semigroups in Y.

(c) $Y \subset D(P(t))$ for any $t \in I$ and the mapping $I \ni t \mapsto P(t)|_Y \in \mathcal{L}(Y, X)$ is continuous in the norm topology.

Then, there is a propagator $U : \Delta \to \mathcal{L}(X)$ with $\|U(t, s)\|_{\mathcal{L}(X)} \leq M e^{\beta(t-s)}$ and

$$\left(\frac{\partial}{\partial t}\right)^+ U(t, s)y|_{t=s} = P(s)y \quad \text{and} \quad \frac{\partial}{\partial s} U(t, s)y = -U(t, s)P(s)y$$

for $(t, s) \in \Delta, y \in Y$.

1.6.3 is a consequence of the next result 1.6.4. Differentiability of the propagator with respect to t cannot be shown because Y cannot be shown to be an invariant subspace of the propagator. However, if we do not work in only two spaces Y and X, but in a scale of Banach spaces, a loss in the highest order does not matter, and one can show the following proposition.

1.6.4 Proposition. Suppose that $m \in \mathbb{N}$ and that $(X^k, \|\cdot\|_k)_k$ is a scale of Banach spaces, i.e., $X^k, k \in \mathbb{N}_0$ are Banach spaces with dense and continuous embeddings $X^k \hookrightarrow X^l$ for $k \geq l$. Moreover, assume that $K_1, K_2 \in \mathbb{N}_0$ with $K_1 \leq K_2$ and

(a) $I \ni t \mapsto A(t) \in \bigcap\limits_{k=K_1}^{K_2} \mathcal{L}(X^{k+m}, X^k)$ is continuous in the (uniform) operator topology.

(b) $(T_t(\tau))_{\tau \geq 0} \subset \bigcap\limits_{k=K_1}^{K_2+m} \mathcal{L}(X^k)$ for $t \in I$ is a family of linear operators such that

- $\{(T_t(\tau))_{\tau \geq 0} : t \in I\}$ is a Kato-stable family of C_0-semigroups in X^k with constants (M_k, β_k) for $K_1 \leq k \leq K_2 + m$.
- $[0, \infty) \ni \tau \mapsto T_t(\tau)x \in X^k$ is differentiable for $K_1 \leq k \leq K_2$ and $x \in X^{k+m}$ with $\frac{d}{d\tau} T_t(\tau)x = A(t)T_t(\tau)x = T_t(\tau)A(t)x$.

Then there exists a family $U(t, s) \in \bigcap\limits_{k=K_1}^{K_2} \mathcal{L}(X^k), (t, s) \in \Delta$ such that the following statements hold:

(a) $U : \Delta \to \mathcal{L}(X^k)$ is a propagator in X^k for $K_1 \leq k \leq K_2$.

(b) $\|U(t, s)\|_{\mathcal{L}(X^k)} \leq M_k e^{\beta_k(t-s)}$ for $K_1 \leq k \leq K_2, (t, s) \in \Delta$.

(c) $\left(\dfrac{\partial}{\partial t}\right)^{+} U(t,s)x|_{t=s} = A(s)x$ and $\dfrac{\partial}{\partial s}U(t,s)x = -U(t,s)A(s)x$ with derivatives in X^k for $x \in X^{k+m}$ and $K_1 \leq k \leq K_2$, .

(d) If $K_2 \geq K_1 + m$, then $\triangle \ni (t,s) \mapsto U(t,s) \in \mathscr{L}(X^{k+m}, X^k)$ is strongly continuously differentiable for $K_1 \leq k \leq K_2 - m$ with strong derivatives

$$\frac{\partial}{\partial t}U(t,s) = A(t)U(t,s) \quad \text{and} \quad \frac{\partial}{\partial s}U(t,s) = -U(t,s)A(s).$$

PROOF: Let $t_j^{(l)} := t_0 + \frac{j}{l}(t_1 - t_0), j = 0, \ldots, l, l \in \mathbb{N}$, and

$$A_l(t) := \begin{cases} A(t_j^{(l)}) & , t_j^{(l)} \leq t < t_{j+1}^{(l)}, j = 0, \ldots, l-1 \\ A(b) & , t = b \end{cases}.$$

Since $I \ni t \mapsto A(t) \in \mathscr{L}(X^{k+m}, X^k)$ is continuous for $K_1 \leq k \leq K_2$ we have

$$\|A(t) - A_l(t)\|_{\mathscr{L}(X^{k+m}, X^k)} \xrightarrow[l\to\infty]{} 0 \quad \text{uniformly for } t \in I.$$

Moreover, let

$$U_l(t,s) := \begin{cases} T_{t_i^{(l)}}(t - t_i^{(l)}) \displaystyle\prod_{\mu=j+1}^{i-1} T_{t_\mu^{(l)}}(\frac{t_1 - t_0}{l})T_{t_j^{(l)}}(t_{j+1}^{(l)} - s) & , \begin{array}{l} t_i^{(l)} \leq t \leq t_{i+1}^{(l)} \\ t_j^{(l)} \leq s \leq t_{j+1}^{(l)}, j < i \end{array} \\ T_{t_j^{(l)}}(t - s) & , t_j^{(l)} \leq s \leq t \leq t_{j+1}^{(l)} \end{cases}.$$

Then we clearly have for $l \in \mathbb{N}$

- $U_l : \triangle \to \mathscr{L}(X^k)$ is a propagator in X^k for $K_1 \leq k \leq K_2 + m$.

- $\|U_l(t,s)\|_{\mathscr{L}(X^k)} \leq M_k e^{\beta_k(t-s)}$ for $(t,s) \in \triangle$ and $K_1 \leq k \leq K_2 + m$.

- for $K_1 \leq k \leq K_2, l \in \mathbb{N}, (t_0, s_0) \in \triangle$, and $x \in X^{k+m}$

  - $\{t \in I : t \geq s_0, t \neq t_j^{(l)}, j = 0, \ldots, l\} \ni t \mapsto U_l(t, s_0)x \in X^k$
    is differentiable with derivative $\dfrac{\partial}{\partial t}U_l(t, s_0) = A_l(t)U_l(t, s_0)$,

  - $\{s \in I : s \leq t_0, s \neq t_j^{(l)}, j = 0, \ldots, l\} \ni s \mapsto U_l(t_0, s)x \in X^k$
    is differentiable with derivative $\dfrac{\partial}{\partial s}U_l(t_0, s) = -U_l(t_0, s)A_l(s).$

Now let $K_1 \leq k \leq K_2$ and $\varepsilon > 0$. Then, for $y \in X^{k+m}, l, l' \in \mathbb{N}, (t, s) \in \Delta$

$$\|U_l(t, s)y - U_{l'}(t, s)y\|_k = \left\| \int_s^t \frac{\partial}{\partial \tau} \left(U_{l'}(t, \tau) U_l(\tau, s)y \right) d\tau \right\|_k$$

$$\leq \int_s^t \left\| U_{l'}(t, \tau) \left(A_l(\tau) - A_{l'}(\tau) \right) U_l(\tau, s)y \right\|_k d\tau$$

$$\leq M_k M_{k+m} e^{(\beta_k + \beta_{k+m})(t_1 - t_0)} \|y\|_{k+m} \int_s^t \|A_l(\tau) - A_{l'}(\tau)\|_{\mathscr{L}(X^{k+m}, X^k)} d\tau .$$

Hence there is an $L \in \mathbb{N}$ with $\|U_l(t, s)y - U_{l'}(t, s)y\|_k < \varepsilon$ $(t, s) \in \Delta$, $l, l' \geq L$. Thus $\left(U_l(\cdot, \cdot)x \right)_{l \in \mathbb{N}} \subset C(\Delta, X^k)$ is a Cauchy sequence for $x \in X^k$ due to 1.1.3. Let $U(t, s)x := \lim_{l \to \infty} U_l(t, s)x$ for $(t, s) \in \Delta, x \in X$. Then

$$U(t, s) \in \bigcap_{k=K_1}^{K_2} \mathscr{L}(X^k)$$

by the principle of uniform boundedness. We clearly have $U(s, s) = \text{Id}$ for $s \in I$ and (b). Furthermore, for $t_0 \leq s \leq q \leq t \leq t_1$ and $x \in X^k, K_1 \leq k \leq K_2$

$$U(t, q)U(q, s)x \xleftarrow[\infty \leftarrow l]{X^k} U_l(t, q)U_l(q, s)x = U_l(t, s)x \xrightarrow[l \to \infty]{X^k} U(t, s)x,$$

which implies (a). Now fix $K_1 \leq k \leq K_2, x \in X^{k+m}, t_0 \leq s < t \leq t_1$, and $\tau \in I$, then

$$\|U_l(t, s)x - T_\tau(t - s)x\|_k = \left\| - \int_s^t \frac{\partial}{\partial q} [U_l(t, q)T_\tau(q - s)x] \, dq \right\|_k$$

$$\leq \int_s^t \|U_l(t, q)(A(\tau) - A_l(q))T_\tau(q - s)x\|_k \, dq$$

$$\leq M_k M_{k+m} e^{(\beta_k + \beta_{k+m})(t_1 - t_0)} \|x\|_{k+m} \int_s^t \|A(\tau) - A_l(q)\|_{\mathscr{L}(X^{k+m}, X^k)} \, dq$$

With $l \to \infty$ we obtain

$$\left\| \frac{U(t, s)x - x}{t - s} - \frac{T_\tau(t - s)x - x}{t - s} \right\|_k$$

$$\leq \frac{M_k M_{k+m}}{t - s} e^{(\beta_k + \beta_{k+m})(t_1 - t_0)} \|x\|_{k+m} \int_s^t \|A(\tau) - A(q)\|_{\mathscr{L}(X^{k+m}, X^k)} \, dq$$

Choosing $\tau = s$ and letting $t \searrow s$ we obtain

$$\left(\frac{\partial}{\partial t} \right)^+ U(t, s)x \bigg|_{t=s} = A(s)x$$

and choosing $\tau = t$ and letting $s \nearrow t$ we obtain

$$\left(\frac{\partial}{\partial s}\right)^{-} U(t,s)x\Big|_{s=t} = \lim_{s \nearrow t} \frac{U(t,t)x - U(t,s)x}{t-s} = \lim_{s \nearrow t} \frac{x - U(t,s)x}{t-s} = -A(t)x\,.$$

Therefore

$$\left(\frac{\partial}{\partial s}\right)^{+} U(t,s)x = \lim_{h \searrow 0} U(t,s+h)\frac{x - U(s+h,s)x}{h}\,.$$

$$= -U(t,s)\left(\frac{\partial}{\partial \tau}\right)^{+} U(\tau,s)x\Big|_{\tau=s} = -U(t,s)A(s)x$$

and

$$\left(\frac{\partial}{\partial s}\right)^{-} U(t,s)x = \lim_{h \searrow 0} U(t,s)\frac{x - U(s,s-h)x}{h}$$

$$= U(t,s)\left(\frac{\partial}{\partial \sigma}\right)^{-} U(s,\sigma)x\Big|_{\sigma=s} = -U(t,s)A(s)x\,.$$

Therefore, $[t_0,t] \ni s \mapsto U(t,s)x \in X^k$ is differentiable with derivative $\frac{\partial}{\partial s}U(t,s)x = -U(t,s)A(s)x$. Moreover, if $K_2 \geq K_1+m$ and $K_1 \leq k \leq K_2-m$, then

$$\left(\frac{\partial}{\partial t}\right)^{+} U(t,s)x = \lim_{h \searrow 0} \frac{U(t+h,t) - \mathrm{Id}}{h}U(t,s)x$$

$$= \left(\frac{\partial}{\partial \tau}\right)^{+} U(\tau,t)U(t,s)x\Big|_{\tau=t} = A(t)U(t,s)x$$

and

$$\left(\frac{\partial}{\partial t}\right)^{-} U(t,s)x = \lim_{h \searrow 0} \frac{U(t,t-h) - \mathrm{Id}}{h}U(t-h,s)x$$

$$= -\left(\frac{\partial}{\partial \sigma}\right)^{-} U(t,\sigma)U(t,s)x\Big|_{\sigma=t} = A(t)U(t,s)x\,.$$

This shows that $[s,t_1] \ni t \mapsto U(t,s)x \in X^k$ is differentiable with derivative $\frac{\partial}{\partial t}U(t,s)x = A(t)U(t,s)x$. With (a) we obtain (d). $\square$

Our next goal will be a result showing differentiability of $U(t,s)$ with respect to t without a loss of order. To this end we will need some preparations. The following estimate is called Gronwall's lemma.

1.6.5 Lemma. Let $\alpha, \beta \geq 0$ and $u \in C(I,\mathbf{R})$ with $u(t) \leq \alpha + \int_{t_0}^{t} \beta u(\tau)d\tau$ for $t \in I$. Then $u(t) \leq \alpha e^{\beta(t-t_0)}$ for $t \in I$.

PROOF: Let $v(t) := \alpha + \beta \int_{t_0}^{t} u(\tau)d\tau$. Then $v'(t) = \beta u(t) \leq \beta v(t)$ and

$$\frac{d}{dt}\left(v(t)e^{-\beta(t-t_0)}\right) = v'(t)e^{-\beta(t-t_0)} - \beta v(t)e^{-\beta(t-t_0)} \leq 0.$$

Hence $v(t)e^{-\beta(t-t_0)} \leq v(t_0) = \alpha$ and $u(t) \leq v(t) \leq \alpha e^{\beta(t-t_0)}$. $\qquad\square$

Then, we will need the following existence and uniqueness result for solutions of special types of integral equations. These equations are also called Volterra equations.

1.6.6 Proposition. Let X be a Banach space and let $U : \Delta \to \mathscr{L}(X)$ and $B : I \to \mathscr{L}(X)$ be strongly continuous. Then the following holds:

(a) There is a unique, strongly continuous mapping $W : \Delta \to \mathscr{L}(X)$ with

$$W(t,s)x = U(t,s)x + \int_{s}^{t} U(t,\tau)B(\tau)W(\tau,s)x d\tau \text{ for } (t,s) \in \Delta, x \in X.$$

$$(1.6.1)$$

(b) W is also the unique, strongly continuous mapping $W : \Delta \to \mathscr{L}(X)$ with

$$W(t,s)x = U(t,s)x + \int_{s}^{t} W(t,\tau)B(\tau)U(\tau,s)x d\tau \text{ for } (t,s) \in \Delta, x \in X.$$

$$(1.6.2)$$

(c) If U is a propagator on X, so is W.

PROOF: For $j = 1, 2$ and strongly continuous mappings $U_j : \Delta \to \mathscr{L}(X)$ and $B_j : I \to \mathscr{L}(X)$ we introduce the notation

$$(U_1 B_1 U_2)(t,s)x := \int_{s}^{t} U_1(t,\tau)B_1(\tau)U_2(\tau,s)x d\tau \quad \text{for } (t,s) \in \Delta, x \in X.$$

It is easy to check that $(U_1 B_1 U_2) : \Delta \to \mathscr{L}(X)$ is strongly continuous and that the notation is associative, i.e., $((U_1 B_1 U_2)B_2 U_3) = (U_1 B_1(U_2 B_2 U_3))$.
Let $M, b \geq 0$ with $\|U(t,s)\|_{\mathscr{L}(X)} \leq M, \|B(t)\|_{\mathscr{L}(X)} \leq b$ for $(t,s) \in \Delta$, let $W_0 := \text{Id}$ and inductively $W_{k+1} := (UBW_k)$ for $k \in \mathbb{N}_0$. Then it is easy to check that $\|W_k(t,s)\|_{\mathscr{L}(X)} \leq M^{k+1}b^k \frac{(t-s)^k}{k!}$. Hence $W(t,s) := \sum_{k=0}^{\infty} W_k(t,s) \in \mathscr{L}(X)$ exists and defines a strongly continuous mapping $\Delta \ni (t,s) \mapsto W(t,s) \in \mathscr{L}(X)$ satisfying $W = U + UBW$. Let $\widetilde{W} : \Delta \to \mathscr{L}(X)$ be strongly continuous with $\widetilde{W} = U + UB\widetilde{W}$, let $x \in X$, and let $v(t) := W(t,s)x - \widetilde{W}(t,s)x$ for $t \in [s, t_1]$. Then

$$\|v(t)\| = \left\| \int_{s}^{t} U(t,\tau)B(\tau)v(\tau)d\tau \right\| \leq Mb \int_{s}^{t} \|v(s)\| ds$$

Hence Gronwall's lemma 1.6.5 implies $\|v(t)\| = 0$ and we have proved the uniqueness statement. This proves (a). For the proof of (b) note inductively that $W_{k+1} = U(BU)^k = (UB)^k U = W_k BU$. Hence W satisfies $W = U + WBU$. To prove the uniqueness statement in (b) let $\widetilde{W} : \Delta \to \mathscr{L}(X)$ be strongly continuous with $\widetilde{W} = U + \widetilde{W}BU$. Then the part proved so far shows that there is a unique, strongly continuous $V : \Delta \to \mathscr{L}(X)$ with $V = \widetilde{W} - \widetilde{W}BV$ that also satisfies $V = \widetilde{W} - VB\widetilde{W}$. Hence the uniqueness statement and $U = \widetilde{W} - \widetilde{W}BU$ imply $V = U$ and hence $U = \widetilde{W} - UB\widetilde{W}$. This shows $\widetilde{W} = U + UB\widetilde{W}$ and the uniqueness statement in (b) is implied by (a). For the proof of (c) note inductively that $W_k(s, s) = \mathrm{Id}$ and $W_k(t, s) = \sum_{l=0}^{k} W_{k-l}(t, q)W_l(q, s)$ for $t_0 \leq s \leq q \leq t \leq t_1$. Therefore $W(s, s) = \mathrm{Id}$ and

$$
\begin{aligned}
W(t, s)x &= \lim_{k \to \infty} \sum_{k=0}^{\infty} \sum_{l=0}^{k} W_{k-l}(t, q)W_l(q, s)x \\
&= \lim_{k \to \infty} \left(\sum_{k=0}^{\infty} W_k(t, q) \right) \left(\sum_{k=0}^{\infty} W_k(q, s) \right) x = W(t, q)W(q, s)x \,.
\end{aligned}
$$

$\square$

1.6.7 Lemma. Let X, Y be Banach spaces with $Y \hookrightarrow X$ densely and continuously and assume that there is an isometric isomorphism $S : Y \to X$. Assume that $A : D(A) \to X$ generates a quasi-contractive C_0-semigroup $(T(t))_{t \geq 0} \subset \mathscr{L}(X)$ and assume that $A_1 := SAS^{-1} : D(A_1) \to X$ with domain $D(A_1) := \{x \in X : S^{-1}x \in D(A), AS^{-1}x \in Y\}$ generates a quasi-contractive C_0-semigroup with constant $\gamma_1 \geq 0$.
Then $T(t)(Y) \subset Y$ and $(T(t)|_Y)_{t \geq 0} \subset \mathscr{L}(Y)$ generates a quasi-contractive C_0-semigroup with constant $\gamma_1 \geq 0$. Its infinitesimal generator is given by $A_Y := A|_{D(A_Y)}$ with $D(A_Y) := \{y \in Y \sqcap D(A) \,.\, Ay \subset Y\}$.

PROOF: Note that $D(A_1) = \{x \in X : S^{-1}x \in D(A_Y)\} = S(D(A_Y))$. Since $D(A_1) \subset X$ densely, this shows $D(A_Y) \subset Y$ densely. Obviously, A_Y is a closed operator in Y. Moreover, for $y \in D(A_Y)$ and $\lambda > \gamma_1$ we have

$$
\|\lambda y - A_Y y\|_Y = \|(\lambda \mathrm{Id} - A_1)Sy\|_X \geq (\lambda - \gamma_1) \|Sy\|_X = (\lambda - \gamma_1) \|y\|_Y \,.
$$

Therefore, A_Y is quasi-dissipative in Y. Let $\lambda > \gamma_1$ and $y \in Y$. Then there is an $x \in D(A_1) = S(D(A_Y))$ with $Sy = (\lambda \mathrm{Id} - A_1)x$. Then $z := S^{-1}x \in D(A_Y)$ and $y = (\lambda \mathrm{Id} - A_Y)z$. Therefore, the Lumer-Phillips theorem 1.1.12 shows that A_Y generates a C_0-semigroup $(T_Y(t))_{t \geq 0}$.

Let $y \in D(A_Y)$ and $u(t) := T_Y(t)y - T(t)y$ for $t \geq 0$. Then $u : [0, \infty) \to X$ is differentiable with $u'(t) = A_Y T_Y(t)y - AT(t)y = Au(t)$ and $u(0) = 0$. Therefore, 1.1.10 shows $u(t) = 0$ for $t \geq 0$. Since $D(A_Y) \subset Y$ densely, this proves $T_Y(t) = T(t)|_Y$ and completes the proof of the lemma. □

Finally, we can state an existence theorem for propagators with differentiability with respect to t without accepting a loss of order. It is an important result for applications.

1.6.8 Theorem. Suppose that X, Y are Banach spaces with $Y \hookrightarrow X$ densely such that the following statements hold:

- $P(t) : D(P(t)) \to X$ for $t \in I$ generates a C_0-semigroup $(e^{\tau P(t)})_{\tau \geq 0}$ and $\{(e^{\tau P(t)})_{\tau \geq 0} : t \in I\}$ is Kato-stable with stability constants (M, β).

- $Y \subset D(P(t))$ for $t \in I$ and $I \ni t \mapsto P(t) \in \mathscr{L}(Y, X)$ is strongly continuous.

- There are a dense subspace $D \subset Y$, an isomorphism $S : Y \to X$, and a strongly continuous mapping $I \ni t \mapsto Q(t) \in \mathscr{L}(X)$ such that D is a core for $P(t), t \in I$, and

$$P(t)S^{-1}z = S^{-1}P(t)z + S^{-1}Q(t)z \quad \text{for } z \in D, t \in I. \tag{1.6.3}$$

Then there is a propagator $U : \Delta \to \mathscr{L}(X) \cap \mathscr{L}(Y)$ such that

(a) $\Delta \ni (t, s) \mapsto U(t, s) \in \mathscr{L}(Y, X)$ is strongly continuously differentiable with strong derivatives

$$\frac{\partial}{\partial t}U(t, s) = P(t)U(t, s), \quad \frac{\partial}{\partial s}U(t, s) = -U(t, s)P(s) \quad \text{for } (t, s) \in \Delta.$$

(b) $\|U(t, s)\|_{\mathscr{L}(X)} \leq Me^{\beta(t-s)}$ and $\|U(t, s)\|_{\mathscr{L}(Y)} \leq M_Y e^{\beta_Y(t-s)}$ for $(t, s) \in \Delta$ and suitable $M_Y \geq 1, \beta_Y \geq 0$.

We will not prove this result due to Kobayashi (cf. [92], [93], cf. also [144, theorem 7.3]) because its proof is rather long and tedious. Instead, we will prove 1.6.8 with the assumption of strong continuity of $t \mapsto A(t) \in \mathscr{L}(Y, X)$ substituted by (uniform) continuity in the operator norm of $t \mapsto A(t) \in \mathscr{L}(Y, X)$ for the sake of simplicity. Then, we can use the results obtained in 1.6.3 and the proof is much shorter. Moreover, since we apply this result only for quasi-contractive semigroups $(e^{\tau P(t)})_{\tau \geq 0}$ we will give the proof only in this case which simplifies the proof once more.

PROOF: We will show that assumption 1.6.3(b) is satisfied. By using the norm $\|y\|_Y^{\sim} := \|Sy\|_X$ for $y \in Y$ on Y we can assume that $S : Y \to X$ is an isometric isomorphism. We will show that $S^{-1}(D(P(t))) \subset D(P(t))$, $P(t)S^{-1}(D(P(t))) \subset Y$, and

$$SP(t)S^{-1} = P(t) + Q(t) \qquad \text{on } D(P(t)), t \in I. \tag{1.6.4}$$

Take $z \in D(P(t)), t \in I$, then there is a sequence $(z_k)_k \subset D$ with $z_k \xrightarrow[k\to\infty]{X} z$ and $P(t)z_k \xrightarrow[k\to\infty]{X} P(t)z$. Thus $S^{-1}z_k \xrightarrow[k\to\infty]{Y} S^{-1}z \in Y \subset D(P(t))$, and

$$P(t)S^{-1}z_k = S^{-1}P(t)z_k + S^{-1}Q(t)z_k \xrightarrow[k\to\infty]{Y} S^{-1}P(t)z + S^{-1}Q(t)z .$$

This implies $S^{-1}z \in D(P(t))$ and $P(t)S^{-1}z = S^{-1}P(t)z + S^{-1}Q(t)z \in Y$. Application of S yields (1.6.4). Hence, due to lemma 1.6.7 and proposition 1.1.15 $(e^{\tau P(t)}|_Y)_{\tau \geq 0}$ generates a quasi-contractive C_0-semigroup in X, i.e. (b) in 1.6.3 is satisfied and 1.6.3 shows that there is a propagator $U : \triangle \to \mathscr{L}(X)$ satisfying all assertions except the strong continuity of $U : \triangle \to \mathscr{L}(Y)$ and the differentiability of $t \mapsto U(t, s)y \in X$ for $y \in Y$. Due to 1.6.6 there is a unique, strongly continuous solution $W : \triangle \to \mathscr{L}(X)$ of

$$W(t,s)x = U(t,s)x + \int_s^t W(t,\tau)Q(\tau)U(\tau,s)x\,d\tau . \tag{1.6.5}$$

Let $V(t,s) := U(t,s)S^{-1}$. Then $s \mapsto V(t,s)y \in X$ is differentiable for any $y \in Y$ with

$$\frac{\partial}{\partial s}V(t,s)y = -U(t,s)P(s)S^{-1}y = -V(t,s)(P(s)+Q(s))y .$$

Let $P_l(t), U_l(t,s)$ as in the proof of 1.6.4. Then $t \mapsto U_l(t,s) \in \mathscr{L}(Y,X)$ is strongly differentiable except for a finite number of values of t with derivative $\frac{\partial}{\partial t}U_l(t,s)y = P_l(t)U_l(t,s)y$ for $y \in Y$. Hence

$$S^{-1}U_l(t,s)y - V(t,s)y = V(t,\tau)U_l(\tau,s)y|_{\tau=s}^{\tau=t} = \int_s^t \frac{\partial}{\partial \tau}[V(t,\tau)U_l(\tau,s)y]d\tau$$

$$= \int_s^t V(t,\tau)(P_l(\tau) - P(\tau) - Q(\tau))U_l(\tau,s)y\,d\tau .$$

Since $\|U_l(\tau,s)\|_{\mathscr{L}(X)} \leq e^{\beta(\tau-s)}$ and $P_l(\tau) \xrightarrow[l\to\infty]{} P(\tau)$ uniformly for $\tau \in I$ in $\mathscr{L}(Y,X)$, we obtain with $l \to \infty$

$$S^{-1}U(t,s)y - V(t,s)y = -\int_s^t V(t,\tau)Q(\tau)U(\tau,s)y\,d\tau .$$

Since $Y \subset X$ densely, this shows

$$V(t,s)x = S^{-1}U(t,s)x + \int_s^t V(t,\tau)Q(\tau)U(\tau,s)x d\tau \quad \text{for } x \in X.$$

Moreover, (1.6.5) implies

$$S^{-1}W(t,s)x = S^{-1}U(t,s)x + \int_s^t S^{-1}W(t,\tau)Q(\tau)U(\tau,s)x\tau \quad \text{for } x \in X.$$

Thus, 1.6.6 shows $S^{-1}W(t,s) = V(t,s) = U(t,s)S^{-1}$. Hence $U(t,s)(Y) \subset Y$ and $U(t,s) \in \mathcal{L}(Y)$ with

$$\|U(t,s)\|_{\mathcal{L}(Y)} \le \|W(t,s)\|_{\mathcal{L}(X)} \le M_Y e^{\beta_Y(t-s)}$$

for suitable $M_Y \ge 1, \beta_Y \ge 0$. Finally, for $y \in Y, (t,s) \in \Delta$ we have

$$\left(\frac{\partial}{\partial t}\right)^+ U(t,s)y = \lim_{h \searrow 0} \frac{U(t+h,t) - \mathrm{Id}}{h} U(t,s)y$$

$$= \left(\frac{\partial}{\partial \tau}\right)^+ U(\tau,t)U(t,s)y\Big|_{\tau=t} = P(t)U(t,s)y$$

and

$$\left(\frac{\partial}{\partial t}\right)^- U(t,s)y = \lim_{h \searrow 0} \frac{U(t,t-h) - \mathrm{Id}}{h} U(t-h,s)y$$

$$= -\left(\frac{\partial}{\partial \sigma}\right)^- U(t,\sigma)U(t,s)y\Big|_{\sigma=t} = P(t)U(t,s)y.$$

This implies the assertion. $\qquad \square$

Once a propagator is constructed, that is differentiable with respect to both parameters, the corresponding (inhomogeneous) Cauchy problem can be uniquely solved.

1.6.9 Proposition. Assume that X,Y are Banach spaces with $Y \hookrightarrow X$, that $U : \Delta \to \mathcal{L}(X) \cap \mathcal{L}(Y)$ is a propagator, and that $I \ni t \mapsto A(t) \in \mathcal{L}(Y,X)$ is strongly continuous. Moreover, assume that $\Delta \ni (t,s) \mapsto U(t,s) \in \mathcal{L}(Y,X)$ is strongly continuously differentiable with derivatives

$$\frac{\partial}{\partial t}U(t,s) = A(t)U(t,s) \quad \text{and} \quad \frac{\partial}{\partial s}U(t,s) = -U(t,s)A(s) \quad \text{for } (t,s) \in \Delta.$$

Then, for $u_0 \in Y$ and $f \in C(I,X) \cap L^1(I,Y)$, there exists a unique solution $u \in C(I,Y) \cap C^1(I,X)$ of

$$\frac{du}{dt}(t) = A(t)u(t) + f(t), \; t \in I, \quad u(t_0) = u_0.$$

This solution is given by Duhamel's principle

$$u(t) = U(t,t_0)u_0 + \int_{t_0}^t U(t,\tau)f(\tau)d\tau. \tag{1.6.6}$$

PROOF: Due to the uniform boundedness principle there is an $M \geq 0$ with $\|A(t)\|_{\mathscr{L}(Y,X)} \leq M$ and $\|U(t,s)\|_{\mathscr{L}(Z)} \leq M$ for $(t,s) \in \Delta, Z = X, Y$. First we assume $f \in C(I,Y)$. Define u by (1.6.6). Then one can verify directly that $u \in C(I,Y)$ and $u \in C^1(I,X)$ with $u'(t) = A(t)u(t)$. For $f \in C(I,X) \cap L^1(I,Y)$ choose $f_k \in C(I,Y)$ with $f_k \underset{k \to \infty}{\longrightarrow} f$ in $C(I,X) \cap L^1(I,Y)$ and define u_k by (1.6.6). Then

$$\|u_k(t) - u_{k'}(t)\|_Y \leq \left\| \int_{t_0}^t U(t,\tau)(f_k(\tau) - f_{k'}(\tau))d\tau \right\|_Y \leq M \|f_k - f_{k'}\|_{L^1(I,Y)}$$

and

$$\|u_k'(t) - u_{k'}'(t)\|_X \leq M \|u_k(t) - u_{k'}(t)\|_Y + \|f_k - f_{k'}\|_{C(I,X)} \,.$$

Hence $(u_k)_k \subset C(I,Y)$ and $(u_k)_k \subset C^1(I,X)$ are Cauchy sequences and we have proved the existence of a solution.

To prove also the uniqueness assume that $v \in C(I,Y) \cap C^1(I,X)$ satisfies

$$\frac{dv}{dt}(t) = A(t)v(t) + f(t), \; t \in I, \quad v(t_0) = u_0.$$

Define $w := u - v$, then $w \subset C(I,Y) \cap C^1(I,X)$ with

$$\frac{dw}{dt}(t) = A(t)w(t), \; t \in I, \quad w(t_0) = 0.$$

Thus

$$\begin{aligned}
w(t) &= \int_{t_0}^t \frac{\partial}{\partial \tau} U(t,\tau)w(\tau)d\tau \\
&= \int_{t_0}^t \left(-U(t,\tau)A(\tau)w(\tau) + U(t,\tau)A(\tau)w(\tau)\right) d\tau = 0
\end{aligned}$$

for $t \in I$, which proves the assertion. $\square$

1.6.10 Bibliographical remarks. The generalizations of C_0-semigroup theory to time-dependent evolution equations have been mainly developed by Kato, cf. e.g. [78], [79], [83], [77]. 1.6.3 is due to [78], cf. also [122, 5.3] or [144, theorem 7.1]). Our proof of 1.6.4 uses the proof of 1.6.3 simultaneously in the whole scale. The result on Volterra integral equations and the "calculus" for Volterra integral operators are taken from [87], cf. also [79]. 1.6.8 is a generalization of a theorem of Kato due to Kobayashi ([92], [93], cf. also [144, theorem 7.3]). The proof of this result is due to Dorroh [40]. Proposition 1.6.9 is taken from [79]. Further results on abstract, time-dependent evolution equations have been obtained e.g. by [35], [51], [60], [91], [115]. There are many applications of 1.6.8 to differential evolution equations, cf. e.g. the previously mentioned papers or [106], [159], [158].

Chapter 2

Well-posedness of the time-dependent linear Cauchy problem

In the second chapter we will study well-posedness of the linear, time-dependent Cauchy problem

$$\frac{du}{dt}(t) = A(t)u(t), \qquad u(t_0) = u_0$$

for operators $A(t) \in \bigcap_{k \in \mathbb{N}_0} \mathscr{L}(X^{k+m}, X^k)$ of order m in a scale of Banach spaces $(X^k)_k$ (i.e., X^k for $k \in \mathbb{N}_0$ is a Banach space with $X^k \hookrightarrow X^l$ for $k \geq l$). Here well-posedness roughly means that for sufficiently smooth initial values u_0 there are unique solutions depending continuously on u_0.

We give a precise definition of well-posedness of the linear Cauchy problem in scales of Banach spaces in section 2.1, where we also use a growth restriction. Moreover, we indicate several simple properties of well-posed Cauchy problems in scales of Banach spaces. In section 2.2 we introduce scales of Banach spaces generated by a finite family of closed operators, we construct regularizing operators if the scale generating operators have suitable properties and we give a construction of spectrally invariant operator algebras. Moreover, in section 2.3 we show how several estimates (in particular dissipativity estimates) can be obtained with commutator methods. Then, in section 2.4, for several scales of this type well-posedness of the Cauchy problem is characterized by means of estimates for the operator $A(t)$. The proofs of the characterizations are finally given in section 2.5. In this section we also give sufficient conditions for operators $A(t)$ that are only defined in a part of the scale.

2.1 Properties of well-posed linear Cauchy problems in scales of Banach spaces

There are several slightly different but similar concepts to define well-posedness of the abstract Cauchy problem

$$\frac{du}{dt}(t) = A(t)u(t), \qquad u(t_0) = u_0 \, ,$$

where $A(t)$ are densely defined, closed linear operators in a Banach space X, cf. e.g. [96, II §3], [44, 7.1], and also [89], [112]. Here well-posedness always roughly means that for initial values u_0 in a dense subset there are unique solutions of this Cauchy problem that depend continuously on these values. One main goal in the theory of abstract evolution equations is to give conditions for the operator $A(t)$ that imply well-posedness or even characterize it. Many approaches to this question are formulated within the setting of unbounded, densely defined operators in Banach spaces (cf. e.g. [122], [52], [78], [79], [83], [144], [25]).
But in pseudodifferential analysis it is often not natural to consider pseudodifferential operators $A(t)$ as densely defined, closed operators in one Banach space X. One often considers these operators as operators of fixed order in a scale of Banach spaces, e.g. in a scale of Sobolev spaces. The operators often are not closed in the sense of unbounded operators in Banach spaces. Hence one has to choose an appropriate closure to apply the results mentioned in section 1.1 or section 1.6, which can be a difficult question.
These difficulties motivate the different approach we want to describe in this chapter. We consider operators of fixed order in a scale of Banach spaces and we do no longer ask about closedness (in the sense of unbounded operators). This corresponds to the natural situation for pseudodifferential operators in a scale of function spaces. But now we have to adapt the notion of well-posedness of the Cauchy problem to this situation. Here the Cauchy problem should be called well-posed, if it is well-posed everywhere in the scale. Moreover, we also will have to use growth restrictions for the solutions. This leads to the following definitions.
    2.1.1 Definition. A scale of Banach spaces $(X^k)_k$ is a family of Banach spaces $(X^k, \|\cdot\|_k)$ for $k \in \mathbb{N}_0$ such that $X^k \hookrightarrow X^l$ densely and continuously for $k \geq l$. In this case we define also

$$X^\infty := \bigcap_{k \in \mathbb{N}_0} X^k$$

Throughout this section, $(X^k)_k$ will always denote a scale of Banach spaces.

2.1.2 Lemma. For a scale of Banach space $(X^k)_k$ we have $X^\infty \subset X^l$ densely for $l \in \mathbf{N}_0$.

PROOF: We only have to prove the lemma for $l = 0$.
Moreover, we can assume that $\|x\|_k \leq \|x\|_{k+1}$ for $x \in X^{k+1}$ because we can substitute the norms $\|x\|_k$ by the equivalent norms $\|x\|_k^{\sim} := \max_{0 \leq j \leq k} \|x\|_j$.
Now, for $x_0 \in X^0$ and $\varepsilon > 0$ we can find inductively $x_k \in X^k$ with $\|x_{k+1} - x_k\|_k < \frac{\varepsilon}{2^{k+1}}$. Thus

$$\|x_{j+l} - x_j\|_j \leq \frac{\varepsilon}{2^j} \qquad \text{for } l \in \mathbf{N}, j \in \mathbf{N}_0 \tag{2.1.1}$$

and $\|x_{j+l} - x_j\|_k \leq \|x_{j+l} - x_j\|_j \leq \frac{\varepsilon}{2^j}$. This shows that $(x_j)_{j \geq k}$ is a Cauchy sequence in X^k for $k \in \mathbf{N}_0$. Hence, we can find a $z \in X^\infty$ with $x_j \xrightarrow[j \to \infty]{} z$ in X^k for every $k \in \mathbf{N}_0$.
Finally, with $l \to \infty$ in (2.1.1) we obtain $\|z - x_0\|_0 \leq \varepsilon$. This proves the assertion. $\qquad\qquad\square$

2.1.3 Definition. Let $m \in \mathbf{N}$ and $I \ni t \mapsto A(t) \in \bigcap_{k \in \mathbf{N}_0} \mathscr{L}(X^{k+m}, X^k)$ be a strongly continuous family of linear operators.

(a) We call the Cauchy problem for $A(t) \in \mathscr{L}(X^{k+m}, X^k)$ in the scale $(X^k)_k$ well-posed with exponential growth, if for every $k \in \mathbf{N}_0$ there exists a constant $\beta_k \geq 0$ such that the following properties are satisfied:

   (i) For $k \in \mathbf{N}_0, u_0 \in X^{k+m}$ and $s \in I$ there exists a unique function $u = u(\cdot, s, u_0) \in C([s, t_1], X^{k+m}) \cap C^1([s, t_1], X^k)$ with

$$\frac{du}{dt}(t) = A(t)u(t), \ t \in [s, t_1], \quad \text{and} \quad u(s) = u_0.$$

   (ii) $\|u(t, s, u_0)\|_k \leq e^{\beta_k(t-s)} \|u_0\|_k$ for $t \in [s, t_1], u_0 \in X^{k+m}$.

(b) An exponentially growing evolution operator or propagator $U(t, s)$ for $A(t)$ in the scale $(X^k)_k$ is a family of operators

$$U(t, s) \in \bigcap_{k \in \mathbf{N}_0} \mathscr{L}(X^k), \quad (t, s) \in \Delta := \{(t, s) \in I^2 : t \geq s\},$$

such that

(i) $U(s, s) = \mathrm{Id}$ and $U(t, s) = U(t, q)U(q, s)$ for $t_0 \leq s \leq q \leq t \leq t_1$.

(ii) $\Delta \ni (t, s) \mapsto U(t, s) \in \mathscr{L}(X^k)$ is strongly continuous for $k \in \mathbb{N}_0$.

(iii) $\Delta \ni (t, s) \mapsto U(t, s) \in \mathscr{L}(X^{k+m}, X^k)$ is strongly continuously differentiable for $k \in \mathbb{N}_0$ with strong derivatives

$$\frac{\partial}{\partial t}U(t, s) = A(t)U(t, s), \quad \frac{\partial}{\partial s}U(t, s) = -U(t, s)A(s) \text{ for } (t, s) \in \Delta.$$

(iv) for $k \in \mathbb{N}_0$ there is a $\beta_k \geq 0$ with $\|U(t, s)\|_{\mathscr{L}(X^k)} \leq e^{\beta_k(t-s)}$ for $(t, s) \in \Delta$.

The notions of well-posedness and the existence of evolution operators are equivalent.

2.1.4 Theorem.   Let $m \in \mathbb{N}$ and $I \ni t \mapsto A(t) \in \bigcap_{k \in \mathbb{N}_0} \mathscr{L}(X^{k+m}, X^k)$ be a strongly continuous family of linear operators. Then the following two statements are equivalent:

(a) The Cauchy problem for $A(t) \in \mathscr{L}(X^{k+m}, X^k)$ in the scale $(X^k)_k$ is well-posed with exponential growth.

(b) There exists an exponentially growing evolution operator $U(t, s)$ for $A(t)$ in the scale $(X^k)_k$.

Moreover, in this case we have:
For $k \in \mathbb{N}_0, u_0 \in X^{k+m}, s \in I$, and $f \in C([s, t_1], X^k) \cap L^1([s, t_1], X^{k+m})$ there is a unique function $u \in C([s, t_1], X^{k+m}) \cap C^1([s, t_1], X^k)$ with

$$\frac{du}{dt}(t) = A(t)u(t) + f(t), \ t \in [s, t_1], \quad u(s) = u_0. \tag{2.1.2}$$

The function u is given by Duhamel's principle

$$u(t) = U(t, s)u_0 + \int_s^t U(t, \tau)f(\tau)d\tau, \ t \in [s, t_1].$$

PROOF: (b) $\Rightarrow$ (a) and the statement about inhomogeneous Cauchy problems are a consequence of 1.6.9. To prove the remaining part (a) $\Rightarrow$ (b) let $U(t,s)x := u(t,s,x)$ for $(t,s) \in \Delta, x \in X^m$. Due to the uniqueness of $u(\cdot,s,x)$ clearly $U(t,s)$ is linear, moreover

$$\|U(t,s)x\|_k = \|u(t,s,x)\|_k \le e^{\beta_k(t-s)} \|x\|_k \quad \text{for } (t,s) \in \Delta, x \in X^{k+m}.$$

Hence, we have a unique extension $U(t,s) \in \bigcap_{k \in \mathbb{N}_0} \mathscr{L}(X^k)$ of $U(t,s)$ with

$$\|U(t,s)\|_{\mathscr{L}(X^k)} \le e^{\beta_k(t-s)} \quad \text{for } (t,s) \in \Delta. \tag{2.1.3}$$

The uniqueness of $u(\cdot,s,x)$ implies $U(s,s) = \text{Id}$ and $U(t,s) = U(t,q)U(q,s)$ for $t_0 \le s \le q \le t \le t_1$ (because $u(t,s,x) = u(t,q,u(q,s,x))$). Due to (2.1.3) there are $b_k \ge 0$ with $\|U(t,s)\|_{\mathscr{L}(X^k)} \le e^{\beta_k(t-s)} \le b_k$ for $k \in \mathbb{N}, (t,s) \in \Delta$. Therefore, we have to establish the strong continuity in (ii) only on a dense set (due to 1.1.3). Fix $k \in \mathbb{N}_0$ and $x \in X^{k+m}$, then for $(\tilde{t},\tilde{s}) \in \Delta$

$$\|U(\tilde{t},\tilde{s})x - x\|_k = \|u(\tilde{t},\tilde{s},x) - u(\tilde{s},\tilde{s},x)\|_k = \left\| \int_{\tilde{s}}^{\tilde{t}} A(\tau)u(\tau,\tilde{s},x)d\tau \right\|_k$$

$$\le \int_{\tilde{s}}^{\tilde{t}} \|A(\tau)U(\tau,\tilde{s})x\|_k \, d\tau \le c_k b_{k+m} \|x\|_{k+m} (\tilde{t} - \tilde{s})$$

for a suitable constant $c_k > 0$. Fix $(t,s) \in \Delta$. If $t = s$, then

$$\|U(t+h_1, s+h_2)x - x\|_k \le c_k b_{k+m} \|x\|_{k+m} (h_1 - h_2)$$

for $h_1 \ge h_2$, which implies the strong continuity in $(s,s) = (t,s)$.
Now suppose $t > s$ and let $\varepsilon > 0$, then there is a $\delta > 0$ with $t + h_1 > s + h_2$ and

$$\|U(t+h_1, s)x - U(t,s)x\|_k = \|u(t+h_1, s, x) - u(t,s,x)\|_k < \frac{\varepsilon}{2}$$

for $|h_1|, |h_2| < \delta$ with $t + h_1, s + h_2 \in I$. Note that

$$\|U(t+h_1, s+h_2)x - U(t+h_1, s)x\|_k \le b_k c_k b_{k+m} \|x\|_{k+m} |h_2| < \frac{\varepsilon}{2}$$

for suitable $\delta_1 < \delta$ and $|h_2| < \delta_1$. Thus, for $|h_1|, |h_2| < \delta_1$ with $t + h_1, s + h_2 \in I$

$$\|U(t+h_1, s+h_2)x - U(t,s)x\|_k$$
$$\le \|U(t+h_1, s+h_2)x - U(t+h_1, s)x\|_k + \|U(t+h_1, s)x - U(t,s)x\|_k < \varepsilon.$$

This proves (ii). For $k \in \mathbb{N}_0$ and $x \in X^{k+m}$ $[s, t_1] \ni t \mapsto U(t, s)x \in X^k$ is differentiable with $\partial_t U(t, s)x = A(t)U(t, s)x$. Due to the strong continuity of $A(t)$ and $U(t, s)$ the mapping $\triangle \ni (t, s) \mapsto \frac{\partial}{\partial t}U(t, s)x = A(t)U(t, s)x \in X^k$ is continuous. Finally, for $(t, s), (t, s+h) \in \triangle$ with $t \geq s+h \geq s$ we have

$$\frac{U(t, s+h)x - U(t, s)x}{h} = U(t, s+h)\frac{x - U(s+h, s)x}{h} \xrightarrow[h \to 0]{X^k} U(t, s)(-A(s)x)$$

and for $(t, s), (t, s+h) \in \triangle$ with $t \geq s \geq s+h$ we have

$$\left\| \frac{U(t, s+h)x - U(t, s)x}{h} + U(t, s)A(s)x \right\|_k$$

$$\leq b_k \frac{1}{h} \int_{s+h}^{s} \|A(\tau)U(\tau, s+h)x - A(s)x\|_k \, d\tau \xrightarrow[h \to 0]{} 0$$

because $\triangle \ni (t, s) \mapsto A(t)U(t, s)x \in X^k$ is continuous. This shows that $[t_0, t] \ni s \mapsto U(t, s)x \in X^k$ is differentiable with $\partial_s U(t, s)x = -U(t, s)A(s)x$. Thus, we obtain the assertion with the strong continuity of $U(t, s)$ and $A(s)$. $\square$

2.1.5 Remark.

(a) If the Cauchy problem for $-A(t_0 + t_1 - t)$ is also well-posed, then we can construct the evolution operator for $t, s \in I$, and (i),...,(iii) in 2.1.3(b) remain valid for all $t, s, q \in I$, and we can substitute (iv) by

$$\|U(t, s)\|_{\mathscr{L}(X^k)} \leq e^{\beta_k |t-s|} \quad \text{for } t, s \in I.$$

Moreover, we can also find past solutions for (2.1.2), i.e., (2.1.2) remains true if we substitute $[s, t_1]$ by $[t_0, t_1]$.

PROOF: Due to 2.1.4 we only have to show that for $s \in I$, $k \in \mathbb{N}_0$ and $u_0 \in X^{k+m}$ then there are unique $u \in C([t_0, s], X^{k+m}) \cap C^1([t_0, s], X^k)$ with $u'(t) = A(t)u(t)$, $u(s) = u_0$, and $\|u(t)\|_k \leq e^{\beta_k(s-t)}$ for $t \in [t_0, s]$. To this end, let $v \in C([t_0 + t_1 - s], t_1], X^{k+m}) \cap C([t_0 + t_1 - s], X^k)$ with $v'(t) = -A(t_0 + t_1 - t)v(t)$ and $v(t_0 + t_1 - s) = u_0$. Then $u(t) := u(t_0 + t_1 - t)$ has the desired properties. $\square$

(b) As a corollary we obtain an existence and uniqueness theorem for linear differential equations in the Fréchet space X^∞, if the Cauchy problem for $A(t)$ is well-posed in the scale $(X^k)_k$ with exponential growth:

For $u_0 \in X^\infty, t_0 \in I$, and $f \in C([t_0, t_1], X^\infty)$ there is a unique function $u \in C^1([t_0, t_1], X^\infty)$ with

$$\frac{du}{dt}(t) = A(t)u(t) + f(t) \quad \text{for } t \in [t_0, t_1] \text{ and } u(t_0) = u_0$$

in X^∞.

(c) Linear differential equations in Fréchet spaces in general do not even have a local solution. Consider for example the Fréchet space $C^\infty(0,1)$ with the topology of uniform convergence of any derivative on compact subsets. Then $Af := -f'$ for $f \in C^\infty(0,1)$ defines a linear, continuous operator $A : C^\infty(0,1) \to C^\infty(0,1)$, but there exists no $\varepsilon > 0$ and a function $u \in C^1([0, \varepsilon], C^\infty(0,1))$ with $u'(t) = Au(t)$ and $u(0) = u_0$. If there would exist a function of that type, then $v(t, x) := u(t)(x)$ for $t \in [0, \varepsilon], x \in (0,1)$ would define a continuously differentiable function $v \in [0, \varepsilon] \times (0,1) \to \mathbb{C}$ with $v_t(t, x) = -v_x(t, x)$ and $v(0, x) = \frac{1}{x}$. But a solution of this partial differential equation is given by $v(t, x) = \frac{1}{x-t}$, which is a contradiction.

Sometimes (in particular in connection with quasi-linear equations) one has to distinguish between real and complex scales. Here the following remark is useful.

2.1.6 Remark.

(a) Let X be a real vector space. Its complexification $X_\mathbb{C}$ is defined by $X_\mathbb{C} := X \oplus iX$ with the canonic operations on $X_\mathbb{C}$. If $(X, \|\cdot\|)$ is a real Banach space we can define a complex Banach space structure on $X_\mathbb{C}$ by $\|x + iy\|^\mathbb{C} := \sqrt{\|x\|^2 + \|y\|^2}$ for $x + iy \in X_\mathbb{C}$, and if $(X, \langle\cdot,\cdot\rangle)$ is a real Hilbert space we obtain a complex Hilbert space structure on $X_\mathbb{C}$ by the scalar product $\langle x + iy, u + iv\rangle^\mathbb{C} := \langle x, u\rangle + \langle y, v\rangle + i(\langle y, u\rangle - \langle x, v\rangle)$. We have the canonic embedding $X \ni x \mapsto x + i0 \in X \oplus iX$. For a linear operator $A : Y \to X$, were X, Y are real vector spaces, we define its complexification by $A^\mathbb{C}(x + iy) := Ax + iAy$ for $x + iy \in Y_\mathbb{C}$. In particular, this defines complexifications of densely defined operators $(A, D(A))$ with domain $D(A)$ in a real Banach space X. Clearly, A is quasi-dissipative if and only if $A^\mathbb{C}$ is quasi-dissipative.

(b) Let $(X^k, \|\cdot\|_k)_k$ be a scale of real Banach spaces, and

$$I \ni t \mapsto A(t) \in \bigcap_{k \in \mathbb{N}_0} \mathscr{L}(X^{k+m}, X^k)$$

be strongly continuous. Then the Cauchy problem for $A(t)$ in the scale $(X^k)_k$ is well-posed with exponential growth if and only if the (complex-ified) Cauchy problem for $A(t)^{\mathbb{C}}$ in the scale $(X_{\mathbb{C}}^k)_k$ is well-posed with exponential growth.

PROOF: If the Cauchy problem for $A(t) \in \mathcal{L}(X^{k+m}, X^k)$ in the scale $(X^k)_k$ is well-posed with exponential growth and evolution operators $U(t, s)$, then clearly the complexified Cauchy problem is also well-posed with exponential growth and evolution operators $U(t, s)^{\mathbb{C}}$.

Now assume conversely that the complexified Cauchy problem is well-posed with exponential growth. Then, for $u_0 \in X^{k+m}$ and $s \in I$ there is a unique $u \in C([s, t_1], X_{\mathbb{C}}^{k+m}) \cap C^1([s, t], X_{\mathbb{C}}^k)$ with $u'(t) = A(t)^{\mathbb{C}} u(t)$ for $t \in [s, t_1]$, $u(s) = u_0$, and $\|u(t)\|_k^{\mathbb{C}} \le e^{\beta_k(t-s)} \|u_0\|_k^{\mathbb{C}}$. Denote by $R : X_{\mathbb{C}}^k \ni (x + iy) \mapsto x \in X^k$ the canonic projection on X^k. Then $v := Ru \in C([s, t_1], X^{k+m}) \cap C^1([s, t_1], X^k)$ with

$$v'(t) = (Ru)'(t) = R(A(t)^{\mathbb{C}} u(t)) = A(t) Ru(t) = A(t) v(t)$$

and

$$\|v(t)\|_k = \|Ru(t)\|_k \le \|u(t)\|_k^{\mathbb{C}} \le e^{\beta_k(t-s)} \|u_0\|_k^{\mathbb{C}} = e^{\beta_k(t-s)} \|u_0\|_k .$$

This implies well-posedness of the Cauchy problem for $A(t)$ in the scale $(X^k)_k$ with exponential growth. □

If the Cauchy problem in a scale of Banach spaces is well-posed, it is well-posed in corresponding interpolated scales as well. This will be useful in applications.

2.1.7 Proposition. Suppose that the Cauchy problem for $A(t)$ in the scale of Banach spaces $(X^k)_k$ is well-posed with exponential growth.
Let $[\cdot, \cdot] = (\cdot, \cdot)_{\theta, q}$ be the real interpolation functor or $[\cdot, \cdot] = (\cdot, \cdot)_{[\theta]}$ be the complex interpolation functor (cf. section 1.5) with $\theta \in (0, 1)$ and $1 \le q < \infty$.
Let $X^{k+\theta} := [X^k, X^{k+1}]$ for $k \in \mathbb{N}_0$.
Then the Cauchy problem for $A(t) \in \mathcal{L}(X^{\theta+k+m}, X^{\theta+k})$ in the scale $(X^{\theta+k})_k$ is also well-posed with exponential growth.

PROOF: This is a consequence of 2.1.4 and 1.5.6.   □

Perturbations of order zero do not disturb well-posedness.

2.1.8 Proposition. Let $m \in \mathbb{N}$ and assume that the Cauchy problem for $A(t)$ in the scale $(X^k)_k$ is well-posed with exponential growth. Moreover, suppose that

$$I \ni t \mapsto B(t) \in \bigcap_{k \in \mathbb{N}_0} \mathcal{L}(X^k) \text{ is strongly continuous.}$$

Then the Cauchy problem for $A(t) + B(t) \in \mathscr{L}(X^{k+m}, X^k)$ in the scale $(X^k)_k$ is also well-posed with exponential growth.

PROOF: If $\{U(t,s) : (t,s) \in \Delta\}$ is the evolution operator for $A(t)$ as in 2.1.4, then, due to 1.6.6, there is a propagator $W : \Delta \to \bigcap_{k \in \mathbb{N}_0} \mathscr{L}(X^k)$ satisfying $\|W(t,s)\|_{\mathscr{L}(X^k)} \leq e^{\tilde{\beta}_k(t-s)}$ for suitable $\tilde{\beta}_k \geq 0$ and

$$W(t,s)x = \int_s^t U(t,\tau)B(\tau)W(\tau,s)x\,d\tau \quad \int_s^t W(t,\tau)B(\tau)U(\tau,s)x\,d\tau \; .$$

For $k \in \mathbb{N}_0$ and $x \in X^{k+m}$ the first equality shows that $t \mapsto W(t,s)x \in X^k$ is differentiable with derivative $A(t)W(t,s)x + B(t)W(t,s)x$ and the second integral equation shows that $s \mapsto W(t,s)x \in X^k$ is differentiable with derivative $-W(t,s)A(s)x - W(t,s)B(s)x$. This proves the assertion. $\qquad\square$

Sometimes it is also important to have information about parameter dependence of the solutions of the Cauchy problem in scales of spaces.

2.1.9 Proposition. Let $(X^k, \|\cdot\|_k)_k$ be a scale of Banach spaces, Ω be a metric space, $I \subset \mathbb{R}$ a compact interval, and suppose that for any $\omega \in \Omega$ the Cauchy problem for $A_\omega(t) \in \mathscr{L}(X^{k+m}, X^k)$ in the scale $(X^k)_k$ is well-posed with exponential growth and corresponding evolution operators $U_\omega(t,s)$. Suppose further that $\Omega \ni \omega \mapsto A_\omega(t)x \in X^k$ is uniformly continuous for $t \in I$, where $x \in X^{k+m}$ and $k \in \mathbb{N}_0$, and that for $k \in \mathbb{N}_0$ there are $a_k, u_k \geq 0$ with $\|A_\omega(t)\|_{\mathscr{L}(X^{k+m}, X^k)} \leq a_k$ and $\|U_\omega(t,s)\|_{\mathscr{L}(X^k)} \leq u_k$ for $t \in I, (t,s) \in \Delta, \omega \in \Omega$. Then, for $k \in \mathbb{N}_0$ $\Omega \ni \omega \mapsto U_\omega(t,s) \in \mathscr{L}(X^k)$ is uniformly strongly continuous for $(t,s) \in \Delta$.

PROOF: Let $(\omega_l)_l \subset \Omega$ with $\omega_l \xrightarrow[l\to\infty]{} \omega \in \Omega$ and let $k \in \mathbb{N}_0, \varepsilon > 0$. Then $K := \{U_\omega(t,s)y : (t,s) \in \Delta\} \subset X^{k+m}$ is a compact subset for any $y \in X^{k+m}$. Hence, for $(t,s) \in \Delta$ we have

$$\|U_\omega(t,s)y - U_{\omega_l}(t,s)y\|_k = \left\| \int_s^t \frac{\partial}{\partial \tau} \Big(U_{\omega_l}(t,\tau)U_\omega(\tau,s)y\Big)d\tau \right\|_k$$

$$= \left\| \int_s^t U_{\omega_l}(t,\tau)\Big(A_\omega(\tau) - A_{\omega_l}(\tau)\Big)U_\omega(\tau,s)y\,d\tau \right\|_k$$

$$\leq \int_I u_k \sup_{z \in K} \left\|\Big(A_\omega(\tau) - A_{\omega_l}(\tau)\Big)z\right\|_k d\tau \; .$$

Note that due to the compactness of K the last integrand can be approximated by a sequence of measurable functions and thus is measurable. The compactness shows also that $\sup_{z \in K} \left\|\Big(A_\omega(\tau) - A_{\omega_l}(\tau)\Big)z\right\|_k \xrightarrow[l\to\infty]{} 0$ uniformly for $t \in I$, and the dominated convergence theorem shows that there is an $L \in \mathbb{N}$ with

$$\|U_{\omega_l}(t,s)y - U_\omega(t,s)y\|_k < \varepsilon \text{ for } (t,s) \in \Delta, l \geq L.$$

Hence 1.1.3 implies the assertion.   □

Our main goal in this chapter will be characterizations of well-posedness in several types of scales. It is easy to give a necessary condition for well-posedness of the Cauchy problem in scales of Banach spaces. This will be done in the next proposition. In section 2.4 we will show that for special settings in scales of Banach spaces these conditions are also sufficient and characterize well-posedness. Clearly, only sufficient conditions for well-posedness are interesting in applications.

2.1.10 Proposition. Suppose that the Cauchy problem for $A(t)$ in the scale of Banach spaces $(X^k)_k$ is well-posed with exponential growth.
Then, for $k \in \mathbb{N}_0$ there exist constants $\beta_k \geq 0$ with

$$\|\lambda x - A(t)x\|_k \geq (\lambda - \beta_k) \|x\|_k \quad \text{for } x \in X^{k+m}, t \in I, \lambda > \beta_k .$$

PROOF: This is a consequence of 1.1.11 and 2.1.3.   □

We conclude this section with some properties of time-independent, well-posed Cauchy problems.

2.1.11 Definition. Let $m \in \mathbb{N}$, $A \in \bigcap_{k \in \mathbb{N}_0} \mathscr{L}(X^{k+m}, X^k)$, and suppose that the Cauchy problem for $[-T,T] \ni t \mapsto A \in \mathscr{L}(X^{k+m}, X^k)$ is well-posed with exponential growth in the scale $(X^k)_k$ for any $T > 0$. Then we call the time-independent Cauchy problem for $A \in \mathscr{L}(X^{k+m}, X^k)$ in the scale $(X^k)_k$ well-posed with exponential growth.

2.1.12 Proposition. For $A \in \bigcap_{k \in \mathbb{N}_0} \mathscr{L}(X^{k+m}, X^k)$, the following statements are equivalent:

(a) The Cauchy problem for $A \in \mathscr{L}(X^{k+m}, X^k)$ is well-posed with exponential growth in the scale $(X^k)_k$.

(b) There is a family of linear operators $\{T(t) : t \geq 0\} \subset \bigcap_{k \in \mathbb{N}_0} \mathscr{L}(X^k)$ with:

    (i) $(T(t))_{t \geq 0} \subset \mathscr{L}(X^k)$ is a quasi-contractive C_0-semigroup in X^k for any $k \in \mathbb{N}_0$.

(ii) $[0, \infty) \ni t \mapsto T(t) \in \mathscr{L}(X^{k+m}, X^k)$ is strongly continuously differentiable for any $k \in \mathbb{N}_0$ with derivative $\frac{d}{dt}T(t) = AT(t) = T(t)A$.

(c) $A|_{X^{k+m}}$ is closable and the closure A_k of $A|_{X^{k+m}}$ generates a quasi-contractive C_0-semigroup in X^k for any $k \in \mathbb{N}_0$.

(d) For any $k \in \mathbb{N}_0$ there are $\beta_k \geq 0$ with $\|\lambda x - Ax\|_k \geq (\lambda - \beta_k)\|x\|_k$ for $x \in X^{k+m}, \lambda > \beta_k$, and $(\lambda_k \text{Id} - A)(X^{k+m}) \subset X^k$ densely for a $\lambda_k > \beta_k$.

PROOF: First assume (a) and let $\{U(t, s) : (t, s) \in \mathbb{R}^2, t \geq s\}$ be the corresponding evolution operator. Fix $t_0, s \in \mathbb{R}$ and $x \in X^m$, and let $u(t) := U(t + s, t_0 + s)x, t \in [t_0, \infty)$.
Then $u \in C([t_0, \infty), X^m) \cap C^1([t_0, \infty), X^0)$ satisfies $\frac{du}{dt}(t) = Au(t)$ and $u(t_0) = x$.
Hence $U(t+s, t_0+s)x = u(t) = U(t, t_0)x$ and $U(t+s, t_0+s) = U(t, t_0)$ because $X^m \subset X^0$ densely. Therefore $T(t) := U(t + s, s), t \geq 0, s \in \mathbb{R}$, is well-defined, and we have

- $T(0) = \text{Id}$ and $T(t)T(s) = U(t + s, s)U(s, 0) = U(t + s, 0) = T(t + s)$.

- $[0, \infty) \ni t \mapsto T(t) \in \mathscr{L}(X^k)$ strongly continuously for $k \in \mathbb{N}_0$.

- For $x \in X^{k+m}, t \geq 0$ we have
$$\frac{d}{dt}T(t)x = \frac{d}{dt}U(t, 0)x = AU(t, 0)x = AT(t)x \text{ and}$$
$$\frac{d}{dt}T(t)x = \frac{d}{dt}U(0, -t)x = U(0, -t)Ax = T(t)Ax \text{ in } X^k .$$

This implies (a) $\Rightarrow$ (b). (b) $\Rightarrow$ (c) is a consequence of 1.1.14 and the equivalence of (c) and (d) is a consequence of the Lumer-Phillips theorem 1.1.12. Now assume (c), hence the closure A_k of $A|_{X^{k+m}}$ in X^k generates a C_0-semigroup $T_k(t) \in \mathscr{L}(X^k)$ in X^k with $\|T_k(t)\|_{\mathscr{L}(X^k)} \leq e^{\beta_k t}$ for any $k \in \mathbb{N}_0$. In particular, for $x \in X^\omega$ and $k \in \mathbb{N}_0$ there is a unique $u \in C^1([0, \infty), X^k)$ with $u(0) = x, u(t) \in D(A_k)$ and $u'(t) = Au(t)$ for $t \in [0, \infty)$, namely $u(t) = T_k(t)x$ (cf. 1.1.10). Hence for $k \geq l$ this implies $T_k(t)x = T_l(t)x$ for $x \in X^\infty$ and thus $T_k(t)x = T_l(x)$ for $x \in X^k$ by continuity. Therefore $T(t)x := T_k(t)x$ for $x \in X^k$ is well-defined and the properties of $T_k(t)$ imply (c) $\Rightarrow$ (b). Finally, (b) $\Rightarrow$ (a) follows by setting $U(t, s) := T(t - s)$ for $t, s \in \mathbb{R}, t \geq s$. $\square$

2.1.13 Proposition. Suppose that $(X^k)_k$ is a scale of Hilbert spaces and that the Cauchy problem for $\pm iA \in \mathscr{L}(X^{k+m}, X^k)$ is well-posed with exponential growth in the scale $(X^k)_k$. If A is a symmetric operator in X^0 with domain X^m, then every power of $A|_{X^\infty}$ is essentially selfadjoint in X^0 on X^∞.

PROOF: 2.1.12 implies the existence of a C_0-group $(T(t))_t \subset \bigcap_{k\in\mathbb{N}_0} \mathscr{L}(X^k)$ with

$$\frac{d}{dt}T(t)x = iAT(t)x = iT(t)Ax \quad \text{for } t \in \mathbb{R}, x \in X^{k+m}$$

in X^k for any $k \in \mathbb{N}_0$. Due to 1.2.6 we therefore only have to show that $T(t)$ is unitary in X for $t \in \mathbb{R}$. Now, for $x \in X^m, t \in \mathbb{R}$

$$
\begin{aligned}
\frac{d}{dt}\langle T(t)x, T(t)x \rangle &= \langle iAT(t)x, T(t)x \rangle + \langle T(t)x, iAT(t)x \rangle \\
&= \langle iAT(t)x, T(t)x \rangle - \langle iAT(t)x, T(t)x \rangle = 0 ,
\end{aligned}
$$

hence $\langle T(t)x, T(t)x \rangle = \langle T(0)x, T(0)x \rangle = \langle x, x \rangle$ for $t \in \mathbb{R}, x \in X^m$. This proves the assertion because $X^m \subset X^0$ densely. $\qquad\square$

2.1.14 Bibliographical remarks. For other notions of well-posedness of abstract Cauchy problems we refer to (cf. [96, II §3], [89, 1.4], [112, 3.10]). In this setting, perturbations of order zero (i.e., by bounded operators) (cf. e.g. [124, 6.2], [78, 4.5], [83, 1.7]) do also not disturb well-posedness. More counterexamples on linear differential equations in Fréchet spaces can be found in [61].

2.2 Scales of Banach spaces generated by families of closed operators

In the sequel we do not work with arbitrary scales of Banach spaces but focus on special types of scales. Starting with a family of closed operators in a Banach space one can define joint iterated domains in the same manner as the usual Sobolev spaces are defined with derivatives.

2.2.1 Definition. Suppose that $Z = \{Z_1, \ldots, Z_M\}$ is a family of closed operators in a Banach space X with domains $D(Z_j)$ for $j = 1, \ldots, M$. Define

- $X_Z^0 := X$

- $X_Z^1 := D(Z_1) \cap \ldots \cap D(Z_M)$

- $X_Z^{k+1} := \{x \in X_Z^k : Z_j x \in X_Z^k, j = 1, \ldots, M\}$ for $k \in \mathbb{N}$

- $X_Z^\infty := \bigcap_{k\in\mathbb{N}_0} X_Z^k$

We write also $X^k_\Lambda := X^k_{\{\Lambda\}}$ for $k \in \mathbb{N}_0 \cup \{\infty\}$, if $Z = \{\Lambda\}$ consists of one operator. Let

$$\|x\|_{X^0_Z} := \|x\| \text{ for } x \in X^0_Z \quad \text{and} \quad \|x\|_{X^{k+1}_Z} := \left(\|x\|^2_{X^k_Z} + \sum_{j=1}^{M} \|Z_j x\|^2_{X^k_Z} \right)^{1/2}$$

for $x \in X^{k+1}_Z$. Moreover, if $(X, \langle \cdot, \cdot \rangle)$ is a Hilbert space, then let

$$\langle x, y \rangle_{X^0_Z} := \langle x, y \rangle \text{ for } x, y \in X^0_Z \text{ and } \langle x, y \rangle_{X^{k+1}_Z} := \langle x, y \rangle_{X^k_Z} + \sum_{j=1}^{M} \langle Z_j x, Z_j y \rangle_{X^k_Z}$$

for $x, y \in X^{k+1}_Z$, $k \in \mathbb{N}_0$. If no confusion can occur we write $\|\cdot\|_k := \|\cdot\|_{X^k_Z}$ resp., $\langle \cdot, \cdot \rangle_k := \langle \cdot, \cdot \rangle_{X^k_Z}$ for simplicity.

The following remark gives a wide class of examples of spaces of this type.

2.2.2 Remark. Let $\pi : G \to \mathscr{L}(H)$ be a strongly continuous, unitary representation of a Lie group G on a Hilbert space H. Then Goodman [53] has shown that there is a selfadjoint operator Λ in H with

$$H^k_\Lambda = H^k(\pi) := \{x \in H : [G \ni g \mapsto \pi(g)x] \in C^k(G, H)\} \quad \text{for } k \in \mathbb{N}_0 .$$

Several other examples will be discussed in detail in chapter 4.

2.2.3 Proposition. If $Z = \{Z_1, \ldots, Z_M\}$ is a family of closed operators in a Banach space X, then $(X^k_Z, \|\cdot\|_k)$ is a Banach space for any $k \in \mathbb{N}_0$. Moreover, if X is a Hilbert space, then $(X^k_Z, \langle \cdot, \cdot \rangle_k, \|\cdot\|_k)$ is a Hilbert space space.

PROOF: We only have to prove the completeness of X^k_Z. Assume that X^k_Z is a Banach space for a given $k \in \mathbb{N}_0$ and let $(x_l)_l \subset X^{k+1}_Z$ be a Cauchy sequence. Then there are $x, y_j \in X^k_Z$ with $x_l \xrightarrow[l \to \infty]{X^k_Z} x$ and $Z_j x_l \xrightarrow[l \to \infty]{X^k_Z} y_j$ for $j = 1, \ldots, M$. Hence the closedness of Z_j shows $x \in D(Z_j)$ and $Z_j x = y_j \in X^k_Z$. Therefore $x \in X^{k+1}_Z$ and $x_l \xrightarrow[l \to \infty]{X^{k+1}_Z} x$. $\qquad \square$

Note, that $(X_Z^k)_k$ is not a scale of Banach spaces in the sense of 2.1.1, because $X_Z^{k+1} \subset X_Z^k$ is not dense in general. However, we will use and study conditions for Z implying that $(X_Z^k)_k$ is a scale of Banach spaces in the sense of 2.1.1. Scales that are generated by one operator will be of particular interest, and several scales generated by a family of closed operators can also be written as scales generated by only one operator. Nevertheless, in some applications it will be easier to check the assumptions if we consider the scale as generated by a family of operators rather than by only one operator.

Now we will construct "mollifiers" in these scales. To this end we have to introduce a notion of commutativity for generators of bounded C_0-semigroups

2.2.4 Lemma. Let $Z = \{Z_1, \ldots, Z_M\}$ be a family of infinitesimal generators of bounded C_0-semigroups in a Banach space X. Then $(\lambda \mathrm{Id} - Z_j)^{-1} \in \mathcal{L}(X)$ exists for $\lambda > 0, j = 1, \ldots, M$ and Z is called *commuting*, if one of the following equivalent conditions is satisfied:

(a) $e^{tZ_j} e^{tZ_l} = e^{tZ_l} e^{tZ_j}$ for any $t \geq 0$, $j, l = 1, \ldots, M$.

(b) $(\lambda \mathrm{Id} - Z_j)^{-1}(\lambda \mathrm{Id} - Z_l)^{-1} = (\lambda \mathrm{Id} - Z_l)^{-1}(\lambda \mathrm{Id} - Z_j)^{-1}$ for any $\lambda > 0$, $j, l = 1, \ldots, M$.

PROOF: Since $(\lambda \mathrm{Id} - Z_j)^{-1} x = \int_0^\infty e^{-\lambda t} e^{tZ_j} x \, dt$ for $\lambda > 0$ we have (a)$\Rightarrow$(b). Assume (b). Since $Z_{j,\lambda} := \lambda Z_j (\lambda \mathrm{Id} - Z_j)^{-1} = \lambda^2 (\lambda \mathrm{Id} - Z_j)^{-1} - \lambda \mathrm{Id} \in \mathcal{L}(X)$, the set $\{e^{tZ_{j,\lambda}} : t \geq 0, j = 1 \ldots, M\} \subset \mathcal{L}(X)$ consists of mutually commuting operators for $\lambda > 0$. This implies (a) because $e^{tZ_j} x = \lim_{\lambda \to \infty} e^{tZ_{j,\lambda}} x$ for $t \geq 0$ and $x \in X$ (cf. the proof of 1.1.4. $\qquad\square$

2.2.5 Lemma. Let $Z = \{Z_1, \ldots, Z_M\}$ be a commuting family of infinitesimal generators of bounded C_0-semigroups in a Banach space X. Then

(a) $(\lambda \mathrm{Id} - Z_j)^{-1} \in \mathcal{L}(X_Z^k)$.

(b) $(\lambda \mathrm{Id} - Z_j)^{-1} Z_i x = Z_i (\lambda \mathrm{Id} - Z_j)^{-1} x$ for $x \in X_Z^1$.

(c) $Z_j Z_i x = Z_i Z_j x$ for $x \in X_Z^2$.

(d) For $k \in \mathbb{N}_0$ there are $d_k \geq 0$ with $\left\| (\mathrm{Id} - \frac{1}{l} Z_j)^{-1} \right\|_{\mathcal{L}(X_Z^k)} \leq d_k$ for $l \in \mathbb{N}$.

(e) $(\mathrm{Id} - \frac{1}{l} Z_j)^{-1} \xrightarrow[l \to \infty]{} \mathrm{Id}$ strongly in X_Z^k.

PROOF: For $x \in X_{\bar{Z}}^1$ and $y = (\lambda\mathrm{Id} - Z_i)x$ 2.2.4(b) implies

$$(\lambda\mathrm{Id} - Z_j)^{-1}x = (\lambda\mathrm{Id} - Z_i)^{-1}(\lambda\mathrm{Id} - Z_j)^{-1}(\lambda\mathrm{Id} - Z_i)x$$

for $j, i = 1 \ldots, M$, thus $(\lambda\mathrm{Id} - Z_j)^{-1}(X_{\bar{Z}}^1) \subset X_{\bar{Z}}^1$, which proves (a) for $k = 1$ due to the closed graph theorem. Application of $(\lambda\mathrm{Id} - Z_i)$ yields

$$\lambda(\lambda\mathrm{Id} - Z_j)^{-1}x - Z_i(\lambda\mathrm{Id} - Z_j)^{-1}x = \lambda(\lambda\mathrm{Id} - Z_j)^{-1}x - (\lambda\mathrm{Id} - Z_j)^{-1}Z_ix.$$

This proves (b). Let $y \in X_{\bar{Z}}^2$ and apply (b) with $x = (\lambda\mathrm{Id} - Z_j)y \in X_{\bar{Z}}^1$, then

$$(\lambda\mathrm{Id} - Z_j)^{-1}Z_i(\lambda\mathrm{Id} - Z_j)y = Z_iy \in D(Z_j)$$

and application of $(\lambda\mathrm{Id} - Z_j)$ shows $\lambda Z_iy - Z_iZ_jy = \lambda Z_iy - Z_jZ_iy$, which proves (c). Now assume (a) for a $k \in \mathbb{N}$ and take $x \in X_{\bar{Z}}^{k+1} \subset X_{\bar{Z}}^k$. Then $(\lambda\mathrm{Id} - Z_j)^{-1}x \in X_{\bar{Z}}^k$ and

$$Z_i(\lambda\mathrm{Id} - Z_j)^{-1}x = (\lambda\mathrm{Id} - Z_j)^{-1}Z_ix \in X_{\bar{Z}}^k$$

for $i = 1, \ldots, M$ due to the induction hypothesis because $Z_ix \in X_{\bar{Z}}^k$. This shows $(\lambda\mathrm{Id} - Z_j)^{-1}x \in X_{\bar{Z}}^{k+1}$, and the closed graph theorem implies (a). Moreover, we clearly have (d) for $k = 0$. Therefore we can assume (d) inductively for $k \in \mathbb{N}_0$, then for $x \in X_{\bar{Z}}^{k+1}$ and suitable $d_{k+1} \geq 0$

$$\left\|\left(\mathrm{Id} - \frac{1}{l}Z_j\right)^{-1}x\right\|_{k+1}^2 = \left\|\left(\mathrm{Id} - \frac{1}{l}Z_j\right)^{-1}x\right\|_k^2 + \sum_{i=1}^M \left\|\left(\mathrm{Id} - \frac{1}{l}Z_j\right)^{-1}Z_ix\right\|_k^2$$

$$\leq d_{k+1}^2 \|x\|_{k+1}^2$$

This proves (d). Finally, $\left\|\left(\mathrm{Id} - \frac{1}{l}Z_j\right)^{-1}x - x\right\| \xrightarrow[l\to\infty]{} 0$ for $x \in X, j = 1, \ldots, M$, due to 1.1.3 because there is a $c > 0$ with

$$\left\|\left(\mathrm{Id} - \frac{1}{l}Z_j\right)^{-1}w - w\right\| = \frac{1}{l}\left\|\left(\mathrm{Id} - \frac{1}{l}Z_j\right)^{-1}Z_jw\right\| \leq \frac{c}{l}\|Z_jw\| \xrightarrow[l\to\infty]{} 0$$

for $w \in D(Z_j)$. Hence we can assume (e) inductively for a $k \in \mathbb{N}_0$, then for $x \in X_{\bar{Z}}^{k+1}$

$$\left\|\left(\mathrm{Id} - \frac{1}{l}Z_j\right)^{-1}x - x\right\|_{k+1}^2$$

$$= \left\|\left(\mathrm{Id} - \frac{1}{l}Z_j\right)^{-1}x - x\right\|_k^2 + \sum_{i=1}^M \left\|\left(\mathrm{Id} - \frac{1}{l}Z_j\right)^{-1}Z_ix - Z_ix\right\|_k^2 \xrightarrow[l\to\infty]{} 0$$

$\square$

2.2.6 Definition. Let $Z = \{Z_1, \ldots, Z_M\}$ be a commuting family of infinitesimal generators of bounded C_0-semigroups. Then, due to 2.2.5 the definition

$$Z^\alpha x := Z_1^{\alpha_1} Z_2^{\alpha_2} \cdots Z_M^{\alpha_M} x \qquad \text{for } x \in X_Z^{|\alpha|}, \alpha = (\alpha_1 \ldots \alpha_M) \in \mathbb{N}_0^M$$

is well-defined with $Z^\alpha Z^\beta x = Z^{\alpha+\beta} x$ for $x \in X_Z^{|\alpha+\beta|}, \alpha, \beta \in \mathbb{N}_0^M$. Clearly,

$$\sqrt{\sum_{|\alpha| \le k} \|Z^\alpha x\|^2} \qquad \text{for } x \in X_Z^k$$

defines an equivalent norm on X_Z^k for any $k \in \mathbb{N}_0$.

2.2.7 Remark. Hence, if a family $Z = \{Z_1, \ldots, Z_M\}$ of infinitesimal generators of bounded C_0-semigroups is commuting, then the $Z_j, j = 1, \ldots, M$, mutually commute on X_Z^2, but the opposite conclusion is wrong.
Take $X := L^2(0,1), Z_1 f := f'$ for $f \in D(Z_1) := \{f \in W^1(0,1) : f(0) = f(1)\}$ and $Z_2 f := -if''$ for $f \in D(Z_2) := \{f \in W^2(0,1) : f(0) = f(1) = 0\}$. Here $W^k(0,1)$ denotes the Sobolev space of order k on $(0,1)$, cf. 4.5. iZ_1 and iZ_2 are selfadjoint, i.e., Z_1 and Z_2 generate C_0-semigroups of contractions. Clearly we have $X_{\{Z_1,Z_2\}}^1 = D(Z_1) \cap D(Z_2) = D(Z_2)$. We will show that

$$(\mathrm{Id} - Z_1)^{-1}(X_{\{Z_1,Z_2\}}^1) = (\mathrm{Id} - Z_1)^{-1}(D(Z_2)) \not\subset D(Z_2) = X_{\{Z_1,Z_2\}}^1,$$

which shows in view of 2.2.5(a) that $\{Z_1, Z_2\}$ is not commuting.
For $u \in C^\infty[0,1]$ we can verify directly that

$$(\mathrm{Id} - Z_1)^{-1}u(x) = e^x \left(\frac{e}{e-1} \int_0^1 e^{-y}u(y)dy - \int_0^x e^{-y}u(y)dy \right) \qquad \text{for } x \in [0,1].$$

Now choose a $u \in C_c^\infty(0,1) \subset D(Z_2)$ with $\int_0^1 e^{-y}u(y)dy \ne 0$, then we have $(\mathrm{Id} - Z_1)^{-1}u(0) \ne 0$, i.e., $(\mathrm{Id} - Z_1)^{-1}u \notin D(Z_2)$.

2.2.8 Examples. Let $1 < p < \infty$. Suppose that $\Omega \subset \mathbb{R}^n$ is an open subset and $\Phi_j : \mathbb{R} \times \Omega \longrightarrow \Omega, j = 1, \ldots, M$, is a commuting family of smooth flows, i.e., there exist functions $a_j = (a_{j,k})_k \in C^\infty(\Omega, \mathbb{R}^n), j = 1, \ldots, M$, with:

- $\partial_t \Phi_j(t,x) = a_j(\Phi_j(t,x))$ for $j = 1, \ldots, M, t \in \mathbb{R}, x \in \Omega$

- $\Phi_j(0,x) = x$ for $j = 1 \ldots, M, x \in \Omega$

- $\Phi_j(s, \Phi_j(t,x)) = \Phi_j(s+t, x)$ for $j = 1, \ldots, M, s, t \in \mathbb{R}, x \in \Omega$

- $\Phi_j(t, \Phi_k(s,x)) = \Phi_k(s, \Phi_j(t,x))$ for $j, k = 1, \ldots, M, s, t \in \mathbb{R}, x \in \Omega$

Then $U_j(t)f(x) := f(\Phi_j(t,x)) \left| \det \frac{\partial \Phi_j}{\partial x}(t,x) \right|^{\frac{1}{p}}$, $f \in L^p(\Omega)$, $j = 1, \ldots, M$, $t \in \mathbb{R}$, $x \in \Omega$, defines a commuting family $\{U_1(t), \ldots, U_M(t)\}$ of isometric C_0-groups in $L^p(\Omega)$, and their infinitesimal generators satisfy

$$A_j(f)(x) = \sum_{k=1}^{n} a_{j,k}(x) \partial_k f(x) + \frac{1}{p} \mathrm{div}\, a_j(x) f(x), f \in C_c^\infty(\Omega), j = 1, \ldots, M, x \in \Omega.$$

Similarly, we can construct examples of commuting groups on manifolds Ω. Proofs of theses statements are left as an exercise.

2.2.9 Lemma. Let $Z = \{Z_1, \ldots, Z_M\}$ be a commuting family of infinitesimal generators of bounded C_0-semigroups in a Banach space X, and let

$$J_l := \left(\mathrm{Id} - \frac{1}{l} Z_1 \right)^{-1} \cdots \left(\mathrm{Id} - \frac{1}{l} Z_M \right)^{-1} \quad \text{for } l \in \mathbb{N}.$$

Then the following statements hold:

(a) $(J_l)_{l \in \mathbb{N}} \subset \mathcal{L}(X_Z^k, X_Z^{k+1})$ for $k \in \mathbb{N}_0$.

(b) For $k \in \mathbb{N}_0$ there are $j_k \geq 0$ with $\|J_l\|_{\mathcal{L}(X_Z^k)} \leq j_k$ for $l \in \mathbb{N}$.

(c) For $k \in \mathbb{N}_0$ there are $j_k' \geq 0$ with $\|J_l - \mathrm{Id}\|_{\mathcal{L}(X_Z^{k+1}, X_Z^k)} \leq \frac{j_k'}{l}$ for $l \in \mathbb{N}$.

(d) $J_l \xrightarrow[l \to \infty]{} \mathrm{Id}$ strongly in $\mathcal{L}(X_Z^k)$ for $k \in \mathbb{N}_0$.

(e) $(X_Z^k)_k$ is a scale of Banach spaces.

PROOF: Since

$$J_l = \left(\mathrm{Id} - \frac{1}{l} Z_i \right)^{-1} \prod_{\mu \neq i} \left(\mathrm{Id} - \frac{1}{l} Z_\mu \right)^{-1}$$

for $i = 1, \ldots, M$, we clearly have (a) for $k = 0$ with the closed graph theorem. Now assume (a) for a $k \in \mathbb{N}_0$ and let $x \in X_Z^{k+1} \subset X_Z^k$, then $J_l x \in X_Z^{k+1}$ and $Z_i J_l x = J_l Z_i x \in X_Z^{k+1}$ for $i = 1, \ldots, M$, hence $J_l x \in X_Z^{k+2}$ and (a) is implied by the closed graph theorem. (b) is a consequence of 2.2.5(d). Furthermore, for $x \in X_Z^1$ we have with suitable constants $d_0, d_0' \geq 0$

$$\|J_l x - x\| \leq \sum_{j=1}^{M} \left\| \prod_{i=1}^{M-j} \left(\mathrm{Id} - \frac{1}{l} Z_i \right)^{-1} \left[\left(\mathrm{Id} - \frac{1}{l} Z_{M-j+1} \right)^{-1} x - x \right] \right\|$$

$$\leq d_0 \sum_{j=1}^{M} \left\| \left(\mathrm{Id} - \frac{1}{l} Z_{M-j+1} \right)^{-1} x - x \right\| \leq \frac{d_0'}{l} \sum_{j=1}^{M} \|Z_j x\|$$

This proves (c) for $k = 0$. Assume (c) for a $k \in \mathbb{N}_0$, then for $x \in X_Z^{k+1}$ with suitable $j'_{k+1} \geq 0$

$$\|J_l x - x\|_{k+1}^2 = \|J_l x - x\|_k^2 + \sum_{i=1}^{M} \|J_l Z_i x - Z_i x\|_k^2 \leq \frac{(j'_{k+1})^2}{l^2} \|x\|_{k+1}^2 .$$

Hence we have proved (c). Furthermore, for $x \in X_Z^k$ and $k \in \mathbb{N}_0$ we have due to 2.2.5(d), 2.2.5(e)

$$\|J_l x - x\|_k \leq \sum_{j=1}^{M} \left\| \prod_{i=1}^{M-j} \left(\mathrm{Id} - \frac{1}{l} Z_i \right)^{-1} \left[\left(\mathrm{Id} - \frac{1}{l} Z_{M-j+1} \right)^{-1} x - x \right] \right\|_k$$

$$\leq \sum_{j=1}^{M} \tilde{d}_k \left\| \left(\mathrm{Id} - \frac{1}{l} Z_{M-j+1} \right)^{-1} x - x \right\|_k \xrightarrow[l \to 0]{} 0$$

with suitable $\tilde{d}_k \geq 0$. This proves (d). Finally, (a) and (d) imply (e). $\qquad \square$

2.2.10 Lemma. Let $A : D(A) \to H$ be a densely defined, closed operator in a Hilbert space H and let $\lambda \in \rho(A)$.
Then $\bar{\lambda} \in \rho(A^*)$ and $(\bar{\lambda}\mathrm{Id} - A^*)^{-1} = \left[(\lambda\mathrm{Id} - A)^{-1} \right]^*$.

PROOF: Let $x \in D(A^*)$ with $(\bar{\lambda}\mathrm{Id} - A^*)x = 0$, then for $y \in D(A)$ we have $0 = \langle (\bar{\lambda}\mathrm{Id} - A^*)x, y \rangle = \langle x, (\lambda\mathrm{Id} - A)y \rangle$ and we have $x \in \mathcal{R}(\lambda\mathrm{Id} - A)^{\perp} = \{0\}$. Hence $\bar{\lambda}\mathrm{Id} - A^* : D(A^*) \to H$ is injective. Moreover, for $x \in D(A^*)$ we have

$$\begin{aligned} \langle x, y \rangle &= \langle x, (\lambda\mathrm{Id} - A)(\lambda\mathrm{Id} - A)^{-1}y \rangle = \langle (\bar{\lambda}\mathrm{Id} - A^*)x, (\lambda\mathrm{Id} - A)^{-1}y \rangle \\ &= \langle \left[(\lambda\mathrm{Id} - A)^{-1} \right]^* (\bar{\lambda}\mathrm{Id} - A^*)x, y \rangle \end{aligned}$$

for $y \in X$, which proves $\left[(\lambda\mathrm{Id} - A)^{-1} \right]^* (\bar{\lambda}\mathrm{Id} - A^*)x = x$ for $x \in H$. Finally, for $x \in H$ and $y \in D(A)$

$$\langle x, y \rangle = \langle x, (\lambda\mathrm{Id} - A)^{-1}(\lambda\mathrm{Id} - A)y \rangle = \langle \left[(\lambda\mathrm{Id} - A)^{-1} \right]^* x, (\lambda\mathrm{Id} - A)y \rangle .$$

This shows $\left[(\lambda\mathrm{Id} - A)^{-1} \right]^* x \in D(A^*)$ with $(\bar{\lambda}\mathrm{Id} - A^*) \left[(\lambda\mathrm{Id} - A)^{-1} \right]^* x = x$, which proves the assertion.

$\qquad \square$

2.2.11 Proposition. Suppose that $Z = \{Z_1, \ldots, Z_M\}$ is a commuting family of infinitesimal generators of bounded C_0-semigroups in a Hilbert space H. Moreover, assume that there exist $\alpha_j \neq 0$ with $Z_j^* = \alpha_j Z_j$ for $j = 1, \ldots, M$. Define

$$J_l^* := \left(\mathrm{Id} - \frac{1}{l} Z_M^* \right)^{-1} \cdots \left(\mathrm{Id} - \frac{1}{l} Z_1^* \right)^{-1} \quad \text{for } l \in \mathbb{N}.$$

Then

(a) $Z^* = \{Z_1^*, \ldots, Z_M^*\}$ is a commuting family of infinitesimal generators of bounded C_0-semigroups.

(b) $H_Z^k = H_{Z^*}^k$ topologically for $k \in \mathbb{N}_0$.

(c) $(J_l)_{l \in \mathbb{N}}, (J_l^*)_{l \in \mathbb{N}} \subset \mathscr{L}(H_Z^k, H_Z^{k+1})$ for $k \in \mathbb{N}_0$.

(d) $(J_l)_{l \in \mathbb{N}}, (J_l^*)_{l \in \mathbb{N}} \subset \mathscr{L}(H_Z^k)$ boundedly for $k \in \mathbb{N}_0$.

(e) $J_l \xrightarrow[l \to \infty]{} \mathrm{Id}$ and $J_l^* \xrightarrow[l \to \infty]{} \mathrm{Id}$ strongly in $\mathscr{L}(H_Z^k)$ for $k \in \mathbb{N}_0$.

(f) For $k \in \mathbb{N}_0$ there are $j_k' > 0$ with $\|J_l - \mathrm{Id}\|_{\mathscr{L}(H_Z^{k+1}, H_Z^k)} \leq \frac{j_k'}{l}$ and $\|J_l^* - \mathrm{Id}\|_{\mathscr{L}(H_Z^{k+1}, H_Z^k)} \leq \frac{j_k'}{l}$ for $l \in \mathbb{N}$.

(g) $\langle J_l u, v \rangle_0 = \langle u, J_l^* v \rangle_0$ for $u, v \in H_Z^k$, $k \in \mathbb{N}_0$.

PROOF: Since $\|A\| = \|A^*\|$ for $A \in \mathscr{L}(H)$, lemma 2.2.10 and the Hille-Yosida theorem 1.1.4 shows that $(Z_j^*, D(Z_j))$ generates a bounded C_0-semigroup and

$$\left(\lambda \mathrm{Id} - Z_j^* \right)^{-1} = \left(\lambda \mathrm{Id} - Z_j \right)^{-1*} \quad \text{for } j = 1, \ldots, M, \lambda > 0.$$

This implies that Z^* is a commuting family of infinitesimal generators of C_0-semigroups of contractions and the assertion is a consequence of 2.2.9. $\square$

2.2.12 Definition.

(a) If $\mathcal{A}$ is an algebra with unit $e \in \mathcal{A}$, then $\mathcal{A}^{-1}$ denotes the set of all invertible elements of $\mathcal{A}$.

(b) Let $\mathcal{B}$ be a Banach algebra with unit $e \in \mathcal{B}$. A linear operator $\delta : D(\delta) \to \mathcal{B}$ with domain $D(\delta) \subset \mathcal{B}$ is called derivation, if its domain $D(\delta)$ is an algebra with $e \in D(\delta)$, $\delta(e) = 0$, and $\delta(ab) = a\delta(b) + \delta(a)b$ for $a, b \in D(\delta)$. The last equation is also called Leibniz formula. If $\mathcal{B}$ is a C^*-algebra, then a derivation δ is called $*$-derivation, if $a \in D(\delta)$ implies $a^* \in D(\delta)$ and $\delta(a^*) = \delta(a)^*$.

2.2.13 Proposition. Let $\mathcal{D} = \{\delta_j : D(\delta_j) \to X, j = 1, \ldots, M\}$ be a finite set of closed derivations in a Banach algebra $\mathcal{B}$. Let $\mathcal{B}_{\mathcal{D}}^k$ be defined as in 2.2.1 and let $\|\cdot\|_{\mathcal{B}_{\mathcal{D}}^0} := \|\cdot\|_{\mathcal{B}}$ and

$$\|x\|_{\mathcal{B}_{\mathcal{D}}^{k+1}} := \|x\|_{\mathcal{B}_{\mathcal{D}}^k} + \sum_{j=1}^{M} \|\delta_j(x)\|_{\mathcal{B}_{\mathcal{D}}^k} \qquad \text{for } x \in \mathcal{B}_{\mathcal{D}}^{k+1}, k \in \mathbb{N}_0.$$

Then:

(a) $(\mathcal{B}_{\mathcal{D}}^k, \|\cdot\|_{\mathcal{B}_{\mathcal{D}}^k})$ is a Banach algebra with $\|ab\|_{\mathcal{B}_{\mathcal{D}}^k} \leq \|a\|_{\mathcal{B}_{\mathcal{D}}^k} \|b\|_{\mathcal{B}_{\mathcal{D}}^k}$ and $\|e\|_{\mathcal{B}_{\mathcal{D}}^k} = 1$ for $a, b \in \mathcal{B}_{\mathcal{D}}^k, k \in \mathbb{N}_0$.

(b) For $a \in \mathcal{B}_{\mathcal{D}}^k$ with $\|a\|_{\mathcal{B}} < 1$ we have $e - a \in (\mathcal{B}_{\mathcal{D}}^k)^{-1}$ for $k \in \mathbb{N}_0 \cup \{\infty\}$.

(c) $\mathcal{B}_{\mathcal{D}}^k \cap (\overline{\mathcal{B}_{\mathcal{D}}^k})^{-1} = (\mathcal{B}_{\mathcal{D}}^k)^{-1}$ for $k \in \mathbb{N}_0 \cup \{\infty\}$.

PROOF: (a) can be easily proved with the Leibniz formula. The Leibniz formula implies further $\delta(a^l) = \sum_{\nu=0}^{l-1} a^\nu \delta(a) a^{l-1-\nu}$ for $a \in D(\delta)$. Using this identity one can verify inductively that for any $r > 0, k \in \mathbb{N}_0, a \in \mathcal{B}_{\mathcal{D}}^k$ with $\|a\| \leq r$ there are constants $c_k(a)$ with $\|a^l\|_{\mathcal{B}_{\mathcal{D}}^k} \leq c_k(a) l^{2^k-1} r^{l-2^k+1}$. This implies $\lim_{l \to \infty} \|a^l\|_{\mathcal{B}_{\mathcal{D}}^k}^{1/l} \leq r$. Hence $e - a \in (\mathcal{B}_{\mathcal{D}}^k)^{-1}$, if $\|a\|_{\mathcal{B}} < 1$. In particular, we have proved (b). Finally, let $a \in \mathcal{B}_{\mathcal{D}}^k \cap (\overline{\mathcal{B}_{\mathcal{D}}^k})^{-1}$ for $k \in \mathbb{N}_0 \cup \{\infty\}$ and let $b := a^{-1} \in \overline{\mathcal{B}_{\mathcal{D}}^k}$. Then there is a sequence $(b_l)_l \subset \mathcal{B}_{\mathcal{D}}^k$ with $b_l \xrightarrow[l \to \infty]{\mathcal{B}} b$. Let $x_l := e - b_l a$. Since $\|x_l\|_{\mathcal{B}} < 1$ for large l and $x_l \in \mathcal{B}_{\mathcal{D}}^k$ the part proved so far shows $e - x_l \in (\mathcal{B}_{\mathcal{D}}^k)^{-1}$ for large l. Therefore $e - x_l = b_l a$ implies $(e - x_l)^{-1} b_l a = e$ for large l and $(e - x_l)^{-1} b_l \in \mathcal{B}_{\mathcal{D}}^k$ is a left inverse of a in $\mathcal{B}_{\mathcal{D}}^k$. Similarly, we can also find a right inverse, and we have proved the proposition. $\square$

2.2.14 Definition. Let $\Lambda : D(\Lambda) \to X$ be a closed operator in a Banach space X. Let

$$D(\delta_\Lambda) := \{A \in \mathscr{L}(X) : A(D(\Lambda)) \subset D(\Lambda), \exists\, a \geq 0 \text{ with}$$
$$\|\Lambda A x - A \Lambda x\| \leq a \|x\| \text{ for } x \in D(\Lambda)\}$$

and let $\delta_\Lambda(A) \in \mathscr{L}(X)$ be the unique extension of $\Lambda A - A\Lambda : D(\Lambda) \to X$ for $A \in D(\delta_\Lambda)$. $\delta_\Lambda : D(\delta_\Lambda) \to \mathscr{L}(X)$ is called the derivation implemented by Λ. Then $\text{ad}\Lambda := \delta_\Lambda$ is a closed derivation. Moreover, if X is a Hilbert space and Λ is skew-selfadjoint, then δ_Λ is a $*$-derivation. We will also use the notation $\text{ad}^0(\Lambda) := \text{Id}, \text{ad}^1(\Lambda) := \text{ad}\Lambda$ and $\text{ad}^k(\Lambda) := \delta_\Lambda^k$ for $k \in \mathbb{N}$.

PROOF: Clearly, $\delta_\Lambda : D(\delta_\Lambda) \to \mathscr{L}(X)$ is a derivation. Let $(A_l)_l \subset D(\delta_\Lambda)$ with $A_l \xrightarrow{l\to\infty} A \in \mathscr{L}(X)$ and $\delta_\Lambda(A_l) \xrightarrow{l\to\infty} B \in \mathscr{L}(X)$. For $x \in D(\Lambda)$ we have $A_l x \in D(\Lambda), A_l x \xrightarrow{l\to\infty} Ax$, and $\Lambda A_l x = \delta_\Lambda(A_l)x + A_l \Lambda x \xrightarrow{l\to\infty} Bx + A\Lambda x$. This shows $Ax \in D(\Lambda)$ and $\Lambda Ax - A\Lambda x = Bx$ for $x \in D(\Lambda)$. Therefore $A \in D(\delta_\Lambda)$ with $\delta_\Lambda(A) = B$.

If X is a Hilbert space and Λ is skew-selfadjoint, then for $A \in D(\delta_\Lambda)$ and $x, y \in D(\Lambda)$

$$\langle \Lambda x, A^* y \rangle = \langle A\Lambda x, y \rangle = \langle \Lambda A x - \delta_\Lambda(A)x, y \rangle = \langle x, -A^*\Lambda y - \delta_\Lambda(A)^* y \rangle$$

Hence $A^* y \in D(\Lambda^*) = D(\Lambda)$ with $\Lambda A^* y = A^*\Lambda y + \delta_\Lambda(A)^* y$. This proves $A^* \in D(\delta_\Lambda)$ with $\delta_\Lambda(A^*) = \delta_\Lambda(A)^*$.   □

2.2.15 Proposition. Let $Z = \{Z_1, \ldots, Z_M\}$ be a family of closed operators in a Banach space X and let $\mathcal{D} := \{\delta_{Z_1}, \ldots, \delta_{Z_M}\}$ be the family of derivations implemented by the elements of Z. Let $(\Psi_Z^k, \|\cdot\|_{\Psi_Z^k}) := (\mathscr{L}(X)_{\mathcal{D}}^k, \|\cdot\|_{\mathscr{L}(X)_{\mathcal{D}}^k})$ for $k \in \mathbb{N}_0$ and $\Psi_Z^\infty := \bigcap_{k \in \mathbb{N}_0} \Psi_Z^k$. Then:

(a) $\Psi_Z^k \times X_Z^k \ni (A, x) \mapsto Ax \in X_Z^k$ is a continuous bilinear map.

(b) $\Psi_Z^\infty \times X_Z^\infty \ni (A, x) \mapsto Ax \in X_Z^\infty$ is a bilinear map.

(c) For $A \in \Psi_Z^k$ with $\|A\|_{\mathscr{L}(X)} < 1$ we have $\mathrm{Id} - A \in (\Psi_Z^k)^{-1}$ for $k \in \mathbb{N}_0 \cup \{\infty\}$.

(d) $\Psi_Z^k \cap (\overline{\Psi_Z^k})^{-1} = (\Psi_Z^k)^{-1}$ for $k \in \mathbb{N}_0 \cup \{\infty\}$.

(e) If X is a Hilbert space and Z_j is skew-selfadjoint for $j = 1, \ldots, M$, then $\Psi_Z^k \cap (\mathscr{L}(X))^{-1} = (\Psi_Z^k)^{-1}$ for $k \in \mathbb{N}_0 \cup \{\infty\}$.

PROOF: We only have of prove (a) and (e). Assume (a) for a $k \in \mathbb{N}_0$ and let $A \in \Psi_Z^{k+1}, x \in X_Z^{k+1}$. Then $Ax \in X_Z^k$ and $Z_j Ax = \delta_{Z_j}(A)x + AZ_j x \in X_Z^k$ for $j = 1, \ldots, M$ by induction because $\delta_{Z_j}(A) \in \Psi_Z^k$. We get $Ax \in X_Z^{k+1}$ and it is easy to check the continuity. Finally, (e) follows from the spectral invariance of C^*-algebras $\mathcal{A} \subset \mathscr{L}(X)$, i.e., $\mathcal{A} \cap \mathscr{L}(X)^{-1} = \mathcal{A}^{-1}$, and 2.2.13, because $\overline{\Psi_Z^k} \subset \mathscr{L}(X)$ is a C^*-subalgebra for any $k \in \mathbb{N}_0$.   □

2.2.16 Proposition. Let $\Lambda : D(\Lambda) \to X$ be the generator of a C_0-group $(T(t))_{t\in\mathbf{R}} \subset \mathscr{L}(X)$ of operators in a Banach space X. Then, for $A \in \mathscr{L}(X)$, the following statements are equivalent:

(a) $[\mathbf{R} \ni t \mapsto T(t)AT(-t) \in \mathscr{L}(X)] \in C^\infty(\mathbf{R}, \mathscr{L}(X))$.

(b) $A \in \bigcap\limits_{k\in\mathbb{N}} D(\delta_\Lambda^k)$.

(c) $A(D(\Lambda^\infty)) \subset D(\Lambda^\infty)$ and for $k \in \mathbb{N}_0$ there are constants $a_k \geq 0$ with $\left\|\operatorname{ad}^k(\Lambda)(A)x\right\| \leq a_k \left\|x\right\|$ for $x \in D(\Lambda^\infty)$.

PROOF: Throughout this proof let $\alpha_B(t) := T(t)BT(-t)$ for $B \in \mathscr{L}(X)$. To prove (a) $\Rightarrow$ (b) we will show inductively that $A \in D(\delta_\Lambda^k)$ for any $k \in \mathbb{N}_0$ with $\alpha_A^{(k)}(t) = \alpha_{\delta_\Lambda^k(A)}(t)$ for $t \in \mathbf{R}$. First, for $x \in D(\Lambda)$

$$\frac{T(h)Ax - Ax}{h} = \frac{T(h)AT(-h) - A}{h}T(h)x + A\frac{T(h)x - x}{h} \xrightarrow[h\to 0]{} \alpha_A'(0)x + A\Lambda x .$$

Hence $A(D(\Lambda)) \subset D(\Lambda)$ and $A \in D(\delta_\Lambda)$ with $\delta_\Lambda(A) = \alpha_A'(0)$. Moreover,

$$\alpha_A'(t) = \lim_{h\to 0} T(t)\frac{T(h)AT(-h) - A}{h}T(-t) = T(t)\alpha_A'(0)T(-t) = \alpha_{\delta_\Lambda(A)}(t) .$$

Assume the induction hypothesis for a $k \in \mathbb{N}$. Then $\alpha_{\delta_\Lambda^k(A)} = \alpha_A^{(k)} : \mathbf{R} \to \mathscr{L}(X)$ is smooth and the induction hypothesis for $k = 1$ shows $\delta_\Lambda^k(A) \in D(\delta_\Lambda)$ with $\delta_\Lambda(\delta_\Lambda^k(A)) = \alpha_{\delta_\Lambda^k(A)}'(0)$. Hence $A \in D(\delta_\Lambda^{k+1})$ and for the $k+1$-th derivative of α_A we have $\alpha_A^{(k+1)}(t) = (\alpha_A^{(k)})'(t) = \alpha_{\delta_\Lambda^k(A)}'(t) = \alpha_{\delta_\Lambda(\delta_\Lambda^k(A))}(t) = \alpha_{\delta_\Lambda^{k+1}(A)}(t)$.

To prove (b) $\Rightarrow$ (c) we will show that for any $A \in \mathscr{L}(X)$ and $k \in \mathbb{N}$

$$A \in D(\delta_\Lambda^k) \implies A(D(\Lambda^j)) \subset D(\Lambda^j) \text{ and } \left\|\operatorname{ad}^j(\Lambda)(A)x\right\| \leq a_j\left\|x\right\|$$
$$\text{for } x \in D(\Lambda^k) \text{ and suitable } a_j \geq 0 \text{ for } 1 \leq j \leq k .$$

This clearly holds for $k = 1$. Now suppose this inductively for a $k \in \mathbb{N}$. Let $A \in D(\delta_\Lambda^{k+1})$ and $x \in D(\Lambda^{k+1})$. Then $x, \Lambda x \in D(\Lambda^k)$ and $A, \delta_\Lambda(A) \in D(\delta_\Lambda^k)$, hence $Ax, A\Lambda x, \delta_\Lambda(A)x \in D(\Lambda^k)$. This implies $\Lambda Ax = \delta_\Lambda(A)x + A\Lambda x \in D(\Lambda^k)$ and $Ax \in D(\Lambda^{k+1})$. Moreover,

$$\left\|\operatorname{ad}^{k+1}(\Lambda)(A)x\right\| = \left\|\operatorname{ad}^k(\Lambda)(\delta_\Lambda(A))x\right\| \leq a_k\left\|\delta_\Lambda(A)x\right\| \leq a'\left\|x\right\|$$

for $x \in D(\Lambda^{k+1})$. Finally assume (c) and let $\widetilde{\delta}^k(A) \in \mathscr{L}(X)$ be the extension of $\operatorname{ad}^k(\Lambda)(A)$ for $k \in \mathbb{N}_0$. Note that $\widetilde{\delta}^1(\widetilde{\delta}^k(A)) = \widetilde{\delta}^{k+1}(A)$. Fix $x \in D(\Lambda^\infty)$.

Then

$$\frac{\alpha_A(t+h)x - \alpha_A(t)x}{h}$$

$$= \frac{T(t+h) - T(t)}{h} AT(-t-h)x + T(t)A\frac{T(-t-h) - T(-t)}{h}x$$

$$\xrightarrow[h\to 0]{} T(t)\Lambda AT(-t)x - T(t)A\Lambda T(-t)x = \alpha_{\tilde\delta^1(A)}(t)x$$

Hence $\alpha_A(\cdot)x : \mathbb{R} \to X$ is differentiable with $(\alpha_A(t)x)'(t) = \alpha_{\tilde\delta^1(A)}(t)x$. Inductively, this shows that $\alpha_A(\cdot)x : \mathbb{R} \to X$ is k-times differentiable with derivative $(\alpha_A(t)x)^{(k)}(t) = \alpha_{\tilde\delta^k(A)}(t)x$ for any $k \in \mathbb{N}$. Hence $\alpha_A(\cdot)x : \mathbb{R} \to X$ is smooth for any $x \in D(\Lambda^\infty)$. Since $D(\Lambda^\infty) \subset X$ densely, this shows that $\alpha_A(\cdot)x : \mathbb{R} \to X$ is smooth for any $x \in X$ and the uniform boundedness principle implies (a). $\qquad\square$

2.2.17 Bibliographical remarks. Spaces generated by one or a family of closed operators have been often considered in literature, cf. e.g. [55], [56], [101], [151], [9], [111]. Lemma 2.2.4 is due to [110, p. 24], [28, 3.6]). In [128, VIII.5] a counterexample due to Nelson of operators is given which commute in the resolvent sense but do not commute on joint cores. The equivalence of (a) and (b) is due to [101, A.3.8].

2.3   Commutator estimates and scales of Banach spaces

If one works in a scale of Banach spaces generated by one or a family of closed operators one is interested in extending certain inequalities from one part of the scale to others. This can be done effectively with commutator estimates and we will describe several results of this type in this section. In particular, we are interested in extending quasi-dissipativity estimates from one part of the scale to others. This will be needed several times in later sections. The first lemma shows how quasi-dissipativity estimates can be extended from the base space to the whole scale, if first order commutator estimates are known.

2.3.1 Lemma. Let X be a Banach space, $Z = \{Z_1, \dots, Z_M\}$ be a family of closed operators, $m \in \mathbb{N}$, and

$$\mathcal{A} \subset \bigcap_{k \in \mathbb{N}_0} \mathscr{L}(X_Z^{k+m}, X_Z^k)$$

be a family of linear operators such that for $k \in \mathbb{N}_0$ there are $\beta \geq 0$ and $a_k \geq 0$ with

$$\|Z_j A x - A Z_j x\|_k \leq a_k \|x\|_{k+1} \quad \text{and} \quad \|\lambda x - A x\| \geq (\lambda - \beta) \|x\|$$

for $x \in X_Z^\infty, j = 1, \ldots, M, A \in \mathcal{A}, \lambda > \beta$.

Then, for $k \in \mathbb{N}_0$ there are $\beta_k \geq 0$ with

$$\|\lambda x - A x\|_k \geq (\lambda - \beta_k) \|x\|_k \qquad \text{for } x \in X_Z^\infty, A \in \mathcal{A}, \lambda > \beta_k. \tag{2.3.1}$$

PROOF: Suppose (2.3.1) inductively for a $k \in \mathbb{N}_0$, then

$$(\lambda - \beta_k) \|x\|_{k+1} = \sqrt{((\lambda - \beta_k) \|x\|_k)^2 + \sum_{j=1}^M ((\lambda - \beta_k) \|Z_j x\|_k)^2}$$

$$\leq \sqrt{\|\lambda x - A x\|_k^2 + \sum_{j=1}^M \|\lambda Z_j x - A Z_j x\|_k^2}$$

$$\leq \sqrt{\|\lambda x - A x\|_k^2 + \sum_{j=1}^M (\|Z_j (\lambda x - A x)\|_k + \|Z_j A x - A Z_j x\|_k)^2}$$

$$\leq \sqrt{\|\lambda x - A x\|_k^2 + \sum_{j=1}^M \|Z_j (\lambda x - A x)\|_k^2} + \sqrt{\sum_{j=1}^M \|Z_j A x - A Z_j x\|_k^2}$$

$$\leq \|\lambda x - A x\|_{k+1} + \sqrt{M} a_k \|x\|_{k+1}$$

for $\lambda > \beta_k, x \in X_Z^\infty, A \in \mathcal{A}$. This implies the assertion. $\qquad \square$

In the case of Hilbert spaces one can also use the following result effectively. Note that only double commutator estimates and no single commutator estimates are assumed. This will be essential for applications to second-order pseudodifferential operators (cf. 4.2.17).

2.3.2 Lemma. Let $Z = \{Z_1, \ldots, Z_M\}$ be a commuting family of infinitesimal generators of bounded C_0-semigroups in a Hilbert space H and assume that there is a $\gamma \in \mathbb{R}$ with $Z_j^* = \gamma Z_j$ for $j = 1, \ldots, M$. Let

$$\mathcal{A} \subset \bigcap_{k \in \mathbb{N}_0} \mathcal{L}(H_Z^{k+m}, H_Z^k)$$

be a family of linear operators for an $m \in \mathbb{N}$ such that any $A \in \mathcal{A}$ is symmetric on H_Z^m in H. Moreover, assume that for $k \in \mathbb{N}_0$ there are $\beta \geq 0$ and $a_k \geq 0$ with

$$|\langle [Z^\alpha, [Z^\alpha, A]] x, x \rangle| \leq a_k \|x\|_k^2 \quad \text{and} \quad \operatorname{Re} \langle A x, x \rangle \leq \beta \|x\|^2$$

for $\alpha \in \mathbb{N}_0^M, |\alpha| \leq k, A \in \mathcal{A}, x \in H_Z^\infty$, where $[B, C] := BC - CB$ denotes the commutator of operators $B, C : H_Z^\infty \to H_Z^\infty$.
Then, for $k \in \mathbb{N}_0$ there are $\beta_k \geq 0$ with

$$\text{Re} \langle Ax, x \rangle_k \leq \beta_k \|x\|_k^2 \qquad \text{for } x \in H_Z^\infty, A \in \mathcal{A}. \qquad (2.3.2)$$

PROOF: For $\alpha \in \mathbb{N}_0^M$ with $|\alpha| \leq k$ we have

$$\begin{aligned}
2\text{Re} \langle Z^\alpha Ax, Z^\alpha x \rangle &= 2\text{Re} \langle AZ^\alpha x, Z^\alpha x \rangle + 2\text{Re} \langle [Z^\alpha, A]x, Z^\alpha x \rangle \\
&\leq 2\beta \|Z^\alpha x\|^2 + \langle [Z^\alpha, A]x, Z^\alpha x \rangle + \langle Z^\alpha x, [Z^\alpha, A]x \rangle \\
&= 2\beta \|Z^\alpha x\|^2 + \gamma^{|\alpha|} (\langle Z^\alpha [Z^\alpha, A]x, x \rangle - \langle [Z^\alpha, A]Z^\alpha x, x \rangle) \\
&\leq 2\beta \|Z^\alpha x\|^2 + |\langle [Z^\alpha, [Z^\alpha, A]]x, x \rangle| \leq (2\beta + a_k) \|x\|_k^2
\end{aligned}$$

This implies the assertion. $\qquad\qquad\qquad\qquad\qquad\qquad\qquad\qquad\qquad\qquad\square$

If the scale generating operator is of positive type these results can be improved.

2.3.3 Proposition. Let $\Lambda : D(\Lambda) \to X$ be an operator of positive type in a complex Banach space X, and let $s \in \mathbb{R}, 0 < t \leq 1, m' \in \mathbb{R}$ be fixed. Suppose that a linear operator $C : D(\Lambda^\infty) \to D(\Lambda^\infty)$ satisfies

- $\|\Lambda^{s/2} Cx\| \leq c_s \|\Lambda^{(s+m')/2} x\|$ for $x \in D(\Lambda^\infty)$ and a suitable $c_s \geq 0$.

- $\|\Lambda^{s/2}[\Lambda, C]x\| \leq d_s \|\Lambda^{(s+m'+1)/2} x\|$ for $x \in D(\Lambda^\infty)$ and a suitable $d_s \geq 0$.

Then there is a constant $c_{s,t} \geq 0$ with $\|\Lambda^{(s-t)/2} Cx\| \leq c_{s,t} \|\Lambda^{(s+m'-t)/2} x\|$ for $x \in D(\Lambda^\infty)$.

PROOF:
Step 1: m'=1 :
Let $\alpha := \frac{t}{2}$. Then (cf. 1.4.7)

$$\Lambda^{-\alpha} x = \frac{\sin(\alpha\pi)}{\pi} \int_0^\infty \lambda^{-\alpha} (\lambda \text{Id} + \Lambda)^{-1} x d\lambda, \ x \in D(\Lambda^\infty).$$

Here $[\lambda \mapsto \lambda^{-\alpha}(\lambda \text{Id} + \Lambda)^{-1} x] \in L^1((0, \infty), D(\Lambda^r))$ is Bochner-integrable for $r \geq 0$ and $x \in D(\Lambda^\infty)$. Hence

$$\begin{aligned}
\Lambda^{s/2}[\Lambda^{-\alpha}, C]x &= \Lambda^{s/2}\Lambda^{-\alpha} Cx - \Lambda^{s/2} C\Lambda^{-\alpha} x \\
&= \frac{\sin(\alpha\pi)}{\pi} \int_0^\infty \lambda^{-\alpha}\Lambda^{s/2}[(\lambda \text{Id} + \Lambda)^{-1}, C]x d\lambda \\
&= -\frac{\sin(\alpha\pi)}{\pi} \int_0^\infty \lambda^{-\alpha}(\lambda \text{Id} + \Lambda)^{-1}\Lambda^{s/2}[\Lambda, C](\lambda \text{Id} + \Lambda)^{-1} x d\lambda
\end{aligned}$$

This implies

$$\left\|\Lambda^{s/2}[\Lambda^{-\alpha}, C]x\right\|$$

$$\leq \frac{\sin(\alpha\pi)}{\pi} \int_0^\infty \lambda^{-\alpha} \left\|(\lambda\mathrm{Id} + \Lambda)^{-1}\Lambda^{s/2}[\Lambda, C](\lambda\mathrm{Id} + \Lambda)^{-1}x\right\| d\lambda$$

$$\leq \frac{\sin(\alpha\pi)}{\pi} \int_0^\infty \lambda^{-\alpha}\frac{M}{1+\lambda} \left\|\Lambda^{s/2}[\Lambda, C](\lambda\mathrm{Id} + \Lambda)^{-1}x\right\| d\lambda$$

$$\leq d_s M \frac{\sin(\alpha\pi)}{\pi} \int_0^\infty \frac{\lambda^{-\alpha}}{1+\lambda} \left\|(-\lambda\mathrm{Id} + (\lambda\mathrm{Id} + \Lambda))(\lambda\mathrm{Id} + \Lambda)^{-1}\Lambda^{s/2}x\right\| d\lambda$$

$$\leq d_s M \frac{\sin(\alpha\pi)}{\pi} \int_0^\infty \frac{\lambda^{-\alpha}}{1+\lambda} \left(\left\|\lambda(\lambda\mathrm{Id} + \Lambda)^{-1}\Lambda^{s/2}x\right\| + \left\|\Lambda^{s/2}x\right\|\right) d\lambda$$

$$\leq d_s M(M+1)\frac{\sin(\alpha\pi)}{\pi} \int_0^\infty \frac{\lambda^{-\alpha}}{1+\lambda} d\lambda \left\|\Lambda^{s/2}x\right\| \leq d_s' \left\|\Lambda^{s/2}x\right\|$$

for suitable $d_s' \geq 0$. Thus, for a constant $c_{s,t} \geq 0$

$$\left\|\Lambda^{(s-t)/2}Cx\right\| \leq \left\|\Lambda^{s/2}C\Lambda^{-t/2}x\right\| + \left\|\Lambda^{s/2}[\Lambda^{-\alpha}, C]x\right\| \leq c_{s,t}\left\|\Lambda^{(s+1-t)/2}x\right\|.$$

Step 2: m' arbitrary: Apply step 1 to $C\Lambda^{(1-m')/2}$. $\qquad\square$

2.3.4 Definition. If $\Lambda : D(\Lambda) \to X$ is a densely defined, closed linear operator in a Banach space X, $D \subset D(\Lambda)$ is a dense subset with $\Lambda(D) \subset D$, and $A : D \to D$ is a linear operator, we define

$$\mathrm{ad}^0(\Lambda)(A) := A, \quad \mathrm{ad}^{j+1}(\Lambda)(A) = [\Lambda, \mathrm{ad}^j(\Lambda)(A)] = \Lambda\mathrm{ad}^j(\Lambda)(A) - \mathrm{ad}^j(\Lambda)(A)\Lambda$$

for $j \in \mathbb{N}_0$.

2.3.5 Corollary. Let $Z : D(Z) \to X$ be an operator of positive type in a complex Banach space X such that $Z^2 : D(Z^2) \to X$ is also of positive type. Moreover, let $A, C : D(Z^\infty) \to D(Z^\infty)$ and $m, m' \in \mathbb{N}_0$.

(a) If $r \in \mathbb{R}$ and $c_{r,0}, c_{r,1}, c_{r+1,1} \geq 0$ with $\|Z^r Cx\| \leq c_{r,0}\|Z^{r+m'}x\|$, $\|Z^r[Z, C]x\| \leq c_{r,1}\|Z^{r+m'}x\|$, and $\|Z^{r+1}[Z, C]x\| \leq c_{r+1,1}\|Z^{r+1+m'}x\|$ for $x \in D(Z^\infty)$, then there is a $c_{r-1,0}$ with $\|Z^{r-1}Cx\| \leq c_{r-1,0}\|Z^{r-1+m'}x\|$ for $x \in D(Z^\infty)$.

(b) If $\|Z^k\mathrm{ad}^j(Z)(A)x\| \leq b_{k,j}\|Z^{k+m}x\|$ for $x \in D(Z^\infty), k, j \in \mathbb{N}_0$, and suitable constants $b_{k,j}$, then, for $k, j \in \mathbb{N}_0$ there are $b_{-k,j}$ with $\|Z^{-k}\mathrm{ad}^j(Z)(A)x\| \leq b_{-k,j}\|Z^{-k+m}x\|$ for $x \in D(Z^\infty), k, j \in \mathbb{N}_0$.

PROOF: Let $\Lambda := Z^2$, then

$$\left\|\Lambda^{r/2}Cx\right\| = \|Z^rCx\| \le c_{r,0}\left\|Z^{r+m'}x\right\| = c_{r,0}\left\|\Lambda^{(r+m')/2}x\right\| \text{ and}$$
$$\left\|\Lambda^{r/2}[\Lambda,C]x\right\| = \left\|Z^r[Z^2,C]x\right\| \le \left\|Z^{r+1}[Z,C]x\right\| + \|Z^r[Z,C]Zx\|$$
$$\le (c_{r+1,1}+c_{r,1})\left\|\Lambda^{(r+1+m')/2}x\right\|.$$

Hence $\|Z^{r-1}Cx\| = \left\|\Lambda^{(r-1)/2}Cx\right\| \le c_{r-1,0}\left\|\Lambda^{(r+m'-1)/2}x\right\| = c_{r-1,0}\left\|Z^{r-1+m'}x\right\|$
with a constant $c_{r-1,0} \ge 0$ due to 2.3.3. This proves (a). (b) is a consequence (a). $\qquad\square$

In the final part of this section we have to work in Hilbert spaces. We will show how commutator estimates for Λ and A can be extended to commutator estimates for $\Lambda^{1/2}$ and A.

2.3.6 Proposition. Assume that Λ is a strictly positive, selfadjoint operator in a complex Hilbert space H. Let $D : H_\Lambda^\infty \longrightarrow H_\Lambda^\infty$, let $s,m \in \mathbb{R}$, and assume $\left\|\Lambda^{s/2}Dx\right\| \le c\left\|\Lambda^{(s+m)/2}x\right\|$ and $\left\|\Lambda^{s/2-1/4}[\Lambda,D]x\right\| \le c'\left\|\Lambda^{(s+m)/2+1/4}x\right\|$ for constants $c,c' \ge 0$ and all $x \in H_\Lambda^\infty$.
Then $\left\|\Lambda^{s/2}[\Lambda^{1/2},D]x\right\| \le d\left\|\Lambda^{(s+m)/2}x\right\|$ for $x \in H_\Lambda^\infty$ with $d := \frac{c'}{2}$.

PROOF: Step 1: m=0: Due to 1.4.7 we have

$$\Lambda^{s/2}[\Lambda^{1/2},D]x = \frac{1}{\pi}\int_0^\infty \lambda^{-1/2}\Lambda^{s/2}[(\lambda\mathrm{Id}+\Lambda)^{-1}\Lambda,D]x d\lambda$$

for $x \in H_\Lambda^\infty$. Now

$$[(\lambda\mathrm{Id}+\Lambda)^{-1}\Lambda,D]x = [(\lambda\mathrm{Id}+\Lambda)^{-1}(-\lambda\mathrm{Id}+(\lambda\mathrm{Id}+\Lambda)),D]x$$
$$= [(-\lambda)(\lambda\mathrm{Id}+\Lambda)^{-1}+\mathrm{Id},D]x = \lambda(\lambda\mathrm{Id}+\Lambda)^{-1}[\Lambda,D](\lambda\mathrm{Id}+\Lambda)^{-1}x$$

and therefore

$$\Lambda^{s/2}[\Lambda^{1/2},D]x = \frac{1}{\pi}\int_0^\infty \lambda^{-1/2}\lambda(\lambda\mathrm{Id}+\Lambda)^{-1}\Lambda^{s/2}[\Lambda,D](\lambda\mathrm{Id}+\Lambda)^{-1}x d\lambda.$$

This implies for $x,y \in H_\Lambda^\infty$

$$\left|\left\langle\Lambda^{s/2}[\Lambda^{1/2},D]x,y\right\rangle\right|^2$$
$$= \frac{1}{\pi^2}\left|\int_0^\infty \lambda^{1/2}\left\langle(\lambda\mathrm{Id}+\Lambda)^{-1}\Lambda^{s/2}[\Lambda,D](\lambda\mathrm{Id}+\Lambda)^{-1}x,y\right\rangle d\lambda\right|^2$$

$$\leq \frac{1}{\pi^2}\left(\int_0^\infty \lambda^{1/2}\left\|\Lambda^{s/2-1/4}[\Lambda,D](\lambda\mathrm{Id}+\Lambda)^{-1}x\right\|\left\|\Lambda^{1/4}(\lambda\mathrm{Id}+\Lambda)^{-1}y\right\|d\lambda\right)^2$$

$$\leq \frac{c'^2}{\pi^2}\left(\int_0^\infty \lambda^{1/2}\left\|\Lambda^{s/2+1/4}(\lambda\mathrm{Id}+\Lambda)^{-1}x\right\|\left\|\Lambda^{1/4}(\lambda\mathrm{Id}+\Lambda)^{-1}y\right\|d\lambda\right)^2$$

$$\leq \frac{c'^2}{\pi^2}\left(\int_0^\infty \lambda^{1/2}\left\|\Lambda^{s/2+1/4}(\lambda\mathrm{Id}+\Lambda)^{-1}x\right\|^2 d\lambda\right)\left(\int_0^\infty \lambda^{1/2}\left\|\Lambda^{1/4}(\lambda\mathrm{Id}+\Lambda)^{-1}y\right\|^2 d\lambda\right)$$

$$= \frac{c'^2}{\pi^2}\left\langle\int_0^\infty \lambda^{1/2}(\lambda\mathrm{Id}+\Lambda)^{-2}\Lambda^{(s+1)/2}xd\lambda,\Lambda^{s/2}x\right\rangle\left\langle\int_0^\infty \lambda^{1/2}(\lambda\mathrm{Id}+\Lambda)^{-2}\Lambda^{1/2}yd\lambda,y\right\rangle$$

$$= \frac{c'^2}{4}\left\langle\Lambda^{1/2}\Lambda^{(s-1)/2}x,\Lambda^{s/2}x\right\rangle\left\langle\Lambda^{1/2}\Lambda^{-1/2}y,y\right\rangle = \frac{c'^2}{4}\left\|\Lambda^{s/2}x\right\|^2\|y\|^2.$$

This shows $\left\|\Lambda^{s/2}[\Lambda^{1/2},D]x\right\| = \displaystyle\sup_{y\in H_\Lambda^\infty,\|y\|\leq 1}\left|\left\langle\Lambda^{s/2}[\Lambda^{1/2},D]x,y\right\rangle\right| \leq \dfrac{c'}{2}\left\|\Lambda^{s/2}x\right\|.$

Step 2: Apply step 1 with $D' := D\Lambda^{-m/2}$. □

2.3.7 Corollary. Assume that Λ is a strictly positive, selfadjoint operator in a complex Hilbert space H. Let $D : H_\Lambda^\infty \longrightarrow H_\Lambda^\infty$, let $r,m \in \mathbb{R}$, and assume that there are constants a_r,b_r,c_r, with

- $\left\|\Lambda^{r/2}Dx\right\| \leq a_r\left\|\Lambda^{(r+m)/2}x\right\|$ for $x \in H_\Lambda^\infty$,

- $\left\|\Lambda^{r/2}[\Lambda,D]x\right\| \leq b_r\left\|\Lambda^{(r+m+1)/2}x\right\|$ for $x \in H_\Lambda^\infty$,

- $\left\|\Lambda^{r/2}[\Lambda,[\Lambda,D]]x\right\| \leq c_r\left\|\Lambda^{(r+m+2)/2}x\right\|$ for $x \in H_\Lambda^\infty$.

Then there exists a $c_r' \geq 0$ with $\left\|\Lambda^{r/2}[\Lambda^{1/2},D]x\right\| \leq c_r'\left\|\Lambda^{(r+m)/2}x\right\|$ for $x \in H_\Lambda^\infty$.

PROOF: Let $m' = m+1, s = r, C = [\Lambda,D]$, and $t = \frac{1}{2}$, then 2.3.3 implies $\left\|\Lambda^{r/2-1/4}[\Lambda,D]x\right\| \leq \tilde{c}_r\left\|\Lambda^{(r+m)/2+1/4}x\right\|$. Hence 2.3.6 implies the assertion. □

Now we can use all these results to prove the following proposition. Its importance lies in the fact that one often chooses $Z = \sqrt{P}$ as a scale generating operator, where P is a suitable second-order differential operator (cf. e.g. sections 4.5 and 4.7). The following proposition shows how estimates for the non-local operator Z can be obtained from estimates for the differential operator P, which are much easier to verify.

2.3.8 Proposition. Let $Z : D(Z) \to H$ be a strictly positive, selfadjoint operator in a complex Hilbert space H. Moreover, let $A : H_Z^\infty \to H_Z^\infty$ such that for $k,j \in \mathbb{N}_0$ there are constants $a_{2k,j} \geq 0$ with

$$\left\|Z^{2k}\mathrm{ad}^j(Z^2)(A)x\right\| \leq a_{2k,j}\left\|Z^{2k+m+j}x\right\|$$

for $x \in H_Z^\infty, A \in \mathcal{A}$. Then, for $k \in \mathbb{Z}, j \in \mathbb{N}_0$ there are $c_{k,j} \geq 0$ with

$$\left\| Z^k \mathrm{ad}^j(Z)(A)x \right\| \leq c_{k,j} \left\| Z^{k+m}x \right\| \quad \text{for } x \in H_Z^\infty, A \in \mathcal{A}.$$

PROOF: We will show that for $k, l, j \in \mathbb{N}_0$ there are $b_{k,l,j} \geq 0$ with

$$\left\| Z^k \mathrm{ad}^l(Z^2)\mathrm{ad}^j(Z)(A)x \right\| \leq b_{k,l,j} \left\| Z^{k+m+l}x \right\| \quad \text{for } x \in H_Z^\infty. \quad (2.3.3)$$

1.5.10 proves (2.3.3) for $j = 0$. Now assume (2.3.3) inductively for a $j \in \mathbb{N}_0$. Then 2.3.7 with $\Lambda := Z^2, D := \mathrm{ad}^l(Z^2)\mathrm{ad}^j(Z)(A), r = k, m'' = m + l$ implies the existence of a constant $b_{k,l,j+1} \geq 0$ with

$$\left\| Z^k \left[Z, \mathrm{ad}^l(Z^2)\mathrm{ad}^j(Z)(A) \right] x \right\| \leq b_{k,l,j+1} \left\| Z^{k+m+l}x \right\| \quad \text{for } x \in H_Z^\infty, A \in \mathcal{A}.$$

This implies (2.3.3) because $\left[Z, \mathrm{ad}^l(Z^2)\mathrm{ad}^j(Z)(A) \right] x = \mathrm{ad}^l(Z^2)\mathrm{ad}^{j+1}(Z)(A)x$. In particular, (2.3.3) with $l = 0$ implies the assertion for $k \in \mathbb{N}_0$. The assertion for $-k \in \mathbb{N}$ follows with 2.3.5. $\qquad\square$

Combining this result with 2.3.1 and 2.3.2 we obtain the following propositions.

2.3.9 Proposition. Let $Z : D(Z) \to H$ be a strictly positive, selfadjoint operator in a complex Hilbert space H, and let

$$\mathcal{A} \subset \bigcap_{k \in \mathbb{N}_0} \mathscr{L}(H_Z^{k+m}, H_Z^k)$$

be a family of operators such that for $k, j \in \mathbb{N}_0$ there are $\beta \geq 0$ and $a_{2k,j} \geq 0$ with

$$\left\| Z^{2k}(\mathrm{ad}^j(Z^2)(A)x) \right\| \leq a_{2k,j} \left\| x \right\|_{2k+j+1} \quad \text{and} \quad \mathrm{Re}\,\langle Ax, x \rangle \leq \beta \left\| x \right\|^2$$

for $x \in H_Z^\infty, A \in \mathcal{A}$. Then, for $k \in \mathbb{N}_0$ there are $\beta_k \geq 0$ with

$$\mathrm{Re}\,\langle Ax, x \rangle_k \leq \beta_k \left\| x \right\|_k^2 \quad \text{for } x \in H_Z^\infty, A \in \mathcal{A}.$$

PROOF: Due to 2.3.8 for $k \in \mathbb{N}_0$ there are $c_{k,1} \geq 0$ with

$$\left\| Z^k(ZA - AZ)x \right\| = \left\| Z^k \mathrm{ad}(Z)(A)x \right\| \leq c_{k,1} \left\| x \right\|_{k+1}.$$

Hence 2.3.1 implies the assertion. $\qquad\square$

2.3.10 Proposition. Let $Z : D(Z) \to H$ be a strictly positive, selfadjoint operator in a complex Hilbert space H, and let

$$\mathcal{A} \subset \bigcap_{k \in \mathbb{N}_0} \mathscr{L}(H_Z^{k+m}, H_Z^k)$$

be a family of symmetric operators in H on H_Z^m such that for $k, j \in \mathbb{N}_0$ there are $\beta \geq 0, a_{2k,j} \geq 0$ with

$$\left\| Z^{2k}(\text{ad}^j(Z^2)(A)x) \right\| \leq a_{2k,j} \left\| x \right\|_{2k+j+2} \quad \text{and} \quad \text{Re}\,\langle Ax, x \rangle \leq \beta \left\| x \right\|^2$$

for $x \in H_Z^\infty, A \in \mathcal{A}$. Then, for $k \in \mathbb{N}_0$ there are $\beta_k \geq 0$ with

$$\text{Re}\,\langle Ax, x \rangle_k \leq \beta_k \left\| x \right\|_k^2 \quad \text{for } x \in H_Z^\infty, A \in \mathcal{A}.$$

PROOF: For $k \in \mathbb{Z}$ there are $c_{k,2} \geq 0$ with $\left\| Z^k \text{ad}^2(Z)(A)x \right\| \leq c_{k,2} \left\| Z^{k+2} x \right\|$ (due to 2.3.8). Hence

$$\left| \langle [Z^k, [Z^k, A]]x, x \rangle \right| = \left| \langle Z^{-k}[Z^k, [Z^k, A]]x, Z^k x \rangle \right|$$

$$\leq c \sum_{j,l=0}^{k-1} \left| \langle Z^{-k+(j+l)} \text{ad}^2(Z)(A) Z^{2(k-1)-(j+l)} x, Z^k x \rangle \right| \leq c \max_{\mu=-k}^{k} c_{\mu,2} \left\| Z^k x \right\|^2$$

Hence 2.3.2 implies the assertion. □

We conclude this section with some remarks on C^∞-elements.

2.3.11 Proposition. Let Λ be an operator of positive type in a complex Banach space X, and let

$$\Psi_\varepsilon^k := \left\{ A \in \mathscr{L}(X) : A(D(\Lambda^\infty)) \subset D(\Lambda^\infty) \text{ and } \left\| \text{ad}^j(\Lambda^\varepsilon)(A)x \right\| \leq a_j \left\| x \right\| \right.$$
$$\left. \text{for } x \in D(\Lambda^\infty), 0 \leq j \leq k, \text{ and suitable } a_j \geq 0 \right\}$$

for $0 < \varepsilon \leq 1, k \in \mathbb{N}_0$. Then $\Psi_{\varepsilon'}^k \subset \Psi_\varepsilon^k$ for $k \in \mathbb{N}_0$ and $0 < \varepsilon < \varepsilon' \leq 1$.

PROOF: Since we can substitute Λ by $\Lambda^{\varepsilon'}$ (due to 1.4.7) we only have to prove the assertion for $\varepsilon' = 1$ and $0 < \varepsilon < 1$. First, it is easy to show inductively for $k \in \mathbb{N}_0$ that

$$[\Lambda, \text{ad}^k(\Lambda^\varepsilon)(A)]x = \text{ad}^k(\Lambda^\varepsilon)([\Lambda, A])x \quad \text{for } A \in \Psi_1^k, x \in D(\Lambda^\infty). \quad (2.3.4)$$

Assume the assertion now for a $k \in \mathbb{N}_0$ and let $A \in \Psi_1^{k+1}$, then

$$\Lambda(\lambda \text{Id} + \Lambda)^{-1} = (-\lambda \text{Id} + (\lambda \text{Id} + \Lambda))(\lambda \text{Id} + \Lambda)^{-1} = -\lambda(\lambda \text{Id} + \Lambda)^{-1} + \text{Id} \quad (2.3.5)$$

and 1.4.7 implies

$$\left\|\mathrm{ad}^{k+1}(\Lambda^\varepsilon)(A)x\right\| = \left\|[\Lambda^\varepsilon, \mathrm{ad}^k(\Lambda^\varepsilon)(A)]x\right\|$$

$$\leq \frac{\sin(\varepsilon\pi)}{\pi} \int_0^\infty \lambda^{\varepsilon-1} \left\|[\Lambda(\lambda\mathrm{Id}+\Lambda)^{-1}, \mathrm{ad}^k(\Lambda^\varepsilon)(A)]x\right\| d\lambda$$

$$\underset{(2.3.5)}{=} \frac{\sin(\varepsilon\pi)}{\pi} \int_0^\infty \lambda^\varepsilon \left\|[(\lambda\mathrm{Id}+\Lambda)^{-1}, \mathrm{ad}^k(\Lambda^\varepsilon)(A)]x\right\| d\lambda$$

$$= \frac{\sin(\varepsilon\pi)}{\pi} \int_0^\infty \lambda^\varepsilon \left\|(\lambda\mathrm{Id}+\Lambda)^{-1}[\Lambda, \mathrm{ad}^k(\Lambda^\varepsilon)(A)](\lambda\mathrm{Id}+\Lambda)^{-1}x\right\| d\lambda$$

$$\underset{(2.3.4)}{\leq} \frac{\sin(\varepsilon\pi)}{\pi} M \int_0^\infty \frac{\lambda^\varepsilon}{1+\lambda} \left\|\underbrace{\mathrm{ad}^k(\Lambda^\varepsilon)([\Lambda, A])}_{\in \Psi_1^k}(\lambda\mathrm{Id}+\Lambda)^{-1}x\right\| d\lambda$$

$$\leq \frac{\sin(\varepsilon\pi)}{\pi} M \int_0^\infty \frac{\lambda^\varepsilon}{1+\lambda} \widetilde{a}_k \left\|(\lambda\mathrm{Id}+\Lambda)^{-1}x\right\| d\lambda$$

$$\leq \frac{\sin(\varepsilon\pi)}{\pi} M^2 \widetilde{a}_k \int_0^\infty \frac{\lambda^\varepsilon}{1+\lambda}\frac{1}{1+\lambda} d\lambda \|x\| \leq a_{k+1}\|x\|$$

for a suitable $a_{k+1} \geq 0$. □

2.3.12 Corollary. Let Λ be a strictly positive, selfadjoint operator in a Hilbert space H. Let

$$\Psi_\varepsilon^\infty := \{A \in \mathscr{L}(X) : \mathbf{R} \ni t \mapsto e^{it\Lambda^\varepsilon} A e^{-it\Lambda^\varepsilon} \in \mathscr{L}(X) \text{ is smooth}\} \quad \text{for } \varepsilon > 0.$$

Then $\Psi_{\varepsilon'}^\infty \subset \Psi_\varepsilon^\infty$ for $0 < \varepsilon < \varepsilon'$.

PROOF: We can assume $\varepsilon' = 1$. Then the assertion is implied by 2.3.11 and 2.2.16 with $Z = i\Lambda^\varepsilon$. □

2.3.13 Bibliographical remarks. The proof of 2.3.6 is motivated from a result due to Kato, cf. Massey [106, Proposition 1]). For more information on C^∞-elements we refer to Gramsch/Kalb [55], Gramsch/Ueberberg/Wagner [56], Lauter [101], Cordes [32] and the references given therein.

2.4 Characterization of well-posedness of the Cauchy problem

Now we can formulate the main theorems of this chapter and give character-izations of well-posedness of the Cauchy problem in scales of Banach spaces generated by one or a family of special closed operators in several settings. Their proofs will be given in section 2.5. We find a correspondence between the order of the operators $A(t)$ we can consider in the scale and the quality of the scale generating operator. General speaking, if we want to consider opera-tors $A(t)$ of higher order with respect to the scale generating operators, these should have better properties as generators. We start with scales generated by quasi-contractive C_0-semigroups. Here we have to assume that the operators $A(t)$ are relatively bounded with bound 0 with respect to this generator.

2.4.1 Theorem. Suppose that $\Lambda : D(\Lambda) \to X$ generates a quasi-contractive C_0-semigroup in a Banach space X. Moreover, assume that

- $I \ni t \mapsto A(t) \in \bigcap_{k \in \mathbb{N}_0} \mathscr{L}(X_\Lambda^{k+1}, X_\Lambda^k)$ is (uniformly) continuous.

- $A(t)$ is relatively bounded by Λ with bound 0 in X_Λ^k for any $k \in \mathbb{N}_0$, i.e., for $\varepsilon > 0, k \in \mathbb{N}_0$ there are $\beta_{\varepsilon,k} \geq 0$ with $\|A(t)x\|_k \leq \varepsilon \|\Lambda x\|_k + \beta_{\varepsilon,k} \|x\|_k$ for $t \in I$ and $x \in X_\Lambda^{k+1}$.

Then the following statements are equivalent:

(I) The Cauchy problem for $A(t) \in \mathscr{L}(X_\Lambda^{k+1}, X_\Lambda^k)$ in the scale $(X_\Lambda^k)_k$ is well-posed with exponential growth.

(II) For $k \in \mathbb{N}_0$ there are $\beta_k \geq 0$ with $\|\lambda x - A(t)x\|_k \geq (\lambda - \beta_k) \|x\|_k$ for $x \in X_\Lambda^\infty, \lambda > \beta_k, t \in I$.

2.4.2 Remark. In 2.4.1 each of the following two conditions implies that $A(t) \in \mathscr{L}(X_\Lambda^{k+1}, X_\Lambda^k)$ is relatively bounded by Λ with bound 0 in X_Λ^k for any $k \in \mathbb{N}_0$:

- (II) is satisfied and for $k \in \mathbb{N}_0$ $A(t)^2 \in \mathscr{L}(X_\Lambda^{k+1}, X_\Lambda^k)$ boundedly for $t \in I$.

- For $k \in \mathbb{N}_0$ there is a Banach space Y_k with $X_\Lambda^{k+1} \hookrightarrow Y_k$ compactly, $Y_k \hookrightarrow X_\Lambda^k$ continuously, and $A(t) \in \mathscr{L}(Y_k, X_\Lambda^k)$ boundedly for $t \in I$.

PROOF: The first assertion follows because $A(t) - \beta_k$ is dissipative in X_Λ^k and

$$
\begin{aligned}
\varepsilon \, \|A(t)x\|_k &\leq \|x + \varepsilon(A(t) - \beta_k)x\|_k + (1 + \varepsilon\beta_k)\,\|x\|_k \\
&\leq \|(\mathrm{Id} - \varepsilon(A(t) - \beta_k))(x + \varepsilon(A(t) - \beta_k)x)\|_k + (1 + \varepsilon\beta_k)\,\|x\|_k \\
&= \|x - \varepsilon^2(A(t) - \beta_k)^2 x\|_k + (1 + \varepsilon\beta_k)\,\|x\|_k \\
&\leq \varepsilon^2 \left(\|A(t)^2 x\|_k + 2\beta_k\,\|A(t)x\|_k + \beta_k^2\,\|x\|_k \right) + (2 + \varepsilon\beta_k)\,\|x\|_k \\
&\leq \varepsilon^2 (a_k + 2\beta_k a_k)\,\|x\|_{k+1} + (2 + \varepsilon\beta_k + \varepsilon^2\beta_k^2)\,\|x\|_k \\
&\leq \varepsilon^2 (a_k + 2\beta_k a_k)\,\|\Lambda x\|_k + c'_{k,\varepsilon}\,\|x\|_k \; .
\end{aligned}
$$

To prove the second assertion it is enough to show that for any $\varepsilon > 0$ there is a $c_\varepsilon > 0$ with $\|x\|_{Y_k} \leq \varepsilon\,\|x\|_{k+1} + c_\varepsilon\,\|x\|_k$ for $x \in X_\Lambda^{k+1}$. Assume that there are an $\varepsilon > 0$ and $\tilde{x}_j \in X_\Lambda^{k+1}$ with $\|\tilde{x}_j\|_{Y_k} > \varepsilon\,\|\tilde{x}_j\|_{k+1} + j\,\|\tilde{x}_j\|_k$. Then $(x_j)_j := (\frac{\tilde{x}_j}{\|\tilde{x}_j\|_{k+1}})_j \subset X_\Lambda^{k+1}$ boundedly and $\|x_j\|_{Y_k} > \varepsilon + j\,\|x_j\|_k$. Due to the compactness of $X_\Lambda^{k+1} \hookrightarrow Y_k$ we can assume that $x_j \to y \in Y_k$ in Y_k for $j \to \infty$, hence $\|y\|_k \xleftarrow[\infty \leftarrow j]{} \|x_j\|_k \leq \frac{1}{j}\,\|x_j\|_{Y_k} \xrightarrow[j \to \infty]{} 0$. Thus $y = 0$ and $\varepsilon \leq \|x_j\|_{Y_k} \xrightarrow[j \to \infty]{} 0$, which is a contradiction. $\square$

The next result considers operators $A(t)$ of order greater or equal 1 with respect to the scale generating operator, but now we have to assume that this operator generates a bounded analytic semigroup with an angle of holomorphy depending on the order of the operators $A(t)$.

2.4.3 Theorem.   Suppose that $\Lambda : D(\Lambda) \to X$ generates a bounded analytic semigroup with angle $\theta > \frac{m}{m+1}\frac{\pi}{2}, m \in \mathbb{N}$, in a complex Banach space X. Moreover, assume that

- $I \ni t \mapsto A(t) \in \bigcap_{k \in \mathbb{N}_0} \mathscr{L}(X_\Lambda^{k+m}, X_\Lambda^k)$ is strongly continuous.

- for $k \in \mathbb{N}_0$ there are $d_k \geq 0$ with $\|[\Lambda, A(t)]x\|_k \leq d_k\,\|x\|_{k+1}, x \in X_\Lambda^\infty, t \in I$.

Then the following statements are equivalent:

(I) The Cauchy problem for $A(t) \in \mathscr{L}(X_\Lambda^{k+m}, X_\Lambda^k)$ in the scale $(X_\Lambda^k)_k$ is well-posed with exponential growth.

(II) There is a $\beta \geq 0$ with $\|\lambda x - A(t)x\| \geq (\lambda - \beta)\,\|x\|$ for $x \in X_\Lambda^\infty, \lambda > \beta, t \in I$.

2.4.4 Remark. Many generators of bounded analytic semigroups generate semigroups with an angle of holomorphy of $\frac{\pi}{2}$ (cf. [62], [119]) and hence arbitrary values of m are allowed in theorem 2.4.3. Moreover, symmetric Markov semigroups in L^2 can always be extended to bounded analytic semigroups in L^p and an estimate for the sector of holomorphy is available (cf. [105], [153]).

The next type of scales we consider are generated by commuting families of isometric C_0-groups. This case will be reduced to 2.4.3 and no order restriction will be needed.

2.4.5 Theorem. Suppose that $Z = \{Z_1, \ldots, Z_M\}$ is a commuting family of infinitesimal generators of isometric C_0-groups in a complex Banach space X, $m \in \mathbb{N}$. Moreover, assume that

- there exist $c_1, c_2 \geq 0$ with $c_1 \|x\|_2 \leq \|x\| + \|\Lambda x\| \leq c_2 \|x\|_2$ for $x \in X_Z^2$,
  where $\Lambda := \displaystyle\sum_{j=1}^{M} Z_j^2$.

- $I \ni t \mapsto A(t) \in \displaystyle\bigcap_{k \in \mathbb{N}_0} \mathscr{L}(X_Z^{k+m}, X_Z^k)$ is strongly continuous.

- for $k \in \mathbb{N}_0$ there exist $d_k \geq 0$ with $\left\| [Z_j, A(t)]x \right\|_k \leq d_k \|x\|_{k+1}$, $x \in X_Z^\infty$, $t \in I$, $j = 1, \ldots, M$.

Then the following two statements are equivalent:

(I) The Cauchy problem for $A(t) \in \mathscr{L}(X_Z^{k+m}, X_Z^k)$ in the scale $(X_Z^k)_k$ is well-posed with exponential growth.

(II) There is a $\beta \geq 0$ with $\|\lambda x - A(t)x\| \geq (\lambda - \beta) \|x\|$ for $x \in X_Z^\infty$, $\lambda > \beta, t \in I$.

If one considers scales of Hilbert spaces, this last theorem can be improved, in particular commutator estimates are no longer necessary, which is crucial for some applications (cf. section 4.2). Moreover, we can substitute the dissipativity assumption (II) in Hilbert spaces by an equivalent estimate with scalar products (cf. 1.1.7). This estimate is much easier to check in applications.

2.4.6 Theorem. Suppose that $Z = \{Z_1, \ldots, Z_M\}$ is a commuting family of infinitesimal generators of C_0-semigroups of contractions in a Hilbert space H. Moreover, suppose $Z_j^* = \alpha_j Z_j$ for $j = 1, \ldots, M$ and suitable $\alpha_j \neq 0$. Let $m \in \mathbb{N}$ and

$$I \ni t \mapsto A(t) \in \bigcap_{k \in \mathbb{N}_0} \mathscr{L}(H_Z^{k+m}, H_Z^k)$$

be strongly continuous. Then the following two statements are equivalent:

(I) The Cauchy problem for $A(t) \in \mathscr{L}(H_Z^{k+m}, H_Z^k)$ in the scale $(H_Z^k)_k$ is well-posed with exponential growth.

(II) For $k \in \mathbb{N}_0$ there is a $\beta_k \geq 0$ with $\mathrm{Re}\, \langle A(t)x, x \rangle_k \leq \beta_k \|x\|_k^2$ for $x \in H_Z^\infty$, $t \in I$.

2.4.7 Remark. Note that (in contrast to semigroup theorems) condition (II) in 2.4.1, 2.4.3, 2.4.5, 2.4.6 is "additive" (due to 1.1.7), i.e., if the Cauchy problem for $A_1(t)$ and $A_2(t)$ is well-posed with exponential growth, so it is for $A_1(t) + A_2(t)$. Thus, in applications it is sufficient to check the assumptions for each term of an evolution equation separately.

Theorems 2.4.1, 2.4.3, 2.4.5, and 2.4.6 will be proved in section 2.5 (cf. 2.5.3, 2.5.7, 2.5.14, 2.5.2). For the proof of the Hilbert space result 2.4.6 we will use a functional analytic "mollifier" argument, whereas for the proofs of the Banach space results we use a functional analytic parabolic regularization technique motivated from [132], [13], [131]. More precisely, in this case we apply Kato's theory for time-dependent evolution equations and perturbation arguments to solve the Cauchy problem for $-\varepsilon(-\Lambda)^q + A(t)$ and show that the corresponding solutions converge for $\varepsilon \to 0$.

As remarked in 2.1.12 these theorems have an immediate consequence to the question whether an operator A in a scale of Banach spaces is the pregenerator of a C_0-semigroup. We obtain:

2.4.8 Corollary. Suppose that $A = A(t)$ satisfies the assumptions of 2.4.1, 2.4.3, 2.4.5, or 2.4.6 and that the Cauchy problem for A is well-posed in the scale $(X_Z^k)_k$. Then the closure of $A|_{X_Z^{k+m}}$ in X_Z^k generates a C_0-semigroup in X_Z^k for any $k \in \mathbb{N}_0$.

2.4.9 Bibliographical remarks. 2.4.8 is connected to results of Robinson [132], [131], [130] and Batty and Robinson [13] who gave several conditions for being a pregenerator using mapping properties of operators in a scale generated by a reference operator. They used weak regularity conditions (that is, they considered operators A that operate only in a "small" part of the scale) but they had to use strong additional assumptions like multi-commutator estimates or relative boundedness. We assume always the mapping properties of A to hold in the complete scale (which seems natural for pseudodifferential operators) but we have simpler additional assumptions. More precisely, in 2.4.1 resp., 2.4.6 we do not use any commutator estimate in contrast to [131, 2.1] resp., [130]. In 2.4.3 we use much weaker commutator conditions than [132, 4.3]. Batty and Robinson used scales generated by only one generator of a C_0-group of isometries ([13, 2.1, 3.3], [132, 1.1]), whereas in 2.4.5 we obtain a related result also for scales that are generated by a commuting family of generators of C_0-groups of isometries. This enables applications to scales of L^p Sobolev spaces on $\mathbf{R}^n$. Note in particular that we treated time-dependent equations, whereas Batty and Robinson worked only in time-independent situations.

2.5 Sufficient conditions for well-posedness of the Cauchy problem

In this section we will prove the theorems of section 2.4 and give moreover sufficient conditions for solvability of the Cauchy problem for operators $A(t)$ that are only defined in a part of the scale. The necessary parts of the statements are consequences of 2.1.10, so we only have to prove the sufficient parts. To this end we will use always the following principle.

2.5.1 Proposition.
Assumptions:

(i) $(X^k, \|\cdot\|_k)_k$ is a scale of Banach spaces.

(ii) $K_1, K_2 \in \mathbb{N}_0, K_1 \leq K_2, m, m' \in \mathbb{N}, m' \geq m.$

(iii) $I \ni t \mapsto A(t) \in \bigcap\limits_{k=K_1}^{K_2+m'} \mathscr{L}(X^{k+m}, X^k), \ I \ni t \mapsto A_l(t) \in \bigcap\limits_{k=K_1}^{K_2+m'} \mathscr{L}(X^{k+m'}, X^k)$
are strongly continuous for $l \in \mathbb{N}$.

(iv) $A_l(t) \underset{l\to\infty}{\longrightarrow} A(t)$ uniformly for $t \in I$ in $\mathscr{L}(X^{k+m'}, X^k)$ for $K_1 \leq k \leq K_2+m'$.

(v) For $l \in \mathbb{N}, (t,s) \in \Delta$ there is a propagator $U_l : \Delta \to \bigcap\limits_{k=K_1}^{K_2+2m'} \mathscr{L}(X^k)$ with:

- $\Delta \ni (t,s) \mapsto U_l(t,s) \in \mathscr{L}(X^{k+m'}, X^k)$ is strongly continuously differentiable for $K_1 \leq k \leq K_2 + m'$ with strong derivatives

$$\frac{\partial}{\partial t}U_l(t,s) = A_l(t)U_l(t,s), \frac{\partial}{\partial s}U_l(t,s) = -U_l(t,s)A_l(s) \text{ for } (t,s) \in \Delta.$$

- For $K_1 \leq k \leq K_2 + 2m'$ there are $u_k \geq 0$ with $\|U_l(t,s)\|_{\mathscr{L}(X^k)} \leq u_k$.

Assertion:

There is a (unique) propagator $U : \Delta \to \bigcap\limits_{k=K_1}^{K_2+m'} \mathscr{L}(X^k)$ with

(a) $\Delta \ni (t,s) \mapsto U(t,s) \in \mathscr{L}(X^{k+m}, X^k)$ is strongly continuously differentiable for $K_1 \leq k \leq K_2$ with strong derivatives

$$\frac{\partial}{\partial t}U(t,s) = A(t)U(t,s), \quad \frac{\partial}{\partial s}U(t,s) = -U(t,s)A(s) \quad \text{for } (t,s) \in \Delta.$$

(b) $\|U(t,s)\|_{\mathscr{L}(X^k)} \leq u_k \quad$ for $(t,s) \in \Delta, K_1 \leq k \leq K_2 + m'$.

PROOF: Fix $K_1 \leq k \leq K_2 + m'$ and $\varepsilon > 0$. For $l,l' \in \mathbb{N}, (t,s) \in \Delta, y \in X^{k+m}$

$$\|U_l(t,s)y - U_{l'}(t,s)y\|_k = \left\| \int_s^t \frac{\partial}{\partial \tau}\Big(U_{l'}(t,\tau)U_l(\tau,s)y\Big)d\tau \right\|_k$$

$$= \left\| \int_s^t U_{l'}(t,\tau)\Big(A_l(\tau) - A_{l'}(\tau)\Big)U_l(\tau,s)y\,d\tau \right\|_k$$

$$\leq \int_s^t u_k \|A_l(\tau) - A_{l'}(\tau)\|_{\mathscr{L}(X^{k+m'},X^k)} u_{k+m'}\|y\|_{k+m'}d\tau$$

$$\leq u_k u_{k+m'}(t-s)\sup_{\tau \in I} \|A_l(\tau) - A_{l'}(\tau)\|_{\mathscr{L}(X^{k+m'},X^k)} \|y\|_{k+m'} .$$

Thus there is an $L \in \mathbb{N}$ with $\|U_l(t,s)y - U_{l'}(t,s)y\|_k < \varepsilon$ $(t,s) \in \Delta, l,l' \geq L$. Hence $\Big(U_l(\cdot,\cdot)x\Big)_{l\in\mathbb{N}} \subset C(\Delta, X^k)$ is a Cauchy sequence for any $x \in X^k$ due to 1.1.3. Define

$$U(t,s)x := \lim_{l\to\infty} U_l(t,s)x \text{ for } (t,s) \in \Delta.$$

Then $U(t,s) \in \mathscr{L}(X^k)$ for $(t,s) \in \Delta$ by the principle of uniform boundedness and $U_l(t,s) \xrightarrow[l\to\infty]{} U(t,s)$ strongly uniformly in $\mathscr{L}(X^k)$ for $(t,s) \in \Delta$. Therefore we have proved (b). Furthermore, for $t_0 \le s \le q \le t \le t_1$,

$$U(t,q)U(q,s)x \xleftarrow[\infty\leftarrow l]{X^k} U_l(t,q)U_l(q,s)x = U_l(t,s)x \xrightarrow[l\to\infty]{X^k} U(t,s)x,$$

which implies that U is a propagator. Now assume $K_1 \le k \le K_2$, then for $x \in X^{k+m'}$

$$\frac{\partial}{\partial t}\Big(U_l(t,s)x\Big) = A_l(t)U_l(t,s)x \xrightarrow[l\to\infty]{X^k} A(t)U(t,s)x$$

$$\frac{\partial}{\partial s}\Big(U_l(t,s)x\Big) = -U_l(t,s)A_l(s)x \xrightarrow[l\to\infty]{X^k} -U(t,s)A(s)x$$

uniformly for $(t,s) \in \Delta$. Therefore $\Delta \ni (t,s) \mapsto U(t,s)x \in X^k$ is differentiable with derivatives $\frac{\partial}{\partial t}U(t,s)x = A(t)U(t,s)x$ and $\frac{\partial}{\partial s}U(t,s)x = -U(t,s)A(s)x$. Hence for $t_0 \le s \le t \le t_1, s \le t+h \le t_1, h \neq 0, x \in X^{k+m'}$ with suitable $a_k \ge 0$ we have

$$\left\| \frac{U(t+h,s)x - U(t,s)x}{h} \right\|_k = \left\| \frac{1}{h}\int_t^{t+h} A(\tau)U(\tau,s)x\,d\tau \right\|_k \le a_k u_{k+m}\|x\|_{k+m}.$$

Thus $\left\{ \frac{U(t+h,s)-U(t,s)}{h} : t_0 \le s \le t \le t_1, s \le t+h \le t_1, h \neq 0 \right\} \subset \mathscr{L}(X^{k+m}, X^k)$ boundedly and therefore $[s,t_1] \ni t \mapsto U(t,s)x \in X^k$ is differentiable. Similarly, $[t_0,t] \ni s \mapsto U(t,s)x \in X^k$ is also differentiable for $x \in X^{k+m}$, and the strong continuity of U implies (a). □

First we will prove 2.4.6. We will show the following more precise statement that gives sufficient conditions for well-posedness for operators $A(t)$ that are only defined in a part of the scale.

2.5.2 Theorem.
Assumptions:

(i) $Z = \{Z_1, \ldots, Z_M\}$ is a commuting family of infinitesimal generators of C_0-semigroups of contractions in a Hilbert space H such that $Z_j^* = \alpha_j Z_j$ with suitable $\alpha_j \neq 0$.

(ii) $m \in \mathbb{N}, K_1, K_2 \in \mathbb{N}_0, K_1 \le K_2$, are fixed.

(iii) $I \ni t \mapsto A(t) \in \bigcap_{k=K_1}^{K_2+2m+2} \mathscr{L}(H_Z^{k+m}, H_Z^k)$ is strongly continuous.

(iv) For $K_1 \le k \le K_2 + 2m + 2$ there are $\beta_k \ge 0$ with

$$\mathrm{Re}\,\langle A(t)x, x\rangle_k \le \beta_k \|x\|_k^2 \quad \text{for } t \in I, x \in H_Z^{k+m}.$$

Assertion:

There is a (unique) propagator $U(t,s) : \Delta \rightarrow \bigcap\limits_{k=K_1}^{K_2+m+1} \mathcal{L}(H_Z^k)$ with:

(a) $\Delta \ni (t,s) \mapsto U(t,s) \in \mathcal{L}(H_Z^{k+m}, H_Z^k)$ is strongly continuously differentiable for $K_1 \leq k \leq K_2$ with strong derivatives

$$\frac{\partial}{\partial t}U(t,s) = A(t)U(t,s), \quad \frac{\partial}{\partial s}U(t,s) = -U(t,s)A(s) \quad \text{for } (t,s) \in \Delta.$$

(b) For $K_1 \leq k \leq K_2 + m + 1, (t,s) \in \Delta$ we have $\|U(t,s)\|_{\mathcal{L}(H_Z^k)} \leq e^{\beta_k(t-s)}$.

PROOF: Let J_l, J_l^* be as in 2.2.11, $m' := m+1$, and $A_l(t) := (J_l^*)^m A(t) (J_l)^m$. Then, due to 2.5.1, we have to show that

$\bullet$ $I \ni t \mapsto A_l(t) \in \bigcap\limits_{k=K_1}^{K_2+m'} \mathcal{L}(H_Z^{k+m'}, H_Z^k), l \in \mathbb{N}$ is strongly continuous.

$\bullet$ $A_l(t) \xrightarrow[l \to \infty]{} A(t)$ uniformly for $t \in I$ in $\mathcal{L}(H_Z^{k+m'}, H_Z^k), K_1 \leq k \leq K_2 + m'$.

$\bullet$ for $l \in \mathbb{N}$ there is a propagator $U_l : \Delta \rightarrow \bigcap\limits_{k=K_1}^{K_2+2m'} \mathcal{L}(H_Z^k)$ with:

(i) $\Delta \ni (t,s) \mapsto U_l(t,s) \in \mathcal{L}(H_Z^{k+m'}, H_Z^k)$ is strongly continuously differentiable for $K_1 \leq k \leq K_2 + m'$ with strong derivatives

$$\frac{\partial}{\partial t}U_l(t,s) = A_l(t)U_l(t,s), \frac{\partial}{\partial s}U_l(t,s) = -U_l(t,s)A_l(s) \text{ for } (t,s) \in \Delta.$$

(ii) For $K_1 \leq k \leq K_2 + 2m'$ there are $u_k \geq 0$ with $\|U_l(t,s)\|_{\mathcal{L}(H_Z^k)} \leq u_k$ for $(t,s) \in \Delta, l \in \mathbb{N}$.

Proof: Due to 2.2.11 clearly $I \ni t \mapsto A_l(t) \in \mathcal{L}(H_Z^{k+m'}, H_Z^k)$ is strongly continuous and $A_l(t) \xrightarrow[l \to \infty]{} A(t)$ uniformly for $t \in I$ in $\mathcal{L}(H_Z^{k+m'}, H_Z^k)$ for $K_1 \leq k \leq K_2 + m'$. Since $I \ni t \mapsto A_l(t) \in \mathcal{L}(H_Z^k)$ is strongly continuous for $K_1 \leq k \leq K_2 + 2m'$, the theory of linear differential equations in Banach spaces (cf. [96, II, §2]) implies the existence of a propagator $U_l : \Delta \rightarrow \mathcal{L}(H_Z^k)$ for $l \in \mathbb{N}, K_1 \leq k \leq K_2 + 2m'$ such that $I^2 \ni (t,s) \mapsto U_l(t,s) \in \mathcal{L}(H_Z^k)$ is strongly continuously differentiable with strong derivatives

$$\frac{\partial}{\partial t}U_l(t,s) = A_l(t)U_l(t,s), \quad \frac{\partial}{\partial s}U_l(t,s) = -U_l(t,s)A_l(s) \text{ for } t, s \in I.$$

Finally, for $K_1 \leq k \leq K_2 + 2m'$ and $x \in H_Z^k$

$$\operatorname{Re} \langle A_l(t)x, x \rangle_k = \operatorname{Re} \langle A(t) (J_l)^m x, (J_l)^m x \rangle_k \leq \beta_k \, \|(J_l)^m x\|_k^2 \leq \beta_k \, \|x\|_k^2 \ .$$

Thus $\|U_l(t,s)\|_{\mathscr{L}(H_Z^k)} \leq e^{\beta_k(t-s)} \leq u_k$ for $(t,s) \in \Delta, K_1 \leq k \leq K_2 + 2m', l \in \mathbb{N}$ due to 1.1.10, which implies the assertion. $\qquad\square$

Now we prove the Banach space results. The main idea in the proof of 2.4.1, 2.4.3, and 2.4.5 is to apply 2.5.1 using a perturbation result with regularized operators $A_l(t) = -\frac{1}{l}(-\Lambda)^q + A(t)$ with a suitable q. In the situation of 2.4.1 this can be done rather easily with $q = 1$ using 1.6.4 and the perturbation result 1.1.15.

2.5.3 Theorem.
Assumptions:

(i) Let $\Lambda : D(\Lambda) \to X$ be the generator of a quasi-contractive C_0-semigroup in a Banach space X, $m \in \mathbb{N}, K_1, K_2 \in \mathbb{N}_0, K_1 \leq K_2$, be fixed.

(ii) $I \ni t \mapsto A(t) \in \bigcap\limits_{k=K_1}^{K_2+3} \mathscr{L}(X_\Lambda^{k+1}, X_\Lambda^k)$ is continuous (in the uniform operator topology).

(iii) $A(t)$ is relatively bounded with bound 0 by Λ in X_Λ^k for $K_1 \leq k \leq K_2 + 3$, i.e., for any $\varepsilon > 0$ and $K_1 \leq k \leq K_2 + 3$ there is a $\beta_{\varepsilon,k} \geq 0$ with $\|A(t)x\|_k \leq \varepsilon \|\Lambda x\|_k + \beta_{\varepsilon,k} \|x\|_k$ for $t \in I$ and $x \in X_\Lambda^{k+1}$.

(iv) For $K_1 \leq k \leq K_2 + 3$ there are $\beta_k \geq 0$ with $\|\lambda x - A(t)x\|_k \geq (\lambda - \beta_k) \|x\|_k$ for $x \in X_\Lambda^\infty, \lambda > \beta_k, t \in I$.

Assertion:

There is a (unique) propagator $U : \Delta \to \bigcap\limits_{k=K_1}^{K_2+1} \mathscr{L}(X_\Lambda^k)$ with:

(a) $\Delta \ni (t,s) \mapsto U(t,s) \in \mathscr{L}(X_\Lambda^{k+1}, X_\Lambda^k)$ is strongly continuously differentiable for $K_1 \leq k \leq K_2$ with strong derivatives

$$\frac{\partial}{\partial t}U(t,s) = A(t)U(t,s), \quad \frac{\partial}{\partial s}U(t,s) = -U(t,s)A(s) \quad \text{for } (t,s) \in \Delta.$$

(b) For $K_1 \leq k \leq K_2 + 1$ we have $\|U(t,s)\|_{\mathscr{L}(X_\Lambda^k)} \leq e^{\beta_k(t-s)}$ for $(t,s) \in \Delta$.

PROOF: Let $A_l(t) := \frac{1}{l}\Lambda + A(t), m' = 1$. Clearly, we have (iii), (iv) of 2.5.1. We only have to verify (v) of 2.5.1, then we have proved the assertion. Due to 1.1.15 $(A_l(t), X_\Lambda^{k+1})$ generates a quasi-contractive C_0-semigroup $(T_{l,t}(\tau))_{\tau \geq 0} \subset \mathscr{L}(X_\Lambda^k)$ for $K_1 \leq k \leq K_2 + 3$. Hence assertion 1.6.4(b) is satisfied, and 1.6.4 implies the assertion. $\square$

For the proof of 2.4.3 the application of the perturbation result 1.1.15 is much more complicated than in the case of 2.4.1 because we have to use powers of the scale generating operator in order to obtain relative boundedness of $A(t)$ with respect to this power. But semigroups generated by powers $-(-\Lambda)^n$ of generators of analytic semigroups in many cases fail to be quasi-contractive on $[0, \infty)$, even if this is the case for the semigroup generated by Λ. Thus the perturbation result 1.1.15 cannot be applied directly. Moreover, there is also a perturbation result for analytic semigroups (cf. [52, 1.6.6]), but the growth of the perturbed semigroup is difficult to control. So we decided to apply nevertheless the perturbation result for quasi-contractive semigroups. To this end we have to renormalize the spaces in such a way that the semigroups generated by $-(-\Lambda)^n$ turn out to be quasi-contractive. But now the quasi-dissipativity of $A(t)$ creates a second problem: Quasi-dissipativity clearly is not invariant under renormalizations. However, using commutator estimates between Λ and $A(t)$ we will show that in our situation the $A(t)$ remain quasi-dissipative and finally the application of 1.1.15 will be possible. Then we can conclude similarly to 2.5.3.

2.5.4 Corollary. Let $\Lambda : D(\Lambda) \to X$ be the generator of a bounded analytic semigroup with angle $\theta > \frac{m}{m+1}\frac{\pi}{2}$ with an $m \in \mathbb{N}$ in a complex Banach space X. Then $-(-\Lambda)^{m+1} : X_\Lambda^{k+m+1} \to X_\Lambda^k$ is sectorial in X_Λ^k for $k \in \mathbb{N}_0$ with angle

$$\theta_m := \pi - (m+1)\left(\pi - \left(\theta + \frac{\pi}{2}\right)\right) > \pi - \frac{m+1}{2}\pi + (m+1)\frac{m}{m+1}\frac{\pi}{2} = \frac{\pi}{2}.$$

Moreover, for $0 < \theta' < \theta_m$ there is an $M_{m,\theta'} > 0$ with

$$\left\|\Lambda^j(\lambda\mathrm{Id} + (-\Lambda)^{m+1})^{-1}\right\|_{\mathscr{L}(X_\Lambda^k)} \leq \frac{M_{m,\theta'}}{|\lambda|^{\frac{m+1-j}{m+1}}}, \qquad \text{for } 0 \leq j \leq m+1.$$

In particular, $-(-\Lambda)^{m+1} : X_\Lambda^{k+m+1} \to X_\Lambda^k$ generates a bounded analytic semigroup $(e^{-z(-\Lambda)^{m+1}})_z$ with angle $\theta_m - \frac{\pi}{2} > 0$ in X_Λ^k for $k \in \mathbb{N}_0$.

PROOF: By an easy induction $\Lambda : X_\Lambda^{k+1} \to X_\Lambda^k$ is sectorial in X_Λ^k with angle $\theta + \frac{\pi}{2}$ for any $k \in \mathbb{N}_0$. Then the assertion is a consequence of 1.3.6. $\square$

We will use the following renormalization procedure.

2.5.5 Proposition. Let $\Lambda : D(\Lambda) \to X$ be the generator of a bounded analytic semigroup with angle $\theta > \frac{m}{m+1}\frac{\pi}{2}$ with an $m \in \mathbb{N}$ in a complex Banach space X and let $M_{k,m} \geq 1$ with $\left\| e^{-\tau(-\Lambda)^{m+1}} \right\|_{\mathscr{L}(X_\Lambda^k)} \leq M_{k,m}$ for $\tau \geq 0$ and

$$\|x\|_k^{(m)} := \sup_{\tau \geq 0} \left\| e^{-\tau(-\Lambda)^{m+1}} x \right\|_k e^{-\tau \log(M_{k,m})} \qquad \text{for } x \in X_\Lambda^k, k \in \mathbb{N}_0.$$

Then $\|\cdot\|_k^{(m)}$ is a norm on X_Λ^k, and for $x \in X_\Lambda^k$ the following statements hold:

- $\|x\|_k \leq \|x\|_k^{(m)} \leq M_{k,m} \|x\|_k.$

- $\|x\|_k^{(m)} = \sup_{0 \leq \tau \leq 1} \left\| e^{-\tau(-\Lambda)^{m+1}} x \right\|_k e^{-\tau \log(M_{k,m})}.$

- $\left\| e^{-\tau(-\Lambda)^{m+1}} x \right\|_k^{(m)} \leq e^{\tau \log(M_{k,m})} \|x\|_k^{(m)}.$

PROOF: Clearly $\|\cdot\|_k^{(m)}$ is a norm on X_Λ^k, and we have

$$\|x\|_k \leq \sup_{\tau \geq 0} \left\| e^{-\tau(-\Lambda)^{m+1}} x \right\|_k e^{-\tau \log(M_{k,m})} \leq \sup_{\tau \geq 0} M_{k,m} \|x\|_k = M_{k,m} \|x\|_k .$$

Moreover, for $\tau \geq 1$ we have $\left\| e^{-\tau(-\Lambda)^{m+1}} x \right\|_k e^{-\tau \log(M_{k,m})} \leq \|x\|_k$, hence

$$\|x\|_k^{(m)} = \sup_{\tau \geq 0} \left\| e^{-\tau(-\Lambda)^{m+1}} x \right\|_k e^{-\tau \log(M_{k,m})} = \sup_{0 \leq \tau \leq 1} \left\| e^{-\tau(-\Lambda)^{m+1}} x \right\|_k e^{-\tau \log(M_{k,m})}.$$

Finally

$$\left\| e^{-\tau(-\Lambda)^{m+1}} x \right\|_k^{(m)} = \sup_{s \geq 0} \left\| e^{-(s+\tau)(-\Lambda)^{m+1}} x \right\|_k e^{-(s+\tau) \log(M_{k,m})} e^{\tau \log(M_{k,m})}$$

$$\leq \sup_{s \geq 0} \left\| e^{-s(-\Lambda)^{m+1}} x \right\|_k e^{-s \log(M_{k,m})} e^{\tau \log(M_{k,m})} = e^{\tau \log(M_{k,m})} \|x\|_k^{(m)} .$$

$\square$

2.5.6 Proposition. Suppose that $S : D(S) \to X$ is the infinitesimal generator of a C_0-semigroup $(e^{\tau S})_{\tau \geq 0}$ in a Banach space X with $\left\| e^{\tau S} \right\|_{\mathscr{L}(X)} \leq M$ for $\tau \geq 0$ and a suitable $M \geq 1$.
Then, for every $\varepsilon > 0$ and $n \in \mathbb{N}_0$ there is a $c_{n,\varepsilon} > 0$ with

$$\sum_{j=0}^{n} \|S^j x\| \leq \varepsilon \|S^{n+1} x\| + c_{n,\varepsilon} \|x\| \qquad \text{for } x \in D(S^{n+1}). \tag{2.5.1}$$

PROOF: For $x \in D(S^2)$ and $\tau \geq 0$ we have $e^{\tau S}x = x + \tau Sx + \int_0^\tau (\tau - s)e^{sS}S^2 x \, ds$. This implies

$$\|Sx\| \leq \frac{1}{\tau}\|e^{\tau S}x\| + \frac{1}{\tau}\|x\| + \frac{1}{\tau}\int_0^\tau (\tau - s)\|e^{sS}S^2 x\| \, ds \leq \frac{2M}{\tau}\|x\| + \frac{M\tau}{2}\|S^2 x\|.$$

Take $\tau = \frac{2\varepsilon}{M}$, then $\|Sx\| \leq \varepsilon\|S^2 x\| + \frac{M^2}{\varepsilon}\|x\|$. Now we assume inductively that there are $d_{n,\varepsilon} > 0$ with $\|S^n x\| \leq \varepsilon\|S^{n+1}x\| + d_{n,\varepsilon}\|x\|$ for an $n \in \mathbb{N}$ and $x \in D(S^{n+1})$. Then for $x \in D(S^{n+2})$

$$\|S^{n+1}x\| = \|S(S^n x)\| \leq \frac{\varepsilon}{2}\|S^2(S^n x)\| + d_{1,\varepsilon/2}\|S^n x\|$$

$$\leq \frac{\varepsilon}{2}\|S^{n+2}x\| + d_{1,\varepsilon/2}\underbrace{\frac{1}{2d_{1,\varepsilon/2}}}_{=:\varepsilon'}\|S^{n+1}x\| + d_{1,\varepsilon/2}d_{n,\varepsilon'}\|x\|.$$

Hence $\frac{1}{2}\|S^{n+1}x\| \leq \frac{\varepsilon}{2}\|S^{n+2}x\| + d_{1,\varepsilon/2}d_{n,\varepsilon'}\|x\|$. This proves that for any $\varepsilon > 0$, $n \in \mathbb{N}$ there is a $d_{n,\varepsilon}$ with $\|S^n x\| \leq \varepsilon\|S^{n+1}x\| + d_{n,\varepsilon}\|x\|$ for $x \in D(S^{n+1})$. Next suppose inductively that for any $\varepsilon > 0, n \in \mathbb{N}$ and an $m \in \mathbb{N}$ there is a $d_{n,m,\varepsilon}$ with

$$\|S^n x\| \leq \varepsilon\|S^{n+m}x\| + d_{n,m,\varepsilon}\|x\| \text{ for } x \in D(S^{n+m}).$$

Then

$$\|S^n x\| \leq \varepsilon\|S^{n+m}x\| + d_{n,m,\varepsilon}\|x\| \leq \varepsilon\|S^{n+m+1}x\| + (\varepsilon d_{n+m,1} + d_{n,m,\varepsilon})\|x\|$$

for $x \in D(S^{n+m+1})$. This shows that for any $\varepsilon > 0$ and any $n, m \in \mathbb{N}$ there is a $d_{n,m,\varepsilon}$ with

$$\|S^n x\| \leq \varepsilon\|S^{n+m}x\| + d_{n,m,\varepsilon}\|x\| \tag{2.5.2}$$

for $x \in D(S^{n+m})$. Finally, we can prove the assertion. For $n = 1$ we have for $x \in D(S^2)$

$$\|x\| + \|Sx\| \leq \varepsilon\|S^2 x\| + (d_{1,\varepsilon} + 1)\|x\|.$$

Now we assume (2.5.1) for an $n \in \mathbb{N}$, then for $x \in D(S^{n+2})$

$$\sum_{j=0}^{n+1}\|S^j x\| = \|x\| + \sum_{j=0}^{n}\|S^j Sx\| \underset{(2.5.1)}{\leq} \|x\| + \frac{\varepsilon}{2}\|S^{n+1}(Sx)\| + c_{n,\varepsilon/2}\|Sx\|$$

$$\underset{(2.5.2)}{\leq} \|x\| + \frac{\varepsilon}{2}\|S^{n+2}x\| + c_{n,\varepsilon/2}\left(\underbrace{\frac{\varepsilon}{2c_{n,\varepsilon/2}}}_{=:\varepsilon'}\|S^{n+2}x\| + d_{1,n+1,\varepsilon'}\|x\|\right)$$

$$= \varepsilon\|S^{n+2}x\| + \left(1 + c_{n,\varepsilon/2}d_{1,n+1,\varepsilon'}\right)\|x\|.$$

This proves the assertion.   □

Now we can prove the sufficient part of 2.4.3. Again we prove a more precise result including operators $A(t)$ that are only defined in a part of the scale.

2.5.7 Theorem.

Assumptions:

(i) Let $\Lambda : D(\Lambda) \to X$ be the generator of a bounded analytic semigroup with angle $\theta > \frac{m}{m+1}\frac{\pi}{2}$ with an $m \in \mathbb{N}$ in a complex Banach space X and with bounded imaginary powers, $K_1, K_2 \in \mathbb{N}_0, K_1 \leq K_2$, be fixed.

(ii) $I \ni t \mapsto A(t) \in \bigcap\limits_{k=K_1}^{K_2+3m+3} \mathscr{L}(X_\Lambda^{k+m}, X_\Lambda^k)$ is strongly continuous.

(iii) For $K_1 \leq k \leq K_2+3m+2$ there are $d_k \geq 0$ with $\|[\Lambda, A(t)]x\|_k \leq d_k \|x\|_{k+1}$ for $x \in X_\Lambda^\infty, t \in I$.

(iv) For $K_1 \leq k \leq K_2 + 3m + 2$ there are constants $\beta_k \geq 0$ such that $\|\lambda x - A(t)x\|_k \geq (\lambda - \beta_k) \|x\|_k$ for $x \in X_\Lambda^\infty, \lambda > \beta_k, t \in I$.

Assertion:

There is a (unique) propagator $U : \Delta \to \bigcap\limits_{k=K_1}^{K_2+m+1} \mathscr{L}(X_\Lambda^k)$ with:

(a) $\Delta \ni (t, s) \mapsto U(t, s) \in \mathscr{L}(X_\Lambda^{k+m}, X_\Lambda^k)$ is strongly continuously differentiable for $K_1 \leq k \leq K_2$ with strong derivatives

$$\frac{\partial}{\partial t}U(t, s) = A(t)U(t, s), \quad \frac{\partial}{\partial s}U(t, s) = -U(t, s)A(s) \quad \text{for } (t, s) \in \Delta.$$

(b) For every $K_1 \leq k \leq K_2 + m + 1$ we have $\|U(t, s)\|_{\mathscr{L}(X_\Lambda^k)} \leq e^{\beta_k(t-s)}$ for $(t, s) \in \Delta$.

PROOF:

Step 1: For $K_1 \leq k \leq K_2 + 2m + 2$ there are $c_{k,m}, c'_{k,m} \geq 0$ with

$$\left\|[(\tau\mathrm{Id} + (-\Lambda)^{m+1})^{-1}, A(t)]x\right\|_k \leq \frac{c_{k,m}}{\tau} \|x\|_k \qquad \tau \geq 1, x \in X_\Lambda^\infty \quad \text{and}$$

$$\left\|[e^{-\tau(-\Lambda)^{m+1}}, A(t)]x\right\|_k \leq c_{k,m} \|x\|_k \qquad 0 \leq \tau \leq 1, x \in X_\Lambda^\infty.$$

Proof: Take $\frac{\pi}{2} < \theta' < \theta'' < \theta_m, \lambda \in \Sigma_{\theta''}, |\lambda| \geq 1, K_1 \leq k \leq K_2 + 2m + 2$, and $x \in X_\Lambda^\infty$, then with 2.5.4

$$
\begin{aligned}
& \left\|[(\lambda \mathrm{Id} + (-\Lambda)^{m+1})^{-1}, A(t)]x\right\|_k \\
= \ & \left\|-(\lambda \mathrm{Id} + (-\Lambda)^{m+1})^{-1}[(-\Lambda)^{m+1}, A(t)](\lambda \mathrm{Id} + (-\Lambda)^{m+1})^{-1}x\right\|_k \\
\leq \ & \frac{M_{\theta''}}{|\lambda|} d_k \left\|(\lambda \mathrm{Id} + (-\Lambda)^{m+1})^{-1}x\right\|_{k+m+1} \\
\leq \ & \frac{M'_{\theta''}}{|\lambda|} \sum_{j=0}^{m+1} \left\|\Lambda^j (\lambda \mathrm{Id} + (-\Lambda)^{m+1})^{-1}x\right\|_k \leq \frac{c_{k,m}}{|\lambda|} \|x\|_k
\end{aligned}
$$

for suitable $M_{\theta''}, M'_{\theta''}, c_{k,m} \geq 0$. Now fix a $0 < \tau \leq 1$ and choose the integration paths

$$
\begin{aligned}
\gamma_1 &: (-\infty, -1/\tau] \ni r \mapsto -re^{-i\theta'} \\
\gamma_2 &: [-\theta', \theta'] \ni \varphi \mapsto \frac{1}{\tau} e^{i\varphi} \\
\gamma_3 &: [1/\tau, \infty) \ni r \mapsto re^{i\theta'}
\end{aligned}
$$

1.3.2 and 2.5.4 show that for $x \in X_\Lambda^\infty$ in X_Λ^k for $K_1 \leq k \leq K_2 + 2m + 2$

$$
e^{-\tau(-\Lambda)^{m+1}} x = \frac{1}{2\pi i} \int_{\gamma_1 \gamma_2 \gamma_3} e^{\tau\lambda} (\lambda \mathrm{Id} + (-\Lambda)^{m+1})^{-1} x \, d\lambda .
$$

Thus for suitable $d_{k,m}, d'_{k,m}, c'_{k,m} \geq 0$

$$
\begin{aligned}
& \left\|[e^{-\tau(-\Lambda)^{m+1}}, A(t)]x\right\|_k = \left\|\frac{1}{2\pi i} \int_{\gamma_1 \gamma_2 \gamma_3} e^{\tau\lambda}[(\lambda \mathrm{Id} + (-\Lambda)^{m+1})^{-1}, A(t)]x \, d\lambda\right\|_k \\
\leq \ & \frac{1}{2\pi} \left[2 \int_{\frac{1}{\tau}}^\infty \left|e^{re^{-i\theta'}\tau}\right| \frac{c_{k,m}}{|r|} dr + \int_{-\theta'}^{\theta'} \left|e^{\frac{1}{\tau}e^{i\varphi}\tau}\right| \frac{c_{k,m}}{|1/\tau|} \frac{1}{\tau} d\varphi \right] \|x\|_k \\
\leq \ & d_{k,m} \left[\int_{\frac{1}{\tau}}^\infty e^{\cos(\theta')r\tau} \frac{dr}{r} + \int_{-\theta'}^{\theta'} e^{\cos(\varphi)} d\varphi \right] \|x\|_k \\
\leq \ & d'_{k,m} \left[\int_1^\infty e^{\cos(\theta')r'} \frac{dr'}{r'} + 1 \right] \|x\|_k = c'_{k,m} \|x\|_k
\end{aligned}
$$

Step 2: For $K_1 \leq k \leq K_2 + 2m + 2$ there are $\beta_{k,m} \geq 0$ with

$$
\|\lambda x - A(t)x\|_k^{(m)} \geq (\lambda - \beta_{k,m}) \|x\|_k^{(m)} \qquad \text{for } x \in X_\Lambda^\infty, \lambda > \beta_{k,m}, t \in I .
$$

Proof: For $0 \leq \tau \leq 1$ we have due to step 1

$$\left\| e^{-\tau(-\Lambda)^{m+1}}(\lambda x - A(t)x) \right\|_k e^{-\tau \log(M_{k,m})}$$

$$\geq \underbrace{\left(\left\| (\lambda \mathrm{Id} - A(t)) \left(e^{-\tau(-\Lambda)^{m+1}} x \right) \right\|_k}_{\geq (\lambda - \beta_k) \left\| e^{-\tau(-\Lambda)^{m+1}} x \right\|_k} - \underbrace{\left\| [e^{-\tau(-\Lambda)^{m+1}}, A(t)]x \right\|_k}_{\leq c_{k,m} \|x\|_k \leq c_{k,m} \|x\|_k^{(m)}} \right) \underbrace{e^{-\tau \log(M_{k,m})}}_{\leq 1} .$$

Thus

$$\|\lambda x - A(t)x\|_k^{(m)} = \sup_{0 \leq \tau \leq 1} \left\| e^{-\tau(-\Lambda)^{m+1}}(\lambda x - A(t)x) \right\|_k e^{-\tau \log(M_{k,m})}$$

$$\geq \sup_{0 \leq \tau \leq 1} (\lambda - \beta_k) \left\| e^{-\tau(-\Lambda)^{m+1}} x \right\|_k e^{-\tau \log(M_{k,m})} - c_{k,m} \|x\|_k^{(m)}$$

$$\geq (\lambda - \beta_k) \|x\|_k^{(m)} - c_{k,m} \|x\|_k^{(m)} =: (\lambda - \beta_{k,m}) \|x\|_k^{(m)} .$$

Step 3: For $0 < \varepsilon < 1$, $K_1 \leq k \leq K_2 + 2m + 2$ $(-\varepsilon(-\Lambda)^{m+1} + A(t), X_\Lambda^{k+m+1})$ generates a C_0-semigroup in X_Λ^k with

$$\left\| e^{\tau(-\varepsilon(-\Lambda)^{m+1}+A(t))} x \right\|_k^{(m)} \leq e^{\tau(\beta_{k,m}+\log(M_{k,m}))} \|x\|_k^{(m)} \quad \text{for } x \in X_\Lambda^k, \tau \geq 0,$$

where $M_{k,m}$ is as in 2.5.5 and $\beta_{k,m}$ is as step 2.
Proof: For $0 < \delta < 1, 0 < \varepsilon < 1$ we have with suitable $a_{k,m}, b_{\varepsilon,\delta,k,m}, b'_{\varepsilon,\delta,k,m} \geq 0$

$$\|A(t)x\|_k^{(m)} \leq a_{k,m} \left(\sum_{j=0}^m \|\Lambda^j x\|_k \right) \leq a_{k,m} \left(\frac{\varepsilon \delta}{a_{k,m}} \|\Lambda^{m+1} x\|_k + b_{\varepsilon,\delta,k,m} \|x\|_k \right)$$

$$\leq \delta \left\| -\varepsilon(-\Lambda)^{m+1} x \right\|_k^{(m)} + b'_{\varepsilon,\delta,k,m} \|x\|_k^{(m)} .$$

Moreover, note that

$$\left\| e^{\tau(-\varepsilon(-\Lambda)^{m+1})} x \right\|_k^{(m)} = \left\| e^{(\varepsilon\tau)(-(-\Lambda)^{m+1})} x \right\|_k^{(m)} \leq e^{\tau \log(M_{k,m})} \|x\|_k^{(m)} .$$

Hence step 2 and 1.1.15 imply step 3.
Step 4: Let $A_l(t) := -\frac{1}{l}(-\Lambda)^{m+1} + A(t), t \in I, l \in \mathbb{N}, m' = m + 1$. We clearly have (iii), (iv) of 2.5.1. Therefore, due to 2.5.1, we have to show that for $l \in \mathbb{N}$ there exists a propagator $U_l : \triangle \to \bigcap_{k=K_1}^{K_2+2m+2} \mathcal{L}(X_\Lambda^k)$ such that for $l \in \mathbb{N}$:

- $\triangle \ni (t, s) \mapsto U_l(t, s) \in \mathcal{L}(X_\Lambda^{k+m+1}, X_\Lambda^k)$ is strongly continuously differentiable for $K_1 \leq k \leq K_2 + m + 1$ with strong derivatives

$$\frac{\partial}{\partial t} U_l(t, s) = A_l(t) U_l(t, s) \quad \text{and} \quad \frac{\partial}{\partial s} U_l(t, s) = -U_l(t, s) A_l(s).$$

- For $K_1 \le k \le K_2 + 2m + 2$ there are $u_k \ge 0$ with $\|U_l(t,s)\|_{\mathscr{L}(X_\Lambda^k)} \le u_k$ for $(t,s) \in \Delta, l \in \mathbb{N}$.

Proof:

Let $(X, \|\cdot\|_X) = (X_\Lambda^k, \|\cdot\|_k^{(m)})$, $(Y, \|\cdot\|_Y) = (X_\Lambda^{k+m+1}, \|\cdot\|_{k+m+1}^{(m)})$. We apply 1.6.8 with $K_1 \le k \le K_2 + 2m + 2$ and $P(t) := A_l(t)$. Due to step 3 $P(t)$ with domain Y is a family of infinitesimal generators of C_0-semigroups in X, and $\left\| e^{\tau P(t)} x \right\|_X \le e^{\tau \gamma_{k,m}} \|x\|_X$, where $\gamma_{k,m} = \beta_{k,m} + \log M_{k,m}$ as in step 3. Moreover, $I \ni t \mapsto P(t) \in \mathscr{L}(Y, X)$ is strongly continuous.

Let $S := \mathrm{Id} + \frac{1}{l}(-\Lambda)^{m+1}$, $Q(t) := -\frac{1}{l}[A(t), (-\Lambda)^{m+1}](\mathrm{Id} + \frac{1}{l}(-\Lambda)^{m+1})^{-1}$, and $D := X_\Lambda^{k+m+1}$. Then D is a core of $(P(t), Y)$ (due to step 3), $S : Y \to X$ is an isomorphism, $\{Q(t) : t \in I\} \subset \mathscr{L}(X)$ boundedly, and $I \ni t \mapsto Q(t)x \in X$ continuously for $x \in X_Z^\infty$. Thus $I \ni t \mapsto Q(t) \in \mathscr{L}(X)$ is strongly continuous. Moreover, for $z \in D$ we have

$$P(t)S^{-1}z - S^{-1}P(t)z = -S^{-1}(P(t)S - SP(t))S^{-1}z$$
$$= -S^{-1}\left(A(t)\frac{1}{l}(-\Lambda)^{m+1} - \frac{1}{l}(-\Lambda)^{m+1}A(t) \right) S^{-1}z = S^{-1}Q(t)z \,.$$

Hence we can apply 1.6.8, and the proof is complete. $\qquad\square$

In the last part of this section we show how one can deduce 2.4.5 from 2.4.3. We assume always that

- $Z = \{Z_1, \dots, Z_M\}$ is a commuting family of infinitesimal generators of C_0-groups of isometries in a complex Banach space X.

- $\Lambda := \sum_{j=1}^{M} Z_j^2$ and there are $c_1, c_2 > 0$ with

$$c_1 \|x\|_2 \le \|x\| + \|\Lambda x\| \le c_2 \|x\|_2 \qquad \text{for } x \in X_Z^2 \,.$$

2.5.8 Lemma. For $k \in \mathbb{N}_0, q \in \mathbb{N}$ there are $c_{k,q}, d_{k,q} \ge 0$ with

$$c_{k,q} \|x\|_{k+2q} \le \left(\sum_{j=0}^{q} \|\Lambda^j x\|_k^2 \right)^{1/2} \le d_{k,q} \|x\|_{k+2q} \qquad \text{for } x \in X_Z^{k+2q} \,.$$

PROOF: For $q = 1$ we have with suitable constants $b, c, c', c'' \geq 0$

$$\|x\|_{k+2}^2 \;\leq\; b \sum_{|\alpha| \leq k} \|Z^\alpha x\|_2^2 \leq c \sum_{|\alpha| \leq k} (\|Z^\alpha x\| + \|\Lambda Z^\alpha x\|)^2$$

$$\leq\; 2c \sum_{|\alpha| \leq k} \left(\|Z^\alpha x\|^2 + \|Z^\alpha \Lambda x\|^2 \right) = c' \left(\|x\|_k^2 + \|\Lambda x\|_k^2 \right) \leq c'' \|x\|_{k+2}^2$$

for $x \in X_Z^{k+2}$. If we assume the assertion inductively for a $q \in \mathbb{N}$ we obtain

$$\|x\|_{k+2q+2}^2 \;\leq\; a_k \sum_{|\alpha| \leq 2q} \|Z^\alpha x\|_{k+2}^2 \leq a_k' \sum_{|\alpha| \leq 2q} \left(\|Z^\alpha x\|_k^2 + \|\Lambda Z^\alpha x\|_k^2 \right)$$

$$\leq\; a_k'' \left(\|x\|_{k+2q}^2 + \|\Lambda x\|_{k+2q}^2 \right) \leq b_k \sum_{j=0}^{q+1} \|\Lambda^j x\|_k^2 \leq b_k' \|x\|_{k+2q+2}^2$$

for $x \in X_Z^{k+2q+2}$ and suitable constants a_k, a_k', a_k'', b_k. □

2.5.9 Lemma. $X_Z^2 = \displaystyle\bigcap_{j=1}^{M} D(Z_j^2)$.

PROOF: Take an $x \in \displaystyle\bigcap_{j=1}^{M} D(Z_j^2)$ and let J_l be as in 2.2.9. Then $J_l^2 x \in X_Z^2$ and $J_l^2 x \xrightarrow[l\to\infty]{X} x$ due to 2.2.9. Moreover

$$\left\| \Lambda(J_l^2 x - x) \right\| \leq \sum_{j=1}^{M} \left\| Z_j^2 (J_l^2 x - x) \right\| = \sum_{j=1}^{M} \left\| J_l^2 Z_j^2 x - Z_j^2 x \right\| \xrightarrow[l\to\infty]{} 0 \,.$$

Therefore $(J_l^2 x)_l \subset X_Z^2$ is a Cauchy sequence with respect to $\|\cdot\| + \|\Lambda\cdot\|$ and thus in X_Z^2, as well. Hence there is an $y \in X_Z^2$ with $J_l^2 x \xrightarrow[l\to\infty]{X_Z^2} y$. This implies $J_l^2 x \xrightarrow[l\to\infty]{X} y$ and thus $x = y \in X_Z^2$. □

2.5.10 Lemma. Let $x \in X_Z^k$ and $\Lambda x \in X_Z^k$ for $k \geq 2$, then $x \in X_Z^{k+2}$.

PROOF: $J_l^2 x \in X_Z^{k+2}$, $J_l^2 x \xrightarrow[l\to\infty]{X_Z^k} x$, and $\left\| \Lambda(J_l^2 x - x) \right\|_k = \left\| J_l^2 \Lambda x - \Lambda x \right\|_k \xrightarrow[l\to\infty]{} 0$ due to 2.2.9. Hence $(J_l^2 x)_l \subset X_Z^{k+2}$ is a Cauchy sequence due to 2.5.8. Thus there is an $y \in X_Z^{k+2}$ with $J_l^2 x \xrightarrow[l\to\infty]{X_Z^{k+2}} y$, which shows $x = y \in X_Z^{k+2}$. □

The square of a generator of an isometric C_0-group is dissipative and sectorial with angle π (cf. 1.3.8). Moreover, commutativity of generators of isometric C_0-groups implies commutativity of the analytic semigroups generated by their squares.

2.5.11 Proposition. $(Z_j^2, D(Z_j^2)), j = 1, \ldots, M,$ is sectorial with angle π. Moreover,

$$(\lambda\mathrm{Id}-Z_j^2)^{-1}(\lambda\mathrm{Id}-Z_l^2)^{-1} = (\lambda\mathrm{Id}-Z_l^2)^{-1}(\lambda\mathrm{Id}-Z_j^2)^{-1} \text{ for } \lambda \in \Sigma_\pi, j, l = 1, \ldots, M.$$

PROOF: 1.3.8 shows that Z_j^2 is sectorial with angle π. To prove the commutativity relation we only have to show that

$$(\lambda\mathrm{Id} - Z_j)^{-1}(\mu\mathrm{Id} - Z_l)^{-1} = (\mu\mathrm{Id} - Z_l)^{-1}(\lambda\mathrm{Id} - Z_j)^{-1} \qquad (2.5.3)$$

for $\mathrm{Re}\,\lambda \neq 0, \mathrm{Re}\,\mu \neq 0, j, l = 1, \ldots, M.$ Take an $x \in X_Z^2$, then, due to 2.2.5 $y := (\lambda\mathrm{Id} - Z_j)^{-1}(\mu\mathrm{Id} - Z_l)^{-1}x \in X_Z^2$, thus with 2.2.5

$$x = (\mu\mathrm{Id} - Z_l)(\lambda\mathrm{Id} - Z_j)y = (\lambda\mathrm{Id} - Z_j)(\mu\mathrm{Id} - Z_l)y,$$

and $y = (\mu\mathrm{Id} - Z_l)^{-1}(\lambda\mathrm{Id} - Z_j)^{-1}x.$ This proves (2.5.3) on X_Z^2 and with the density of $X_Z^2 \subset X$ the proof is complete. $\qquad\square$

2.5.12 Proposition. $\Lambda : X_Z^{k+2} \to X_Z^k$ is sectorial with angle π in X_Z^k for $k \in \mathbb{N}_0.$

PROOF:

Step 1: k=0

Due to 2.5.11 and 1.3.2 $(Z_j^2, D(Z_j^2)), j = 1, \ldots, M,$ generate bounded analytic semigroups $(e^{zZ_j^2})_z$ of angle $\frac{\pi}{2}$ and we have $e^{zZ_j^2}e^{z'Z_l^2} = e^{z'Z_l^2}e^{zZ_j^2}$ for $z, z' \in \Sigma_{\pi/2},$ $j, l = 1, \ldots, M,$ where we have used the integral representation in 1.3.2 and the commutativity of the resolvents. Let $T(z) := e^{zZ_1^2} \cdots e^{zZ_M^2}.$ Then $(T(z))_z$ is a bounded analytic semigroup with angle $\frac{\pi}{2}.$ Let $(S, D(S))$ be its infinitesimal generator, which is sectorial of angle π due to 1.3.2. Now the commutativity shows that $e^{tZ_j^2}(D(Z_l^2)) \subset D(Z_l^2),$ thus (due to 2.5.9)

$$T(t)(X_Z^2) = T(t)\left(\bigcap_{j=1}^M D(Z_j^2)\right) \subset \bigcap_{j=1}^M D(Z_j^2) = X_Z^2.$$

and 1.1.14 shows that X_Z^2 is a core of $S.$ Finally, for $x \in X_Z^2$ we have

$$\frac{d}{dt}T(t)x = (Z_1^2 + \ldots + Z_M^2)T(t)x = \Lambda T(t)x,$$

hence $\Lambda = S|_{X_Z^2}.$ But Λ is closed, so $\Lambda = S$ and we have finished step 1.

Step 2: Assume the assertion inductively for a $k \in \mathbb{N}_0$: Let $\lambda \in \Sigma_\theta, 0 < \theta < \pi$, and an $x \in X_Z^{k+1} \subset X_Z^k$. Inductively there is an $y \in X_Z^{k+2}$ with $(\lambda \mathrm{Id} - \Lambda)y = x$. In particular, $\Lambda y = \lambda y - x \in X_Z^{k+1}$, i.e., $y \in X_Z^{k+3}$ due to 2.5.10. Therefore the operator $\lambda \mathrm{Id} - \Lambda : X_Z^{k+3} \to X_Z^{k+1}$ is surjective. Moreover, inductively there is an $M_\theta \geq 0$ with

$$\|y\|_k \leq \frac{M_\theta}{|\lambda|} \|(\lambda \mathrm{Id} - \Lambda)y\|_k \qquad \text{for } y \in X_Z^{k+2} .$$

Hence

$$\|x\|_{k+1}^2 \leq \frac{M_\theta^2}{|\lambda|^2}\left(\|(\lambda \mathrm{Id} - \Lambda)x\|_k^2 + \sum_{j=1}^M \|(\lambda \mathrm{Id} - \Lambda)Z_j x\|_k^2 \right) = \frac{M_\theta^2}{|\lambda|^2} \|(\lambda \mathrm{Id} - \Lambda)x\|_{k+1}^2 .$$

This shows that (Λ, X_Z^{k+3}) is sectorial with angle π in X_Z^{k+1}.   □

2.5.13 Lemma. For $k \in \mathbb{N}_0$ and $D(\Lambda) := X_Z^{k+2}$ we have

$$\left(X_Z^k\right)_\Lambda^q = X_Z^{k+2q} \text{ topologically for } q \in \mathbb{N}_0 .$$

PROOF: This is a consequence of 2.5.8 and 2.5.10.   □

Finally, we can prove the sufficient part of 2.4.5.

2.5.14 Theorem.
Assumptions:

(i) Let $Z = \{Z_1, \ldots, Z_M\}$ be a commuting family of infinitesimal generators of isometric C_0-groups in a complex Banach space X, $m \in \mathbb{N}$. Moreover, let $\Lambda := \sum_{j=1}^M Z_j^2$, and $K_1, K_2 \in \mathbb{N}_0$ with $K_1 \leq K_2$.

(ii) There exist $c_1, c_2 \geq 0$ with $c_1 \|x\|_2 \leq \|x\| + \|\Lambda x\| \leq c_2 \|x\|_2$ for $x \in X_Z^2$.

(iii) $I \ni t \mapsto A(t) \in \bigcap_{k=K_1}^{K_2+7m+6} \mathscr{L}(X_Z^{k+m}, X_Z^k)$ is strongly continuous.

(iv) For $K_1 \leq k \leq K_2 + 7m + 5$ there are constants $d_k \geq 0$ with

$$\|[Z_j, A(t)]x\|_k \leq d_k \|x\|_{k+1}, x \in X_Z^\infty, t \in I, j = 1, \ldots, M .$$

(v) For $K_1 \leq k \leq K_2 + 7m + 5$ there are $\beta_k \geq 0$ with

$$\|\lambda x - A(t)x\|_k \geq (\lambda - \beta_k) \|x\|_k \quad \text{for } x \in X_Z^\infty, \lambda > \beta_k, t \in I.$$

Assertion:

There is a (unique) propagator $U : \Delta \to \bigcap\limits_{k=K_1}^{K_2+m} \mathscr{L}(X_Z^k)$ satisfying the following
properties:

(a) $\Delta \ni (t, s) \mapsto U(t, s) \in \mathscr{L}(X_Z^{k+m}, X_Z^k)$ is strongly continuously differentiable for $K_1 \leq k \leq K_2$ with strong derivatives

$$\frac{\partial}{\partial t} U(t, s) = A(t)U(t, s), \quad \frac{\partial}{\partial s} U(t, s) = -U(t, s)A(s) \quad \text{for } (t, s) \in \Delta.$$

(b) For $K_1 \leq k \leq K_2 + m$ we have $\|U(t, s)\|_{\mathscr{L}(X_Z^k)} \leq e^{\beta_k(t-s)}$ for $(t, s) \in \Delta$.

PROOF: Let $K_1 \leq k \leq K_2 + m$ and $Y := X_Z^k$. Then (Λ, X_Z^{k+2}) generates a
bounded analytic semigroup with angle $\frac{\pi}{2}$ in Y and with $Y_\Lambda^l = X_Z^{k+2l}$ topologically (due to 2.5.12, 2.5.13). Moreover,

- $I \ni t \mapsto A(t) \in \bigcap\limits_{l=0}^{3m+3} \mathscr{L}(Y_\Lambda^{l+m}, Y_\Lambda^l) = \bigcap\limits_{l=0}^{3m+3} \mathscr{L}(X_Z^{k+2l+2m}, X_Z^{k+2l})$ is strongly
continuous.

- For $0 \leq l \leq 3m + 2$ we have

$$\|[\Lambda, A(t)]x\|_{Y_\Lambda^l} \leq \sum_{j=1}^M \left(\|Z_j[Z_j, A(t)]x\|_{k+2l} + \|[Z_j, A(t)]Z_j x\|_{k+2l} \right)$$

$$\leq \sum_{j=1}^M (d_{k+2l+1} + d_{2k+l}) \|x\|_{k+2(l+1)} \leq d_k' \|x\|_{Y_\Lambda^{l+1}}$$

for $x \in Y_\Lambda^\infty = X_Z^\infty$ and a suitable $d_k' \geq 0$.

- For every $0 \leq l \leq 3m + 2$ there are constants $\widetilde{\beta}_l \geq 0$ such that
$\|\lambda x - A(t)x\|_{Y_\Lambda^l} \geq (\lambda - \widetilde{\beta}_l) \|x\|_{Y_\Lambda^l}$ for $x \in Y_\Lambda^{l+1}, \lambda > \beta_l$.
Proof: $\|\lambda x - A(t)x\|_Y = \|\lambda x - A(t)x\|_k \geq (\lambda - \beta_k) \|x\|_k = (\lambda - \beta_k) \|x\|_Y$,
therefore the assertion is implied by 2.3.1.

Hence 2.5.7 implies the existence of $U(t,s) \in \bigcap_{l=0}^{m+1} \mathscr{L}(Y_\Lambda^l), (t,s) \in \Delta$, with

- $\Delta \ni (t,s) \mapsto U(t,s) \in \mathscr{L}(Y_\Lambda^m, Y)$ is strongly continuously differentiable with derivatives

$$\frac{\partial}{\partial t}U(t,s) = A(t)U(t,s) \quad \text{and} \quad \frac{\partial}{\partial s}U(t,s) = -U(t,s)A(s).$$

- $\|U(t,s)\|_{\mathscr{L}(Y_\Lambda^l)} \le e^{\tilde{\beta}_l(t-s)}$ for $0 \le l \le m+1$.

This implies that $U : \Delta \to \mathscr{L}(X_Z^k)$ is a propagator for $K_1 \le k \le K_2 + m$ with (b) and shows moreover that $\Delta \ni (t,s) \mapsto U(t,s) \in \mathscr{L}(X_Z^{k+2m}, X_Z^k)$ is strongly differentiable for $K_1 \le k \le K_2$ with derivative

$$\frac{\partial}{\partial t}U(t,s) = A(t)U(t,s) \quad \text{and} \quad \frac{\partial}{\partial s}U(t,s) = -U(t,s)A(s).$$

Therefore, for $t_0 \le s \le t, t+h \le t_1$ with $h \neq 0$ and $x \in X^{k+2m}$ we have with suitable $a_k, u_{k+m} \ge 0$

$$\left\| \frac{U(t+h,s)x - U(t,s)x}{h} \right\|_k \le \left| \frac{1}{h} \int_t^{t+h} \|A(\tau)U(\tau,s)x\|_k \, d\tau \right| \le a_k u_{k+m} \|x\|_{k+m}.$$

This shows $\left\{ \frac{U(t+h,s)-U(t,s)}{h} : t_0 \le s \le t, t+h \le t_1, h \neq 0 \right\} \subset \mathscr{L}(X_Z^{k+m}, X_Z^k)$ boundedly and hence implies that $[s,t_1] \ni t \mapsto U(t,s)x \in X_Z^k$ is differentiable with derivative $A(t)U(t,s)x$. Similarly one can also show that $[t_0,t] \ni s \mapsto U(t,s)x \in X^k$ is differentiable for $x \in X^{k+m}$ with derivative $-U(t,s)A(s)x$. This proves the assertion. $\qquad \Box$

2.5.15 Bibliographical remarks. 2.5.6 is an extension of [132, 2.2]. 2.5.11 and 2.5.12 can be essentially found in [110, A-II,1.13]).

Chapter 3

Quasilinear Evolution Equations

This chapter is devoted to the study of semilinear evolution equations of the form

$$u'(t) = A(t)u(t) + f(t, u(t))$$

and quasilinear evolution equations of the form

$$u'(t) = A(t, u(t))u(t) + f(t, u(t))$$

in scales of Banach spaces, where $A(t)$ resp., $A(t, v)$ are linear operators. Although semilinear evolution equations are special cases of quasilinear ones, we will start this chapter with a discussion of semilinear evolution equations in scales of Banach spaces in section 3.1. We do this for two reasons. On the one hand, the proofs are much easier and clearer than in the case of quasilinear evolution equations; on the other hand, we also get more precise results in this special case. Then, in section 3.2 we collect some commutator estimates that we will need in the sequel. In 3.3 we give a general uniqueness and existence result for quasilinear evolution equations provided that the Cauchy problem for related time-dependent linear evolutions equations is well-posed. These assumptions will always be satisfied in the situations described in chapter 2. Finally, in 3.4 we will prove a regularity result for these equations in scales of Hilbert spaces.

3.1 Semilinear Evolution Equations

In [139], [52, II, 2.4], [127, section X.13], [25, 4.3] the semilinear Cauchy problem

$$u'(t) = Au(t) + f(t, u(t)), \qquad u(t_0) = u_0$$

for a generator $A : D(A) \to X$ of a C_0-semigroup in a Banach space X is solved. We will apply these methods to semilinear evolution equations in scales of Banach spaces substituting the theory of C_0-semigroups by the theory of linear evolution equations in scales of Banach spaces.

In this section we will always assume that $(X^k)_k$ is a scale of Banach spaces.

3.1.1 Proposition. Let $\triangle \ni (t, s) \mapsto U(t, s) \in \mathcal{L}(X)$ be a propagator and $f : I \times X \to X$ be continuous such that

- $\|U(t, s)\|_{\mathcal{L}(X)} \leq \tilde{U}$ for $(t, s) \in \triangle$ and a suitable constant $\tilde{U}$.

- $\|f(s, 0)\| \leq \tilde{F}$ for $s \in I$.

- for any $R > 0$ there is an $F(R) > 0$ with

$$\|f(t, u) - f(t, v)\| \leq F(R) \|u - v\|$$

for $t \in I, u, v \in X$ with $\|u\| \leq R, \|v\| \leq R$.

Then the following statements hold:

(a) Let $T_0(M) := \min\left(t_1 - t_0, \dfrac{1}{2\tilde{U}F(2\tilde{U}M + \tilde{U}\tilde{F}) + 2} \right)$. Then, for $x \in X$ with $\|x\| \leq M$, there is a unique $u \in C([t_0, t_0 + T_0(M)], X)$ with

$$u(t) = U(t, t_0)x + \int_{t_0}^t U(t, s)f(s, u(s))ds .$$

We have $\|u(t)\| \leq 2\tilde{U}M + \tilde{U}\tilde{F}$ for $t \in [t_0, t_0 + T_0(M)]$.

(b) There is a mapping $T : X \to (t_0, \infty]$ such that for $x \in X$ there exists a unique function $u \in C([t_0, T(x)) \cap I, X)$ with

$$u(t) = U(t, t_0)x + \int_{t_0}^t U(t, s)f(s, u(s))ds \quad \text{for } t \in [t_0, T(x)) \cap I .$$

Moreover, either $T(x) > t_1$ and there is an $L \geq 0$ with $\|u(t)\| \leq L$ for $t \in I$ or $T(x) \leq t_1$ and

$$\|u(t)\| \xrightarrow[t \to T(x)-]{} \infty \quad \text{with} \quad 2\tilde{U}F(2\tilde{U}\|u(t)\| + \tilde{U}\tilde{F}) > \frac{1}{T(x) - t} - 2 .$$

PROOF: Let $K := 2\tilde{U}M + \tilde{U}\tilde{F}$, $E := \{u \in C([t_0, t_0 + T_0(M)], X) : \|u(t)\| \leq K\}$, and $d(u, v) := \sup_{t \in [t_0, t_0 + T_0(M)]} \|u(t) - v(t)\|$ for $u, v \in E$. Moreover, let

$$(\Phi u)(t) := U(t, t_0)x + \int_{t_0}^{t} U(t, s)f(s, u(s))ds \quad \text{for } u \in E, t \in [t_0, t_0 + T_0(M)].$$

Then $\Phi u \in C([t_0, t_0 + T_0(M)], X)$ for $u \in E$ and

$$\|f(s, u(s))\| \leq \|f(s, 0)\| + \|f(s, u(s)) - f(s, 0)\| \leq \tilde{F} + F(K) \underbrace{\|u(s)\|}_{\leq K}$$

$$= \tilde{F} + (2\tilde{U}M + \tilde{U}\tilde{F})F(K) \leq 2(\tilde{U}F(K) + 1)(M + \tilde{F}) \leq \frac{M + \tilde{F}}{T_0(M)}.$$

Thus $\|\Phi u(t)\| \leq \tilde{U}\|x\| + \int_{t_0}^{t}\tilde{U}\|f(s, u(s))\| \leq \tilde{U}M + (t - t_0)\tilde{U}\frac{M + \tilde{F}}{T_0(M)} \leq K$ Finally,

$$\|\Phi u(t) - \Phi v(t)\| \leq \int_{t_0}^{t}\tilde{U}F(K)\|u(s) - v(s)\|\,ds \leq \frac{1}{2}d(u, v),$$

which implies (a) by the Banach fix point theorem .
For the proof of (b) let $u \in C([t_0, T(x)) \cap I, X)$ be the maximal solution (i.e., the solution with a maximal domain, which always exists due to Zorn's lemma) and assume that $T(x) \leq t_1$ and that there is a $t \in [t_0, T(x))$ with

$$2\tilde{U}F(2\tilde{U}\|u(t)\| + \tilde{U}\tilde{F}) \leq \frac{1}{T(x) - t} - 2$$

Then, with $M := \|u(t)\|$ we have $T(x) - t \leq T_0(M)$.
Let $v \in C([t, t + T_0(M)] \cap I, X)$ be the solution of

$$v(s) - U(s, t)u(t) + \int_{t}^{s} U(s, \sigma)f(\sigma, v(\sigma))d\sigma.$$

Define $w \in C([t_0, t + T_0(M)] \cap I, X)$ by

$$w(s) := \begin{cases} u(s) & , s \in [t_0, t] \\ v(s) & , s \in [t, t + T_0(M)] \cap I \end{cases}$$

Then $w \in C([t_0, t + T_0(M)] \cap I, X)$ clearly satisfies

$$w(s) = U(s, t_0)x + \int_{t_0}^{s} U(s, \sigma)f(\sigma, w(\sigma))d\sigma \quad \text{for } s \in [t_0, t + T_0(M)] \cap X)$$

which contradicts the maximality of $T(x)$. $\qquad\qquad\qquad\qquad\qquad\square$

Applied to scales of Banach spaces we immediately obtain the following existence and uniqueness result for semilinear evolution equations in scales of Banach spaces. Moreover, this result contains information about the blow-up behaviour.

3.1.2 Theorem. Suppose that

- $m \in \mathbb{N}, K \in \mathbb{N}_0$.

- $I \ni t \mapsto A(t) \in \bigcap_{k \in \mathbb{N}_0} \mathscr{L}(X^{k+m}, X^k)$ is strongly continuous.

- the Cauchy problem for $A(t)$ in the scale $(X^k)_k$ is well-posed with exponential growth and the associated evolution operator $U(t,s)$ satisfies $\|U(t,s)\|_{\mathscr{L}(X^k)} \leq \tilde{U}_k$ for $k \in \mathbb{N}_0, (t,s) \in \Delta$, and suitable constants $\tilde{U}_k$.

- $f : I \times X^{K+m} \to X^{K+m}$ is continuous and $\|f(t,0)\|_{K+m} \leq \tilde{F}_{K+m}$ for $t \in I$.

- for $R > 0$ there is an $F_{K+m}(R)$ with

$$\|f(t,u) - f(t,v)\|_{K+m} \leq F_{K+m}(R) \|u - v\|_{K+m}$$

for $u, v \in X^{K+m}$ with $\|u\|_{K+m} \leq R, \|v\|_{K+m} \leq R$.

Then there is a function $T_{K+m} : X^{K+m} \to (t_0, \infty]$ such that the following conditions are satisfied:

(a) For $x \in X^{K+m}$ there is a unique $u \in C([t_0, T_{K+m}(x)) \cap I, X^{K+m}) \cap C^1([t_0, T_{K+m}(x)) \cap I, X^K)$ with

$$u'(t) = A(t)u(t) + f(t, u(t)), \qquad u(t_0) = x .$$

(b) Either $T_{K+m}(x) > t_1$ and $\|u(t)\|_{K+m} \leq L_{K+m}$ for $t \in I$ and a suitable $L_{K+m} \geq 0$ or $T_{K+m}(x) \leq t_1$ and $\|u(t)\|_{K+m} \xrightarrow[t \to T_{k+m}(x)-]{} \infty$ and

$$2\tilde{U}_{K+m}F_{K+m}(2\tilde{U}_{K+m}\|u(t)\|_{K+m} + \tilde{U}_{K+m}\tilde{F}_{K+m}) > \frac{1}{T_{K+m}(x) - t} - 2$$

for $t \in [t_0, T_{K+m}(x)) \cap I$.

PROOF: Due to 3.1.1 there is a $u \in C([t_0, T_{K+m}(x)) \cap I, X^{K+m})$ with

$$u(t) = U(t, t_0)x + \int_{t_0}^{t} U(t, s)f(s, u(s))ds \,.$$

Hence $f(s, u(s)) \in C([t_0, T_{K+m}(x)) \cap I, X^{K+m})$, thus (due to 2.1.4) we have $u \in C^1([t_0, T_{K+m}(x)) \cap I, X^K)$ with $u'(t) = A(t)u(t) + f(t, u(t))$. ☐

The previous result shows that under the assumptions mentioned above we can solve the semilinear Cauchy problem everywhere in the scale on a time interval depending on the scale index k. But the length of the interval of existence can converge to zero if $k \to \infty$, as the following example shows. Therefore we are interested in conditions implying that this effect cannot occur and showing that the length of the interval of existence is independent of the scale index k. This is what we call regularity.

3.1.3 Example. Let $X^k := C[\frac{1}{k}, 1 - \frac{1}{k}]$, $A(t) := 0$, and $f(t, x) := x^2$ for $t \in [0, 1]$. Trivially, the Cauchy problem for $A(t)$ in the scale $(X^k)_k$ is well-posed with exponential growth and with evolution operators $U(t, s) := 0$, but there is no $\varepsilon > 0$ and no $u \in C^1([0, \varepsilon), X^\infty)$ with

$$u'(t) = A(t)u(t) + f(t, u(t)), t \in [0, \varepsilon), \quad u(0) = u_0 \,,$$

where $u_0(x) := \frac{1}{x}$ for $x \in (0, 1)$.

PROOF: Assume that $u : [0, \varepsilon) \to X^\infty$ is differentiable with $u'(t) = u(t)^2$ for $t \in [0, \varepsilon)$ and $u(0) = u_0$. Let $v(t, x) := u(t)(x)$ for $t \in [0, \varepsilon), x \in (0, 1)$. Since

$$\frac{v(t + h, x) - v(t, x)}{h} = \frac{u(t + h)(x) - u(t)(x)}{h} \xrightarrow[h \to 0]{} u(t)^2(x) = v(t, x)^2$$

uniformly for $x \in [\frac{1}{k}, 1 - \frac{1}{k}]$ for any $k \in \mathbb{N}$, the mapping $[0, \varepsilon) \ni t \mapsto v(t, x)$ is differentiable for any $x \in (0, 1)$ with

$$v_t(t, x) = v(t, x)^2, t \in [0, \varepsilon), \qquad v(0, x) = \frac{1}{x}$$

Hence the formula of separation of variables for ordinary differential equations shows $v(t, x) = \frac{1}{x-t}$ for $t \in [0, \min(\varepsilon, x)), x \in (0, 1)$. This is a contradiction. ☐

Our next aim will be a regularity result for semilinear evolution equation in scales of Hilbert spaces. Therefore, we will need the following result on integral inequalities.

3.1.4 Lemma. Let $I = [t_0, t_1]$, $J \subset \mathbb{R}$ be intervals, $\varphi \in C(J, (0, \infty))$ non-decreasing, $w, v \in C(I, J)$, and $\alpha \geq 0$ with

$$0 \leq v(t) \leq \alpha + \int_{t_0}^{t} \varphi(v(s))ds \quad \text{and} \quad w(t) = \alpha + \int_{t_0}^{t} \varphi(w(s))ds \quad \text{for } t \in I.$$

Then $v(t) \leq w(t)$ for $t \in I$.

PROOF: There are open intervals $I \subset I' \subset \mathbb{R}$, $J \subset J' \subset \mathbb{R}$ and an non-decreasing function $\Phi \in C(J', (0, \infty))$ with $\Phi|_J = \varphi$. Let $w_\varepsilon \in C^1(I_\varepsilon, J')$ be the solution with maximal domain of

$$w_\varepsilon'(t) = \Phi(w_\varepsilon(t)) \quad \text{for } t \in I_\varepsilon, \qquad w_\varepsilon(t_0) = \alpha + \varepsilon$$

for $\varepsilon \in [0, 1]$. Then $I \subset I_0$ and the theorem on continuous parameter dependence for ordinary differential equations (cf. [8, II,8.3]) shows that there is an $\varepsilon_0 > 0$ with $I \subset I_\varepsilon$ for $0 < \varepsilon < \varepsilon_0$ and $w_\varepsilon \xrightarrow[\varepsilon \to 0]{} w_0$ uniformly on I. Therefore we only have to show $v(t) < w_\varepsilon(t)$ for $t \in I$ and $0 < \varepsilon < \varepsilon_0$. Assume that this is wrong, then $t_2 := \inf\{t \in I : v(t) \geq w_\varepsilon(t)\} \in (t_0, t_1]$, $v(t_2) = w_\varepsilon(t_2)$, and $v(t) < w_\varepsilon(t)$ for $t \in [t_0, t_2)$. Thus

$$v(t_2) \leq \alpha + \int_{t_0}^{t_2} \Phi(v(s))ds < w_\varepsilon(t_0) + \int_{t_0}^{t_2} \Phi(w_\varepsilon(s))ds = w_\varepsilon(t_2),$$

which is a contradiction. $\qquad\square$

As a corollary we have Gronwall's inequality.

3.1.5 Lemma. Let $I \subset \mathbb{R}$ be an interval with left boundary point $t_0 \in I$, $v : I \to [0, \infty)$ be continuous, and $\alpha, \beta \geq 0$ with

$$v(t) \leq \alpha + \int_{t_0}^{t} \beta v(s)ds \quad \text{for } t \in I.$$

Then $v(t) \leq \alpha e^{\beta(t - t_0)}$ for $t \in I$.

PROOF: Let $\varphi(x) := \beta x$ and $w(t) = \alpha e^{\beta(t - t_0)}$, and apply 3.1.4. $\qquad\square$

With these inequalities we can prove now a regularity result for semilinear evolution equations in scales of Banach spaces assuming estimates of Nash-Moser type .

3.1.6 Theorem.   Assume that

- $K \in \mathbb{N}$ is fixed.

- $I \ni t \mapsto A(t) \in \bigcap\limits_{k \in \mathbb{N}_0} \mathscr{L}(X^{k+m}, X^k)$ is strongly continuous with $m \in \mathbb{N}$ fixed.

- the Cauchy problem for $A(t) \in \mathscr{L}(X^{k+m}, X^k)$ in the scale $(X^k)_k$ is well-posed with exponential growth with corresponding evolution operators $(U(t,s))_{(t,s)\in\Delta} \subset \bigcap\limits_{k \in \mathbb{N}_0} \mathscr{L}(X^k)$.

- for any $k \geq K$ we have $f : I \times X^k \longrightarrow X^k$ continuously.

- for $k \geq K$ and $R > 0$ there is an $F_k(R) > 0$ with

$$\|f(t,u) - f(t,v)\|_k \leq F_k(R) \|u - v\|_k$$

for $u, v \in X^k$ with $\|u\|_k \leq R, \|v\|_k \leq R, t \in I$.

- for $k \geq K$ and $R > 0$ there is an $\widetilde{F}_k(R) > 0$ with

$$\|f(t,u)\|_{k+1} \leq \widetilde{F}_k(R) \|u\|_{k+1} \quad \text{for } u \in X^{k+1} \text{ with } \|u\|_k \leq R, t \in I .$$

Then the following statements hold:

(a) Let $k \geq K + m, x \in X^{k+1}, [t_0, t_2] \subset I$, and $u \in C([t_0, t_2], X^k)$ with

$$u(t) = U(t, t_0)x + \int_{t_0}^t U(t,s)f(s, u(s))ds . \tag{3.1.1}$$

Then $u \in C([t_0, t_2], X^{k+1})$ and (3.1.1) is also valid in X^{k+1}.

(b) Let $x \in X^\infty$. Then there is a $t_2 \in I$ with $t_0 < t_2$ and a unique function $u \in C^1([t_0, t_2], X^\infty)$ with

$$u'(t) = A(t)u(t) + f(t, u(t)) \text{ for } t \in [t_0, t_2], \quad u(t_0) = x .$$

PROOF: We only have to prove (a). Let $I_{k'} \subset I$ be the maximal interval of existence for the solution $u \in C(I_{k'}, X^{k'})$ of (3.1.1) with left boundary point t_0 for $k' = k$ and $k' = k+1$. Assume that there is a $t_3 \in I_k$ with $I_{k+1} \subset [t_0, t_3]$. Then there is an $R_k > 0$ with $\|u(t)\|_k \leq R_k$ for $t \in [t_0, t_3]$. Hence

$$\|u(t)\|_{k+1} \leq \|U(t,t_0)x\|_{k+1} + \int_{t_0}^t \|U(t,s)f(s,u(s))\|_{k+1}\, ds$$

$$\leq \tilde{u}_{k+1} \|x\|_{k+1} + \tilde{u}_{k+1}\tilde{F}_k(R_k) \int_{t_0}^t \|u(s)\|_{k+1}\, ds$$

for $t \in I_{k+1}$ and suitable $\tilde{a}_{k+1} \geq 0$. Therefore

$$\|u(t)\|_{k+1} \leq \|x\|_{k+1}\tilde{u}_{k+1} e^{\tilde{u}_{k+1}\tilde{F}_k(R_k)(t-t_0)} \leq R_{k+1}$$

for $t \in I_{k+1}$ due to 3.1.5. This contradicts 3.1.1.   □

In the last part of this section we give some results on global existence. Since the case of ordinary differential equations is included in the previous statements (if $A(t) = 0$), these results can be only as good as general global existence results for ordinary differential equations. However, we will prove a global existence result with the greatest generality consistent with the theory of ordinary differential equations. In particular, this includes superlinear growth. Therefore we recall the following simple lemma.

3.1.7 Lemma. Let $\varphi : [0, \infty) \to (0, \infty)$ be continuous and $t_0 \in \mathbb{R}, x_0 \geq 0$. Then the initial value problem

$$x' = \varphi(x) \qquad x(t_0) = x_0 \tag{3.1.2}$$

has a unique solution on $[t_0, \infty)$ if and only if

$$\int_{x_0}^\infty \frac{dx}{\varphi(x)} = \infty . \tag{3.1.3}$$

PROOF: (3.1.2) has a unique solution $u : [t_0, \infty) \mapsto [0, \infty)$ if and only if any solution is bounded on bounded intervals. Thus the lemma is a consequence of the formula of separation of variables

$$t - t_0 = \int_{x_0}^{u(t)} \frac{dx}{\varphi(x)} .$$

□

3.1.8 Example. (3.1.3) is satisfied by $\varphi(x) = ax + b$ or $\varphi(x) = ax \ln(x+1) + b$ with $a, b > 0$.

3.1.9 Proposition. Let $\Delta \ni (t, s) \mapsto U(t, s) \in \mathscr{L}(X)$ be a propagator and $f : I \times X \to X$ be continuous such that

- $\|U(t, s)\|_{\mathscr{L}(X)} \leq \tilde{u}$ for $(t, s) \in \Delta$ and a suitable constant $\tilde{u}$.

- $\|f(s, 0)\| \leq \tilde{F}$ for $s \in I$.

- for any $R > 0$ there is an $F(R) > 0$ with

$$\|f(t, u) - f(t, v)\| \leq F(R) \|u - v\|$$

  for $t \in I, u, v \in X$ with $\|u\| \leq R, \|v\| \leq R$.

- there is a non-decreasing and continuous function $\varphi : [0, \infty) \to (0, \infty)$ with

$$\|f(t, x)\| \leq \varphi(\|x\|) \text{ for } t \in I, x \in X, \qquad \text{and} \qquad \int_0^\infty \frac{dx}{\varphi(x)} = \infty.$$

Then, for $x_0 \in X$, there is a unique $u \in \mathcal{C}(I, X)$ with

$$u(t) = U(t, t_0)x_0 + \int_{t_0}^t U(t, s)f(s, u(s))ds \quad \text{for } t \in I.$$

PROOF: Let $u \in \mathcal{C}(I', X)$ be the solution with maximal domain $I' \subset I$ and left boundary point $t_0 \in I'$ and let $w : [t_0, \infty) \to \mathbb{R}$ be the solution of

$$w'(t) = \tilde{u}\varphi(w(t)) \qquad w(t_0) = \tilde{u}\,\|x_0\|\ .$$

Then for $v(t) := \|u(t)\|$ we have

$$
\begin{aligned}
v(t) &= \left\| U(t, t_0)x_0 + \int_{t_0}^t U(t, s)f(s, u(s))ds \right\| \leq \tilde{u}\,\|x_0\| + \int_{t_0}^t \tilde{u}\,\|f(s, u(s))\|\,ds \\
&\leq \tilde{u}\,\|x_0\| + \int_{t_0}^t \tilde{u}\varphi(\|u(s)\|)ds \leq \tilde{u}\,\|x_0\| + \int_{t_0}^t \tilde{u}\varphi(v(s))ds
\end{aligned}
$$

for $t \in I'$. Hence 3.1.4 implies $\|u(t)\| \leq w(t)$ for $t \in I'$, which proves $I' = I$.
$\square$

3.1.10 Theorem. Suppose that

- $m \in \mathbf{N}, K \in \mathbf{N}_0$

- $I \ni t \mapsto A(t) \in \bigcap_{k \in \mathbf{N}_0} \mathscr{L}(X^{k+m}, X^k)$ is strongly continuous.

- the Cauchy problem for $A(t)$ in the scale $(X^k)_k$ is well-posed with exponential growth and the associated evolution operator $U(t, s)$ satisfies $\|U(t,s)\|_{\mathscr{L}(X^k)} \leq \tilde{u}_k$ for $k \in \mathbf{N}_0, (t,s) \in \triangle$, and suitable constants $\tilde{u}_k \geq 0$.

- $f : I \times X^{K+m} \to X^{K+m}$ is continuous and $\|f(t,0)\|_{K+m} \leq \tilde{F}_{K+m}$ for $t \in I$.

- for $R > 0$ there is an $F_{K+m}(R)$ with

$$\|f(t,u) - f(t,v)\|_{K+m} \leq F_{K+m}(R) \|u - v\|_{K+m}$$

for $u, v \in X^{K+m}$ with $\|u\|_{K+m} \leq R, \|v\|_{K+m} \leq R$.

- there is a non-decreasing and continuous $\varphi_{K+m} : [0, \infty) \to (0, \infty)$ with $\|f(t,x)\|_{K+m} \leq \varphi_{K+m}(\|x\|_{K+m})$ for $t \in I, x \in X^{K+m}$, and

$$\int_0^\infty \frac{dx}{\varphi_{K+m}(x)} = \infty.$$

Then, for $x_0 \in X^{K+m}$ there is a unique $u \in C^1(I, X^K) \cap C(I, X^{K+m})$ with

$$\frac{du}{dt}(t) = A(t)u(t) + f(t, u(t)) \text{ for } t \in I, \qquad u(t_0) = x_0.$$

PROOF: 3.1.10 is implied by 3.1.9 in the same manner as 3.1.2 is implied by 3.1.1. □

3.1.11 Bibliographical remarks.

Semilinear evolution equations in the setting of C_0-semigroups have been considered first by Segal [139], cf. also [52, II, 2.4], [127, section X.13], [25, 4.3]. In these references also regularity and global existence results for linearly bounded f (i.e., $\|f(t,x)\| \leq a_t \|x\| + b_t$) have been obtained. Lemma 3.1.4 can be found e.g. in [57, III4.4].

3.2   Commutator estimates and quasilinear evolution equations

As in the linear case (cf. section 2.3) one can also use commutator estimates in the quasilinear case to reduce several assumptions to the base space. We will collect some results in this section that will be useful later on.

3.2.1 Lemma. Let $(X, \|\cdot\|)$ be a Banach space, $Z = \{Z_1, \ldots, Z_M\}$ a commuting family of generators of bounded C_0-semigroups, and $K \in \mathbb{N}_0, m \in \mathbb{N}$, Moreover, assume that $I \ni t \mapsto A(t, v) \in \mathscr{L}(X_Z^{k+m}, X_Z^k)$ is strongly continuous for $v \in X_Z^k, k \geq K$, and suppose that for $R > 0$ and $k \geq K$ there are $d_{k,R}, \beta_R \geq 0$ with (using the notation of 2.2.6)

$$\|Z^\alpha A(t, v)x - A(t, v)Z^\alpha x\| \leq d_{k,R} \|x\|_k \quad \text{for } v \in X_Z^k, \|v\|_k \leq R,$$
$$\|\lambda x - A(t, v)x\| \geq (\lambda - \beta_R) \|x\| \quad \text{for } v \in X_Z^K, \|v\|_K \leq R, \lambda > \beta_R,$$

$t \in I, x \in X_Z^\infty, \alpha \in \mathbb{N}_0^M, |\alpha| \leq k$. Then, for $R > 0$ and $k \geq K$ there are $\beta_{k,R} \geq 0$ with

$$\|\lambda x - A(t, v)x\|_k \geq (\lambda - \beta_{k,R}) \|x\|_k$$

for $t \in I, v \in X_Z^k, \|v\|_k \leq R, x \in X_Z^\infty, \lambda > \beta_{k,R}$.

PROOF: For $k \in \mathbb{N}_0$ and $0 \leq |\alpha| \leq k$ there are constants $\gamma_{\alpha,k} \geq 0$ with

$$\|x\|_k = \sqrt{\sum_{|\alpha| \leq k} \gamma_{\alpha,k} \|Z^\alpha x\|^2} \qquad \text{for } x \in X_Z^k, k \in \mathbb{N}_0.$$

Therefore

$$(\lambda - \beta_R) \|x\|_k = \sqrt{\sum_{|\alpha| \leq k} \gamma_{\alpha,k}((\lambda - \beta_R) \|Z^\alpha x\|)^2}$$

$$\leq \sqrt{\sum_{|\alpha| \leq k} \gamma_{\alpha,k} (\|Z^\alpha(\lambda x - A(t, v)x)\| + \|[Z^\alpha, A(t, v)]x\|)^2}$$

$$\leq \sqrt{\sum_{|\alpha| \leq k} \gamma_{\alpha,k} \|Z^\alpha(\lambda x - A(t, v)x)\|^2} + \sqrt{\sum_{|\alpha| \leq k} \gamma_{\alpha,k} \|[Z^\alpha, A(t, v)]x\|^2}$$

$$\leq \|\lambda x - A(t, v)x\|_k + \beta'_{k,R} \|x\|_k$$

for $t \in I, v \in X_Z^k, \|v\|_k \leq R, k \geq K, x \in X_Z^\infty, \lambda > \beta_R$ and suitable $\beta'_{k,R} \geq 0$. This implies the assertion with $\beta_{k,R} := \beta_R + \beta'_{k,R}$.

$\square$

The next result does not only assume a weak commutator estimate as in 3.2.1, but even a stronger commutator estimate of Nash-Moser type. This lemma will be needed for applications of regularity results; the estimate can be verified with Gagliardo-Moser-Nirenberg estimates.

3.2.2 Lemma. Let $(H, \|\cdot\|)$ be a Hilbert space, $Z = \{Z_1, \ldots, Z_M\}$ a commuting family of generators of bounded C_0-semigroups, $K \in \mathbb{N}_0, m \in \mathbb{N}$, and $I \subset \mathbb{R}$ a compact interval. Moreover, assume that $I \ni t \mapsto A(t, v) \in \mathscr{L}(H_Z^{k+m}, H_Z^k)$ is strongly continuous for $v \in H_Z^k, k \geq K$, and suppose that for $R > 0$ and $k \geq K$ there are $d'_{k,R}, \beta_R > 0$ with

$$\|Z^\alpha A(t, v)x - A(t, v)Z^\alpha x\| \leq d'_{k,R}(\|v\|_{k+m} \|x\|_k + \|x\|_{k+m})$$

for $t \in I, v \in H_Z^{k+m}, \|v\|_k \leq R, x \in H_Z^\infty, \alpha \in \mathbb{N}_0^M, |\alpha| \leq k + m$, and

$$\mathrm{Re} \langle A(t, v)x, x \rangle \leq \beta_R \|x\|^2 \qquad \text{for } t \in I, v \in H^K, \|v\|_K \leq R, x \in H_Z^\infty.$$

Moreover, let $J_l := \left(\mathrm{Id} - \frac{1}{l}Z_1\right)^{-1} \cdots \left(\mathrm{Id} - \frac{1}{l}Z_M\right)^{-1}$ and $S_l := J_l^{k'}$ for $l \in \mathbb{N}$ and a fixed $k' \geq m$. Then, for $R > 0$ and $k \geq K + 1$ there are $\gamma_{k,R} \geq 0$ with

$$\mathrm{Re} \langle S_l A(t, u)u, S_l u \rangle_{k+m} \leq \gamma_{k,R} \|u\|_{k+m}^2$$

for $l \in \mathbb{N}, t \in I, u \in H_Z^{k+m}, \|u\|_k \leq R$.

PROOF:
Step 1: For $R > 0, k \geq K$ there is a $\beta_{k,R} > 0$ with

$$\mathrm{Re} \langle A(t, v)x, x \rangle_{k+m} \leq \beta_{k,R}(\|x\|_{k+m}^2 + \|x\|_k \|v\|_{k+m} \|x\|_{k+m})$$

for $t \in I, v \in H_Z^{k+m}, \|v\|_k \leq R, x \in H_Z^{k+2m}$.
Proof: For $k \in \mathbb{N}_0$ and $0 \leq |\alpha| \leq k$ there are $\gamma_{\alpha,k} \geq 0$ with

$$\langle x, y \rangle_k = \sum_{|\alpha| \leq k} \gamma_{\alpha,k} \langle Z^\alpha x, Z^\alpha y \rangle$$

for $x, y \in H_Z^k$. Hence, for $t \in I, v \in H_Z^{k+m}, \|v\|_k \leq R, x \in H_Z^\infty$

$$\mathrm{Re} \langle A(t, v)x, x \rangle_{k+m} = \sum_{|\alpha| \leq k+m} \gamma_{\alpha,k+m} \mathrm{Re} \langle Z^\alpha A(t, v)x, Z^\alpha x \rangle$$

$$= \sum_{|\alpha| \leq k+m} \gamma_{\alpha,k+m} \mathrm{Re} \langle A(t, v)Z^\alpha x + Z^\alpha A(t, v)x - A(t, v)Z^\alpha x, Z^\alpha x \rangle$$

$$\leq \sum_{|\alpha| \leq k+m} (\beta_R \|Z^\alpha x\|^2 + d'_{k,R}(\|v\|_{k+m} \|x\|_k + \|x\|_{k+m}) \|x\|_{k+m})$$

$$\leq \beta_{k,R}(\|x\|_{k+m}^2 + \|x\|_k \|v\|_{k+m} \|x\|_{k+m})$$

for suitable $\beta_{k,R} \geq 0$. Since $H_Z^\infty \subset H_Z^{k+2m}$ densely this implies the assertion of step 1.

Step 2: For $k \geq K+1, R > 0$ there is a $g_{k,R} \geq 0$ with

$$\|S_l A(t,u)u - A(t,u)S_l u\|_{k+m} \leq g_{k,R}\|u\|_{k+m}$$

for $t \in I, u \in H_Z^{k+m}, \|u\|_k \leq R, l \in \mathbb{N}$.

Proof: For $k \geq K+1, \mu = 1,\ldots,M$, and $u \in H_Z^{k+m}, \|u\|_k \leq R, x \in H_Z^\infty$ we have

$$\|Z_\mu A(t,u)x - A(t,u)Z_\mu x\|_{k+m-1} \leq c \sum_{|\alpha| \leq k+m-1} \|Z^\alpha (Z_\mu A(t,u)x - A(t,u)Z_\mu x)\|$$

$$\leq c \sum_{|\alpha| \leq k+m-1} [\|Z^{\alpha+e_\mu}A(t,u)x - A(t,u)Z^{\alpha+e_\mu}x\| + \|A(t,u)Z^\alpha(Z_\mu x) - Z^\alpha A(t,u)Z_\mu x\|]$$

$$\leq \tilde{d}_{k,R}(\|u\|_{k+m}\|x\|_k + \|x\|_{k+m}) \tag{3.2.1}$$

with suitable $c, \tilde{d}_{k,R} \geq 0$, and for $k \geq K+1, \mu = 1,\ldots,M, u \in H_Z^{k+m}$, $\|u\|_k \leq R, x \in H_Z^\infty$

$$\left(\text{Id} - \frac{1}{l}Z_\mu\right)^{-1} A(t,u)x - A(t,u)\left(\text{Id} - \frac{1}{l}Z_\mu\right)^{-1}x$$

$$= \frac{1}{l}\left(\text{Id} - \frac{1}{l}Z_\mu\right)^{-1}(Z_\mu A(t,u) - A(t,u)Z_\mu)\left(\text{Id} - \frac{1}{l}Z_\mu\right)^{-1}x .$$

Moreover, for $u \in H_Z^{k+m}, \|u\|_k \leq R, k \geq K+1, x \in H_Z^\infty, \mu = 1,\ldots,M$ we have

$$J_l A(t,u)x - A(t,u)J_l x$$

$$= \left(\text{Id} - \frac{1}{l}Z_\mu\right)^{-1}\left[\prod_{\substack{j=1 \\ j\neq\mu}}^{M}\left(\text{Id} - \frac{1}{l}Z_j\right)^{-1} A(t,u) - A(t,u)\prod_{\substack{j-1 \\ j\neq\mu}}^{M}\left(\text{Id} - \frac{1}{l}Z_j\right)^{-1}\right]x$$

$$+ \left[\left(\text{Id} - \frac{1}{l}Z_\mu\right)^{-1} A(t,u) - A(t,u)\left(\text{Id} - \frac{1}{l}Z_\mu\right)^{-1}\right]\prod_{\substack{j=1 \\ j\neq\mu}}^{M}\left(\text{Id} - \frac{1}{l}Z_j\right)^{-1}x$$

$$= \frac{1}{l}\left(\text{Id} - \frac{1}{l}Z_\mu\right)^{-1}\left[\sum_{\substack{j=1 \\ j\neq\mu}}^{M}\prod_{\substack{p=1 \\ p\neq\mu}}^{j}\left(\text{Id} - \frac{1}{l}Z_\rho\right)^{-1}(Z_j A(t,u) - A(t,u)Z_j)\right.$$

$$\times \prod_{\substack{p=j \\ p\neq\mu}}^{M}\left(\text{Id} - \frac{1}{l}Z_\rho\right)^{-1}x + (Z_\mu A(t,u) - A(t,u)Z_\mu)\prod_{j=1}^{M}\left(\text{Id} - \frac{1}{l}Z_j\right)^{-1}x\right] .$$

Hence

$$\|J_l A(t,u)x - A(t,u)J_l x\|_{k+m-1}$$

$$\underset{(3.2.1)}{\leq} \tilde{d}'_{k,R}\left(\sum_{\substack{j=1\\j\neq\mu}}^{M}\left(\|u\|_{k+m}\left\|\prod_{\substack{\rho=j\\\rho\neq\mu}}^{M}\left(\mathrm{Id}-\frac{1}{l}Z_\rho\right)^{-1}x\right\|_k\right.\right.$$

$$\left.\left.+\left\|\prod_{\substack{\rho=j\\\rho\neq\mu}}^{M}\left(\mathrm{Id}-\frac{1}{l}Z_\rho\right)^{-1}x\right\|_{k+m}\right)+\|u\|_{k+m}\|J_l x\|_k+\|J_l x\|_{k+m}\right)$$

$$\leq \tilde{d}''_{k,R}(\|u\|_{k+m}\|x\|_k+\|x\|_{k+m})$$

for suitable $\tilde{d}'_{k,R}, \tilde{d}''_{k,R}$, and

$$\left\|Z_\mu(J_l A(t,u)x - A(t,u)J_l x)\right\|_{k+m-1}$$

$$\leq e'_{k,R}\left(\sum_{\substack{j=1\\j\neq\mu}}^{M}\left(\|u\|_{k+m}\left\|\prod_{\substack{\rho=j\\\rho\neq\mu}}^{M}\left(\mathrm{Id}-\frac{1}{l}Z_\rho\right)^{-1}x\right\|_k\right.\right.$$

$$\left.\left.+\left\|\prod_{\substack{\rho=j\\\rho\neq\mu}}^{M}\left(\mathrm{Id}-\frac{1}{l}Z_\rho\right)^{-1}x\right\|_{k+m}\right)+\|u\|_{k+m}\|J_l x\|_k+\|J_l x\|_{k+m}\right)$$

$$\leq e''_{k,R}(\|u\|_{k+m}\|x\|_k+\|x\|_{k+m})$$

for $u\in H_Z^{k+m},\|u\|_k\leq R, x\in H_Z^\infty, k\geq K+1,\mu=1,\ldots,M, l\in\mathbb{N}$, and suitable $e'_{k,R}, e''_{k,R}$. Therefore

$$\left\|J_l^{k'}A(t,u)x - A(t,u)J_l^{k'}x\right\|_{k+m}$$

$$\leq c\left\|J_l^{k'}A(t,u)x - A(t,u)J_l^{k'}x\right\|_{k+m-1}$$

$$+c\sum_{\mu=1}^{M}\left\|Z_\mu(J_l^{k'}A(t,u)x - A(t,u)J_l^{k'}x)\right\|_{k+m-1}$$

$$\leq c\sum_{\nu=0}^{k'-1}\left\|J_l^{k'-1-\nu}(J_l A(t,u)-A(t,u)J_l)J_l^\nu x\right\|_{k+m-1}$$

$$+c\sum_{\mu=1}^{M}\sum_{\nu=0}^{k'-1}\left\|J_l^{k'-1-\nu}Z_\mu(J_l A(t,u)-A(t,u)J_l)J_l^\nu x\right\|_{k+m-1}$$

$$\leq f'_{k,R}\sum_{\nu=0}^{k'-1}(\|u\|_{k+m}\|J_l^\nu x\|_k+\|J_l^\nu x\|_{k+m})\leq f_{k,R}(\|u\|_{k+m}\|x\|_k+\|x\|_{k+m})$$

with suitable constants $c, f'_{k,R}, f_{k,R} \geq 0$. Hence we have shown

$$\|S_l A(t,u)x - A(t,u)S_l x\|_{k+m} \leq f_{k,R}(\|u\|_{k+m}\|x\|_k + \|x\|_{k+m})$$

for $l \in \mathbb{N}, u \in H_Z^{k+m}, \|u\|_k \leq R, k \geq K+1, x \in H_Z^\infty$. Since $H_Z^\infty \subset H_Z^{k+m}$ densely, this holds by continuity for $x \in H_Z^{k+m}$ and implies the assertion of step 2.

Step 3 : Proof of the assertion:

$$\begin{aligned}
&\mathrm{Re} \, \langle S_l A(t,u)u, S_l u \rangle_{k+m} \\
\leq \;& \mathrm{Re} \, \langle A(t,u)S_l u, S_l u \rangle_{k+m} + \|S_l A(t,u)u - A(t,u)S_l u\|_{k+m} \|S_l u\|_{k+m} \\
\leq \;& \beta_{k,R} \left(\|S_l u\|_{k+m}^2 + \|S_l u\|_k \|u\|_{k+m} \|S_l u\|_{k+m} \right) + g'_{k,R} \|u\|_{k+m}^2 \\
\leq \;& \gamma_{k,R} \|u\|_{k+m}^2
\end{aligned}$$

for $t \in I, u \in H_Z^{k+m}, \|u\|_k \leq R, l \in \mathbb{N}, k \geq K+1$, and suitable $g'_{k,R}, \gamma_{k,R} \geq 0$. $\square$

Whereas the preceding lemma dealt with quasilinear operators satisfying first order commutator estimates, the following lemma gives a similar estimate for linear operators.

3.2.3 Lemma. Let $(\Lambda, D(\Lambda))$ be a strictly negative, selfadjoint operator in a Hilbert space H and let $\mathcal{A}$ be a family of linear operators $A : H_\Lambda^\infty \to H_\Lambda^\infty$ such that for $k \in \mathbb{N}_0$ there are $a_k \geq 0$ and $\beta \geq 0$ with

$$\|\Lambda^k Ax - A\Lambda^k x\| \leq a_k \|x\|_k \quad \text{and} \quad \mathrm{Re} \, \langle Ax, x \rangle \leq \beta \|x\|^2$$

for $x \in H_\Lambda^\infty, A \in \mathcal{A}$. Moreover, let $S_l := (\mathrm{Id} - \tfrac{1}{l}\Lambda)^{-k'}$ for $l \in \mathbb{N}$ and a fixed $k' \in \mathbb{N}$.

Then, for $k \in \mathbb{N}_0$ there are $\beta_k \geq 0$ with

$$\mathrm{Re} \, \langle S_l Ax, S_l x \rangle_k \leq \beta_k \|x\|_k^2$$

for $l \in \mathbb{N}, x \in H_\Lambda^\infty, A \in \mathcal{A}$.

PROOF: Let $B := S_l^2 A$ for $l \in \mathbb{N}, A \in \mathcal{A}$, then with a suitable $c \geq 0$

$$\begin{aligned}
\|\Lambda^k Bx - B\Lambda^k x\| &= \|\Lambda^k S_l^2 Ax - S_l^2 A\Lambda^k x\| = \|S_l^2(\Lambda^k Ax - A\Lambda^k x)\| \\
&\leq c\|\Lambda^k Ax - A\Lambda^k x\| \leq ca_k \|x\|_k .
\end{aligned}$$

for $x \in H_\Lambda^\infty$. Moreover, for $x \in H_\Lambda^\infty$

$$\left\| \left(\mathrm{Id} - \frac{1}{l}\Lambda \right)^{-1} Ax - A \left(\mathrm{Id} - \frac{1}{l}\Lambda \right)^{-1} x \right\|$$

$$= \left\| \left(\mathrm{Id} - \frac{1}{l}\Lambda \right)^{-1} \left(A \left(\mathrm{Id} - \frac{1}{l}\Lambda \right) - \left(\mathrm{Id} - \frac{1}{l}\Lambda \right) A \right) \left(\mathrm{Id} - \frac{1}{l}\Lambda \right)^{-1} x \right\|$$

$$\leq \frac{c'}{l} \left\| [\Lambda, A] \left(\mathrm{Id} - \frac{1}{l}\Lambda \right)^{-1} x \right\| \leq c'a_1 \left\| \Lambda^{-1} \left(\mathrm{Id} - \left(\mathrm{Id} - \frac{1}{l}\Lambda \right) \right) \left(\mathrm{Id} - \frac{1}{l}\Lambda \right)^{-1} x \right\|_1$$

$$\leq c'' \left\| \left(\mathrm{Id} - \frac{1}{l}\Lambda \right)^{-1} x - x \right\| \leq 2c'' \|x\|$$

with suitable constants $c', c'' > 0$, and thus $\|S_l Ax - AS_l x\| \leq d\|x\|$ for a suitable $d \geq 0$. Hence

$$\begin{aligned}
\mathrm{Re}\,\langle Bx, x \rangle &= \mathrm{Re}\,\langle S_l^2 Ax, x \rangle = \mathrm{Re}\,\langle S_l Ax, S_l x \rangle \\
&= \mathrm{Re}\,\langle AS_l x, S_l x \rangle + \mathrm{Re}\,\langle S_l Ax - AS_l x, S_l x \rangle \\
&\leq \beta \|S_l x\|^2 + d\|x\|\,\|S_l x\| \leq \beta' \|x\|^2
\end{aligned}$$

for a suitable $\beta' \geq 0$ and 2.3.1 shows that for $k \in \mathbb{N}_0$ there are $\beta_k \geq 0$ with

$$\mathrm{Re}\,\langle S_l Ax, S_l x \rangle_k = \mathrm{Re}\,\langle S_l^2 Ax, x \rangle_k = \mathrm{Re}\,\langle Bx, x \rangle_k \leq \beta_k \|x\|_k^2$$

for $x \in H_\Lambda^\infty, l \in \mathbb{N}$.   □

The last estimate in this section gives an estimate in the spirit of the previous estimate but dealing now with linear operators satisfying second-order commutator estimates.

3.2.4 Lemma. Let $(\Lambda, D(\Lambda))$ be a strictly negative, selfadjoint operator in a Hilbert space H and let $\mathcal{A}$ be a family of linear operators $A : H_\Lambda^\infty \to H_\Lambda^\infty$ such that any $A \in \mathcal{A}$ is symmetric in H on H_Λ^∞ and that for $k \in \mathbb{N}_0$ there are constants $a_{k,2} \geq 0, \beta_k'' \geq 0$ with

$$\left\| \Lambda^k [\Lambda, [\Lambda, A]] x \right\| \leq a_{k,2} \|x\|_{k+2} \quad \text{and} \quad \mathrm{Re}\,\langle Ax, x \rangle_k \leq \beta_k'' \|x\|_k^2$$

for $x \in H_\Lambda^\infty, A \in \mathcal{A}$. Moreover, let $S_l := (\mathrm{Id} - \frac{1}{l}\Lambda)^{-k'}$ with $k' \geq 2$. Then, for $k \in \mathbb{N}_0$ there are $\beta_k \geq 0$ with

$$\mathrm{Re}\,\langle S_l Ax, S_l x \rangle_k \leq \beta_k \|x\|_k^2 \qquad \text{for } l \in \mathbb{N}, x \in H_\Lambda^\infty, A \in \mathcal{A}.$$

PROOF: Let $\tilde{A} := S_l[S_l, A]$, then

$$[\Lambda, \tilde{A}]x = [\Lambda, S_l[S_l, A]]x$$

$$= S_l\left[\Lambda, \sum_{j=0}^{k'-1}\left(\mathrm{Id} - \frac{1}{l}\Lambda\right)^{-j}\left[\left(\mathrm{Id} - \frac{1}{l}\Lambda\right)^{-1}, A\right]\left(\mathrm{Id} - \frac{1}{l}\Lambda\right)^{-(k'-1-j)}\right]x$$

$$= \sum_{j=0}^{k'-1} S_l\left(\mathrm{Id} - \frac{1}{l}\Lambda\right)^{-j-1}\left[\Lambda, \left[\frac{1}{l}\Lambda, A\right]\right]\left(\mathrm{Id} - \frac{1}{l}\Lambda\right)^{-(k'-j)}x$$

$$= \sum_{j=0}^{k'-1} S_l\left(\mathrm{Id} - \frac{1}{l}\Lambda\right)^{-j-1}[\Lambda, [\Lambda, A]]\Lambda^{-1}\left(\mathrm{Id} - \left(\mathrm{Id} - \frac{1}{l}\Lambda\right)\right)\left(\mathrm{Id} - \frac{1}{l}\Lambda\right)^{-(k'-j)}x .$$

for $x \in H_\Lambda^\infty$. Hence

$$\left\|[\Lambda, \tilde{A}]x\right\|_k$$

$$\leq \sum_{j=0}^{k'-1} c_k''\left\|[\Lambda, [\Lambda, A]]\Lambda^{-1}\left(\left(\mathrm{Id} - \frac{1}{l}\Lambda\right)^{-1} - \mathrm{Id}\right)\left(\mathrm{Id} - \frac{1}{l}\Lambda\right)^{-(k'-1-j)}x\right\|_k$$

$$\leq \sum_{j=0}^{k'-1} c_k'\left\|\Lambda^{-1}\left(\left(\mathrm{Id} - \frac{1}{l}\Lambda\right)^{-1} - \mathrm{Id}\right)\left(\mathrm{Id} - \frac{1}{l}\Lambda\right)^{-(k'-1-j)}x\right\|_{k+2}$$

$$\leq c_k \|x\|_{k+1} \tag{3.2.2}$$

for $x \in H_\Lambda^\infty$ and suitable $c_k, c_k', c_k'' \geq 0$. Moreover, for $B : H_\Lambda^\infty \to H_\Lambda^\infty$ and $x \in H_\Lambda^\infty$ we have

$$[S_l, B]x = \sum_{j=0}^{k'-1}\left(\mathrm{Id} - \frac{1}{l}\Lambda\right)^{-j}\left[\left(\mathrm{Id} - \frac{1}{l}\Lambda\right)^{-1}, B\right]\left(\mathrm{Id} - \frac{1}{l}\Lambda\right)^{-(k'-1-j)}x$$

$$= \sum_{j=0}^{k'-1}\left(\mathrm{Id} - \frac{1}{l}\Lambda\right)^{-j-1}\left[\frac{1}{l}\Lambda, B\right]\left(\mathrm{Id} - \frac{1}{l}\Lambda\right)^{-(k'-j)}x .$$

Therefore

$$\|[S_l, [S_l, A]]x\|$$

$$\leq \sum_{j_1,j_2=0}^{k'-1}\left\|\left(\mathrm{Id} - \frac{1}{l}\Lambda\right)^{-(j_1+j_2)-2}[\Lambda, [\Lambda, A]]\Lambda^{-2}\times\right.$$

$$\left.\times\left(\left(\mathrm{Id} - \left(\mathrm{Id} - \frac{1}{l}\Lambda\right)\right)\left(\mathrm{Id} - \frac{1}{l}\Lambda\right)^{-1}\right)^2\left(\mathrm{Id} - \frac{1}{l}\Lambda\right)^{-(2k'-2-j_1-j_2)}x\right\|$$

$$\leq c''\sum_{j_1,j_2=0}^{k'-1}\left\|[\Lambda, [\Lambda, A]]\Lambda^{-2}\left(\left(\mathrm{Id} - \frac{1}{l}\Lambda\right)^{-1} - \mathrm{Id}\right)^2\left(\mathrm{Id} - \frac{1}{l}\Lambda\right)^{-(2k'-2-j_1-j_2)}x\right\|$$

$$\le c' \sum_{j_1,j_2=0}^{k'-1} \left\| \Lambda^{-2}\left(\left(\mathrm{Id}-\frac{1}{l}\Lambda\right)^{-1}-\mathrm{Id}\right)^2 \left(\mathrm{Id}-\frac{1}{l}\Lambda\right)^{-(2k'-2-j_1-j_2)} x\right\|_2$$

$$\le c\,\|x\|$$

for $x \in H_\Lambda^\infty$ with suitable $c'', c', c \ge 0$. This implies

$$2\mathrm{Re}\left\langle \widetilde{A}x, x\right\rangle = \langle S_l[S_l, A]x, x\rangle + \langle x, S_l[S_l, A]x\rangle$$
$$= \langle S_l[S_l, A]x, x\rangle - \langle [S_l, A]S_l x, x\rangle = \langle [S_l, [S_l, A]]x, x\rangle \le c\,\|x\|^2$$

and if we apply 2.3.1 with (3.2.2) we obtain constants $\beta_k' \ge 0$ for $k \in \mathbf{N}_0$ such that

$$\mathrm{Re}\,\langle S_l[S_l, A]x, x\rangle_k = \mathrm{Re}\left\langle \widetilde{A}x, x\right\rangle_k \le \beta_k'\,\|x\|_k^2$$

for $l \in \mathbf{N}, k \in \mathbf{N}_0, A \in \mathcal{A}, x \in H_\Lambda^\infty$. Finally

$$\mathrm{Re}\,\langle S_l A x, S_l x\rangle_k = \mathrm{Re}\,\langle A S_l x, S_l x\rangle_k + \mathrm{Re}\,\langle [S_l, A]x, S_l x\rangle_k$$
$$\le \beta_k''\,\|S_l x\|_k^2 + \mathrm{Re}\,\langle S_l[S_l, A]x, x\rangle_k \le \beta_k\,\|x\|_k^2$$

for $l \in \mathbf{N}, k \in \mathbf{N}_0, A \in \mathcal{A}, x \in H_\Lambda^\infty$ and suitable $\beta_k \ge 0$. $\qquad\square$

3.3   A local existence and uniqueness result for quasilinear evolution equations

In this section we will prove a general existence and uniqueness result for quasilinear evolution equations in scales of Banach spaces provided that related linear time-dependent Cauchy problems are well-posed in this scale. To this end we will use a method similar to a construction of Kato who used his theory for linear, time-dependent evolution equations to construct solutions of quasilinear equations (cf. [82], [85]). Roughly speaking he constructed the solutions as a limit of a sequence $(u_l)_l$ of solutions of time-dependent, linear equations defined inductively by

$$u_{l+1}'(t) = A(t, u_l(t))u_{l+1}(t) + f(t, u_l(t))\,.$$

To let this iteration work it is important that the solutions of the linear, time-dependent evolution equations have the same regularity as the coefficients of the equation. But this does not hold for the theory of linear evolution equations in scales of Banach spaces we developed in chapter 2. We had to assume there that the coefficients of $A(t)$ are more regular than the solutions we obtained. So Kato's arguments cannot be applied directly and we have to involve a regularization procedure in addition. In 3.3.9 we will show how this can be achieved. In the proof we will also use an idea from the proof of well-posedness of the Cauchy problem for first-order, symmetric hyperbolic, quasilinear evolution equations (cf. e.g. [126]). But first we will have to collect some definitions and lemmata.

3.3.1 Definition. Let $I \subset \mathbf{R}$ be an interval and let X be a Banach space, then

$$C_w(I, X) := \left\{ u : I \to X : u(t_l) \xrightarrow[l \to \infty]{} u(t) \text{ weakly for } t, t_l \in I \text{ with } t_l \xrightarrow[l \to \infty]{} t \right\} .$$

We call the elements of $C_w(I, X)$ weakly continuous functions. Note that weakly continuous functions are always locally bounded (due to the uniform boundedness principle).

We have to use some results on sequences in reflexive Banach spaces.

3.3.2 Proposition. Let $(X, \|\cdot\|_X), (Y, \|\cdot\|_Y)$ be Banach spaces with $Y \hookrightarrow X$ and let Y be reflexive. Let $(x_l)_l \subset Y$ be a sequence with $\|x_l\|_Y \leq M$ for $l \in \mathbf{N}$ with an $M \geq 0$, and $x_l \xrightarrow[l \to \infty]{} x \in X$ weakly in X.
Then $x \in Y, \|x\|_Y \leq M$, and $x_l \xrightarrow[l \to \infty]{} x$ weakly in Y.

PROOF: The Eberlein-Šmulian theorem (cf. [160, V,Appendix 4]) shows that there is a subsequence $(x_{l_k})_k$ of $(x_l)_l$, and an $y \in Y$ with $\|y\|_Y \leq M$ and $x_{l_k} \xrightarrow[k \to \infty]{} y$ weakly in Y, hence also in X. Thus $x = y \in Y$ with $\|x\|_Y \leq M$ because $x_l \xrightarrow[l \to \infty]{} x$ weakly in X.
Moreover, assume that there is a sequence $(l_n)_n$ with $l_n \xrightarrow[n \to \infty]{} \infty$, an $\varepsilon > 0$, and an $y^* \in Y^*$ with

$$\left| \langle y^*, x_{l_n} \rangle_{Y^*, Y} - \langle y^*, x \rangle_{Y^*, Y} \right| \geq \varepsilon .$$

Then the same argument as before applied to $(x_{l_n})_n$ shows that there is a subsequence of $(x_{l_n})_n$ converging weakly towards x in Y, which is a contradiction. $\square$

3.3.3 Corollary. Let X, Y be Banach spaces with $Y \hookrightarrow X$ and Y reflexive. Moreover, let $I \subset \mathbf{R}$ be an interval and $f : I \to Y$ be locally bounded with $f \in C_w(I, X)$.
Then we have $f \in C_w(I, Y)$.

3.3.4 Lemma. Let $(\alpha_l)_l, (\beta_l)_l \subset [0, \infty)$ be sequences and let $0 \le q < 1$. Then

$$\alpha_{l+1} \le q\alpha_l + \beta_l \text{ for } l \in \mathbb{N} \quad \text{and} \quad \sum_{l=1}^{\infty} \beta_l < \infty \quad \text{imply} \quad \sum_{l=1}^{\infty} \alpha_l < \infty .$$

PROOF:

$$\sum_{l=1}^{\infty} \alpha_l = \alpha_1 + \sum_{l=1}^{\infty} \alpha_{l+1} \le \alpha_1 + q \sum_{l=1}^{\infty} \alpha_l + \sum_{l=1}^{\infty} \beta_l \Longrightarrow \sum_{l=1}^{\infty} \alpha_l \le \frac{1}{1-q} \left(\alpha_1 + \sum_{l=1}^{\infty} \beta_l \right) .$$

$\square$

3.3.5 Definition. Let $(X^k, \|\cdot\|_k)_k$ be a scale of Banach spaces, $I = [t_0, t_1]$ be a compact interval, and let $K_1, K_2, K_3 \in \mathbb{N}_0, m \in \mathbb{N}$ with $K_1 \le K_2$ and $K_3 \ge m$.
We call the Cauchy problem for $A(t)$ in the scale $(X^k)_k$ well-posed in the part (K_1, K_2, K_3), if

$$I \ni t \mapsto A(t) \in \bigcap_{k=K_1}^{K_2+K_3} \mathscr{L}(X^{k+m}, X^k)$$

is strongly continuous and if the existence of $\beta_k \ge 0$ for $K_1 \le k \le K_2 + K_3$ with

$$\|\lambda x - A(t)x\|_k \ge (\lambda - \beta_k) \|x\|_k \quad \text{for } \lambda > \beta_k, x \in X^{k+m}, K_1 \le k \le K_2 + K_3$$

implies the existence of a propagator

$$U(t, s) \in \bigcap_{k=K_1}^{K_2+m} \mathscr{L}(X^k), (t, s) \in \Delta := \{(t, s) \in I^2 : t \ge s\} ,$$

which has the following properties:

(a) $\|U(t, s)\|_{\mathscr{L}(X^k)} \le e^{\beta_k(t-s)}$ for $(t, s) \in \Delta$ and $K_1 \le k \le K_2 + m$.

(b) $\Delta \ni (t, s) \mapsto U(t, s) \in \mathscr{L}(X^{k+m}, X^k)$ is strongly continuously differentiable for $K_1 \le k \le K_2$ with strong derivatives

$$\frac{\partial}{\partial t} U(t, s) = A(t)U(t, s), \qquad \frac{\partial}{\partial s} U(t, s) = -U(t, s)A(s) .$$

3.3.6 Proposition.

(a) Let $Z = \{Z_1, \ldots, Z_M\}$ be a commuting family of infinitesimal generators of C_0-semigroups of contractions in a Hilbert space H with $Z_j^* = \alpha_j Z_j$ for $j = 1, \ldots, M$ with suitable $\alpha_j \neq 0$. Then the Cauchy problem for any $A(t)$ in the scale $(H_Z^k)_k$ is well-posed in the part $(K_1, K_2, 2m + 2)$.

(b) If Λ is the infinitesimal generator of a bounded analytic semigroup with angle $\theta > \frac{m}{m+1} \frac{\pi}{2}$ in a complex Banach space X, then the Cauchy problem for $A(t)$ in the scale $(X_\Lambda^k)_k$ is well-posed in the part $(K_1, K_2, 3m + 3)$, if for $K_1 \leq k \leq K_2 + 3m + 2$ there are $d_k \geq 0$ with $\|[\Lambda, A(t)]x\|_k \leq d_k \|x\|_{k+1}$ for $x \in X_\Lambda^\infty, t \in I$.

PROOF: This is an immediate consequence of 2.5.2 resp., 2.5.7. $\qquad\square$

3.3.7 Definition. A family of operators $(J_l)_{l \in \mathbb{N}_0}$ is called a mollifier in a scale of Banach spaces $(X^k)_k$, if for $k \in \mathbb{N}_0$ are constants $j_k \geq 0$ with

- $(J_l)_l \subset \mathscr{L}(X^k, X^{k+1})$ for any $k \in \mathbb{N}_0$.

- $\|J_l\|_{\mathscr{L}(X^k)} \leq j_k$ for any $k \in \mathbb{N}_0, l \in \mathbb{N}$.

- $\|J_l - \mathrm{Id}\|_{\mathscr{L}(X^{k+1}, X^k)} \leq \frac{j_k}{l^2}$ for any $k \in \mathbb{N}_0, l \in \mathbb{N}$.

Note that these conditions imply $J_l \xrightarrow[l \to \infty]{} \mathrm{Id}$ strongly in $\mathscr{L}(X^k)$ for any $k \in \mathbb{N}_0$.

3.3.8 Proposition. Let $Z = \{Z_1, \ldots, Z_M\}$ be a family of infinitesimal generators of bounded C_0-semigroups in a Banach space X that is commuting modulo operators of order zero.
Then there is a mollifier $(J_l)_l$ in the scale $(X_Z^k)_k$.

PROOF: This is a consequence of 2.2.9. $\qquad\square$

Now we can formulate and prove the announced existence and uniqueness result for quasilinear evolution equations in scales of Banach spaces. Note that in the general case the constructed solution has less regularity than the initial value. However, for reflexive scales of Banach spaces the solutions have the same regularity as the initial values and are weakly continuous in this space. Note moreover, that in contrast to Kato's results (cf. [82], [85]) we do not use any commutator assumption. This will be important for some applications (cf. e.g. 5.2.1 or 5.3.10).

3.3.9 Theorem.

Assumptions:

(i) $(X^k)_k$ is a scale of Banach spaces with a mollifier $(J_l)_{l \in \mathbb{N}_0}$.

(ii) $B_R^k := \{x \in X^k : \|x\|_k \leq R\}$ for $R > 0, k \in \mathbb{N}_0$.

(iii) $I = [t_0, t_1] \subset \mathbb{R}$ is a compact interval and $K \in \mathbb{N}_0, m \in \mathbb{N}$ are fixed.

(iv) There is a family of linear mappings $A(t, v)$ with

- $I \ni t \mapsto A(t, v) \in \mathscr{L}(X^{k+m}, X^k)$ is strongly continuous for $v \in X^k$, $k \geq K$.

- For $R > 0, k \geq K$ there is a constant $\tilde{a}_{k,R} \geq 0$ with

$$\|A(t, v) - A(t, w)\|_{\mathscr{L}(X^{k+m}, X^k)} \leq \tilde{a}_{k,R} \|v - w\|_k \quad \text{for } v, w \in B_R^k, t \in I.$$

(v) For $K \leq K_1 \leq K_2$ there is a $K_3 \geq m$ such that the Cauchy problem for $A(t, v(t))$ in the scale $(X^k)_k$ is well-posed in the part (K_1, K_2, K_3) for any $v \in C(I, X^{K_2+K_3})$

(vi) For $R > 0, k \geq K$ there is a $\beta_{k,R} \geq 0$ with

$$\|\lambda x - A(t, v)x\|_k \geq (\lambda - \beta_{k,R}) \|x\|_k$$

for $x \in X^{k+m}, v \in B_R^k, t \in I, \lambda > \beta_{k,R}$.

(vii) For $k \geq K$ $f : I \times X^k \ni (t, v) \mapsto f(t, v) \in X^k$ is a mapping with

- $I \ni t \mapsto f(t, v) \in X^k$ is continuous for $v \in X^k, k \geq K$.

- For $R > 0, k \geq K$ there is an $\tilde{F}_{k,R} > 0$ with

$$\|f(t, v) - f(t, w)\|_k \leq \tilde{F}_{k,R} \|v - w\|_k \quad \text{for } v, w \in B_R^k.$$

Then, for $k \geq K$, there is a function $T_k : X^{k+2m} \to (t_0, \infty]$ such that the following conditions are satisfied:

For every $x_0 \in X^{k+2m}$ and $I' := [t_0, T_k(x_0)) \cap I$ there is a unique function $u \in C^1(I', X^k) \cap C(I', X^{k+m})$ with

$$\frac{du}{dt}(t) = A(t, u(t))u(t) + f(t, u(t)), t \in I', \qquad u(t_0) = x_0 .$$

Moreover, if the spaces X^k are reflexive for $k \geq K$ we have $u \in C_w(I', X^{k+2m})$ and either $T_k(x_0) > t_1$ or $[t_0, T_k(x_0)) \ni t \mapsto \|u(t)\|_{k+2m}$ is an unbounded mapping.

PROOF: Fix $k \geq K$ and let $K_1 := K, K_2 := k + 2m$ and choose K_3 as in (v). Note that for $R > 0, k' \geq K$ there are $a_{k',R}, F_{k',R}$ with

$$\|A(t,v)\|_{\mathscr{L}(X^{k'+m},X^{k'})} \leq a_{k',R} \quad \text{and} \quad \|f(t,v)\|_{k'} \leq F_{k',R}$$

for $t \in I$ and $v \in B_R^{k'}$. Moreover, let $S_l := J_l^{2m+K_3}$, then for $k' \in \mathbb{N}_0$ there are $s_{k'} \geq 0$ with

- $(S_l)_l \subset \mathscr{L}(X^{k'}, X^{k'+2m+K_3})$ for any $k' \in \mathbb{N}_0$.

- $\|S_l\|_{\mathscr{L}(X^{k'})} \leq s_{k'}$ for any $k' \in \mathbb{N}_0, l \in \mathbb{N}$.

- $S_l \underset{l\to\infty}{\longrightarrow} \mathrm{Id}$ strongly in $\mathscr{L}(X^{k'})$ for any $k' \in \mathbb{N}_0$.

- $\|S_l - \mathrm{Id}\|_{\mathscr{L}(X^{k'+1},X^{k'})} \leq \frac{s_{k'}}{l^2}$ for any $k' \in \mathbb{N}_0, l \in \mathbb{N}$.

Step 1: Define $B_R(\tilde{I}, X^{k'}) := \{u : \tilde{I} \to X^{k'} : \|u(t)\|_{k'} \leq R \text{ for } t \in \tilde{I}\}$ for $k' \in \mathbb{N}_0$, $R > 0$, and compact subintervals $\tilde{I} = [t_0, \tilde{t}_1] \subset I$. Moreover, let $A_{l,v}(t) := A(t, S_l v(t))$ and $f_{l,v}(t) := f(t, S_l v(t))$ for $l \in \mathbb{N}, t \in \tilde{I}, v \in B_R(\tilde{I}, X^{k'})$, $k' \geq K$. Then the following statements hold:

- For $R > 0, k' \in \mathbb{N}_0, l \in \mathbb{N}$ with $k' \leq k + 2m + K_3$ there exist $R_{k,k',l} > 0$ with $S_l x \in B_{R_{k,k',l}}^{k'}$ for $x \in B_R^k$. In particular, $S_l u \in B_{R_{k,k',l}}(\tilde{I}, X^{k'})$ for $u \in B_R(\tilde{I}, X^k)$ and $S_l u \in C(\tilde{I}, X^{k'})$ for $u \in C(\tilde{I}, X^k)$.

- $\tilde{I} \ni t \mapsto A_{l,v}(t) \in \mathscr{L}(X^{k'+m}, X^{k'})$ is strongly continuous for $l \in \mathbb{N}$, $K_1 \leq k' \leq K_2 + K_3, v \in C(\tilde{I}, X^k)$.

- For $v \in B_R(\tilde{I}, X^k), K_1 \leq k' \leq K_2 + K_3, l \in \mathbb{N}, t \in \tilde{I}$ we have

$$\|\lambda x - A_{l,v}(t)x\|_{k'} = \|\lambda x - A(t, S_l v(t))x\|_{k'} \geq (\lambda - \beta_{k',R_{k,k',l}}) \|x\|_{k'}$$

for $x \in X^{k'+m}, l \in \mathbb{N}, t \in \tilde{I}, \lambda > \beta_{k',R_{k,k',l}}$.

- For $v \in B_R(\tilde{I}, X^{k'}), k' \geq K$, we have

$$\|\lambda x - A_{l,v}(t)x\|_{k'} = \|\lambda x - A(t, S_l v(t))x\|_{k'} \geq (\lambda - \beta_{k',s_{k'}R}) \|x\|_{k'}$$

for $x \in X^{k'+m}, l \in \mathbb{N}, t \in \tilde{I}, \lambda > \beta_{k',s_{k'}R}$.

- For $v \in C(\widetilde{I}, X^k)$ there is a propagator $U_{l,v}(t,s) \in \mathcal{L}(X^{k'})$, where $(t,s) \in \widetilde{\Delta} = \{(t,s) \in \widetilde{I}^2 : t \geq s\}, K \leq k' \leq k + 3m$, such that $\widetilde{\Delta} \ni (t,s) \mapsto U_{l,v}(t,s) \in \mathcal{L}(X^{k'+m}, X^{k'})$ is strongly continuously differentiable for $K \leq k' \leq k + 2m, l \in \mathbb{N}$ with strong derivatives

$$\frac{\partial}{\partial t}U_{l,v}(t,s) = A_{l,v}(t)U_{l,v}(t,s), \frac{\partial}{\partial s}U_{l,v}(t,s) = -U_{l,v}(t,s)A_{l,v}(s).$$

- $f_{l,v} \in C(\widetilde{I}, X^{k'})$ for $K \leq k' \leq k + 3m \leq K_2 + K_3, v \in C(\widetilde{I}, X^k)$.

- $\|f_{l,v}(t)\|_{k'} = \|f(t, S_l v(t))\|_{k'} \leq F'_{k',R} := F_{k',s_{k'}R}$ for $l \in \mathbb{N}, v \in B_R(\widetilde{I}, X^{k'})$, $t \in \widetilde{I}, k' \geq K$.

- For $K \leq k' \leq k + 2m, v \in C(\widetilde{I}, X^k), \widetilde{t}_0 \in \widetilde{I}, \widetilde{t}_0 < \widetilde{t}_1, g \in C([\widetilde{t}_0, \widetilde{t}_1], X^{k'+m})$, and $x \in X^{k'+m}$ there is a unique solution $u_{l,v} \in C([\widetilde{t}_0, \widetilde{t}_1], X^{k'+m}) \cap C^1([\widetilde{t}_0, \widetilde{t}_1], X^{k'})$ of

$$\frac{du_{l,v}}{dt}(t) = A_{l,v}(t)u_{l,v}(t) + g(t) \text{ for } t \in [\widetilde{t}_0, \widetilde{t}_1] \text{ and } u_{l,v}(\widetilde{t}_0) = x.$$

This solution is given by

$$u_{l,v}(t) := U_{l,v}(t, \widetilde{t}_0)x + \int_{\widetilde{t}_0}^t U_{l,v}(t, \tau)g(\tau)d\tau \text{ for } t \in [\widetilde{t}_0, \widetilde{t}_1].$$

- For $v \in C(\widetilde{I}, X^k), g \in C(\widetilde{I}, X^{k+m}), x \in X^{k+2m}$, and $u \in C(\widetilde{I}, X^{k+2m}) \cap C^1(\widetilde{I}, X^{k+m})$ with

$$\frac{du}{dt}(t) = A_{l,v}(t)u(t) + g(t), t \in \widetilde{I}, \qquad u(t_0) = x$$

we have $u(t) = U_{l,v}(t, t_0)x + \int_{t_0}^t U_{l,v}(t, \tau)g(\tau)d\tau$ for $t \in \widetilde{I}$:

Let $w(t) := U_{l,v}(t, t_0)x + \int_{t_0}^t U_{l,v}(t, \tau)g(\tau)d\tau$ for $t \in \widetilde{I}$.
Then $w \in C(\widetilde{I}, X^{k+m}) \cap C^1(\widetilde{I}, X^k)$ with

$$\frac{dw}{dt}(t) = A_{l,v}(t)w(t) + g(t), t \in \widetilde{I}, \qquad w(t_0) = x.$$

Since this also holds for u at the place of w the uniqueness result of above implies $u = w$.

- For every $K \leq k' \leq k + 2m$, $R > 0$ there is a constant $\beta'_{k',R}$ with

$$\|U_{l,v}(t,s)\|_{\mathscr{L}(X^{k'})} \leq e^{\beta'_{k',R}(t-s)}$$

for $l \in \mathbb{N}, v \in C(\widetilde{I}, X^k) \cap B_R(\widetilde{I}, X^{k'}), (t,s) \in \widetilde{\Delta}$:
Take $x \in X^{k'+m}$ and define $u(t) := U_{l,v}(t,s)x$ for $(t,s) \in \widetilde{\Delta}$.
Then $u \in C^1([s,\widetilde{t}_1], X^{k'}) \cap C([s,\widetilde{t}_1], X^{k'+m})$ with

$$\frac{du}{dt}(t) = A_{l,v}(t)u(t).$$

Hence the apriori estimate 1.1.10 shows

$$\|U_{l,v}(t,s)x\|_{k'} = \|u(t)\|_{k'} \leq e^{\beta_{k',s_{k'}}R(t-s)} \|u(s)\|_{k'} = e^{\beta_{k',s_{k'}}R(t-s)} \|x\|_{k'}$$

for $t \geq s$. Since $X^{k'+m} \hookrightarrow X^{k'}$ densely, this implies the assertion with
$\beta'_{k',R} := \beta_{k',s_{k'}R}$.

Step 2: Let $u_0(t) := x_0$ and $u_{l+1}(t) := U_{l,u_l}(t,t_0)x_0 + \int_{t_0}^t U_{l,u_l}(t,s)f_{l,u_l}(s)ds$ for
$t \in I, l \in \mathbb{N}$. Then $u_{l+1} \in C(I, X^{k+2m}) \cap C^1(I, X^{k+m})$ with

$$\frac{du_{l+1}}{dt}(t) = A_{l,u_l}(t)u_{l+1}(t) + f_{l,u_l}(t), t \in I, \quad u_{l+1}(t_0) = x_0.$$

Moreover, if $R_0 := \|x_0\|_{k+2m}$, $R' := 2R_0 + 1$, and $t_0 < t_2 \in I$ is chosen with

$$e^{\beta'_{k+2m,R'}(t_2-t_0)}R_0 + (t_2 - t_0)e^{\beta'_{k+2m,R'}(t_2-t_0)}F'_{k+2m,R'} \leq 2R_0 + 1 = R',$$

then $\|u_l(t)\|_{k+2m} \leq R'$ for $t \in [t_0,t_2], l \in \mathbb{N}$.
Proof: Clearly $\|u_0(t)\|_{k+2m} = \|x_0\|_{k+2m} = R_0 \leq R'$. Moreover, inductively for
$t \in [t_0,t_2]$

$$
\begin{aligned}
\|u_{l+1}(t)\|_{k+2m} &\leq \|U_{l,u_l}(t,t_0)x_0\|_{k+2m} + \int_{t_0}^t \|U_{l,u_l}(t,s)f_{l,u_l}(s)\|_{k+2m} ds \\
&\leq e^{\beta'_{k+2m,R'}(t_2-t_0)}R_0 + (t_2 - t_0)e^{\beta'_{k+2m,R'}(t_2-t_0)}F'_{k+2m,R'} \leq R'
\end{aligned}
$$

Step 3: Let $c_k := \widetilde{a}_{k+m,s_{k+m}R'}R' + \widetilde{F}_{k+m,s_{k+m}R'}$ and choose a $t_3 \in (t_0,t_2]$ such
that $q := (t_3 - t_0)e^{\beta'_{k+m,R'}(t_3-t_0)}c_k s_{k+m} < 1$.
Then $(u_l)_l \subset C([t_0,t_3], X^{k+m})$ is a Cauchy sequence.
Proof: In X^{k+m} we have

$$
\begin{aligned}
u'_{l+1}(t) - u'_l(t) &= A_{l,u_l}(t)(u_{l+1}(t) - u_l(t)) + \left(A_{l,u_l}(t) - A_{l-1,u_{l-1}}(t)\right)u_l(t) \\
&\quad + f_{l,u_l}(t) - f_{l-1,u_{l-1}}(t).
\end{aligned}
$$

This implies in X^{k+m}

$$u_{l+1}(t) - u_l(t)$$
$$= \int_{t_0}^t U_{l,u_l}(t,s) \left[\left(A_{l,u_l}(s) - A_{l-1,u_{l-1}}(s) \right) u_l(s) + f_{l,u_l}(s) - f_{l-1,u_{l-1}}(s) \right] ds .$$

Now

$$\|S_l u_l(s) - S_{l-1} u_{l-1}(s)\|_{k+m}$$
$$\leq \|S_l u_l(s) - S_l u_{l-1}(s)\|_{k+m} + \|S_l u_{l-1}(s) - S_{l-1} u_{l-1}(s)\|_{k+m}$$
$$\leq s_{k+m} \|u_l(s) - u_{l-1}(s)\|_{k+m} + \frac{2 s_{k+m}}{(l-1)^2} R' ,$$

hence

$$\|u_{l+1}(t) - u_l(t)\|_{k+m} \leq \int_{t_0}^t e^{\beta'_{k+m,R'}(t_3 - t_0)} c_k \|S_l u_l(s) - S_{l-1} u_{l-1}(s)\|_{k+m} ds$$
$$\leq q \left[\sup_{s \in [t_0,t_3]} \|u_l(s) - u_{l-1}(s)\|_{k+m} + \frac{2}{(l-1)^2} R' \right] .$$

Therefore

$$\|u_{l+1} - u_l\|_{C([t_0,t_3], X^{k+m})} \leq q \|u_l - u_{l-1}\|_{C([t_0,t_3], X^{k+m})} + \frac{2q}{(l-1)^2} R'$$

and 3.3.4 implies $\sum_{l=1}^\infty \|u_{l+1} - u_l\|_{C([t_0,t_3], X^{k+m})} < \infty$, which proves step 3.

Step 4: Let $u := \lim_{l\to\infty} u_l$, then $u \in C([t_0, t_3], X^{k+m}) \cap C^1([t_0, t_3], X^k)$ with

$$u'(t) = A(t, u(t))u(t) + f(t, u(t)) \quad \text{for } t \in [t_0, t_3], \qquad u(t_0) = u_0 .$$

Proof:

$$\left\| \frac{du_l}{dt}(t) - A(t, u(t))u(t) - f(t, u(t)) \right\|_k$$
$$\leq \|(A(t, S_{l-1} u_{l-1}(t)) - A(t, u(t))) u_l(t)\|_k$$
$$\quad + \|A(t, u(t))(u_l(t) - u(t))\|_k + \|f(t, S_{l-1} u_{l-1}(t)) - f(t, u(t))\|_k$$
$$\leq (\tilde{a}'_{k,R'} + \widetilde{F}'_{k,R'}) \|S_{l-1} u_{l-1}(t) - u(t)\|_k + a'_{k,R'} \|u_l(t) - u(t)\|_{k+m} \xrightarrow[l\to\infty]{} 0$$

uniformly for $t \in [t_0, t_3]$ with suitable constants $\tilde{a}'_{k,R'}, \widetilde{F}'_{k,R'}, a'_{k,R'}$. This implies the assertion of step 4 and the existence statement of the assertion (for general Banach spaces X^k).

Step 5: Proof of uniqueness.

Assume that $I' = [t_0, t_1'] \subset I$ is a compact subinterval and $u_j \in C^1(I', X^k) \cap C(I', X^{k+m}), j = 1, 2$, are two functions with

$$\frac{du_j}{dt}(t) = A(t, u_j(t))u_j(t) + f(t, u_j(t)) \text{ for } t \in I', \qquad u_j(t_0) = x_0$$

for $j = 1, 2$. Then $u_1 = u_2$.

Proof:

First there is an $R > 0$ with $u_1(t), u_2(t) \in B_R^k \cap B_R^{k+m}$ for $t \in I'$, hence

$$\|\lambda x - A(t, u_j(t))x\|_k \geq (\lambda - \beta_{k,R}) \|x\|_k \text{ for } \lambda > \beta_{k,R}, x \in X^{k+m}, t \in \tilde{I}, j = 1, 2.$$

Let $\gamma_{k,R} := \beta_{k,R} + \tilde{a}_{k,R}R + \widetilde{F}_{k,R}$, $v(t) := (u_1(t) - u_2(t))e^{-\gamma_{k,R}t}$ and $f(t) := \|v(t)\|_k$ for $t \in I'$. Then

$$\begin{aligned}
v'(t) &= (A(t, u_1(t)) - \gamma_{k,R})v(t) \\
&\quad + e^{-\gamma_{k,R}t}([A(t, u_1(t)) - A(t, u_2(t))]u_2(t) + [f(t, u_1(t)) - f(t, u_2(t))]) .
\end{aligned}$$

Let $r_t(h) := \|v(t-h)\|_k - \|v(t) + v'(t)(-h)\|_k$ for $t \in (t_0, t_1'], h \in (0, t - t_0)$, then

$$\begin{aligned}
f(t-h) &= \|v(t) + v'(t)(-h)\|_k + r_t(h) \\
&\geq \|v(t) + (A(t, u_1(t)) - \gamma_{k,R})v(t)(-h)\|_k \\
&\quad -he^{-\gamma_{k,R}t}(\|(A(t, u_1(t)) - A(t, u_2(t)))u_2(t)\|_k + \|f(t, u_1(t)) - f(t, u_2(t))\|_k) + r_t(h) \\
&\geq h\left\|\left(\frac{1}{h} + \gamma_{k,R}\right)v(t) - A(t, u_1(t))v(t)\right\|_k \\
&\quad -he^{-\gamma_{k,R}t}\left(\tilde{a}_{k,R}\|u_1(t) - u_2(t)\|_k \|u_2(t)\|_{k+m} + \widetilde{F}_{k,R}\|u_1(t) - u_2(t)\|_k\right) + r_t(h) \\
&\geq f(t) + h(\tilde{a}_{k,R}R + \widetilde{F}_{k,R})f(t) - h(\tilde{a}_{k,R}R + \widetilde{F}_{k,R})f(t) + r_t(h) \\
&= f(t) + r_t(h) .
\end{aligned}$$

Hence, for the left upper Dini derivative we have

$$\begin{aligned}
D^- f(t) &= \limsup_{h \to 0+} \frac{f(t-h) - f(t)}{-h} \leq \limsup_{h \to 0+} \left|\frac{r_t(h)}{-h}\right| \\
&= \lim_{h \to 0} \frac{\|v(t-h) - (v(t) + v'(t)(-h))\|_k}{h} = 0 .
\end{aligned}$$

This shows $0 \leq f(t) \leq f(t_0) = 0$, hence $u_1(t) = u_2(t)$ for $t \in I'$.

Step 6: Assume that X^k is reflexive for $k \geq K$. Then, for $R_0 > 0$, there is an $\varepsilon_{k,R_0} > 0$ such that for $x_0 \in X^{k+2m}$ with $\|x_0\|_{k+2m} \leq R_0$ and with $I' := [t_0, t_0 + \varepsilon_{k,R_0}] \cap I$ there is a unique

$$u \in C^1(I', X^k) \cap C(I', X^{k+m}) \cap C_w(I', X^{k+2m}) \qquad \text{with}$$

$$\frac{du}{dt}(t) = A(t, u(t))u(t) + f(t, u(t)), t \in I' \qquad u(t_0) = x_0 .$$

Proof: Since $\|u_l(t)\|_{k+2m} \leq R'$ for $t \in [t_0, t_3]$ and $u_l(t) \xrightarrow[l \to \infty]{} u(t)$ in X^{k+m}, 3.3.2 shows $u(t) \in X^{k+2m}$ with $\|u(t)\|_{k+2m} \leq R'$ for $t \in [t_0, t_3]$. Then 3.3.3 implies $u \in C_w([t_0, t_3], X^{k+2m})$. Here t_3 only depends on R_0 and the constants concerning $A(t, x)$ and $f(t, x)$. This implies step 6.

Step 7: Proof of the assertion for reflexive spaces X^k:

Using step 6 and Zorn's lemma we can choose a maximal interval $I' \subset I$ with left boundary point t_0 such that there is a unique $u \in C^1(I', X^k) \cap C(I', X^{k+m}) \cap C_w(I', X^{k+2m})$ with

$$\frac{du}{dt}(t) = A(t, u(t))u(t) + f(t, u(t)), t \in I' \qquad u(t_0) = x_0 . \qquad (3.3.1)$$

If $I' = I$ choose an arbitrary $T_k(x_0) > t_1$. Otherwise there is a $T_k(x_0) \in [t_0, t_1]$ with $I' = [t_0, T_k(x_0))$. Thus in any case we have $I' = [t_0, T_k(x_0)) \cap I$.

Finally, suppose $T_k(x_0) \leq t_1$ and assume that there is an $R_0 > 0$ with $\|u(t)\|_{k+2m} \leq R_0$ for $t \in [t_0, T_k(x_0))$. Let ε_{k,R_0} be as in step 6 and choose $t_2 \in [t_0, T_k(x_0))$ with $t_2 \geq T_k(x_0) - \frac{\varepsilon_{k,R_0}}{2}$. Then step 6 shows that with $\tilde{I} = [t_2, t_2 + \varepsilon_{k,R_0}] \cap I$ there is a $v \in C^1(\tilde{I}, X^k) \cap C(\tilde{I}, X^{k+m}) \cap C_w(\tilde{I}, X^{k+2m})$ with

$$\frac{dv}{ds}(s) = A(s, v(s))v(s) + f(s, v(s)), s \in [t_2, t_2 + \varepsilon_{k,R_0}] \cap I, \qquad v(t_2) = u(t_2) .$$

Therefore, with $I'' := [t_0, t_2 + \varepsilon_{k,M}] \cap I$ we can extend u to a function

$$u \in C^1(I'', X^k) \cap C(I'', X^{k+m}) \cap C_w(I'', X^{k+2m})$$

solving (3.3.1) on $[t_0, t_2 + \varepsilon_{k,M}] \cap I$. This contradicts the maximality of I'. $\square$

In the remaining part of this section we will show that, in the case of scales of Hilbert spaces, the solutions are even strongly continuous with values in the space of the initial value, if we assume in addition first-order commutator estimates. In the proof an idea from the construction of solutions for quasilinear symmetric hyperbolic evolution equations is used, cf. [148, 16.1]. Note that a form of first-order commutator estimates is also assumed in Kato's result [82]. We will need the following lemmata.

3.3.10 Lemma. Assume that X and Y are Banach spaces with $Y \hookrightarrow X$ and $f \in C^1(I, X)$, i.e., $g := f' \in C(I, X)$, where $I \subset \mathbb{R}$ is an interval. Moreover, assume that $f(I) \subset Y$ and $g \in C(I, Y)$. Then $f \in C^1(I, Y)$ with $f' = g$.

PROOF: We denote integrals in Z by $Z-\int$ for $Z = X, Y$, then

$$
\frac{f(t+h) - f(t)}{h} = \frac{1}{h}X - \int_t^{t+h} f'(\tau)d\tau = \frac{1}{h}X - \int_t^{t+h} g(\tau)d\tau
$$

$$
= \frac{1}{h}Y - \int_t^{t+h} g(\tau)d\tau \xrightarrow[h \to 0]{Y} g(t)
$$

for $t \in I$. $\square$

3.3.11 Lemma. Assume that $(X, \langle \cdot, \cdot \rangle, \|\cdot\|)$ is a Hilbert space, $(x_l)_{l \in \mathbb{N}} \subset X$ with $x_l \xrightarrow[l \to \infty]{} x \in X$ weakly in X and $\|x_l\| \xrightarrow[l \to \infty]{} \|x\|$. Then $x_l \xrightarrow[l \to \infty]{} x$ in X.

PROOF:

$$
\|x_l - x\|^2 = \langle x_l - x, x_l - x \rangle = \|x_l\|^2 - \overline{\langle x_l, x \rangle} - \langle x_l, x \rangle + \|x\|^2
$$

$$
\xrightarrow[l \to \infty]{} \|x\|^2 - \overline{\langle x, x \rangle} - \langle x, x \rangle + \|x\|^2 = 0
$$

$\square$

3.3.12 Theorem.

Assumptions:

(i) $Z = \{Z_1, \ldots, Z_M\}$ is a commuting family of infinitesimal generators of C_0-semigroups of contractions in a Hilbert space H and there exist $\alpha_j \neq 0$ with $Z_j^* = \alpha_j Z_j$ for $j = 1, \ldots, M$.

(ii) $B_R^k := \{x \in H_Z^k : \|x\|_k \leq R\}$ for $R > 0, k \in \mathbb{N}_0$.

(iii) $I = [t_0, t_1] \subset \mathbb{R}$ is a compact interval and $K \in \mathbb{N}_0, m \in \mathbb{N}$ are fixed.

(iv) There is a family of linear mappings $A(t, v)$ with

- $I \ni t \mapsto A(t, v) \in \mathscr{L}(H_Z^{k+m}, H_Z^k)$ is strongly continuous for $v \in H_Z^k$, $k \geq K$.

- For $R > 0, k \geq K$ there is a constant $\tilde{a}_{k,R} \geq 0$ with

$$
\|A(t, v) - A(t, w)\|_{\mathscr{L}(H_Z^{k+m}, H_Z^k)} \leq \tilde{a}_{k,R} \|v - w\|_k \quad \text{for } v, w \in B_R^k, t \in I.
$$

(v) For $R > 0, k \geq K$ there is a $d_{k,R} > 0$ with

$$\|Z^\alpha A(t,v)x - A(t,v)Z^\alpha x\| \leq d_{k,R} \|x\|_k$$

for $t \in I, v \in B_R^k, x \in H_Z^\infty, \alpha \in \mathbb{N}_0^M, |\alpha| \leq k$.

(vi) For $R > 0$ there is a $\beta_{0,R} \geq 0$ with $|\mathrm{Re}\,\langle A(t,v)x, x\rangle| \leq \beta_{0,R} \|x\|^2$ for $x \in H_Z^\infty, v \in B_R^K, t \in I$.

(vii) For $k \geq K$ $f : I \times H_Z^k \ni (t,v) \mapsto f(t,v) \in H_Z^k$ is a mapping with

- $I \ni t \mapsto f(t,v) \in H_Z^k$ is continuous for $v \in H_Z^k, k \geq K$.
- For $R > 0, k \geq K$ there is an $\widetilde{F}_{k,R} > 0$ with

$$\|f(t,v) - f(t,w)\|_k \leq \widetilde{F}_{k,R} \|v - w\|_k \text{ for } v, w \in B_R^k.$$

Then, for any $k \geq K + m$, $u_0 \in H_Z^{k+m}$, and $\widetilde{t}_0 \in I$, there exists an interval $I_\varepsilon = [\widetilde{t}_0 - \varepsilon, \widetilde{t}_0 + \varepsilon] \cap I$ and a unique function $u \in C(I_\varepsilon, H_Z^{k+m}) \cap C^1(I_\varepsilon, H_Z^k)$ with

$$\frac{du}{dt}(t) = A(t, u(t))u(t) + f(t, u(t)), t \in I_\varepsilon, \qquad u(\widetilde{t}_0) = u_0.$$

PROOF:
Step 1: There is an interval $I_\varepsilon = [\widetilde{t}_0 - \varepsilon, \widetilde{t}_0 + \varepsilon] \cap I$ and a unique function $u \in C^1(I_\varepsilon, H_Z^{k-m}) \cap C(I_\varepsilon, H_Z^k) \cap C_w(I_\varepsilon, H_Z^{k+m}) \cap B_R(I_\varepsilon, H_Z^{k+m})$ with

$$\frac{du}{dt}(t) = A(t, u(t))u(t) + f(t, u(t)), \ t \in I_\varepsilon, \quad u(\widetilde{t}_0) = u_0.$$

Here $B_R(I_\varepsilon, H_Z^{k+m}) := \{v : I_\varepsilon \to H_Z^{k+m} : \|v(t)\|_{k+m} \leq R \text{ for } t \in I_\varepsilon\}$ with a suitable $R > 0$.
Proof: This is a consequence of 3.2.1, 3.3.6, 3.3.8, and 3.3.9. The assertion for $t \leq \widetilde{t}_0$ is obtained if these arguments are applied to $-A(\widetilde{t}_0 - t, v), -f(\widetilde{t}_0 - t, v)$.
Step 2: For $k' \geq K + 1, R > 0$ there exists a $g_{k',R} > 0$ with

$$\left\|J_l^{2m} A(t,v)x - A(t,v)J_l^{2m}x\right\|_{k'} \leq g_{k',R} \|x\|_{k'} \text{ for } v \in B_R^{k'}, x \in H_Z^\infty, l \in \mathbb{N},$$

where $J_l := (\mathrm{Id} - \tfrac{1}{l}Z_1)^{-1} \cdots (\mathrm{Id} - \tfrac{1}{l}Z_M)^{-1}.$

Proof: First, for $k' \geq K+1, j = 1, \ldots, M, y \in H_Z^\infty$, and $v \in B_R^{k'}$ we have

$$\left\| Z_j A(t,v)y - A(t,v)Z_j y \right\|_{k'-1} \leq c_{k'} \sum_{|\alpha| \leq k'-1} \left\| Z^\alpha (Z_j A(t,v)y - A(t,v)Z_j y) \right\|$$

$$\leq c_{k'} \sum_{|\alpha| \leq k'-1} \left[\left\| Z^{\alpha+e_j} A(t,v)y - A(t,v)Z^{\alpha+e_j} y \right\| \right.$$

$$\left. + \left\| A(t,v)Z^\alpha (Z_j y) - Z^\alpha A(t,v)(Z_j y) \right\| \right]$$

$$\leq c_{k'} \sum_{|\alpha| \leq k'-1} \left[d_{k',R} \|y\|_{k'} + d_{k'-1,R} \|Z_j y\|_{k'-1} \right] \leq \tilde{d}_{k',R} \|y\|_{k'}$$

for suitable $c_{k'}, \tilde{d}_{k',R} \geq 0$, and for $k' \geq K+1, j = 1, \ldots, M, y \in H_Z^\infty$, and $v \in B_R^{k'}$

$$\left(\mathrm{Id} - \frac{1}{l}Z_j \right)^{-1} A(t,v)y - A(t,v) \left(\mathrm{Id} - \frac{1}{l}Z_j \right)^{-1} y$$

$$= \left(\mathrm{Id} - \frac{1}{l}Z_j \right)^{-1} A(t,v) \left(\mathrm{Id} - \frac{1}{l}Z_j \right) \left(\mathrm{Id} - \frac{1}{l}Z_j \right)^{-1} y$$

$$- \left(\mathrm{Id} - \frac{1}{l}Z_j \right)^{-1} \left(\mathrm{Id} - \frac{1}{l}Z_j \right) A(t,v) \left(\mathrm{Id} - \frac{1}{l}Z_j \right)^{-1} y$$

$$= \frac{1}{l} \left(\mathrm{Id} - \frac{1}{l}Z_j \right)^{-1} (Z_j A(t,v) - A(t,v)Z_j) \left(\mathrm{Id} - \frac{1}{l}Z_j \right)^{-1} y .$$

Moreover, for $y \in H_Z^\infty, v \in B_R^{k'}, k' \geq K+1, \mu = 1, \ldots, M$ we have

$$J_l A(t,v)y - A(t,v)J_l y$$

$$= \left(\mathrm{Id} - \frac{1}{l}Z_\mu \right)^{-1} \left[\prod_{\substack{j=1 \\ j \neq \mu}}^{M} \left(\mathrm{Id} - \frac{1}{l}Z_j \right)^{-1} A(t,v) - A(t,v) \prod_{\substack{j=1 \\ j \neq \mu}}^{M} \left(\mathrm{Id} - \frac{1}{l}Z_j \right)^{-1} \right] y$$

$$+ \left[\left(\mathrm{Id} - \frac{1}{l}Z_\mu \right)^{-1} A(t,v) - A(t,v) \left(\mathrm{Id} - \frac{1}{l}Z_\mu \right)^{-1} \right] \prod_{\substack{j=1 \\ j \neq \mu}}^{M} \left(\mathrm{Id} - \frac{1}{l}Z_j \right)^{-1} y$$

$$= \frac{1}{l} \left(\mathrm{Id} - \frac{1}{l}Z_\mu \right)^{-1} \left[\sum_{\substack{j=1 \\ j \neq \mu}}^{M} \prod_{\substack{p=1 \\ p \neq \mu}}^{j} \left(\mathrm{Id} - \frac{1}{l}Z_\rho \right)^{-1} (Z_j A(t,v) - A(t,v)Z_j) \times \right.$$

$$\left. \times \prod_{\substack{p=j \\ p \neq \mu}}^{M} \left(\mathrm{Id} - \frac{1}{l}Z_\rho \right)^{-1} y + (Z_\mu A(t,v) - A(t,v)Z_\mu) \prod_{j=1}^{M} \left(\mathrm{Id} - \frac{1}{l}Z_j \right)^{-1} y \right] .$$

Hence, for $y \in H_Z^\infty, v \in B_R^{k'}, k' \geq K+1$, we have

$$\left\| J_l A(t,v)y - A(t,v)J_l y \right\|_{k'-1} \leq \tilde{d}''_{k,R} \|y\|_{k'}$$

and

$$\left\|Z_\mu(J_l A(t,v)y - A(t,v)J_l y)\right\|_{k'-1}$$

$$= \left\|\underbrace{\frac{1}{l}Z_\mu\left(\mathrm{Id} - \frac{1}{l}Z_\mu\right)^{-1}}_{=(\mathrm{Id}-\frac{1}{l}Z_\mu)^{-1}-\mathrm{Id}}\left[\sum_{\substack{j=1\\j\neq\mu}}^M\prod_{\substack{p=1\\p\neq\mu}}^j\left(\mathrm{Id} - \frac{1}{l}Z_p\right)^{-1}(Z_j A(t,v) - A(t,v)Z_j)\times\right.\right.$$

$$\left.\left.\times\prod_{\substack{\rho=j\\\rho\neq\mu}}^M\left(\mathrm{Id} - \frac{1}{l}Z_\rho\right)^{-1}y + (Z_\mu A(t,v) - A(t,v)Z_\mu)\prod_{j=1}^M\left(\mathrm{Id} - \frac{1}{l}Z_j\right)^{-1}y\right]\right\|_{k-1}$$

$$\leq \tilde{d}'''_{k,R}\|y\|_{k'}$$

with suitable $\tilde{d}''_{k,R}, \tilde{d}'''_{k,R} \geq 0$. Hence

$$\left\|J_l^{2m}A(t,v)x - A(t,v)J_l^{2m}x\right\|_{k'}$$

$$\leq c_{k'}\left\|J_l^{2m}A(t,v)x - A(t,v)J_l^{2m}x\right\|_{k'-1}$$

$$+c_{k'}\sum_{\mu=1}^M\left\|Z_\mu(J_l^{2m}A(t,v)x - A(t,v)J_l^{2m}x)\right\|_{k'-1}$$

$$\leq c_{k'}\sum_{\nu=0}^{2m-1}\left\|J_l^{2m-1-\nu}(J_l A(t,v) - A(t,v)J_l)J_l^\nu x\right\|_{k'-1}$$

$$+c_{k'}\sum_{\mu=1}^M\sum_{\nu=0}^{2m-1}\left\|J_l^{2m-1-\nu}Z_\mu(J_l A(t,v) - A(t,v)J_l)J_l^\nu x\right\|_{k'-1}$$

$$\leq g_{k',R}\|x\|_{k'}$$

for $t \in I, l \in \mathbb{N}, v \in B_R^{k'}, k' \geq K + 1, x \in H_Z^\infty$ and a suitable constants $c_{k'}, g_{k',R} \geq 0$.

Step 3: Proof of the assertion. We only have to show that $u \in C(I_\varepsilon, H_Z^{k+m})$ (cf. 3.3.10). To this end suppose that $(s_j)_{j\in\mathbb{N}} \subset I_\varepsilon$ with $s_j \xrightarrow[j\to\infty]{} s_0 \in I_\varepsilon$.
Then $u(s_j) \xrightarrow[j\to\infty]{} u(s_0)$ weakly in H_Z^{k+m} due to step 1. Thus, due to 3.3.11 $u(s_j) \xrightarrow[j\to\infty]{} u(s_0)$ in H_Z^{k+m}, if we show $\|u(s_j)\|_{k+m} \xrightarrow[j\to\infty]{} \|u(s_0)\|_{k+m}$.
Now, for $x \in H_Z^\infty, v \in B_R^{k+m}, l \in \mathbb{N}$ we have

$$\left|\mathrm{Re}\left\langle J_l^{2m}A(t,v)x, J_l^{2m}x\right\rangle_{k+m}\right|$$

$$\leq \left|\mathrm{Re}\left\langle A(t,v)J_l^{2m}x, J_l^{2m}x\right\rangle_{k+m}\right| + \left|\mathrm{Re}\left\langle J_l^{2m}A(t,v)x - A(t,v)J_l^{2m}x, J_l^{2m}x\right\rangle_{k+m}\right|$$

$$\leq \beta_{k+m,R}\left\|J_l^{2m}x\right\|_{k+m}^2 + g_{k+m,R}\|x\|_{k+m}\left\|J_l^{2m}x\right\|_{k+m} < \gamma_{k+m,R}\|x\|_{k+m}^2$$

for a suitable constant $\gamma_{k+m,R} \geq 0$, where we have used step 2 and 3.2.1. Thus, by continuity

$$\left| \text{Re} \left\langle J_l^{2m} A(t,v)x, J_l^{2m}x \right\rangle_{k+m} \right| \leq \gamma_{k+m,R} \|x\|_{k+m}^2$$

for $x \in H_Z^{k+m}, v \in B_R^{k+m}, l \in \mathbb{N}, t \in I_\varepsilon$. Moreover, $J_l^{2m}u \in C^1(I_\varepsilon, H_Z^{k+m}) \cap C(I_\varepsilon, H_Z^{k+2m})$ with $\frac{dJ_l^{2m}u}{dt}(t) = J_l^{2m}\frac{du}{dt}(t)$, hence

$$\left| \frac{d}{dt} \left\| J_l^{2m}u(t) \right\|_{k+m}^2 \right|$$

$$= 2 \left| \text{Re} \left\langle \frac{dJ_l^{2m}u}{dt}(t), J_l^{2m}u(t) \right\rangle_{k+m} \right|$$

$$\leq 2 \left| \text{Re} \left\langle J_l^{2m} A(t, u(t))u(t), J_l^{2m}u(t) \right\rangle_{k+m} \right| + 2 \left| \text{Re} \left\langle J_l^{2m} f(t, u(t)), J_l^{2m}u(t) \right\rangle_{k+m} \right|$$

$$\leq 2\gamma_{k+m,R} \|u(t)\|_{k+m}^2 + 2F_{k+m,R} \|u(t)\|_{k+m}$$

$$\leq C_{k+m,R}$$

for $l \in \mathbb{N}, t \in I_\varepsilon$, and suitable constant $F_{k+m,R}, C_{k+m,R}$. This implies

$$\left| \left\| J_l^{2m}u(s_j) \right\|_{k+m}^2 - \left\| J_l^{2m}u(s_0) \right\|_{k+m}^2 \right| \leq C_{k+m,R} \left| s_j - s_0 \right|.$$

Since $J_l^{2m}w_0 \xrightarrow[l\to\infty]{} w_0$ in H_Z^{k+m} for $w_0 \in H_Z^{k+m}$, this shows with $l \to \infty$ that

$$\left| \left\| u(s_j) \right\|_{k+m}^2 - \left\| u(s_0) \right\|_{k+m}^2 \right| \leq C_{k+m,R} \left| s_j - s_0 \right|$$

and we get $\|u(s_j)\|_{k+m} \xrightarrow[j\to\infty]{} \|u(s_0)\|_{k+m}$. This implies the assertion. $\square$

3.4 Regularity for quasilinear evolution equations in scales of Banach spaces

As for semilinear evolution equations questions of regularity are also important for quasilinear ones, i.e., the question whether the lengths of the intervals of existence depend on the scale index k. In this section we will give a condition implying regularity (cf. 3.4.4). We will have to use additional assumptions that can be verified with estimates of Nash-Moser type (cf. section 3.2). For the proof we need the following generalization of Gronwall's inequality. We recall the short proof (cf. [25, 4.2.1]).

3.4.1 Lemma. Let $I = (t_0, t_1), \alpha, \beta \geq 0$, and $u \in L^1(t_0, t_1)$ with $u(t) \geq 0$ and

$$u(t) \leq \alpha + \beta \int_{t_0}^t u(s)ds \quad \text{for almost every } t \in (t_0, t_1).$$

Then $u(t) \leq \alpha e^{\beta(t-t_0)}$ for almost every $t \in I$.

PROOF: Let $v(t) := \alpha + \beta \int_{t_0}^t u(s)ds$ for $t \in I$. Then v is differentiable almost everywhere with $v'(t) = \beta u(t) \leq \beta v(t)$ for almost every $t \in I$. Hence

$$\frac{d}{dt}\left(v(t)\exp(-\beta(t-t_0))\right) = v'(t)\exp(-\beta(t-t_0)) - \beta v(t)\exp(-\beta(t-t_0)) \leq 0$$

for a.e. $t \in I$. This implies $v(t)\exp(-\beta(t-t_0)) \leq v(t_0) = \alpha$ for a.e. $t \in I$. $\square$

3.4.2 Lemma. Let $I \subset \mathbb{R}$ be an interval, $(X, \|\cdot\|)$ be a Banach space, and $f \in C_w(I, X)$. Then f (and $\|f\|$) are measurable.

PROOF: This is a consequence of Pettis' theorem (cf. [160, V4]). $\square$

We combine these two lemmata to obtain the following estimate.

3.4.3 Lemma. Let $I \subset \mathbb{R}$ be an interval with left boundary point $t_0 \in I$, $(X, \|\cdot\|)$ be a Banach space, and let $\alpha, \beta \geq 0$, $u \in C_w(I, X)$ with

$$\|u(t)\|^2 \leq \alpha + \beta \int_{t_0}^t \|u(s)\|^2\, ds \quad \text{for } t \in I.$$

Then $\|u(t)\|^2 \leq \alpha e^{\beta(t-t_0)}$ for $t \in I$.

PROOF: We only have to prove the assertion for compact intervals $I = [t_0, t_1]$. 3.4.1 and 3.4.2 imply that there is a set $N \subset I$ with measure zero such that

$$\|u(t)\|^2 \leq \alpha e^{\beta(t-t_0)} \quad \text{for } t \in I \setminus N.$$

Let $t \in N$, then there is a sequence $(t_j)_j \subset I \setminus N$ with $t_j \underset{j \to \infty}{\longrightarrow} t$. Then

$$\left|\langle x^*, u(t_j)\rangle_{X^*, X}\right| \leq \|x^*\|_{X^*} \|u(t_j)\| \leq \sqrt{\alpha}\, \|x^*\|_{X^*}\, e^{\beta/2(t_j - t_0)}$$

for any $x^* \in X^*$. With $j \to \infty$ we obtain

$$\left|\langle x^*, u(t)\rangle_{X^*, X}\right| \leq \sqrt{\alpha}\, \|x^*\|_{X^*}\, e^{\beta/2(t-t_0)} \quad \text{for any } x^* \in X^*.$$

This implies $\|u(t)\|^2 \leq \alpha e^{\beta(t-t_0)}$. $\square$

Finally, we can prove the following regularity result.

3.4.4 Theorem.   Suppose with the assumptions of 3.3.9 that $(X^k, \langle \cdot, \cdot \rangle_k)$ is a scale of Hilbert spaces and that for $k \geq K, R > 0$ there are $\nu \geq 2m$ and $\gamma_{\nu,k,R}, f_{k,R} \geq 0$ with

$$\text{Re} \, \langle J_l^\nu A(t, u)u, J_l^\nu u \rangle_{k+m} \leq \gamma_{\nu,k,R} \|u\|_{k+m}^2 \quad \text{and} \quad \|f(t, u)\|_{k+m} \leq G_{k,R} \|u\|_{k+m}$$
(3.4.1)

for $u \in X^{k+m}, \|u\|_k \leq R, k \geq K, R > 0$. Then the following statements hold:

(a) Let $k \geq K, u_0 \in X^{k+3m}, [t_0, t_2] \subset I$, and

$$u \in C^1([t_0, t_2], X^k) \cap C([t_0, t_2], X^{k+m}) \cap C_w([t_0, t_2], X^{k+2m}) \qquad \text{with}$$

$$\frac{du}{dt}(t) = A(t, u(t))u(t) + f(t, u(t)), \; t \in [t_0, t_2], \; u(t_0) = u_0 \,. \quad (3.4.2)$$

Then $u \in C^1([t_0, t_2], X^{k+m}) \cap C([t_0, t_2], X^{k+2m}) \cap C_w([t_0, t_2], X^{k+3m})$ and (3.4.2) holds in X^{k+m}.

(b) Let $u_0 \in X^\infty$, then there is a $t_0 < t_2 \in I$ and a unique $u \in C^1([t_0, t_2], X^\infty)$ with

$$\frac{du}{dt}(t) = A(t, u(t))u(t) + f(t, u(t)), \; t \in [t_0, t_2], \qquad u(t_0) = u_0 \,.$$

PROOF: We only have to prove (a). Due to 3.3.9 there is a maximal interval $I_{k'} \subset I$ with left boundary point $t_0 \in I_{k'}$ and a unique $u \in C^1(I_{k'}, X^{k'}) \cap C(I_{k'}, X^{k'+m}) \cap C_w(I_{k'}, X^{k'+2m})$ with (3.4.2) in $X^{k'}$ for $k' = k, k + m$. Assume that there is a $t_3 \in I_k$ with $I_{k+m} \subset [t_0, t_3]$. Then there is an $R_{k+2m} > 0$ with $\|u(t)\|_{k+2m} \leq R_{k+2m}$ for $t \in [t_0, t_3]$, because $u \in C_w([t_0, t_3], X^{k+2m})$. Now $J_l^\nu u \subset C^1(I_{k+m}, X^{k+3m}) \cap C(I_{k+m}, X^{k+4m})$ with

$$\frac{d}{dt} \|J_l^\nu u(t)\|_{k+3m}^2$$

$$= \quad 2\text{Re} \, \langle J_l^\nu A(t, u(t))u(t), J_l^\nu u(t) \rangle_{k+3m} + 2\text{Re} \, \langle J_l^\nu f(t, u(t)), J_l^\nu u(t) \rangle_{k+3m}$$

$$\leq \quad \gamma_k' \|u(t)\|_{k+3m}^2 + G_k' \|u(t)\|_{k+3m}^2 \leq h_k \|u(t)\|_{k+3m}^2$$

with suitable constants $\gamma_k', G_k', h_k \geq 0$ for $t \in I_{k+m}$. Hence

$$\|J_l^\nu u(t)\|_{k+3m}^2 \quad = \quad \|J_l^\nu u(t_0)\|_{k+3m}^2 + \int_{t_0}^t \frac{d}{d\tau} \|J_l^\nu u(\tau)\|_{k+3m}^2 \, d\tau$$

$$\leq \quad \|J_l^\nu u(t_0)\|_{k+3m}^2 + \int_{t_0}^t h_k \|u(\tau)\|_{k+3m}^2 \, d\tau \qquad \text{for } t \in I_{k+m} \,.$$

With $l \to \infty$ we obtain

$$\|u(t)\|_{k+3m}^2 \leq \|u_0\|_{k+3m}^2 + \int_{t_0}^t h_k \|u(\tau)\|_{k+3m}^2 \, d\tau \quad \text{for } t \in I_{k+m}$$

and 3.4.3 implies $\|u(t)\|_{k+3m}^2 \leq \|u_0\|_{k+3m}^2 e^{h_k(t-t_0)} \leq R_{k+3m}$ for $t \in I_{k+m}$ and a suitable R_{k+3m}. This contradicts 3.3.9 and shows $[t_0, t_2] \subset I_{k+m} = I_k$, which proves the assertion. $\qquad \square$

Chapter 4

Applications to linear, time-dependent evolution equations

After developing the theory of linear and quasilinear (abstract) evolution equations in scales of Banach spaces we will give applications to linear and quasilinear pseudodifferential evolution equations in chapter 4 and 5. First, in the present chapter we describe several consequences for linear equations

$$u'(t) = A(t)u(t) + f(t) .$$

After collecting definitions and properties of pseudodifferential operators in section 4.1, we prove in section 4.2 results on well-posedness of the Cauchy problem for several pseudodifferential evolution equations in scales of un-weighted and weighted L^2-Sobolev spaces. The probably most interesting result in this section (cf. 4.2.17) extends a theorem on well-posedness of first-order hyperbolic pseudodifferential evolution equations (cf. [67, 23.1.2]) to equations with second-order operators $A(t)$ including degenerate parabolic equations. Here we make essential use of the Fefferman-Phong inequality. In section 4.3 we regard Schrödinger equations and give conditions for essential selfadjointness of pseudodifferential operators. Then, in section 4.4 and 4.5 we consider evolution equations in scales of k-times continuously differentiable functions and L^p-Sobolev spaces. Moreover, in 4.5 we give a result on well-posedness of a degenerate parabolic second-order boundary value problem that is non-characteristic at the boundary. Finally, following Ali Mehmeti [6], [5] some applications to evolution equations on networks are described in section 4.7.

4.1 Pseudodifferential operators and weighted Sobolev spaces

For applications to pseudodifferential evolution equations in this and the following chapter we will have to use several properties of vector-valued pseudodifferential operators in weighted and unweighted scales of Sobolev spaces. To this end we will review and collect results that we will need in the sequel. Since the theory of pseudodifferential operators is discussed in detail in many modern textbooks and monographs we only collect results without proofs. We refer to [67], [97], [149], [140], [32], [123], [135], [137].

By $(E, \langle \cdot, \cdot \rangle, \|\cdot\|)$ we will always denote a Hilbert space and for $x \in \mathbf{R}^n$ let $\langle x \rangle := (1 + |x|^2)^{1/2}$. Denote by $(L^p(\Omega, E), \|\cdot\|_{L^p(\Omega, E)})$, where $1 \leq p \leq \infty$, the p-integrable functions on a σ-finite measure space (Ω, μ), by $C^k(\Omega, E)$ the k-times continuously differentiable functions from Ω to E, where $k \in \mathbf{N}_0 \cup \{\infty\}$ and $\Omega \subset \mathbf{R}^n$ is an open subset, and by

$$\mathscr{S}(\mathbf{R}^n, E) := \left\{ u \in C^\infty(\mathbf{R}^n, E) : \|u\|_{\mathscr{S}, k} := \sup_{x \in \mathbf{R}^n, |\alpha| \leq k} \left\| \langle x \rangle^k \partial^\alpha u(x) \right\|_E < \infty \right.$$

$$\left. \text{for all } k \in \mathbf{N}_0 \right\}$$

the Fréchet space of rapidly decreasing functions. Here we use the multi-index notation $\partial^\alpha := \partial_{x_1}^{\alpha_1} \cdots \partial_{x_n}^{\alpha_n}$ for $\alpha = (\alpha_1, \ldots, \alpha_n) \in \mathbf{N}_0^n$. Its dual space $\mathscr{S}'(\mathbf{R}^n, E) := \mathscr{L}(\mathscr{S}(\mathbf{R}^n, \mathbf{C}), E)$ equipped with the topology of point-wise convergence is called the space of E-valued tempered distributions. Let $L^p(\Omega) := L^p(\Omega, \mathbf{C}), \mathscr{S}(\mathbf{R}^n) := \mathscr{S}(\mathbf{R}^n, \mathbf{C}), \mathscr{S}'(\mathbf{R}^n) := \mathscr{S}'(\mathbf{R}^n, \mathbf{C})$.
The Fourier transform

$$\widehat{u}(\xi) := \mathcal{F}[u](\xi) := \int_{\mathbf{R}^n} e^{-ix \cdot \xi} u(x) dx \quad \text{for } u \in \mathscr{S}(\mathbf{R}^n, E)$$

defines a topological isomorphism $\mathcal{F} : \mathscr{S}(\mathbf{R}^n, E) \to \mathscr{S}(\mathbf{R}^n, E)$ with

$$\check{u}(\xi) := \mathcal{F}^{-1}[u](\xi) = \int_{\mathbf{R}^n} e^{ix \cdot \xi} u(\xi) d\xi \quad \text{for } u \in \mathscr{S}(\mathbf{R}^n, E),$$

where $d\xi := \frac{d\xi}{(2\pi)^n}$. $\mathscr{S}(\mathbf{R}^n, E)$ is a dense subspace of $L^2(\mathbf{R}^n, E)$ and $\mathscr{S}'(\mathbf{R}^n, E)$, and the Fourier transform extends continuously to an isometric isomorphism $\mathcal{F} : L^2(\mathbf{R}^n, E) \longrightarrow L^2(\mathbf{R}_\xi^n, E)$ (where $\mathbf{R}_\xi^n$ carries the measure $d\xi$) and to a topological isomorphism $\mathcal{F} : \mathscr{S}'(\mathbf{R}^n, E) \longrightarrow \mathscr{S}'(\mathbf{R}^n, E)$ by $\mathcal{F}(F)(u) := F(\mathcal{F}(u))$ for $F \in \mathscr{S}'(\mathbf{R}^n, E)$ and $u \in \mathscr{S}(\mathbf{R}^n)$. Moreover, let $D^\alpha := \frac{1}{i^{|\alpha|}} \partial^\alpha$, then $\widehat{D^\alpha u}(\xi) = \xi^\alpha \widehat{u}(\xi)$ for $u \in \mathscr{S}(\mathbf{R}^n, E)$ and $\xi \in \mathbf{R}^n$.

4.1.1 Definition. For $m \in \mathbf{R}, 0 \leq \delta < \rho \leq 1$ let

$$S_{\rho,\delta}^m[E] := \left\{ p \in C^\infty(\mathbf{R}_x^n \times \mathbf{R}_\xi^n, \mathcal{L}(E)) : \forall \alpha, \beta \in \mathbf{N}_0^n \, \exists c_{\alpha\beta} \geq 0 \, \forall x, \xi \in \mathbf{R}^n : \right.$$
$$\left. \left\| \partial_\xi^\alpha \partial_x^\beta p(x, \xi) \right\|_{\mathcal{L}(E)} \leq c_{\alpha\beta} \langle \xi \rangle^{m - \rho|\alpha| + \delta|\beta|} \right\}.$$

Taking the infimum of all possible constants $c_{\alpha\beta}$ yields a Fréchet topology on $S_{\rho,\delta}^m[E]$. Moreover, let $S^{-\infty}[E] := \bigcap_{m \in \mathbf{R}} S_{\rho,\delta}^m[E]$, $S^m[E] := S_{1,0}^m[E]$, $S_{\rho,\delta}^m := S_{\rho,\delta}^m[\mathbf{C}]$, and $S^m := S^m[\mathbf{C}]$ for $m \in \mathbf{R} \cup \{-\infty\}$.

We clearly have the embedding $S_{\rho,\delta}^m \ni p \mapsto p\mathrm{Id}_E \in S_{\rho,\delta}^m[E]$. The elements of $S_{\rho,\delta}^m[E]$ are called *symbols* . To a symbol $p \in S_{\rho,\delta}^m[E]$ we associate the pseudodifferential operator

$$p(X, D_x)u(x) := \int_{\mathbf{R}^n} e^{ix \cdot \xi} p(x, \xi) \hat{u}(\xi) d\xi \quad \text{for } u \in \mathscr{S}(\mathbf{R}^n, E).$$

We write $p(X, D_x) \in \mathrm{Op}S_{\rho,\delta}^m[E]$. Then $S_{\rho,\delta}^m[E] \ni p \mapsto p(X, D_x) \in \mathrm{Op}S_{\rho,\delta}^m[E]$ is an isomorphism and induces a Fréchet topology on $\mathrm{Op}S_{\rho,\delta}^m[E]$. There is a natural embedding $\mathrm{Op}S_{\rho,\delta}^m \hookrightarrow \mathrm{Op}S_{\rho,\delta}^m[E]$.

In particular, $\langle D_x \rangle^m = \langle D_x \rangle^m \mathrm{Id}_E \in \mathrm{Op}S_{\rho,\delta}^m[E]$.

4.1.2 Theorem.

(a) For $P \in \mathrm{Op}S_{\rho,\delta}^m[E], m \in \mathbf{R}, 0 \leq \delta < \rho \leq 1$ the pseudodifferential operator $P : \mathscr{S}(\mathbf{R}^n, E) \to \mathscr{S}(\mathbf{R}^n, E)$ is continuous and has a unique, continuous extension $P : \mathscr{S}'(\mathbf{R}^n, E) \to \mathscr{S}'(\mathbf{R}^n, E)$.

(b) Then, for $m, m' \in \mathbf{R}, 0 \leq \delta < \rho \leq 1$ the following mappings are continuous:

- $\mathrm{Op}S_{\rho,\delta}^m[E] \times \mathrm{Op}S_{\rho,\delta}^{m'}[E] \ni (P, Q) \mapsto PQ \in \mathrm{Op}S_{\rho,\delta}^{m+m'}[E]$

- $\mathrm{Op}S_{\rho,\delta}^m[E] \times \mathrm{Op}S_{\rho,\delta}^{m'}[E] \ni (p(X, D_x), q(X, D_x)) \mapsto$
$$p(X, D_x)q(X, D_x) - (pq)(X, D_x) \in \mathrm{Op}S_{\rho,\delta}^{m+m'-(\rho-\delta)}[E]$$

- $\mathrm{Op}S_{\rho,\delta}^m \times \mathrm{Op}S_{\rho,\delta}^{m'}[E] \ni (P, Q) \mapsto$
$$[P, Q] := PQ - QP \in \mathrm{Op}S_{\rho,\delta}^{m+m'-(\rho-\delta)}[E]$$

- $\mathrm{Op}S_{\rho,\delta}^m[E] \ni P \mapsto [P, \langle D_x \rangle^{m'}] \in \mathrm{Op}S_{\rho,\delta}^{m+m'-(1-\delta)}[E]$

- $\mathrm{Op}S_{\rho,\delta}^m[E] \ni P \mapsto P^* \in \mathrm{Op}S_{\rho,\delta}^m[E]$

- $\mathrm{Op}S_{\rho,\delta}^m[E] \ni p(X, D_x) \mapsto p(X, D_x)^* - p^*(X, D_x) \in \mathrm{Op}S_{\rho,\delta}^{m-(\rho-\delta)}[E]$

Here, the operator P^* be defined by $\langle Pu, v\rangle_{L^2(\mathbf{R}^n, E)} = \langle u, P^*v\rangle_{L^2(\mathbf{R}^n, E)}$ for $u, v \in \mathscr{S}(\mathbf{R}^n, E)$ and $p^*(X, D_x) := q(X, D_x)$ with $q(x, \xi) = p(x, \xi)^*$ for $x, \xi \in \mathbf{R}^n$.

PROOF: [129] or [48]. □

4.1.3 Theorem. Let $m \in \mathbf{R}, 0 \leq \delta < \rho \leq 1$. The Weyl-quantized pseudo-differential operator $p^w(X, D_x)$ with symbol $p \in S^m_{\rho,\delta}[E]$ is defined by

$$p^w(X, D_x)u(x) = \int_{\mathbf{R}^n} \left(\int_{\mathbf{R}^n} e^{i(x-x')\cdot\xi} p\left(\frac{x+x'}{2}, \xi\right) u(x')dx' \right) d\xi$$

for $u \in \mathscr{S}(\mathbf{R}^n, E)$. Then $p^w(X, D_x) \in \mathrm{Op}S^m_{\rho,\delta}[E]$, and the mappings

- $S^m_{\rho,\delta}[E] \ni p \mapsto p^w(X, D_x) \in \mathrm{Op}S^m_{\rho,\delta}[E]$
- $S^m_{\rho,\delta}[E] \ni p \mapsto p^w(X, D_x) - p(X, D_x) \in \mathrm{Op}S^{m-(\rho-\delta)}_{\rho,\delta}[E]$

are continuous. If $p(x, \xi)^* = p(x, \xi) \in S^m_{\rho,\delta}[E]$ for $x, \xi \in \mathbf{R}^n$, then $p^w(X, D_x)$ is symmetric on $\mathscr{S}(\mathbf{R}^n, E)$ in $L^2(\mathbf{R}^n, E)$. Moreover, the real part $\mathrm{Re}\, p(X, D_x) := \frac{1}{2}(p(X, D_x) + p(X, D_x)^*)$ of the operator $p(X, D_x)$ satisfies $\mathrm{Re}\, p(X, D_x) - p^w(X, D_x) \in \mathrm{Op}S^{m-2(\rho-\delta)}_{\rho,\delta}[E]$.

PROOF: [48]. □

Next we define weight functions.

4.1.4 Definition. A function $\gamma \in C^\infty(\mathbf{R}^n)$ is called *weight function* if

- $\gamma(x) \geq c > 0$ for all $x \in \mathbf{R}^n$ and some $c > 0$.

- $\partial^\alpha \gamma$ is bounded for all $\alpha \in \mathbf{N}^n_0 \setminus \{0\}$.

4.1.5 Examples. Choose $\gamma(x) = \langle x\rangle$, $1 + \log\langle x\rangle$, $\langle x'\rangle$, $\langle x'\rangle + (1 + \log\langle x''\rangle)$, $\langle x\rangle^{1/2}$, $(\langle x'\rangle + \langle x''\rangle^2)^{1/3}$, $(\langle x'\rangle + \log\langle x''\rangle)^{1/2}$ for $x = (x', x'') \in \mathbf{R}^n$. Moreover, if γ is a weight function, so is γ^t for $0 \leq t \leq 1$.

PROOF: [138, 1.2]. □

Clearly, we have the following lemma.

4.1.6 Lemma. If γ is a weight function, then the multiplication operator with $\gamma^{-r}, r > 0$, is a pseudodifferential operator in $\mathrm{Op}S^0$.

4.1.7 Proposition. Assume that γ is a weight function, $0 \leq \delta < \rho \leq 1$, $m \in \mathbf{R}$, and $p \in S^m_{\rho,\delta}[E]$. Then

$$S^m_{\rho,\delta}[E] \ni p \mapsto [\gamma, p(X, D_x)] \in \mathrm{Op}S^{m-\rho}_{\rho,\delta}[E]$$

is continuous, where γ denotes the multiplication operator with $\gamma\mathrm{Id}_E$.

PROOF: This is a straightforward extension of [138, 2.4]. □

4.1.8 Definition. For a weight function γ, $r, r' \in \mathbf{R}, 1 < p < \infty$ the vector-valued L^p-Sobolev space of order r with weight r' is defined by

$$H^{r,r'}_{p,\gamma}[E] := \{u \in \mathscr{S}'(\mathbf{R}^n, E) : \langle D_x \rangle^r \gamma^{r'} u \in L^p(\mathbf{R}^n, E)\} .$$

Let $H^{r,r'}_{\gamma}[E] := H^{r,r'}_{2,\gamma}[E]$. With the norm $\|u\|_{H^{r,r'}_{p,\gamma}[E]} := \left\|\langle D_x \rangle^r (\gamma^{r'} u)\right\|_{L^p(\mathbf{R}^n, E)}$ for $u \in H^{r,r'}_{p,\gamma}[E]$ we obtain a Banach space structure on $H^{r,r'}_{p,\gamma}[E]$ and with the scalar product $\langle u, v \rangle_{H^{r,r'}_{\gamma}[E]} := \langle \langle D_x \rangle^r \gamma^{r'} u, \langle D_x \rangle^r \gamma^{r'} v \rangle_{L^2(\mathbf{R}^n, E)}$ we obtain a Hilbert space structure on $H^{r,r'}_{\gamma}[E]$.
We also use the abbreviations $H^{r,r'}_p[E] := H^{r,r'}_{p,\langle \cdot \rangle}[E]$, $H^{r,r'}[E] := H^{r,r'}_2[E]$, $H^r_p[E] := H^{r,0}_p[E]$, $H^r[E] := H^{r,0}[E]$, $H^{r,r'}_{p,\gamma} := H^{r,r'}_{p,\gamma}[\mathbf{C}]$, $H^{r,r'}_{\gamma} := H^{r,r'}_{\gamma}[\mathbf{C}]$, $H^{r,r'}_p := H^{r,r'}_p[\mathbf{C}]$, $H^{r,r'} := H^{r,r'}[\mathbf{C}]$, $H^r_p = H^r_p[\mathbf{C}]$, $H^r := H^r[\mathbf{C}]$. Clearly, $\mathscr{S}(\mathbf{R}^n, E) \subset H^{r,r'}_{p,\gamma}$ densely.

4.1.9 Proposition. Assume that γ is a weight function and $r \in \mathbf{N}_0, r' \in \mathbf{R}$. Define

$$\langle u, v \rangle^{(1)}_{\gamma,r,r'} := \left\langle \gamma^{r'} \langle D_x \rangle^r u, \gamma^{r'} \langle D_x \rangle^r v \right\rangle_{L^2(\mathbf{R}^n, E)}$$

$$\langle u, v \rangle^{(2)}_{\gamma,r,r'} := \sum_{|\alpha| \leq r} \left\langle \partial^\alpha (\gamma^{r'} u), \partial^\alpha (\gamma^{r'} v) \right\rangle_{L^2(\mathbf{R}^n, E)}$$

$$\langle u, v \rangle^{(3)}_{\gamma,r,r'} := \sum_{|\alpha| \leq r} \left\langle \gamma^{r'} \partial^\alpha u, \gamma^{r'} \partial^\alpha v \right\rangle_{L^2(\mathbf{R}^n, E)}$$

for $u, v \in H^{r,r'}_{\gamma}[E]$. Then $\langle \cdot, \cdot \rangle^{(j)}_{\gamma,r,r'}, j = 1, 2, 3$, define equivalent scalar products on $H^{r,r'}_{\gamma}[E]$.

PROOF: The assertion is a straightforward extension of [7, 1.7]. □

4.1.10 Theorem. Let γ be a weight function, $m, r, r' \in \mathbf{R}$, and $p = 2$, $0 \le \delta < \rho \le 1$ or $1 < p < \infty$, $0 \le \delta < 1$, $\rho = 1$, $E = \mathbf{C}$. Then

$$\mathrm{Op}S^m_{\rho,\delta}[E] \hookrightarrow \mathscr{L}(H^{r+m,r'}_{p,\gamma}[E], H^{r,r'}_{p,\gamma}[E]) .$$

PROOF: This is a straightforward extension of [138, 1.7], cf. also [7, 1.6]. $\square$

4.1.11 Lemma. Let Ω be a metric space, γ be a weight function, $m, r, r' \in \mathbf{R}$, and $p = 2$, $0 \le \delta < \rho \le 1$ or $1 < p < \infty$, $0 \le \delta < 1$, $\rho = 1$, $E = \mathbf{C}$. Moreover, let $(p_\omega)_{\omega \in \Omega} \subset S^m_{\rho,\delta}[E]$ be a bounded set such that $\Omega \ni \omega \mapsto p_\omega(x, \xi) \in \mathscr{L}(E)$ is continuous for fixed $x, \xi \in \mathbf{R}^n$.
Then the mapping $\Omega \ni \omega \mapsto p_\omega(X, D_x) \in \mathscr{L}(H^{r+m,r'}_{p,\gamma}[E], H^{r,r'}_{p,\gamma}[E])$ is strongly continuous.

PROOF: It is easy to show that $p_{\omega_k}(X, D_x)u \xrightarrow[k \to \infty]{} p_\omega(X, D_x)u$ in $\mathscr{S}(\mathbf{R}^n, E)$ for any $u \in \mathscr{S}(\mathbf{R}^n, E)$ and sequence $(\omega_k)_{k \in \mathbb{N}} \subset \Omega$ with $\omega_k \xrightarrow[k \to \infty]{} \omega \in \Omega$. Hence 4.1.10 implies the assertion. $\square$

For applications of the results in section 2.4 we are interested in symbol conditions that imply quasi-dissipativeness of the associated pseudodifferential operator. Here two important results are known. First, we have the sharp Gårding inequality.

4.1.12 Theorem. Let $0 \le \delta < \rho \le 1$ and $\mathcal{P} \subset S^{\rho-\delta}_{\rho,\delta}[E]$ be a bounded set of symbols. Let $c \ge 0$ and assume $\mathrm{Re}\, p(x, \xi) := \frac{1}{2}(p(x, \xi) + p(x, \xi)^*) \le c\mathrm{Id}_E$ for $x, \xi \in \mathbf{R}^n$ and $p \in \mathcal{P}$. Then there is a $\beta \ge 0$ with

$$\mathrm{Re}\, \langle p(X, D_x)u, u \rangle_{L^2(\mathbf{R}^n, E)} \le \beta \|u\|^2_{L^2(\mathbf{R}^n, E)} \qquad \text{for } p \in \mathcal{P}, u \in \mathscr{S}(\mathbf{R}^n, E).$$

PROOF: [67, 18.6.14] or [48, 2.91]. $\square$

In the scalar case Fefferman and Phong have extended this inequality to symbols of order $S^{2(\rho-\delta)}_{\rho,\delta}$. This inequality is called Fefferman-Phong inequality.

4.1.13 Theorem. Let $0 \le \delta < \rho \le 1$ and let $\mathcal{P} \subset S^{2(\rho-\delta)}_{\rho,\delta}$ be a bounded family of symbols. If there is a $c \ge 0$ with $p(x, \xi) \le c$ for $x, \xi \in \mathbf{R}^n$ and $p \in \mathcal{P}$, then there is a $\beta \ge 0$ with

$$\mathrm{Re}\, \langle p(X, D_x)u, u \rangle_{L^2(\mathbf{R}^n)} \le \beta \|u\|^2_{L^2(\mathbf{R}^n)} \qquad \text{and}$$
$$\mathrm{Re}\, \langle p^w(X, D_x)u, u \rangle_{L^2(\mathbf{R}^n)} \le \beta \|u\|^2_{L^2(\mathbf{R}^n)}$$

for $p \in \mathcal{P}, u \in \mathscr{S}(\mathbf{R}^n)$.

PROOF: [45], [147], or [67, 18.6]. $\square$

4.1.14 Remark.

(a) Brummelhuis [20] has shown that a generalization of the Fefferman-Phong inequality for symbols $p \in S^2(\mathbb{C}^2)$ does not hold.

(b) The Fefferman-Phong inequality in its original version (cf. [45], [46]) is formulated for non-positive symbols $p \in S^2$. In the sequel a lot of work has been invested in further studies and generalizations of this inequality. Let us mention here [67, 18.6], [66], [147], [20], [142], [29], [103].
There is also a related inequality on compact manifolds, called Melin inequality (which requires some additional assumptions), cf. [107], [65], [67, 22.3], [146, 8§4], and [120] (and the references given therein).

Moreover, symbol conditions implying quasi-dissipativity of pseudodifferential operators can be given for special hypoelliptic symbols. Here we focus on the scalar case.

4.1.15 Definition.
Let $m \geq m' \geq 0$ and $0 \leq \delta < \rho \leq 1$. We write $p \in HS_{\rho,\delta}^{m,m'}$, if there are $p_0 \in S_{\rho,\delta}^m$ and $p_1 \in S_{\rho,\delta}^{m-(\rho-\delta)}$ and constants $c_0 > 0$, $c_{\alpha\beta} > 0$, $\alpha, \beta \in \mathbf{N}_0^n$ with $p = p_0 + p_1$ and

(a) $p_0(x,\xi) \geq c_0 \langle \xi \rangle^{m'}$ for $\xi \in \mathbf{R}^n$.

(b) $|\partial_\xi^\alpha \partial_x^\beta p_0(x,\xi)| \leq c_{\alpha\beta} |p_0(x,\xi)| \langle \xi \rangle^{-\rho|\alpha|+\delta|\beta|}$ for $x, \xi \in \mathbf{R}^n$.

(c) $|\partial_\xi^\alpha \partial_x^\beta p_1(x,\xi)| \leq c_{\alpha\beta} |p_0(x,\xi)| \langle \xi \rangle^{-(\rho-\delta)-\rho|\alpha|+\delta|\beta|}$ for $x, \xi \in \mathbf{R}^n$.

Note that $|\partial_\xi^\alpha \partial_x^\beta p(x,\xi)| \leq c_{\alpha\beta} |p_0(x,\xi)| \langle \xi \rangle^{-\rho|\alpha|+\delta|\beta|}$ for $x, \xi \in \mathbf{R}^n$. Moreover, we call $\mathcal{P} \subset S_{\rho,\delta}^m$ a bounded subset of $HS_{\rho,\delta}^{m,m'}$, if we can choose simultaneous constants for the elements of $\mathcal{P}$.

The following symbols satisfy the conditions of 4.1.15.

4.1.16 Examples.

(a) $p = p_0 + p_1$ with an elliptic $p_0 \in S_{\rho,\delta}^m$ (i.e., $p_0(x,\xi) \geq c_0 \langle \xi \rangle^m$) and arbitrary $p_1 \in S_{\rho,\delta}^{m-(\rho-\delta)}$.

(b) Elliptic symbols of variable order $p(x,\xi) = \langle \xi \rangle^{a(x)}$ with $a \in \mathcal{B}^\infty(\mathbf{R}^n)$ and $0 \leq m' := \inf_{x \in \mathbf{R}^n} \{a(x)\} \leq m := \sup_{x \in \mathbf{R}^n} \{a(x)\}, \rho = 1, 0 < \delta < 1$.

(c) $p(x,\xi) = a(x)\langle D_x \rangle^m + \langle D_x \rangle^{m'}, m > m' \geq 0$, for special $a \in \mathcal{B}^\infty(\mathbf{R}^n)$ with $a(x) > 0$ for $x \neq 0$ and $a(0) = 0$.

PROOF: [17], [97, 2.5.5].   □

$HS_{\rho,\delta}^{m,m'}$ is invariant under several operations.

4.1.17 Lemma. Let $0 \leq m' \leq m, 0 \leq k' \leq k$, and $0 \leq \delta < \rho \leq 1$. Then, for $p \in HS_{\rho,\delta}^{m,m'}$ and $q \in HS_{\rho,\delta}^{k,k'}$ we have

(a) $pq \in HS_{\rho,\delta}^{m+k,m'+k'}$

(b) $p\#q \in HS_{\rho,\delta}^{m+k,m'+k'}$, where $p\#q$ denotes the symbol of $p(X, D_x)q(X, D_x)$.

(c) $p^* \in HS_{\rho,\delta}^{m,m'}$, where p^* denotes the symbol of the formal adjoint $p(X, D_x)^*$.

Here, bounded subsets are transformed into bounded subsets.

PROOF: Let $p = p_0 + p_1, q = q_0 + q_1$ be the decompositions as in definition 4.1.15. Then one can check directly that $pq = s_0 + s_1 \in HS_{\rho,\delta}^{m+k,m'+k'}$ with $s_0 = p_0 q_0$ and $s_1 = p_1 q_0 + p_0 q_1 + p_1 q_1$.

Moreover, let $N \in \mathbb{N}$ with $m + k - N(\rho - \delta) < m' + k' - (\rho - \delta)$ and choose a remainder term $r_N \in S_{\rho,\delta}^{m+k-N(\rho-\delta)} \subset S_{\rho,\delta}^{-(\rho-\delta)+m'+k'}$ (cf. [97, 2.1.7]) with

$$p\#q = pq + \sum_{1<|\alpha|<N} \frac{1}{\alpha!}(\partial_\xi^\alpha p)(D_x^\alpha q) + r_N \ .$$

Let $t_0 := p_0 q_0 \in S_{\rho,\delta}^{m+k}$ and

$$t_1 := p_0 q_1 + p_1 q_0 + p_1 q_1 + \sum_{1<|\alpha|<N} \frac{1}{\alpha!}(\partial_\xi^\alpha p)(D_x^\alpha q) + r_N \in S_{\rho,\delta}^{m+k-(\rho-\delta)} \ .$$

Then clearly $t_0(x, \xi) \geq \tilde{c}_0 \langle\xi\rangle^{m'+k'}$ and $|\partial_\xi^\alpha \partial_x^\beta t_0(x, \xi)| \leq \tilde{c}_{\alpha\beta}|t_0(x, \xi)|\langle\xi\rangle^{-\rho|\alpha|+\delta|\beta|}$ for $x, \xi \in \mathbb{R}^n$ and suitable constants $\tilde{c}_0, \tilde{c}_{\alpha\beta}$. As

$$|\partial_\xi^\alpha \partial_x^\beta r_N(x, \xi)| \leq d_{\alpha\beta}\langle\xi\rangle^{m'+k'}\langle\xi\rangle^{-(\rho-\delta)-\rho|\alpha|+\delta|\beta|} \leq \frac{d_{\alpha\beta}}{\tilde{c}_0}|t_0(x, \xi)|\langle\xi\rangle^{-(\rho-\delta)-\rho|\alpha|+\delta|\beta|}$$

for $x, \xi \in \mathbb{R}^n$ and suitable $d_{\alpha\beta}$, one can verify $p\#q = t_0 + t_1 \in HS_{\rho,\delta}^{m+k,m'+k'}$. Finally, choosing an $N \in \mathbb{N}$ with $m - N(\rho - \delta) < m' - (\rho - \delta)$ and a remainder $r_N \in S_{\rho,\delta}^{m-N(\rho-\delta)} \subset S_{\rho,\delta}^{m'-(\rho-\delta)}$ (cf. [97, 2.1.7]) with

$$p^* = \overline{p} + \sum_{1<|\alpha|<N} \frac{1}{\alpha!}\partial_\xi^\alpha D_x^\alpha \overline{p} + r_N$$

and setting $\tilde{t}_0 = p_0$ and $\tilde{t}_1 = \overline{p_1} + \sum_{1<|\alpha|<N} \frac{1}{\alpha!}\partial_\xi^\alpha D_x^\alpha \overline{p} + r_N$ one can show in the same manner that $p^* = \tilde{t}_0 + \tilde{t}_1 \in HS_{\rho,\delta}^{m,m'}$. $\qquad\square$

4.1.18 Theorem. Let $0 \leq m' \leq m$ and $0 \leq \delta < \rho \leq 1$. Then for $p \in S_{\rho,\delta}^m$ with $-p \in HS_{\rho,\delta}^{m,m'}$ there is a $\beta > 0$ with

$$\operatorname{Re} \langle p(X, D_x)u, u \rangle_{L^2(\mathbf{R}^n)} \leq \beta \|u\|_{L^2(\mathbf{R}^n)}^2 \quad \text{for } u \in \mathscr{S}(\mathbf{R}^n).$$

PROOF: For $q \in S_{\rho,\delta}^m$ with $Q := q(X, D_x) = p(X, D_x) + p(X, D_x)^*$ we have

$$2\operatorname{Re} \langle p(X, D_x)u, u \rangle_{L^2} = \langle p(X, D_x)u + p(X, D_x)^* u, u \rangle_{L^2} = \langle Qu, u \rangle_{L^2}.$$

Moreover, 4.1.17 shows $-q \in HS_{\rho,\delta}^{m,m'}$. As Q is symmetric [129, 4.6] shows that $-Q$ is essentially selfadjoint on $\mathscr{S}(\mathbf{R}^n)$ in $L^2(\mathbf{R}^n)$ and that there is a $\beta > 0$ with $\lambda \in \rho(-\overline{Q})$ for $\lambda < -\beta$, where $\overline{Q}$ denotes the minimal closed extension of Q in $L^2(\mathbf{R}^n)$. Hence the spectral theorem implies the assertion. $\square$

Now we give the definitions and some properties of a further pseudodifferential calculus.

4.1.19 Definition. For $m, m' \in \mathbf{R}$ define

$$\psi c^{(m,m')} := \left\{ p \in C^\infty(\mathbf{R}_x^n \times \mathbf{R}_\xi^n) : \text{For } l \in \mathbf{N}_0 \right.$$

$$|p|_l^{\psi c^{(m,m')}} := \max_{|\alpha|+|\beta| \leq l} \sup_{x, \xi \in \mathbf{R}^n} \left\{ |\partial_\xi^\alpha \partial_x^\beta p(x, \xi)| \langle \xi \rangle^{-m+|\alpha|} \langle x \rangle^{-m'+|\beta|} \right\} < \infty \left. \right\}.$$

As usual $(\psi c^{(m,m')}, \{|\cdot|_l^{\psi c^{(m,m')}} : l \in \mathbf{N}_0\})$ is a Fréchet space and

$$\psi c^{(-\infty,-\infty)} := \bigcap_{m,m' \in \mathbf{R}} \psi c^{(m,m')}.$$

For $p \in \psi c^{(m,m')}$ and $u \in \mathscr{S}(\mathbf{R}^n)$ define

$$\left(p(X, D_x)u \right)(x) := \int_{\mathbf{R}^n} e^{ix \cdot \xi} p(x, \xi) \hat{u}(\xi) d\xi \text{ for } x \in \mathbf{R}^n,$$

We write $p(X, D_x) \in \operatorname{Op}\psi c^{(m,m')}$. Then $\psi c^{(m,m')} \ni p \mapsto p(X, D_x) \in \operatorname{Op}\psi c^{(m,m')}$ is bijective and defines a Fréchet topology on $\operatorname{Op}\psi c^{(m,m')}$.

4.1.20 Theorem. Let $m, m', k, k' \in \mathbf{R}$.

(a) For $p \in \psi c^{(m,m')}$ the mapping $p(X, D_x) : \mathscr{S}(\mathbf{R}^n) \to \mathscr{S}(\mathbf{R}^n)$ is continuous and has a unique extension to a mapping $p(X, D_x) : \mathscr{S}'(\mathbf{R}^n) \to \mathscr{S}'(\mathbf{R}^n)$ that we denote again by $p(X, D_x)$.

(b) The following mappings are continuous:

- $\mathrm{Op}\psi c^{(m,m')} \times \mathrm{Op}\psi c^{(k,k')} \ni (P, Q) \mapsto PQ \in \mathrm{Op}\psi c^{(m+k,m'+k')}$

- $\mathrm{Op}\psi c^{(m,m')} \times \mathrm{Op}\psi c^{(k,k')} \ni (P, Q) \mapsto$
  $$[P, Q] := PQ - QP \in \mathrm{Op}\psi c^{(m+k-1,m'+k'-1)}$$

- $\mathrm{Op}\psi c^{(m,m')} \ni P \mapsto P^* \in \mathrm{Op}\psi c^{(m,m')}$, where P^* is defined as in 4.1.2

Moreover, if $p \in \psi c^{(m,m')}$ is real-valued, then the anti-symmetric part of $p(X, D_x)$ satisfies $p(X, D_x) - p(X, D_x)^* \in \mathrm{Op}\psi c^{(m-1,m'-1)}$.

(c) For $p \in \psi c^{(m,m')}$ the Weyl-quantized pseudodifferential operator is defined by

$$p^w(X, D_x)u(x) = \int_{\mathbf{R}^n} \left(\int_{\mathbf{R}^n} e^{i(x-x')\cdot\xi} p\left(\frac{x + x'}{2}, \xi\right) u(x')dx' \right) d\xi$$

for $u \in \mathscr{S}(\mathbf{R}^n)$. Then $p^w(X, D_x) \in \mathrm{Op}\psi c^{(m,m')}$, and the mappings

- $\psi c^{(m,m')} \ni p \mapsto p^w(X, D_x) \in \mathrm{Op}\psi c^{(m,m')}$
- $\psi c^{(m,m')} \ni p \mapsto p^w(X, D_x) - p(X, D_x) \in \mathrm{Op}\psi c^{(m-1,m'-1)}$

are continuous.
Moreover, if $p \in \psi c^{(m,m')}$ is real-valued, then $p^w(X, D_x)$ is symmetric on $\mathscr{S}(\mathbf{R}^n)$ in $L^2(\mathbf{R}^n)$ and $\mathrm{Re}\, p(X, D_x) - p^w(X, D_x) \in \mathrm{Op}\psi c^{(m-2,m'-2)}$

PROOF: [32, 1.2.1, 1.7.2]   □

Then we have the following mapping properties in weighted L^p-Sobolev spaces.

4.1.21 Theorem. For $m, m', s, t \in \mathbf{R}$ and $1 < p < \infty$ we have

$$\mathrm{Op}\psi c^{(m,m')} \hookrightarrow \mathscr{L}(H_p^{s+m,t+m'}, H_p^{s,t}).$$

PROOF: For $P \in \mathrm{Op}\psi c^{(m,m')}$ we have due to 4.1.20

$$Q := \langle D_x \rangle^s \langle x \rangle^t P \langle x \rangle^{-t-m'} \langle D_x \rangle^{-s-m} \in \mathrm{Op}\psi c^{(0,0)} \hookrightarrow \mathrm{Op}S_{1,0}^0 \,,$$

hence 4.1.10 implies

$$\|Pu\|_{p,s,t} = \left\| Q \langle D_x \rangle^{s+m} \langle x \rangle^{t+m'} u \right\|_{L^p} \le c \left\| \langle D_x \rangle^{s+m} \langle x \rangle^{t+m'} u \right\|_{L^p} = \|u\|_{p,s+m,t+m'}$$

for $u \in \mathscr{S}(\mathbf{R}^n)$. $\square$

In this calculus a version of the Fefferman-Phong inequality is also valid.

4.1.22 Theorem. Let $\mathcal{P} \subset \psi c^{(2,2)}$ boundedly with $\{\mathrm{Im}\, p : p \in \mathcal{P}\} \subset \psi c^{(1,1)}$ boundedly and $\mathrm{Re}\, p(x, \xi) \le c$ for $x, \xi \in \mathbf{R}^n, p \in \mathcal{P}$, and a suitable constant $c > 0$. Then there is a $\beta \ge 0$ with

$$\mathrm{Re}\, \langle p(X, D_x)u, u \rangle_{L^2(\mathbf{R}^n)} \le \beta \|u\|_{L^2(\mathbf{R}^n)}^2 \quad \text{and}$$
$$\mathrm{Re}\, \langle p^w(X, D_x)u, u \rangle_{L^2(\mathbf{R}^n)} \le \beta \|u\|_{L^2(\mathbf{R}^n)}^2$$

for $u \in \mathscr{S}(\mathbf{R}^n)$.

PROOF: [67, 18.6.8]. $\square$

Denote by $\mathcal{B}^k(\mathbf{R}^n)$ for $k \in \mathbf{N}_0 \cup \{\infty\}$ the set of all k-times continuously differentiable functions with bounded derivatives. Clearly, $\mathcal{B}^k(\mathbf{R}^n)$ for $k \in \mathbf{N}_0$ is a Banach space with norm $\|u\|_{\mathcal{B}^k(\mathbf{R}^n)} := \sup_{x \in \mathbf{R}^n, |\alpha| \le k} \|\partial^\alpha u\|_{L^\infty(\mathbf{R}^n)}$ for $u \in \mathcal{B}^k(\mathbf{R}^n)$. Moreover, $\mathcal{C}_0^k(\mathbf{R}^n) := \{u \in C^k(\mathbf{R}^n) : \lim_{|x| \to \infty} \partial^\alpha u(x) = 0, |\alpha| \le k\}$ for $k \in \mathbf{N}_0$ is a Banach space with norm $\|u\|_{\mathcal{C}_0^k(\mathbf{R}^n)} := \sup_{x \in \mathbf{R}^n, |\alpha| \le k} \|\partial^\alpha u\|_{L^\infty(\mathbf{R}^n)}$ for $u \in \mathcal{C}_0^k(\mathbf{R}^n)$. The classical Friedrichs mollifier is given in the following way.

4.1.23 Proposition. Let $\rho \in C_c^\infty(\mathbf{R}^n)$ be a smooth function with compact support and with $\rho(0) = 1$, $0 \le \rho \le 1$, $\int_{\mathbf{R}^n} \rho(x)dx = 1$, and $\rho_\varepsilon(x) := \varepsilon^{-n}\rho(\frac{x}{\varepsilon})$ for $\varepsilon > 0, x \in \mathbf{R}^n$. Then

(a) $\rho_\varepsilon * u \in C^\infty(\mathbf{R}^n)$ with $\partial^\alpha(\rho_\varepsilon * u) = (\partial^\alpha \rho_\varepsilon) * u$ for $\alpha \in \mathbf{N}_0^n, \varepsilon > 0$, and $u \in L^p(\mathbf{R}^n), 1 \le p \le \infty$.

(b) $\{\rho_\varepsilon * u : 0 < \varepsilon < 1\} \subset \mathcal{B}^k(\mathbf{R}^n)$ boundedly for $u \in \mathcal{B}^k(\mathbf{R}^n)$ and $\rho_\varepsilon * u \xrightarrow[\varepsilon \to 0]{} u$ in $\mathcal{B}^k(\mathbf{R}^n)$ for $u \in \mathcal{B}^{k+1}(\mathbf{R}^n), k \in \mathbf{N}_0$.

(c) For $u \in H_p^k, k \in \mathbf{N}_0, 1 < p < \infty$, let $u_\varepsilon(x) := \rho(\varepsilon x)(\rho_\varepsilon * u)(x)$ for $x \in \mathbf{R}^n$, $0 < \varepsilon < 1$. Then $u_\varepsilon \in C_c^\infty(\mathbf{R}^n)$, $u_\varepsilon \xrightarrow[\varepsilon \to 0]{} u$ in H_p^k, and $\|u_\varepsilon\|_{H_p^k} \le c\|u\|_{H_p^k}$ for $0 < \varepsilon < 1$ and a suitable constant $c > 0$.

(d) For $u \in C_0^k(\mathbf{R}^n)$, let $u_\varepsilon(x) := \rho(\varepsilon x)(\rho_\varepsilon * u)(x)$ for $0 < \varepsilon < 1$, $x \in \mathbf{R}^n$. Then $u_\varepsilon \in C_c^\infty(\mathbf{R}^n)$, $u_\varepsilon \xrightarrow[\varepsilon \to 0]{} u$ in $C_0^k(\mathbf{R}^n)$, and $\|u_\varepsilon\|_{C_0^k(\mathbf{R}^n)} \leq c \|u\|_{C_0^k(\mathbf{R}^n)}$ for $0 < \varepsilon < 1$ and a suitable $c > 0$.

PROOF: Clearly, we have (a). $\{\rho_\varepsilon * u : 0 < \varepsilon < 1\} \subset B^k(\mathbf{R}^n)$ is bounded for $u \in B^k(\mathbf{R}^n)$ because $\partial^\alpha(\rho_\varepsilon * u) = (\rho_\varepsilon * (\partial^\alpha u))$. Moreover, let $R > 0$ with $\rho(x) = 0$ for $|x| \geq R$ and $\alpha \in \mathbf{N}_0^n$ with $|\alpha| \leq k$. Then, for $u \in B^{k+1}(\mathbf{R}^n)$ and $\delta > 0$, there is an $\eta > 0$ with $|(\partial^\alpha u)(x+z) - (\partial^\alpha u)(x)| < \delta$ for $x, z \in \mathbf{R}^n$, $|z| < \eta$. Hence

$$\sup_{y \in \mathbf{R}^n} |(\partial^\alpha \rho_\varepsilon * u)(y) - (\partial^\alpha u)(y)|$$

$$\leq \sup_{y \in \mathbf{R}^n} \int_{\mathbf{R}^n} \varepsilon^{-n} \rho\left(\frac{x}{\varepsilon}\right) |(\partial^\alpha u)(y-x) - (\partial^\alpha u)(y)| dx$$

$$\leq \sup_{y \in \mathbf{R}^n} \int_{|x| \leq R} \rho(x) \underbrace{|(\partial^\alpha u)(y - \varepsilon x) - (\partial^\alpha u)(y)|}_{< \delta} dx < \delta \qquad (4.1.1)$$

if $\varepsilon R < \eta$. This proves (b). For $u \in H_p^k$, $k \in \mathbf{N}_0$, $1 < p < \infty$ we have $\|\rho_\varepsilon * u\|_{H_p^k} \leq c \|u\|_{H_p^k}$ for suitable $c > 0$ and $\rho_\varepsilon * u \xrightarrow[\varepsilon \to 0]{} u$ in H_p^k due to [97, 1.2.7] and $\partial^\alpha(\rho_\varepsilon * u) = \rho_\varepsilon * (\partial^\alpha u)$. This proves (c) because $\|\rho(\varepsilon x)u(x)\|_{H_p^k} \leq d \|u\|_{H_p^k}$ for suitable $d > 0$ and $\rho(\varepsilon x)u(x) \xrightarrow[\varepsilon \to 0]{} u(x)$ in H_p^k.

Finally, for every $u \in C_0^k(\mathbf{R}^n)$, $|\alpha| \leq k$, and $\delta > 0$ there exists an $\eta > 0$ such that $|(\partial^\alpha u)(x + z) - (\partial^\alpha u)(x)| < \delta$ for $x, z \in \mathbf{R}^n$, $|z| < \eta$. Hence we can see as in (4.1.1) that $\rho_\varepsilon * u \xrightarrow[\varepsilon \to 0]{} u$ in $C_0^k(\mathbf{R}^n)$. Since $\|\rho(\varepsilon x)u(x)\|_{C_0^k} \leq d \|u(x)\|_{C_0^k}$ for suitable $d > 0$ and $\rho(\varepsilon x)u(x) \xrightarrow[\varepsilon \to 0]{} u(x)$ in $C_0^k(\mathbf{R}^n)$ we obtain the assertion. $\square$

Moreover, we also need the following type of mollifiers:

4.1.24 Proposition.
There is a family $(J_l)_{l \in \mathbf{N}} \subset \mathrm{Op}S^{-\infty} \cap \mathrm{Op}\psi c^{(-\infty, -\infty)}$ such that the following statements hold:

- $J_l : \mathscr{S}'(\mathbf{R}^n) \longrightarrow \mathscr{S}(\mathbf{R}^n)$ continuously for $l \in \mathbf{N}$.

- $(J_l)_l \subset \mathrm{Op}S_{1,0}^0 \cap \mathrm{Op}\psi c^{(0,0)}$ boundedly.

- $J_l \xrightarrow[l \to \infty]{} \mathrm{Id}$ strongly in $\mathscr{L}(H_{p,\gamma}^{s,t})$ and $(J_l)_{l \in \mathbf{N}} \subset \mathscr{L}(H_{p,\gamma}^{s,t})$ boundedly for $s, t \in \mathbf{R}$, weight functions γ, and $1 < p < \infty$.

In particular, $\mathscr{S}(\mathbf{R}^n)$ is dense in arbitrary intersections of the spaces $H_{p,\gamma}^{s,t}$.

PROOF:

Step 1:   Choose a $\varphi \in \mathscr{S}(\mathbf{R}^n)$ with $\varphi(0) = 1$ and define $\varphi_l(x) := \varphi(\frac{x}{l})$ for $x \in \mathbf{R}^n$. Then one immediately verifies that

- $\varphi_l \in \mathscr{S}(\mathbf{R}^n_x)$

- $\forall \alpha \in \mathbf{N}_0^n \, \exists \, c_\alpha \, \forall \, l \in \mathbf{N} : |\partial_x^\alpha \varphi_l(x)| \le c_\alpha \langle x \rangle^{-|\alpha|}$

- $\forall \alpha \in \mathbf{N}_0^n \, \exists \, d_\alpha \, \forall \, l \in \mathbf{N} : |\partial_x^\alpha (\varphi_l(x) - 1)| \le \dfrac{d_\alpha}{l} \langle x \rangle^{1-|\alpha|}$

Step 2: Let $J_l := \varphi_l(X)\varphi_l(D_x) \in \mathrm{Op}S^{-\infty} \cap \mathrm{Op}\psi c^{(-\infty,-\infty)}$. Then we clearly have $J_l \in \mathscr{L}(H_{2,\langle\cdot\rangle}^{s',t'}, H_{2,\langle\cdot\rangle}^{s,t})$ for $s, t, s', t' \in \mathbf{R}$ due to 4.1.21 and $J_l : \mathscr{S}'(\mathbf{R}^n) \longrightarrow \mathscr{S}(\mathbf{R}^n)$ continuously, $l \in \mathbf{N}$, because

$$\mathscr{S}'(\mathbf{R}^n) = \bigcup_{s,t\in\mathbf{R}} H_{2,\langle\cdot\rangle}^{s,t} \quad \text{and} \quad \mathscr{S}(\mathbf{R}^n) = \bigcap_{s,t\in\mathbf{R}} H_{2,\langle\cdot\rangle}^{s,t} \qquad (4.1.2)$$

with the inductive resp., projective topology. Since $(J_l)_{l\in\mathbf{N}} \subset \mathrm{Op}S_{1,0}^0 \cap \mathrm{Op}\psi c^{(0,0)}$ boundedly, we have $(J_l)_{l\in\mathbf{N}} \subset \mathscr{L}(H_{p,\gamma}^{s,t})$ boundedly and hence we only have to show that $J_l u \xrightarrow[l\to\infty]{} u$ in $\mathscr{S}(\mathbf{R}^n)$ for $u \in \mathscr{S}(\mathbf{R}^n)$. But this is a consequence of $J_l \xrightarrow[l\to\infty]{} \mathrm{Id}$ in $\mathrm{Op}\psi c^{(1,1)} \hookrightarrow \mathscr{L}(H_{2,\langle\cdot\rangle}^{s+1,t+1}, H_{2,\langle\cdot\rangle}^{s,t})$ and (4.1.2). $\qquad\square$

We also need the pseudodifferential calculus developed by Shubin-Helffer, cf. [140] and [59].

4.1.25 Definition. For $m \in \mathbf{R}$ define

$$G^m := \Big\{ p \in C^\infty(\mathbf{R}_x^n \times \mathbf{R}_\xi^n) : \forall \alpha, \beta \in \mathbf{N}_0^n \, \exists c_{\alpha\beta} :$$

$$|\partial_\xi^\alpha \partial_x^\beta p(x, \xi)| \le c_{\alpha\beta} \langle (x, \xi) \rangle^{m-|\alpha|-|\beta|} \text{ for } x, \xi \in \mathbf{R}^n \Big\}.$$

Moreover, let $G^{-\infty} := \bigcap_{m\in\mathbf{R}} G^m$. For $p \in G^m$ define

$$\big(p(X, D_x)u\big)(x) := \int_{\mathbf{R}^n} e^{ix\cdot\xi} p(x, \xi)\hat{u}(\xi)d\xi \text{ for } u \in \mathscr{S}(\mathbf{R}^n), x \in \mathbf{R}^n .$$

We write $p(X, D_x) \in \mathrm{Op}G^m$. Note that $G^0 \subset S^0$.

4.1.26 Theorem. Let $m, m', k, k' \in \mathbf{R}$.

(a) For $p \in G^m$ the mapping $p(X, D_x) : \mathscr{S}(\mathbf{R}^n) \to \mathscr{S}(\mathbf{R}^n)$ is continuous and has a unique extension to a mapping $p(X, D_x) : \mathscr{S}'(\mathbf{R}^n) \to \mathscr{S}'(\mathbf{R}^n)$ that we denote again by $p(X, D_x)$.

(b) The following mappings are continuous:

- $\mathrm{Op}G^{m_1} \times \mathrm{Op}G^{m_2} \ni (P, Q) \mapsto PQ \in \mathrm{Op}G^{m_1+m_2}$
- $\mathrm{Op}G^{m_1} \times \mathrm{Op}G^{m_2} \ni (P, Q) \mapsto [P, Q] \in \mathrm{Op}G^{m_1+m_2-2}$
- $\mathrm{Op}G^m \ni P \mapsto P^* \in \mathrm{Op}G^m$, where P^* is defined as in 4.1.2.

Moreover, if $p \in G^m$ is real-valued, then $p(X, D_x) - p(X, D_x)^* \in \mathrm{Op}G^{m-2}$.

(c) For $p \in G^m$ the Weyl-quantized pseudodifferential operator is defined by

$$p^w(X, D_x)u(x) = \int_{\mathbf{R}^n} \left(\int_{\mathbf{R}^n} e^{i(x-x')\cdot\xi} p\left(\frac{x+x'}{2}, \xi\right) u(x')dx' \right) d\xi$$

for $u \in \mathscr{S}(\mathbf{R}^n)$. Then $p^w(X, D_x) \in \mathrm{Op}G^m$, and the mappings

- $G^m \ni p \mapsto p^w(X, D_x) \in \mathrm{Op}G^m$
- $G^m \ni p \mapsto p^w(X, D_x) - p(X, D_x) \in \mathrm{Op}G^{m-2}$

are continuous. Moreover, if $p \in G^m$ is real-valued, then $p^w(X, D_x)$ is symmetric on $\mathscr{S}(\mathbf{R}^n)$ in $L^2(\mathbf{R}^n)$ and $\mathrm{Re}\, p(X, D_x) - p^w(X, D_x) \in \mathrm{Op}G^{m-4}$

(d) If $P = p(X, D_x) \in G^m$ is a globally elliptic differential operator, i.e., there are $c, R > 0$ with $|p(x, \xi)| \geq c(|x| + |\xi|)^m$ for $|x| + |\xi| \geq R$, then there is a $Q \in \mathrm{Op}G^{-m}$ with $QP - \mathrm{Id} \in \mathrm{Op}G^{-\infty}$.

PROOF: [59, 1.2.2, 1.2.6, 1.3.1, 1.5.7] . □

Again, there is a version of the Fefferman-Phong inequality for this calculus.

4.1.27 Theorem. Let $\mathcal{P} \subset G^4$ boundedly with $\{\mathrm{Im}\, p : p \in \mathcal{P}\} \subset G^2$ boundedly and $\mathrm{Re}\, p(x, \xi) \leq c$ for $x, \xi \in \mathbf{R}^n, p \in \mathcal{P}$, and a suitable constant $c > 0$. Then there is a $\beta \geq 0$ with

$$\mathrm{Re}\, \langle p(X, D_x)u, u \rangle_{L^2(\mathbf{R}^n)} \leq \beta \|u\|^2_{L^2(\mathbf{R}^n)} \quad \text{and}$$
$$\mathrm{Re}\, \langle p^w(X, D_x)u, u \rangle_{L^2(\mathbf{R}^n)} \leq \beta \|u\|^2_{L^2(\mathbf{R}^n)}$$

for $u \in \mathscr{S}(\mathbf{R}^n)$.

PROOF: [67, 18.6.8]. □

4.1.28 Definition. For a weight function $\gamma, 1 < p < \infty$, and $k \in \mathbb{N}_0$ let

$$\mathcal{H}_{p,\gamma}^k := \bigcap_{j=0}^{k} H_{p,\gamma}^{j,k-j} \quad \text{and} \quad \|u\|_{\mathcal{H}_{p,\gamma}^k} := \left(\sum_{j=0}^{k} \|u\|_{H_{p,\gamma}^{j,k-j}}^p \right)^{1/p} \quad \text{for } u \in \mathcal{H}_{p,\gamma}^k .$$

Then $(\mathcal{H}_{p,\gamma}^k, \|\cdot\|_{\mathcal{H}_{p,\gamma}^k})$ is a Banach space. Clearly, $\mathcal{H}_\gamma^k := \mathcal{H}_{2,\gamma}^k$ is a Hilbert space with respect to the scalar product $\langle u, v \rangle_{\mathcal{H}_\gamma^k} := \sum_{j=0}^{k} \langle u, v \rangle_{H_\gamma^{j,k-j}}$ for $u, v \in \mathcal{H}_\gamma^k$. We also write $\mathcal{H}_p^k := \mathcal{H}_{p,(\cdot)}^k$ and $\mathcal{H}^k := \mathcal{H}_{(\cdot)}^k$.

4.1.29 Proposition. For $1 < p < \infty$ and $k \in \mathbb{N}_0$ we have

$$\mathcal{H}_p^k = \{u \in L^p(\mathbb{R}^n) : x^\alpha \partial^\beta u \in L^p(\mathbb{R}^n) \text{ for } |\alpha| + |\beta| \leq k\}$$

and $\|u\|_{\mathcal{H}_p^k}' := \left(\displaystyle\sum_{|\alpha|+|\beta|\leq k} \|x^\alpha \partial^\beta u\|_{L^p(\mathbb{R}^n)}^p \right)^{1/p}$ defines an equivalent norm on $\mathcal{H}_p^k$.

PROOF: Let $\mathcal{K}_p^k := \{u \in L^p(\mathbb{R}^n) : x^\alpha \partial^\beta u \in L^p(\mathbb{R}^n) \text{ for } |\alpha| + |\beta| \leq k\}$ and let $u \in \mathcal{H}_p^k$, then

$$x^\alpha \partial^\beta u = \underbrace{x^\alpha \partial^\beta \langle D_x \rangle^{-|\beta|} \langle x \rangle^{-|\alpha|}}_{\in \mathrm{Op}\psi c^{(0,0)}} \underbrace{\langle x \rangle^{|\alpha|} \langle D_x \rangle^{|\beta|} u}_{\in L^p(\mathbb{R}^n)} \in L^p(\mathbb{R}^n)$$

for $|\alpha| + |\beta| \leq k$. This shows $\mathcal{H}_p^k \hookrightarrow \mathcal{K}_p^k$.
Since $(\mathcal{H}_p^k, \|\cdot\|_{\mathcal{H}_p^k}), (\mathcal{K}_p^k, \|\cdot\|_{\mathcal{H}_p^k}')$ are Banach spaces, it suffices to show $\mathcal{K}_p^k \subset \mathcal{H}_p^k$.
Let $u \in \mathcal{K}_p^k$ and let $0 \leq j \leq k$.
If j and $k - j$ are even, then there are $c_{\alpha\beta}$ with

$$\langle x \rangle^j \langle D_x \rangle^{k-j} u = (1 + x^2)^{j/2} (\mathrm{Id} - \Delta)^{(k-j)/2} u = \sum_{|\alpha|+|\beta|\leq k} c_{\alpha\beta} x^\alpha \partial^\beta u \in L^p(\mathbb{R}^n) .$$

If j is odd (thus $j \geq 1$) and $k - j$ is even, then there are $c_{\alpha\beta}$ with

$$\langle x \rangle^j \langle D_x \rangle^{k-j} u = \langle x \rangle^{-1} (1 + x^2) \langle x \rangle^{j-1} \langle D_x \rangle^{k-j} u$$

$$= \underbrace{\langle x \rangle^{-1}}_{\in \mathrm{Op}\psi c^{(0,0)}} \sum_{|\alpha|+|\beta|\leq k-1} c_{\alpha\beta} \underbrace{x^\alpha \partial^\beta u}_{\in L^p(\mathbb{R}^n)}$$

$$+ \sum_{l=1}^{n} \underbrace{\langle x \rangle^{-1} x_l}_{\in \mathrm{Op}\psi c^{(0,0)}} \sum_{|\alpha|+|\beta|\leq k-1} c_{\alpha\beta} \underbrace{x_l x^\alpha \partial^\beta u}_{\in L^p(\mathbb{R}^n)} \in L^p(\mathbb{R}^n) .$$

Moreover, if j is even and $k - j$ is odd (thus $k - j \geq 1$), then

$$\langle x \rangle^j \langle D_k \rangle^{k-j} u = \langle x \rangle^j \langle D_x \rangle^{-1} \langle D_x \rangle^2 \langle D_x \rangle^{k-j-1} u$$

$$= \underbrace{\langle x \rangle^j \langle D_x \rangle^{-1} \langle x \rangle^{-j}}_{\in \mathrm{Op}\psi c(0,0)} \underbrace{\langle x \rangle^j \langle D_x \rangle^{k-j-1} u}_{\in L^p(\mathbf{R}^n)}$$

$$- \sum_{l=1}^n \underbrace{\langle x \rangle^j \langle D_x \rangle^{-1} \partial_l \langle x \rangle^{-j}}_{\in \mathrm{Op}\psi c(0,0)} \underbrace{\langle x \rangle^j \partial_l \langle D_x \rangle^{k-j-1} u}_{\in L^p(\mathbf{R}^n)} \in L^p(\mathbf{R}^n) .$$

Finally, if j and $k - j$ are odd, then

$$\langle x \rangle^j \langle D_x \rangle^{k-j} u$$

$$= \underbrace{\langle x \rangle^j \langle D_x \rangle^{-1} \langle x \rangle^{-j-1}}_{\in \mathrm{Op}\psi c(0,0)} \underbrace{\langle x \rangle^{j-1} \langle D_x \rangle^{k-j-1} u}_{\in L^p(\mathbf{R}^n)}$$

$$+ \sum_{l=1}^n \underbrace{\langle x \rangle^j \langle D_x \rangle^{-1} \langle x \rangle^{-j-1} x_l}_{\in \psi c(0,0)} \underbrace{x_l \langle x \rangle^{j-1} \langle D_x \rangle^{k-j-1} u}_{\in L^p(\mathbf{R}^n)}$$

$$- \sum_{l=1}^n \underbrace{\langle x \rangle^j \langle D_x \rangle^{-1} \partial_l \langle x \rangle^{-j-1}}_{\in \mathrm{Op}\psi c(0,0)} \underbrace{\langle x \rangle^{j-1} \partial_l \langle D_x \rangle^{k-j-1} u}_{\in L^p(\mathbf{R}^n)}$$

$$- \sum_{l,l'=1}^n \underbrace{\langle x \rangle^j \langle D_x \rangle^{-1} \partial_l \langle x \rangle^{-j-1} x_l}_{\in \mathrm{Op}\psi c(0,0)} \underbrace{x_l \langle x \rangle^{j-1} \partial_l \langle D_x \rangle^{k-j-1} u}_{\in L^p(\mathbf{R}^n)} \in L^p(\mathbf{R}^n) .$$

$\square$

4.1.30 Proposition. For $j, k \in \mathbf{N}_0$ and weight functions γ there are $c_{jk} \geq 0$ with

$$\left\| \gamma^j \langle D_x \rangle^k u \right\|_{L^2(\mathbf{R}^n)} \leq c_{jk} \left\| \gamma^{j+k} u \right\|_{L^2(\mathbf{R}^n)}^{\frac{j}{j+k}} \left\| \langle D_x \rangle^{j+k} u \right\|_{L^2(\mathbf{R}^n)}^{\frac{k}{j+k}}$$

$$\left\| \langle D_x \rangle^k \gamma^j u \right\|_{L^2(\mathbf{R}^n)} \leq c_{jk} \left\| \gamma^{j+k} u \right\|_{L^2(\mathbf{R}^n)}^{\frac{j}{j+k}} \left\| \langle D_x \rangle^{j+k} u \right\|_{L^2(\mathbf{R}^n)}^{\frac{k}{j+k}}$$

for $u \in \mathcal{S}(\mathbf{R}^n)$. In particular,

$$\left\| \gamma^{j+k} u \right\|_{L^2(\mathbf{R}^n)}^{\frac{j}{j+k}} \left\| \langle D_x \rangle^{j+k} u \right\|_{L^2(\mathbf{R}^n)}^{\frac{k}{j+k}} \leq \left(\left\| \gamma^{j+k} u \right\|_{L^2(\mathbf{R}^n)} + \left\| \langle D_x \rangle^{j+k} u \right\|_{L^2(\mathbf{R}^n)} \right)$$

and $H_\gamma^{k,0} \cap H_\gamma^{0,k} = \mathcal{H}_\gamma^k$ topologically.

PROOF: We can assume that $j, k \in \mathbb{N}$.
Step 1: There are $c_j > 0$ with

$$\left\|\gamma\langle D_x\rangle^j u\right\|_{L^2}^{j+1} \leq c_j \left\|\gamma^{j+1} u\right\|_{L^2} \left\|\langle D_x\rangle^{j+1} u\right\|_{L^2}^j$$

$$\left\|\gamma^j \langle D_x\rangle u\right\|_{L^2}^{j+1} \leq c_j \left\|\gamma^{j+1} u\right\|_{L^2}^j \left\|\langle D_x\rangle^{j+1} u\right\|_{L^2}$$

for $u \in \mathscr{S}(\mathbf{R}^n)$ and $j \in \mathbb{N}$: For j=1 we obtain

$$
\begin{aligned}
\left\|\gamma\langle D_x\rangle u\right\|_{L^2}^2 &= \langle\gamma\langle D_x\rangle u, \gamma\langle D_x\rangle u\rangle_{L^2} = \langle\langle D_x\rangle u, \gamma^2\langle D_x\rangle u\rangle_{L^2} \\
&= \langle\langle D_x\rangle^2 u, \gamma^2 u\rangle_{L^2} + \langle\langle D_x\rangle u, [\gamma^2, \langle D_x\rangle]u\rangle_{L^2} \\
&\leq \left\|\langle D_x\rangle^2 u\right\|_{L^2} \left\|\gamma^2 u\right\|_{L^2} + \left\|\langle D_x\rangle u\right\|_{L^2} \left\|[\gamma^2, \langle D_x\rangle]u\right\|_{L^2} \\
&\leq c_1 \left\|\langle D_x\rangle^2 u\right\|_{L^2} \left\|\gamma^2 u\right\|_{L^2}
\end{aligned}
$$

for a suitable constant $c_1 \geq 0$ due to 4.1.7 and 4.1.10. This shows step 1 for $j = 1$.
$j \to j+1$: First we have inductively

$$\left\|\gamma\langle D_x\rangle^{j+1} u\right\|_{L^2}^{j+1} \left\|\gamma^{j+1}\langle D_x\rangle u\right\|_{L^2}^{j+1}$$

$$\leq c_j' \left\|\gamma^{j+1}\langle D_x\rangle u\right\|_{L^2} \left\|\langle D_x\rangle^{j+2} u\right\|_{L^2}^j \left\|\gamma^{j+2} u\right\|_{L^2}^j \left\|\langle D_x\rangle^{j+1}(\gamma u)\right\|_{L^2}$$

$$\leq b_j^j \left\|\langle D_x\rangle^{j+2} u\right\|_{L^2}^j \left\|\gamma^{j+2} u\right\|_{L^2}^j \left\|\gamma\langle D_x\rangle^{j+1} u\right\|_{L^2} \left\|\gamma^{j+1}\langle D_x\rangle u\right\|_{L^2}$$

for a constants $c_j', b_j \geq 0$. Here we have used the equivalence of norms 4.1.9.
This shows

$$\left\|\gamma\langle D_x\rangle^{j+1} u\right\|_{L^2} \left\|\gamma^{j+1}\langle D_x\rangle u\right\|_{L^2} \leq b_j \left\|\langle D_x\rangle^{j+2} u\right\|_{L^2} \left\|\gamma^{j+2} u\right\|_{L^2} .$$

Hence we can conclude

$$
\begin{aligned}
\left\|\gamma\langle D_x\rangle^{j+1} u\right\|_{L^2}^{j+2} &= \left\|\gamma\langle D_x\rangle^j(\langle D_x\rangle u)\right\|_{L^2}^{j+1} \left\|\gamma\langle D_x\rangle^{j+1} u\right\|_{L^2} \\
&\leq c_j \left\|\gamma^{j+1}\langle D_x\rangle u\right\|_{L^2} \left\|\langle D_x\rangle^{j+2} u\right\|_{L^2}^j \left\|\gamma\langle D_x\rangle^{j+1} u\right\|_{L^2} \\
&\leq c_j b_j \left\|\gamma^{j+2} u\right\|_{L^2} \left\|\langle D_x\rangle^{j+2} u\right\|_{L^2}^{j+1} .
\end{aligned}
$$

Moreover, in the same manner we obtain

$$
\begin{aligned}
\left\|\gamma^{j+1}\langle D_x\rangle u\right\|_{L^2}^{j+2} &\leq d_j' \left\|\langle D_x\rangle(\gamma^j(\gamma u))\right\|_{L^2}^{j+1} \left\|\langle D_x\rangle(\gamma^{j+1}u)\right\|_{L^2} \\
&\leq d_j'' \left\|\langle D_x\rangle^{j+1}(\gamma u)\right\|_{L^2} \left\|\gamma^{j+2}u\right\|_{L^2}^{j} \left\|\langle D_x\rangle(\gamma^{j+1}u)\right\|_{L^2} \\
&\leq d_j''' \left\|\gamma^{j+2}u\right\|_{L^2}^{j} \left\|\gamma\langle D_x\rangle^{j+1}u\right\|_{L^2} \left\|\gamma^{j+1}\langle D_x\rangle u\right\|_{L^2} \\
&\leq d_j''' b_j \left\|\gamma^{j+2}u\right\|_{L^2}^{j+1} \left\|\langle D_x\rangle^{j+2}u\right\|_{L^2}
\end{aligned}
$$

for suitable constants d_j', d_j'', d_j'''.

step 2: Proof of the proposition: We proceed inductively for $k \in \mathbb{N}$.

$k = 1$: step 1 and equivalence of norms.

$k \to k+1$: We calculate inductively with constants $c_{j(k+1)}$

$$
\begin{aligned}
\left\|\gamma^j\langle D_x\rangle^{k+1}u\right\|_{L^2} &\leq c_{jk} \left\|\gamma^{j+k}\langle D_x\rangle u\right\|_{L^2}^{\frac{j}{j+k}} \left\|\langle D_x\rangle^{j+k+1}u\right\|_{L^2}^{\frac{k}{j+k}} \\
&\leq c_{j(k+1)} \left(\left\|\gamma^{j+k+1}u\right\|_{L^2}^{\frac{j+k}{j+k+1}} \left\|\langle D_x\rangle^{j+k+1}u\right\|_{L^2}^{\frac{1}{j+k+1}}\right)^{\frac{j}{j+k}} \left\|\langle D_x\rangle^{j+k+1}u\right\|_{L^2}^{\frac{k}{j+k}} \\
&= c_{j(k+1)} \left\|\gamma^{j+k+1}u\right\|_{L^2}^{\frac{j}{j+k+1}} \left\|\langle D_x\rangle^{j+k+1}u\right\|_{L^2}^{\frac{k+1}{j+k+1}} .
\end{aligned}
$$

Since $a^t b^{1-t} \leq a+b$ for all $a, b \geq 0, 0 \leq t \leq 1$ we obtain the first estimate. The second estimate follows by the equivalence of norms. □

4.1.31 Theorem. $\mathrm{Op}\psi c^{(m,m')} \hookrightarrow \mathscr{L}(H_p^{k+m+m'}, H_p^k)$ for $m, m', k \in \mathbb{N}_0$.

PROOF: Let $P \in \mathrm{Op}\psi c^{(m,m')}$, then due to 4.1.21

$$
\begin{aligned}
\|Pu\|_{\mathcal{H}_p^k} &= \left(\sum_{j=0}^{k} \|Pu\|_{H_p^{j,k-j}}^p\right)^{1/p} \leq c \left(\sum_{j=0}^{k} \|u\|_{H_p^{j+m,k+m'-j}}^p\right)^{1/p} \\
&\leq c \left(\sum_{j=0}^{k+m+m'} \|u\|_{H_p^{j,k+m+m'-j}}^p\right)^{1/p} = c\|u\|_{\mathcal{H}_p^{k+m+m'}} \quad \text{for } u \in \mathcal{H}_p^{k+m+m'} .
\end{aligned}
$$

□

4.1.32 Lemma. For $P \in \mathrm{Op}\psi c^{(l,4k-l)}, k \in \mathbb{N}_0, 0 \leq l \leq 4k$, there is a $c > 0$ with

$$
|\langle Pu, u\rangle_{L^2(\mathbf{R}^n)}| \leq c\|u\|_{\mathcal{H}_2^{2k}}^2 \quad \text{for } u \in \mathscr{S}(\mathbf{R}^n) .
$$

PROOF: If $l = 2j$ with $0 \le j \le 2k$, then

$$|\langle Pu, u\rangle_{L^2(\mathbf{R}^n)}| = |\langle \underbrace{\langle D_x\rangle^{-j}\langle x\rangle^{-(2k-j)}P}_{\in \mathrm{Op}\psi c(j, 2k-j)} u, \langle D_x\rangle^j \langle x\rangle^{2k-j}u\rangle_{L^2(\mathbf{R}^n)}| \le c\|u\|_{\mathcal{H}_2^{2k}}^2$$

for $u \in \mathscr{S}(\mathbf{R}^n)$. If $l = 2j + 1$ with $0 \le j \le 2k - 1$, then

$$|\langle Pu, u\rangle_{L^2(\mathbf{R}^n)}| = |\langle \underbrace{\langle D_x\rangle^{-(j+1)}\langle x\rangle^{-(2k-(j+1))}P}_{\in \mathrm{Op}\psi c(j, 2k-j)} u, \langle D_x\rangle^{j+1}\langle x\rangle^{2k-(j+1)}u\rangle_{L^2(\mathbf{R}^n)}|$$

$$\le c\|u\|_{\mathcal{H}_2^{2k}}^2$$

$\square$

4.1.33 Theorem. For $m, k \in \mathbb{N}_0$ we have $\mathrm{Op}G^m \hookrightarrow \mathscr{L}(\mathcal{H}_p^{k+m}, \mathcal{H}_p^k)$.

PROOF: If $k + m$ is even, then one can choose $Q \in \mathrm{Op}G^{-(k+m)}, R \in \mathrm{Op}G^{-\infty}$ satisfying $Q(\mathrm{Id} - \Delta + x^2)^{(k+m)/2} + R = \mathrm{Id}$ (using 4.1.26). For $|\alpha| + |\beta| \le k$ and $P \in \mathrm{Op}G^m$

$$\|x^\alpha \partial^\beta Pu\|_{L^p(\mathbf{R}^n)}$$

$$\le \| \underbrace{x^\alpha \partial^\beta PQ}_{\in \mathrm{Op}G^0 \subset \mathrm{Op}S^0} (\mathrm{Id} - \Delta + x^2)^{(k+m)/2}u\|_{L^p(\mathbf{R}^n)} + \| \underbrace{x^\alpha \partial^\beta PR}_{\in \mathrm{Op}G^0 \subset \mathrm{Op}S^0} u\|_{L^p(\mathbf{R}^n)}$$

$$\le c \sum_{|\gamma| + |\delta| \le k+m} \|x^\gamma \partial^\delta u\|_{L^p(\mathbf{R}^n)} + c\|u\|_{L^p(\mathbf{R}^n)},$$

which implies the assertion for even $k + m$.

If $k + m$ is odd, then one can choose $Q \in \mathrm{Op}G^{-(k+m)-1}$ and $R \in \mathrm{Op}G^{-\infty}$ such that $Q(\mathrm{Id} - \Delta + x^2)^{(k+m+1)/2} + R = \mathrm{Id}$. Then, for $|\alpha| + |\beta| \le k$ and $P \in \mathrm{Op}G^m$

$$\|x^\alpha \partial^\beta Pu\|_{L^p(\mathbf{R}^n)}$$

$$\le \|x^\alpha \partial^\beta PQ(\mathrm{Id} - \Delta + x^2)^{1+(k+m-1)/2}u\|_{L^p(\mathbf{R}^n)} + \|\underbrace{x^\alpha \partial^\beta PR}_{\in \mathrm{Op}S^0}u\|_{L^p(\mathbf{R}^n)}$$

$$\le \|\underbrace{x^\alpha \partial^\beta PQ}_{\in \mathrm{Op}S^0}(\mathrm{Id} - \Delta + x^2)^{(k+m-1)/2}u\|_{L^p(\mathbf{R}^n)}$$

$$+ \sum_{j=1}^n \|\underbrace{x^\alpha \partial^\beta PQ\partial_j}_{\in \mathrm{Op}S^0} \partial_j (\mathrm{Id} - \Delta + x^2)^{(k+m-1)/2}u\|_{L^p(\mathbf{R}^n)}$$

$$+ \sum_{j=1}^n \|\underbrace{x^\alpha \partial^\beta PQx_j}_{\in \mathrm{Op}S^0} x_j (\mathrm{Id} - \Delta + x^2)^{(k+m-1)/2}u\|_{L^p(\mathbf{R}^n)} + c\|u\|_{L^p(\mathbf{R}^n)}$$

$$\le c' \sum_{|\gamma| + |\delta| \le k+m} \|x^\gamma \partial^\delta u\|_{L^p(\mathbf{R}^n)}.$$

$\square$

4.1.34 Proposition. For $k \in \mathbb{N}_0$ we have

$$\mathcal{H}_p^{2k} = \{u \in L^p(\mathbf{R}^n) : (\Delta - x^2)^k u \in L^p(\mathbf{R}^n)\} \quad \text{and} \quad \|u\|_{L^p} + \|(\Delta - x^2)^k u\|_{L^p}$$

defines an equivalent norm on $\mathcal{H}_p^{2k}$.

PROOF:
For $u \in \mathcal{H}_p^{2k}$ we have $u, (\Delta - x^2)^k u \in L^p(\mathbf{R}^n)$ with

$$\|u\|_{L^p} + \|(\Delta - x^2)^k u\|_{L^p} \leq c \|u\|_{\mathcal{H}_p^{2k}} .$$

Now, take conversely $u \in L^p(\mathbf{R}^n)$ with $(\Delta - x^2)^k u \in L^p(\mathbf{R}^n)$. Due 4.1.26 there are $Q \in \mathrm{Op}G^{-2k}$ and $R \in \mathrm{Op}G^{-\infty}$ with $\mathrm{Id} = Q(\Delta - x^2)^k + R$. Thus, for $|\alpha| + |\beta| \leq 2k$

$$x^\alpha \partial^\beta u = \underbrace{x^\alpha \partial^\beta Q}_{\in \mathrm{Op}G^0} \underbrace{(\Delta - x^2)^k u}_{\in L^p(\mathbf{R}^n)} + \underbrace{x^\alpha \partial^\beta R}_{\in \mathrm{Op}G^0} \underbrace{u}_{\in L^p(\mathbf{R}^n)} \in L^p(\mathbf{R}^n)$$

with $\|x^\alpha \partial^\beta u\|_{L^p} \leq c \left(\|u\|_{L^p} + \|(\Delta - x^2)^k u\|_{L^p} \right)$. This proves the assertion. $\square$

4.1.35 Bibliographical remarks. As mentioned at the beginning of this section, general references for pseudodifferential operators are for example Hörmander [67], Kumano-go [97], Taylor [149], Shubin [140], Cordes [32], Petersen [123], Saint-Raymond [135], or Schröder [137]. In particular, vector-valued pseudodifferential calculus is discussed by Robert [129] or Folland [48] and pseudodifferential calculus in weighted Sobolev spaces by Schrohe [138] or Alvarez/Hounie [7]. The pseudodifferential calculus defined in 4.1.19 is studied in detail by Cordes [32] and more properties of the pseudodifferential calculus defined in 4.1.25 can be found in the monographs of Shubin [140] or Helffer [59]. Further results on the class $HS_{p,\delta}^{m,m'}$ have been given by Robert [129], Iwasaki [72, 3.7], and Kumano-go/Tsutsumi [98, 4.3]. For results on Fefferman-Phong inequalities and Mellin inequalities we refer to Fefferman/Phong [45], [46], Mellin [107], Hörmander [66], [65], [67, 22.3], Taylor [147], [146, 8§4], Brummelhuis [20], Sung [142], Colombini/Del Santo/Zuily [29], Lerner/Nourrigat [103], Parenti/Parmeggiani [120]. For further results on sharp Gårding inequalities we refer to Hörmander [67, 18.6.14], Folland [48, 2.91], Nagase [108], Jacob [73].

4.2   Pseudodifferential evolution equations in scales of weighted Sobolev spaces

In this section several applications of chapter 2 to systems of differential and pseudodifferential evolution equations in scales of unweighted and weighted Sobolev spaces will be given. $(E, \langle \cdot, \cdot \rangle_E, \|\cdot\|_E)$ will always denote a Hilbert space. The scale of E-valued Sobolev spaces can be written as a scale of Hilbert spaces generated by a strictly positive, selfadjoint operator.

4.2.1 Proposition. $Z := \langle D_x \rangle \mathrm{Id}_E$ with domain $D(Z) := H^1[E]$ is a strictly positive, selfadjoint operator in $H := L^2(\mathbf{R}^n, E)$ with $H_Z^k = H^k[E]$ topologically for $k \in \mathbf{N}_0$.

PROOF: Clearly, $Z : D(Z) \to H$ is strictly positive and densely defined. Moreover, $\mathcal{R}(\lambda \mathrm{Id} + Z) \subset E$ densely for $\lambda > 0$ and hence Z is strictly positive and selfadjoint due to 1.2.5. The ellipticity of $\langle D_x \rangle$ implies $H_Z^k = H^k[E]$ topologically for $k \in \mathbf{N}_0$. □

Next, we give a formulation of 2.4.6 for the vector-valued scale of Sobolev spaces $H^k[E]$ and pseudodifferential operators $A(t)$.

4.2.2 Theorem.   Let $0 \le \delta < \rho \le 1$ and $p \in C(I \times \mathbf{R}_x^n \times \mathbf{R}_\xi^n, \mathcal{L}(E))$ with $\{p(t, \cdot, \cdot) : t \in I\} \subset S_{\rho, \delta}^m[E]$.
Then the Cauchy problem for $p(t, X, D_x) \in \mathcal{L}(H^{k+m}[E], H^k[E])$ in the scale $(H^k[E])_k$ is well-posed with exponential growth if and only if for $k \in \mathbf{N}_0$ there are $\beta_k \ge 0$ with

$$\mathrm{Re} \, \langle p(t, X, D_x)u, u \rangle_{H^k[E]} \le \beta_k \|u\|_{H^k[E]}^2 \qquad \text{for } u \in \mathscr{S}(\mathbf{R}^n, E), t \in I . \quad (4.2.1)$$

PROOF: This is a consequence of 4.2.1, 4.1.11, and 2.4.6. □

Therefore one is interested in finding symbol conditions implying (4.2.1). For systems independent of x the answer is easy. Let $\mathrm{Re} \, A := \frac{1}{2}(A + A^*)$ be the real part and $\mathrm{Im} \, A := \frac{1}{2i}(A - A^*)$ the imaginary part of A, and write $A \le c\mathrm{Id}$ if $A^* = A$ and $\langle Ax, x \rangle \le c\|x\|^2$ for $A \in \mathscr{L}(E)$ and $x \in E$. Then we have the following proposition.

4.2.3 Proposition. Let $p(\xi) \in S_{\rho, \delta}^m[E]$ be a symbol independent of $x \in \mathbf{R}^n$ and $k_0 \in \mathbf{N}_0, \beta_{k_0} \ge 0$. Then the following statements are equivalent:

(a) $\mathrm{Re} \, \langle p(D_x)u, u \rangle_{H^{k_0}[E]} \le \beta_{k_0} \|u\|_{H^{k_0}[E]}^2$ for $u \in \mathscr{S}(\mathbf{R}^n, E)$.

(b) $\mathrm{Re} \, p(\xi) \le \beta_{k_0} \mathrm{Id}_E$ for $\xi \in \mathbf{R}^n$.

PROOF: We have

$$
\mathrm{Re}\,\langle p(D_x)u, u\rangle_{H^{k_0}[E]} = \int_{\mathbf{R}^n} \mathrm{Re}\,\left\langle \langle\xi\rangle^{k_0} p(\xi)\widehat{u}(\xi), \langle\xi\rangle^{k_0}\widehat{u}(\xi)\right\rangle_E d\xi
$$

$$
= \int_{\mathbf{R}^n} \langle\xi\rangle^{2k_0} \langle(\mathrm{Re}\,p(\xi))\widehat{u}(\xi), \widehat{u}(\xi)\rangle_E d\xi .
$$

This proves (b) $\Rightarrow$ (a). Conversely assume that there is a $\xi_0 \in \mathbf{R}^n$ with $(\mathrm{Re}\,p)(\xi_0) \not\leq \beta_{k_0}\mathrm{Id}_E$. Then there is an $y \in E$ with $\langle(\mathrm{Re}\,p)(\xi_0)y, y\rangle_E > \beta_{k_0}\langle y, y\rangle_E$ and due to the continuity of the mapping $\mathbf{R}^n \ni \xi \mapsto \langle(\mathrm{Re}\,p)(\xi)y, y\rangle_E$ there is an open set $U \subset \mathbf{R}^n$ with $\langle(\mathrm{Re}\,p)(\xi)y, y\rangle_E > \beta_{k_0}\langle y, y\rangle_E$ for $\xi \in U$. Now choose $\chi \in C_c^\infty(\mathbf{R}^n, \mathbf{R})$ with $0 \leq \chi \leq 1$ and $\mathrm{supp}(\chi) \subset U$, and $u \in \mathscr{S}(\mathbf{R}^n, E)$ with $\widehat{u}(\xi) = \chi(\xi)y \in \mathscr{S}(\mathbf{R}^n, E)$ for $\xi \in \mathbf{R}^n$. Hence

$$
\mathrm{Re}\,\langle p(D_x)u, u\rangle_{H^{k_0}[E]} = \int_{\mathbf{R}^n} \langle\xi\rangle^{2k_0} \langle(\mathrm{Re}\,p(\xi))\chi(\xi)y, \chi(\xi)y\rangle_E d\xi
$$

$$
= \int_{\mathbf{R}^n} \langle\xi\rangle^{2k_0}\chi(\xi)^2 \langle(\mathrm{Re}\,p(\xi))y, y\rangle_E d\xi > \int_{\mathbf{R}^n} \langle\xi\rangle^{2k_0}\chi(\xi)^2\beta_{k_0}\langle y, y\rangle_E d\xi
$$

$$
= \int_{\mathbf{R}^n} \beta_{k_0}\left\langle \langle\xi\rangle^{k_0}\widehat{u}(\xi), \langle\xi\rangle^{k_0}\widehat{u}(\xi)\right\rangle_E d\xi = \beta_{k_0}\|u\|_{H^{k_0}[E]}^2 .
$$

This is a contradiction. □

Hence it is natural to ask whether also for pseudodifferential systems with symbols depending on x quasi-dissipativity of the system corresponds to semi-boundedness of the real part of the symbol. But here the answer is much more complicated and can be given only in part. In particular, a generalization of 4.2.3 to x-dependent symbols does not hold even for differential systems (cf. 4.2.5).

However, for systems of classical pseudodifferential operators semiboundedness of the principal symbol is a necessary condition for well-posedness. Note that the following proposition includes the case of differential operators.

4.2.4 Proposition. Let $p \in S_{\rho,\delta}^m[E], m > 0, 0 \leq \delta < \rho \leq 1$, and suppose that there is a $p_m \in S_\rho^m[E]$ with $t^{-m}p(x, \eta+t\xi) \xrightarrow[t\to\infty]{} p_m(x, \xi)$ strongly in $\mathscr{L}(E)$ for $x, \xi, \eta \in \mathbf{R}^n$. Assume that Re $\langle p(X, D_x)u, u\rangle_{L^2(\mathbf{R}^n, E)} \leq \beta \|u\|_{L^2(\mathbf{R}^n, E)}^2$ for $u \in \mathscr{S}(\mathbf{R}^n, E)$ and a $\beta \geq 0$.
Then Re $p_m(x, \xi) \leq 0$ for $x, \xi \in \mathbf{R}^n$.

PROOF: Replacing p by $p - \beta \mathrm{Id}_E$ we can assume $\beta = 0$. For $w \in \mathscr{S}(\mathbf{R}^n, E)$ we have

$$\mathcal{F}_{y\to\eta}(e^{ity\xi}w(y))(\eta) = \int_{\mathbf{R}^n} e^{-iy\eta}e^{ity\xi}w(y)dy = \int_{\mathbf{R}^n} e^{-iy(\eta-t\xi)}w(y)dy = \widehat{w}(\eta - t\xi)$$

and

$$e^{-ity\xi}p(Y, D_y)(e^{it\cdot\xi}w(\cdot))(y) = \int_{\mathbf{R}^n} e^{iy\eta}p(y, \eta + t\xi)\widehat{w}(\eta)d\eta \,.$$

Choose $v \in E$ and $\rho \in C_c^\infty(\mathbf{R}^n, \mathbf{R})$ with $\mathrm{supp}(\rho) \subset \{y \in \mathbf{R}^n : |y| \leq 1\}$ and $\int_{\mathbf{R}^n} \rho(y)^2 dy = 1$. Moreover, for fixed $x, \xi \in \mathbf{R}^n$ let $\rho_\varepsilon(y) := \varepsilon^{-n/2}\rho\left(\frac{y-x}{\varepsilon}\right)$ and $u(y) := t^{-m/2}\rho_\varepsilon(y)e^{ity\xi}v \in \mathscr{S}(\mathbf{R}^n, E)$ for $0 < \varepsilon < 1$. Then

$$\begin{aligned}
0 \;\geq\;& \mathrm{Re}\,\langle p(Y, D_y)u, u\rangle_{L^2(\mathbf{R}^n_y, E)} \\
=\;& \mathrm{Re}\,\langle t^{-m}e^{-ity\xi}p(Y, D_y)(e^{ity\xi}\rho_\varepsilon(y)v), \rho_\varepsilon(y)v\rangle_{L^2(\mathbf{R}^n_y, E)} \\
=\;& t^{-m}\mathrm{Re}\,\left\langle \int_{\mathbf{R}^n} e^{iy\eta}p(y, \eta + t\xi)\widehat{\rho_\varepsilon}(\eta)d\eta v, \rho_\varepsilon(y)v\right\rangle_{L^2(\mathbf{R}^n_y, E)} \\
=\;& \mathrm{Re}\,\int_{\mathbf{R}^n}\int_{\mathbf{R}^n} e^{iy\eta}\langle t^{-m}p(y, \eta + t\xi)v, v\rangle_E \widehat{\rho_\varepsilon}(\eta)\rho_\varepsilon(y)d\eta dy \\
\xrightarrow{t\to\infty}\;& \mathrm{Re}\,\int_{\mathbf{R}^n}\int_{\mathbf{R}^n} e^{iy\eta}\widehat{\rho_\varepsilon}(\eta)d\eta\langle p_m(y, \xi)v, v\rangle_E \rho_\varepsilon(y)dy \\
=\;& \mathrm{Re}\,\int_{\mathbf{R}^n} \langle p_m(y, \xi)v, v\rangle_E \rho_\varepsilon(y)^2 dy \\
=\;& \mathrm{Re}\,\int_{\mathbf{R}^n} \langle p_m(x + \varepsilon y', \xi)v, v\rangle_E \rho(y')^2 dy' \\
\xrightarrow{\varepsilon\to 0}\;& \int_{\mathbf{R}^n} \mathrm{Re}\,\langle p_m(x, \xi)v, v\rangle_E \rho(y')^2 dy' = \langle(\mathrm{Re}\,p_m(x, \xi))v, v\rangle_E
\end{aligned}$$

This proves the assertion. □

The converse statement to this proposition does not hold, even for differential operators.

4.2.5 Example. Let $p(x_1, x_2, \xi_1, \xi_2) := \begin{pmatrix} -x_2^2\xi_1^2 & -ix_2\xi_1\xi_2 \\ ix_2\xi_1\xi_2 & -\xi_2^2 \end{pmatrix}$.

Then $\mathrm{Re}\,p(x, \xi) \leq 0$ in $M_2(\mathbf{C})$ but Brummelhuis [20] has shown that for any $\beta > 0$ there are $u, v \in C_c^\infty(\mathbf{R}^2)$ supported in the unit cube with

$$\mathrm{Re}\,\left\langle p(X, D_x)\begin{pmatrix} u \\ v \end{pmatrix}, \begin{pmatrix} u \\ v \end{pmatrix}\right\rangle_{L^2(\mathbf{R}^2, \mathbf{C}^2)} \geq \beta(\|u\|_{L^2(\mathbf{R}^2)}^2 + \|v\|_{L^2(\mathbf{R}^2)}^2)\,.$$

Moreover, in the scalar case a converse to 4.2.4 does not hold for $m > 2$ either (cf. [67, 18.6.12 and section 22.3]).

Nevertheless, one can give several sufficient symbol conditions implying L^2-quasi-dissipativity. As an additional complication, we are not only interested in L^2-quasi-dissipativity but also in H^k-quasi-dissipativity in order to apply 4.2.2. First, many hypoelliptic scalar symbols share this property.

4.2.6 Proposition. Let $0 \le m' \le m, 0 \le \delta < \rho \le 1$, and let $\mathcal{P} \subset S^m_{\rho,\delta}$ boundedly such that $\{-p : p \in \mathcal{P}\} \subset HS^{m,m'}_{\rho,\delta}$ boundedly. Then, for $s \in \mathbf{R}$ there are $\beta_s \ge 0$ with

$$\mathrm{Re}\, \langle p(X, D_x)u, u\rangle_{H^s} \le \beta_s \|u\|^2_{L^2(\mathbf{R}^n)} \qquad \text{for } u \in \mathscr{S}(\mathbf{R}^n), p \in \mathcal{P}.$$

PROOF: We have $\mathrm{Re}\, \langle p(X, D_x)u, u\rangle_{H^s} = \mathrm{Re}\, \langle \langle D_x\rangle^{2s} p(X, D_x)u, u\rangle_{L^2(\mathbf{R}^n)}$. Let $q \in S^{m+2s}_{\rho,\delta}$ with $q(X, D_x) = \langle D_x\rangle^{2s} p(X, D_x)$. Then $-q \in HS^{m+2s,m'+2s}_{\rho,\delta}$ due to 4.1.17 and 4.1.18 implies the assertion. $\square$

The next lemma shows that for special types of second-order pseudodifferential systems L^2-quasi-dissipativity implies H^k-quasi-dissipativity.

4.2.7 Lemma. Let $0 < \rho \le 1$ and $\mathcal{P} \subset \mathrm{Op}S^2_{\rho,0}[E]$ be a family of pseudodifferential operators with $\{P - P^* : P \in \mathcal{P}\} \subset \mathrm{Op}S^1_{\rho,0}[E]$ boundedly and such that there is a $\beta \ge 0$ with

$$\mathrm{Re}\, \langle Pu, u\rangle_{L^2(\mathbf{R}^n,E)} \le \beta \|u\|^2_{L^2(\mathbf{R}^n,E)} \qquad \text{for } u \in \mathscr{S}(\mathbf{R}^n, E), P \in \mathcal{P}.$$

Then, for $k \in \mathbf{N}_0$ there are $\beta_k \ge 0$ with

$$\mathrm{Re}\, \langle Pu, u\rangle_{H^k[E]} \le \beta_k \|u\|^2_{H^k[E]} \quad \text{for } u \in \mathscr{S}(\mathbf{R}^n, E), P \in \mathcal{P}.$$

PROOF: Let $P_1 := \frac{P+P^*}{2}, P_2 := \frac{P-P^*}{2}$. Then the properties of pseudodifferential calculus 4.1.2 show $\langle D_x\rangle^{-k}[\langle D_x\rangle^k, [\langle D_x\rangle^k, P_1]] \in \mathrm{Op}S^k_{\rho,0}[E]$, hence

$$\left| \left\langle [\langle D_x\rangle^k, [\langle D_x\rangle^k, P_1]]u, u \right\rangle_{L^2(\mathbf{R}^n,E)} \right| \le a_k \|u\|^2_{H^k[E]}$$

for $u \in \mathscr{S}(\mathbf{R}^n, E)$ and suitable $a_k \ge 0$ and any $k \in \mathbf{N}_0$. Hence the assertion for P_1 is a consequence of 2.3.2. Moreover, $[\langle D_x\rangle, P_2] \in \mathrm{Op}S^1_{\rho,0}[E]$, therefore the assertion for P_2 (and thus P) is a consequence of 2.3.1 because $\mathrm{Re}\, \langle P_2 u, u\rangle_{L^2(\mathbf{R}^n,E)} = 0$. $\square$

Now, using the Fefferman-Phong inequality we obtain a general result on H^k-quasi-dissipativity for semibounded, scalar second-order symbols.

4.2.8 Corollary. Let $\mathcal{P} \subset S_{1,0}^2$ be a bounded set with $\{\operatorname{Im} p : p \in \mathcal{P}\} \subset S_{1,0}^1$ boundedly and $\operatorname{Re} p(x,\xi) \leq c$ for $x, \xi \in \mathbf{R}^n, p \in \mathcal{P}$, and a suitable $c > 0$. Then, for $k \in \mathbf{N}_0$ there are $\beta_k \geq 0$ with

$$\operatorname{Re}\langle p(X, D_x)u, u\rangle_{H^k} \leq \beta_k \|u\|_{H^k}^2 \qquad \text{for } u \in \mathscr{S}(\mathbf{R}^n), p \in \mathcal{P}.$$

PROOF: This statement follows from 4.1.2, 4.1.12, 4.1.13, and 4.2.7. □

As a second corollary we obtain H^k-quasi-dissipativity of symmetric hyperbolic pseudodifferential systems.

4.2.9 Corollary. Let $\mathcal{P} \subset S_{1,0}^1[E]$ be a bounded set with $\operatorname{Re} p(x,\xi) \leq c\operatorname{Id}$ for $x, \xi \in \mathbf{R}^n, p \in \mathcal{P}$, and a suitable $c \geq 0$. Then, for $k \in \mathbf{N}_0$ there are $\beta_k \geq 0$ with

$$\operatorname{Re}\langle p(X, D_x)u, u\rangle_{H^k[E]} \leq \beta_k \|u\|_{H^k[E]}^2 \qquad \text{for } u \in \mathscr{S}(\mathbf{R}^n, E), p \in \mathcal{P}.$$

The same statement holds if $\mathcal{P} \subset S_{1,0}^1[E]$ is a bounded family of pseudodifferential operators with $\{p(X, D_x) + p(X, D_x)^* : p \in \mathcal{P}\} \subset \operatorname{Op}S_{1,0}^0[E]$ boundedly. Systems of this type are called symmetric hyperbolic pseudodifferential systems .

PROOF: The first statement is implied by 4.1.12 and 4.2.7, the second statement by

$$\left|2\operatorname{Re}\langle Pu, u\rangle_{L^2(\mathbf{R}^n, E)}\right| = \left|\langle (P + P^*)u, u\rangle_{L^2(\mathbf{R}^n, E)}\right| \leq \beta \|u\|_{L^2(\mathbf{R}^n, E)}^2 .$$

□

4.2.10 Examples. 4.2.2 can be applied to the following equations:

(a) Symmetric hyperbolic differential systems $u'(t) = P(t)u(t)$, where

$$P(t) = \sum_{j=1}^n a_j(t, x)\partial_j + b(t, x)$$

with $a_j, b \in \mathcal{B}^\infty(I \times \mathbf{R}^n, \mathscr{L}(E))$ with $a_j(t, x)^* = a_j(t, x)$ for $t \in I$, $x \in \mathbf{R}^n$, $j = 1, \ldots, n$. Then $\{P(t) : t \in I\} \subset \operatorname{Op}S_{1,0}^1[E]$ is a symmetric hyperbolic family of pseudodifferential operators in the sense of 4.2.9 because

$$P(t)^*u(x) \;=\; -\sum_{j=1}^{n} \partial_j(a_j(t,x)^*u(x)) + b(t,x)^*u(x)$$

$$=\; -\sum_{j=1}^{n} a_j(t,x)\partial_j u(x) - \sum_{j=1}^{n}(\partial_j a_j(t,x))u(x) + b(t,x)^*u(x)$$

for $u \in \mathscr{S}(\mathbf{R}^n)$, hence $P(t) + P(t)^* \in \mathrm{Op}S_{1,0}^0[E]$ for $t \in I$.

(b) First order Weyl-quantized pseudodifferential evolution equations
$u'(t) \;=\; ip^w(t, X, D_x)u(t)$, where $p \in C(I \times \mathbf{R}^n \times \mathbf{R}^n, \mathscr{L}(E))$ with
$p(t, x, \xi)^* = p(t, x, \xi)$ for $t \in I, x, \xi \in \mathbf{R}^n$ and $\{p(t, \cdot, \cdot) : t \in I\} \subset S_{1,0}^1[E]$
boundedly.

(c) (weakly) degenerate parabolic pseudodifferential equations of arbitrary
order $u'(t) \;=\; p(t, X, D_x)u(t)$ with $p \in C(I \times \mathbf{R}^n \times \mathbf{R}^n)$ and with
$\{p(t, \cdot, \cdot) : t \in I\} \subset HS_{\rho,\delta}^{m,m'}$ boundedly for $0 \le m' \le m, 0 \le \delta < \rho \le 1$.
In particular, this holds for the examples in 4.1.16.

(d) (strongly) degenerate parabolic pseudodifferential equations of order 2
$u'(t) = p(t, X, D_x)u(t)$ with $p \in C(I \times \mathbf{R}^n \times \mathbf{R}^n)$ and $\{p(t, \cdot, \cdot) : t \in I\} \subset S_{1,0}^2$
boundedly, $\{\mathrm{Im}\, p(t, \cdot, \cdot) : t \in I\} \subset S_{1,0}^1$ boundedly, $\mathrm{Re}\, p(t, x, \xi) \le c$ for
$t \in I, x, \xi \in \mathbf{R}^n$.

(e) strongly degenerate parabolic differential systems of order 2
$u'(t) \;=\; P(t)u(t)$, $P(t) = \sum_{j,l=1}^{n} \partial_j(a_{jl}(t, \cdot)\partial_l) + \sum_{j=1}^{n} b_j(t, \cdot)\partial_j + c(t, \cdot)$
with $a_{jl}, b_j, c \in B^\infty(I \times \mathbf{R}^n, \mathscr{L}(E))$ such that $\sum_{j,l=1}^{n} \langle a_{jl}(t, x)e_l, e_j \rangle \ge 0$,
$a_{jl}(t, x)^* = a_{lj}(t, x)$, and $b_j(t, x)^* = b_j(t, x)$ for $t \in I$, $x \in \mathbf{R}^n$, $e_j \in E$, and
$j = 1, \ldots, n$.

(f) Schrödinger equation

$$\frac{du}{dt}(t) = \pm i\Delta u(t) + Q(t)u(t),\ t \in \mathbf{R},\quad u(t_0) = u_0,$$

where $Q(t) = q(t, X, D_x) \in \mathrm{Op}S_{1,0}^1$ is a symmetric hyperbolic pseudodifferential operator with $q \in C(I \times \mathbf{R}^n \times \mathbf{R}^n)$.

(g) coupled Schrödinger equations

$$\frac{du_j}{dt}(t) = \pm i\Delta u_j(t) + \sum_{k=1}^{N} Q_{jk}(t)u_k(t),\ t \in \mathbf{R},\ u_j(t_0) = u_{0,j},\ j = 1, \ldots, N,$$

where $Q(t) = (Q_{jk}(t))_{j,k=1,...,N} \in OpS_{1,0}^1[\mathbb{C}^N]$ is a symmetric hyperbolic family of matrices of pseudodifferential operators depending continuously on t.

(h) linear Korteweg-de Vries equation

$$\frac{du}{dt}(t) + \partial_x^3 u(t) + \partial_x u(t) = 0, \ t \in \mathbb{R}, \quad u(t_0) = u_0$$

Next we will consider evolution equations in scales of weighted Sobolev spaces. We will need several lemmata concerning commutators between pseudodifferential operators and multiplication operators with weight functions.

4.2.11 Lemma. For a weight function γ, $k \in \mathbb{Z}, m, r, r' \in \mathbb{R}$ we have

$$OpS_{1,0}^m[E] \ni P \mapsto [\gamma^k, P] \in \mathscr{L}(H_\gamma^{r+m-1,r'+k-1}[E], H_\gamma^{r,r'}[E])$$

continuously.

PROOF: Let $\varepsilon := \text{signum}(k)$, then $[\gamma^k, P] = \sum_{j=0}^{|k|-1} \gamma^{j\varepsilon}[\gamma^\varepsilon, P]\gamma^{\varepsilon(|k|-1-j)}$ and the assertion is implied by 4.1.6, 4.1.7, and 4.1.10.   □

4.2.12 Lemma. For a weight function γ, $k \in \mathbb{Z}, m, r, r' \in \mathbb{R}$ we have

$$OpS_{1,0}^m[E] \ni P \mapsto [\gamma^k, [\gamma^k, P]] \in \mathscr{L}(H_\gamma^{r+m-2,r'+2k-2}[E], H_\gamma^{r,r'}[E])$$

continuously.

PROOF: Let $\varepsilon := \text{signum}(k)$, then

$$[\gamma^k, [\gamma^k, P]] = \sum_{j=0}^{|k|-1} \gamma^{j\varepsilon} \left[\gamma^\varepsilon, \sum_{l=0}^{|k|-1} \gamma^{l\varepsilon}[\gamma^\varepsilon, P]\gamma^{\varepsilon(|k|-1-l)}\right] \gamma^{\varepsilon(|k|-1-j)}$$

$$= \sum_{j=0}^{|k|-1}\sum_{l=0}^{|k|-1} \gamma^{(j+l)\varepsilon}[\gamma^\varepsilon, [\gamma^\varepsilon, P]]\gamma^{\varepsilon(2|k|-2-l-j)}.$$

The assertion is implied by 4.1.6, 4.1.7, and 4.1.10.   □

4.2.13 Lemma. For $Q_j \in \mathrm{Op}S_{1,0}^{m_j}, j = 1, 2, P \in \mathrm{Op}S_{1,0}^m[E], r \in \mathbf{Z}, r_1, r_2 \in \mathbb{R}$ we have

$$[\gamma^{-r} Q_1 \gamma^r, [\gamma^{-r} Q_2 \gamma^r, \gamma^{-r} P \gamma^r]] \in \mathscr{L}(H_\gamma^{r_1 + m_1 + m_2 + m - 2, r_2}[E], H_\gamma^{r_1, r_2}[E]) \,.$$

PROOF: Since $[Q_1, [Q_2, P]] \in \mathscr{L}(H_\gamma^{r_1 + m_1 + m_2 + m - 2, r_2}[E], H_\gamma^{r_1, r_2}[E])$ we have to show that the assertion for $r \in \mathbf{Z}$ implies the assertion for $r + \varepsilon, \varepsilon = \pm 1$. Now

$$[\gamma^{-r} \gamma^{-\varepsilon} Q_1 \gamma^\varepsilon \gamma^r, [\gamma^{-r} \gamma^{-\varepsilon} Q_2 \gamma^\varepsilon \gamma^r, \gamma^{-r} \gamma^{-\varepsilon} P \gamma^\varepsilon \gamma^r]]$$

$$= [\gamma^{-r} Q_1 \gamma^r, [\gamma^{-r} Q_2 \gamma^r, \gamma^{-r} P \gamma^r]] + [\gamma^{-r} Q_1 \gamma^r, [\gamma^{-r} Q_2 \gamma^r, \gamma^{-r} \underbrace{\gamma^{-\varepsilon}[P, \gamma^\varepsilon]}_{\in \mathrm{Op}S_{1,0}^m[E]} \gamma^r]]$$

$$+ [\gamma^{-r} Q_1 \gamma^r, [\gamma^{-r} \underbrace{\gamma^{-\varepsilon}[Q_2, \gamma^\varepsilon]}_{\in \mathrm{Op}S_{1,0}^{m_2}} \gamma^r, \gamma^{-r} P \gamma^r]]$$

$$+ [\gamma^{-r} Q_1 \gamma^r, [\gamma^{-r} \underbrace{\gamma^{-\varepsilon}[Q_2, \gamma^\varepsilon]}_{\in \mathrm{Op}S_{1,0}^{m_2}} \gamma^r, \gamma^{-r} \underbrace{\gamma^{-\varepsilon}[P, \gamma^\varepsilon]}_{\in \mathrm{Op}S_{1,0}^m[E]} \gamma^r]]$$

$$+ [\gamma^{-r} \underbrace{\gamma^{-\varepsilon}[Q_1, \gamma^\varepsilon]}_{\in \mathrm{Op}S_{1,0}^{m_1}} \gamma^r, [\gamma^{-r} Q_2 \gamma^r, \gamma^{-r} P \gamma^r]]$$

$$+ [\gamma^{-r} \underbrace{\gamma^{-\varepsilon}[Q_1, \gamma^\varepsilon]}_{\in \mathrm{Op}S_{1,0}^{m_1}} \gamma^r, [\gamma^{-r} Q_2 \gamma^r, \gamma^{-r} \underbrace{\gamma^{-\varepsilon}[P, \dot\gamma^\varepsilon]}_{\in \mathrm{Op}S_{1,0}^m[E]} \gamma^r]]$$

$$+ [\gamma^{-r} \underbrace{\gamma^{-\varepsilon}[Q_1, \gamma^\varepsilon]}_{\in \mathrm{Op}S_{1,0}^{m_1}} \gamma^r, [\gamma^{-r} \underbrace{\gamma^{-\varepsilon}[Q_2, \gamma^\varepsilon]}_{\in \mathrm{Op}S_{1,0}^{m_2}} \gamma^r, \gamma^{-r} P \gamma^r]]$$

$$+ [\gamma^{-r} \underbrace{\gamma^{-\varepsilon}[Q_1, \gamma^\varepsilon]}_{\in \mathrm{Op}S_{1,0}^{m_1}} \gamma^r, [\gamma^{-r} \underbrace{\gamma^{-\varepsilon}[Q_2, \gamma^\varepsilon]}_{\in \mathrm{Op}S_{1,0}^{m_2}} \gamma^r, \gamma^{-r} \underbrace{\gamma^{-\varepsilon}[P, \gamma^\varepsilon]}_{\in \mathrm{Op}S_{1,0}^m[E]} \gamma^r]] \,.$$

This shows the assertion. □

4.2.14 Lemma. For $Q \in \mathrm{Op}S_{1,0}^{m_1}, P \in \mathrm{Op}S_{1,0}^{m_2}[E], r \in \mathbf{Z}, r_1, r_2 \in \mathbb{R}$ we have

$$[\gamma^{-r} Q \gamma^r, \gamma^{-r}[\gamma^r, P]] \in \mathscr{L}(H_\gamma^{r_1 + m_1 + m_2 - 2, r_2}[E], H_\gamma^{r_1, r_2}[E]) \,.$$

PROOF: We have

$$[\gamma^{-r} Q \gamma^r, \gamma^{-r}[\gamma^r, P]] = [\gamma^{-r} Q \gamma^r, P] - [\gamma^{-r} Q \gamma^r, \gamma^{-r} P \gamma^r]$$

$$= \gamma^{-r} Q[\gamma^r, P] + \gamma^{-r}[Q, P]\gamma^r + [\gamma^{-r}, P]Q\gamma^r - \gamma^{-r}[Q, P]\gamma^r$$

$$= \gamma^{-r} Q[\gamma^r, P] - \gamma^{-r}[\gamma^r, P]\gamma^{-r} Q\gamma^r$$

$$= \gamma^{-r} Q[\gamma^r, P] - (\gamma^{-r}[\gamma^r, P]Q - \gamma^{-r}[\gamma^r, P]\gamma^{-r}[\gamma^r, Q])$$

$$= \gamma^{-r}[Q, [\gamma^r, P]] + \underbrace{\gamma^{-r}[\gamma^r, P]\gamma^{-r}[\gamma^r, Q]}_{\in \mathscr{L}(H_\gamma^{r_1 + m_1 + m_2 - 2, r_2}[E], H_\gamma^{r_1, r_2}[E])} \,.$$

Here we have used 4.2.11. Moreover, with $\varepsilon := \operatorname{signum}(r)$

$$\gamma^{-r}[Q,[\gamma^r, P]] = \sum_{j=0}^{|r|-1} \gamma^{-r}[Q, \gamma^{j\varepsilon}[\gamma^\varepsilon, P]\gamma^{\varepsilon(|r|-1-j)}]$$

$$= \sum_{j=0}^{|r|-1} \left(\gamma^{-r}\gamma^{j\varepsilon}[\gamma^\varepsilon, P][Q, \gamma^{\varepsilon(|r|-1-j)}] + \gamma^{-r}\gamma^{j\varepsilon}[Q, [\gamma^\varepsilon, P]]\gamma^{\varepsilon(|r|-1-j)} \right.$$

$$\left. + \gamma^{-r}[Q, \gamma^{j\varepsilon}][\gamma^\varepsilon, P]\gamma^{\varepsilon(|r|-1-j)} \right) \in \mathscr{L}(H_\gamma^{r_1+m_1+m_2-2,r_2}[E], H_\gamma^{r_1,r_2}[E]) \,,$$

where we have used 4.2.11. This proves the assertion.   □

Next we have a simple observation.

4.2.15 Proposition. Assume that $(H_j, \langle \cdot, \cdot \rangle_j, \|\cdot\|_j), j = 1, 2$, are Hilbert spaces and $U : H_1 \to H_2$ is an isometric isomorphism, i.e., $\langle Ux, Uy \rangle_2 = \langle x, y \rangle_1$ for $x, y \in H_1$. Moreover, assume that $Z = \{Z_1, \ldots, Z_M\}$ is a commuting family of infinitesimal generators of unitary C_0-groups in H_1. Define

$$\widetilde{Z}_j x := U Z_j U^{-1} x \text{ for } x \in D(\widetilde{Z}_j) := U(D(Z_j)), j = 1, \ldots, M.$$

Then $\widetilde{Z} := \{\widetilde{Z}_1, \ldots, \widetilde{Z}_M\}$ is a commuting family of infinitesimal generators of unitary C_0-groups in H_2, and we have $(H_2)_{\widetilde{Z}}^k = U((H_1)_Z^k)$ for $k \in \mathbb{N}_0$.

PROOF: Let $T_j(t), t \in \mathbb{R}$, denote the groups generated by $Z_j, j = 1, \ldots, M$. One can check directly that $\widetilde{T}_j(t) := U T_j(t) U^{-1}, t \in \mathbb{R}, j = 1, \ldots, M$, are unitary C_0-semigroups in H_2 with infinitesimal generators $\widetilde{Z}_j, j = 1, \ldots, M$. This implies the assertion.   □

Now we can extend 4.2.2 and 4.2.7 to scales of weighted Sobolev spaces.

4.2.16 Theorem. Let γ be a weight function, $p \in C(I \times \mathbb{R}^n \times \mathbb{R}^n, \mathscr{L}(E))$ with $\{p(t, \cdot, \cdot) : t \in I\} \subset S_{1,0}^2[E]$ boundedly and $\{\operatorname{Im} p(t, \cdot, \cdot) : t \in I\} \subset S_{1,0}^1[E]$ boundedly.
If there is a $\beta \geq 0$ with

$$\operatorname{Re} \langle p(t, X, D_x)u, u \rangle_{L^2(\mathbb{R}^n, E)} \leq \beta \|u\|_{L^2(\mathbb{R}^n, E)}^2 \qquad \text{for } u \in \mathscr{S}(\mathbb{R}^n, E),$$

then the Cauchy problem for $p(t, X, D_x) \in \mathscr{L}(H_\gamma^{k+2,k'}[E], H_\gamma^{k,k'}[E])$ in the scale $(H_\gamma^{k,k'}[E])_k$ is well-posed with exponential growth for any $k' \in \mathbb{Z}$.

PROOF: Let $Z := \gamma^{-k'}\langle D_x\rangle\gamma^{k'}\mathrm{Id}_E$ with domain $H^{1,k'}_\gamma[E]$ in $H := H^{0,k'}_\gamma[E]$. Then Z is selfadjoint in H with $H^k_Z = H^{k,k'}_\gamma[E]$ (due to 4.2.15). Moreover $I \ni t \mapsto p(t,X,D_x) \in \mathscr{L}(H^{k+2}_Z, H^k_Z)$ is strongly continuous. Let $P_1(t) := \gamma^{-k'}\mathrm{Re}\ (p(t,X,D_x))\gamma^{k'}$, $P_2(t) := \gamma^{-k'}[\gamma^{k'},\mathrm{Re}\ (p(t,X,D_x))]$, and $P_3(t) := i\mathrm{Im}\ (p(t,X,D_x))$. Then $\langle P_1(t)u,v\rangle_{H^{0,k'}_\gamma[E]} = \langle u, P_1(t)v\rangle_{H^{0,k'}_\gamma[E]}$ and

$$\mathrm{Re}\ \langle P_1(t)u,u\rangle_{H^{0,k'}_\gamma[E]} = \mathrm{Re}\ \left\langle p(t,X,D_x)\gamma^{k'}u, \gamma^{k'}u\right\rangle_{L^2(\mathbf{R}^n,E)}$$

$$\leq \beta\left\|\gamma^{k'}u\right\|^2_{L^2(\mathbf{R}^n,E)} = \beta\|u\|^2_{H^{0,k'}_\gamma[E]}$$

for $u,v \in \mathscr{S}(\mathbf{R}^n,E)$. Moreover, $[Z^k,[Z^k,P_1(t)]] \in \mathscr{L}(H^{r_1+2k,r_2}_\gamma[E], H^{r_1,r_2}_\gamma[E])$ due to 4.2.13. Hence $Z^{-k}[Z^k,[Z^k,P_1(t)]] \in \mathscr{L}(H^k_Z, H^0_Z)$, and 2.3.2 shows $\mathrm{Re}\ \langle P_1(t)u,u\rangle_{H^{k,k'}_\gamma[E]} \leq \beta_{k,k'}\|u\|^2_{H^{k,k'}_\gamma[E]}$ for $u \in \mathscr{S}(\mathbf{R}^n,E), t \in I$, and suitable constants $\beta_{k,k'} \geq 0$ for any $k \in \mathbf{N}_0$. Moreover, for suitable $\beta'_{k'} \geq 0$

$$2\mathrm{Re}\ \langle P_2(t)u,u\rangle_{H^{0,k'}_\gamma[E]} = 2\mathrm{Re}\ \left\langle[\gamma^{k'},\mathrm{Re}\ (p(t,X,D_x))]u, \gamma^{k'}u\right\rangle_{L^2(\mathbf{R}^n,E)}$$

$$= \left\langle[\gamma^{k'},\mathrm{Re}\ (p(t,X,D_x))]u, \gamma^{k'}u\right\rangle_{L^2(\mathbf{R}^n,E)} + \left\langle\gamma^{k'}u, [\gamma^{k'},\mathrm{Re}\ (p(t,X,D_x))]u\right\rangle_{L^2(\mathbf{R}^n,E)}$$

$$= \left\langle\gamma^{k'}[\gamma^{k'},\mathrm{Re}\ (p(t,X,D_x))]u, u\right\rangle_{L^2(\mathbf{R}^n,E)} - \left\langle[\gamma^{k'},\mathrm{Re}\ (p(t,X,D_x))]\gamma^{k'}u, u\right\rangle_{L^2(\mathbf{R}^n,E)}$$

$$= \left\langle\gamma^{-k'}[\gamma^{k'},[\gamma^{k'},\mathrm{Re}\ (p(t,X,D_x))]]u, \gamma^{k'}u\right\rangle_{L^2(\mathbf{R}^n,E)} \leq \beta'_{k'}\|u\|^2_{H^{0,k'}_\gamma[E]},$$

because $[\gamma^{k'},[\gamma^{k'},\mathrm{Re}\ (p(t,X,D_x))]] \in \mathscr{L}(H^{r_1,r_2+2k'}_\gamma[E], H^{r_1,r_2}_\gamma[E])$ due to 4.2.12. Moreover

$$[Z,P_2(t)] = [\gamma^{-k'}\langle D_x\rangle\gamma^{k'}, \gamma^{-k'}[\gamma^{k'},\mathrm{Re}\ (p(t,X,D_x))]] \in \mathscr{L}(H^{k+1}_Z, H^k_Z)$$

due to 4.2.14, and 2.3.1 shows $\mathrm{Re}\ \langle P_2(t)u,u\rangle_{H^{k,k'}_\gamma[E]} \leq \beta'_{k,k'}\|u\|^2_{H^{k,k'}_\gamma[E]}$ for every $u \in \mathscr{S}(\mathbf{R}^n,E), t \in I$, and suitable constants $\beta'_{k,k'} \geq 0$ for any $k \in \mathbf{N}_0$. $P_3(t) = \frac{1}{2}(p(t,X,D_x)-p(t,X,D_x)^*) = i(\mathrm{Im}\ p)(t,X,D_x)+r(t,X,D_x) \in S^1_{1,0}[E]$ with a suitable $r(t,X,D_x) \in S^{2-1}_{1,0}[E]$ and $P_3(t) + P_3(t)^* = 0$, hence with suitable $\beta''_{k'} \geq 0$

$$2\mathrm{Re}\ \langle P_3(t)u,u\rangle_{H^{0,k'}_\gamma[E]}$$

$$= 2\mathrm{Re}\ \left\langle P_3(t)\gamma^{k'}u, \gamma^{k'}u\right\rangle_{L^2(\mathbf{R}^n,E)} + 2\mathrm{Re}\ \left\langle[\gamma^{k'},P_3(t)]u, \gamma^{k'}u\right\rangle_{L^2(\mathbf{R}^n,E)}$$

$$\leq 0 + \left\|[\gamma^{k'},P_3(t)]u\right\|_{L^2(\mathbf{R}^n,E)}\left\|\gamma^{k'}u\right\|_{L^2(\mathbf{R}^n,E)} \leq \beta''_{k'}\|u\|^2_{H^{0,k'}_\gamma[E]}a$$

for $u \in \mathscr{S}(\mathbf{R}^n,E), t \in I$ due to 4.2.11 and $[Z,P_3(t)] \in \mathscr{L}(H^{k+1,k'}_\gamma[E], H^{k,k'}_\gamma[E])$. Therefore $\mathrm{Re}\ \langle P_3(t)u,u\rangle_{H^{k,k'}_\gamma[E]} \leq \beta''_{k,k'}\|u\|^2_{H^{k,k'}_\gamma[E]}$ for $u \in \mathscr{S}(\mathbf{R}^n,E), t \in I$, and suitable constants $\beta''_{k,k'} \geq 0$ for any $k \in \mathbf{N}_0$ due to 2.3.1.

Since $P_1(t) + P_2(t) + P_3(t) = \mathrm{Re}\ (p(t, X, D_x)) + i\mathrm{Im}\ (p(t, X, D_x)) = p(t, X, D_x)$
this shows $\mathrm{Re}\ \langle P(t)u, u \rangle_{H_\gamma^{k,k'}[E]} \leq \gamma_{k,k'} \|u\|^2_{H_\gamma^{k,k'}[E]}$ for $u \in \mathscr{S}(\mathbf{R}^n, E), t \in I$,
and suitable constants $\gamma_{k,k'} \geq 0$ for any $k \in \mathbf{N}_0$. Therefore 2.4.6 implies the
assertion. $\qquad\qquad\qquad\qquad\qquad\qquad\qquad\qquad\qquad\qquad\qquad\qquad\quad$ $\square$

Combining this result with the Fefferman-Phong inequality 4.1.13 we obtain
the main result of this section that combines and generalizes usual results on
hyperbolic and parabolic evolution equations.

4.2.17 Theorem. Let γ be a weight function, $p \in C(I \times \mathbf{R}^n \times \mathbf{R}^n)$ with
$\{p(t, \cdot, \cdot) : t \in I\} \subset S_{1,0}^2$ boundedly and $\{\mathrm{Im}\ p(t, \cdot, \cdot) : t \in I\} \subset S_{1,0}^1$ boundedly.
If there is a $c \geq 0$ with $\mathrm{Re}\ p(t, x, \xi) \leq c$ for $t \in I, x, \xi \in \mathbf{R}^n$, then for any
$k \in \mathbf{N}_0, k' \in \mathbf{Z}, u_0 \in H_\gamma^{k+2,k'}$, and $f \in C(I, H_\gamma^{k,k'}) \cap L^1(I, H_\gamma^{k+2,k'})$ there is a
unique $u \in C^1(I, H_\gamma^{k,k'}) \cap C(I, H_\gamma^{k+2,k'})$ with

$$u'(t) = p(t, X, D_x)u(t) + f(t), t \in I, \qquad u(t_0) = u_0 . \qquad (4.2.2)$$

In particular, in this case, for $u_0 \in \mathscr{S}(\mathbf{R}^n)$ and $f \in C(I, \mathscr{S}(\mathbf{R}^n))$ there is a
unique $u \in C^1(I, \mathscr{S}(\mathbf{R}^n))$ with (4.2.2).

PROOF: This is a consequence of 4.1.12, 4.1.13, and 4.2.16. $\qquad\qquad\qquad$ $\square$

Now we mention further examples satisfying the conditions of 4.2.16.

4.2.18 Examples. 4.2.16 can be applied further to the following examples.

(a) Symmetric hyperbolic systems as in example 4.2.10 (a), (b).

(b) Strongly degenerate parabolic systems of order 2 as in 4.2.10 (e).

(c) $\dfrac{du}{dt}(t) = \displaystyle\sum_{j=1}^{n-1} \partial_j^2 u(t) + \partial_n u(t), t \geq 0, u(0) = u_0$ . .

In the last part of this section we will discuss pseudodifferential wave equations.
Although wave equations can be reduced to symmetrizable (strict) hyperbolic
systems or even symmetric hyperbolic systems we prefer an approach using a
different reduction method because we can admit pseudodifferential perturbations in low order. In fact, we have:

4.2.19 Theorem. Let γ be a weight function and $p, q \in C(I \times \mathbb{R}_x^n \times \mathbb{R}_\xi^n)$ with $\{p(t, \cdot, \cdot), q(t, \cdot, \cdot) : t \in I\} \subset S_{1,0}^1$ and $\{p(t, \cdot, \cdot) + p(t, \cdot, \cdot)^* : t \in I\} \subset S_{1,0}^0$ boundedly. Then, for $t_0 \in I, k \in \mathbb{N}_0, k' \in \mathbb{Z}, u_0 \in H_\gamma^{k+2,k'}, u_1 \in H_\gamma^{k+1,k'}$, and $f \in C(I, H_\gamma^{k+1,k'})$, there is a unique

$$u \in C(I, H_\gamma^{k+2,k'}) \cap C^1(I, H_\gamma^{k+1,k'}) \cap C^2(I, H_\gamma^{k,k'})$$

with

$$\frac{d^2 u}{dt^2}(t) = \Delta u(t) + p(t, X, D_x)\frac{du}{dt}(t) + q(t, X, D_x)u(t) + f(t), \quad t \in I,$$

$$u(t_0) = u_0, \quad \frac{du}{dt}(t_0) = u_1. \tag{4.2.3}$$

In particular, for $t_0 \in I, u_0, u_1 \in \mathscr{S}(\mathbb{R}^n)$, and $f \in C(I, \mathscr{S}(\mathbb{R}^n))$ there is a unique $u \in C^2(I, \mathscr{S}(\mathbb{R}^n))$ with (4.2.3).

PROOF: With the transformation $v = \begin{pmatrix} u \\ u_t \end{pmatrix}$ and $\mathcal{K}_\gamma^{k,k'} := \begin{array}{c} H_\gamma^{k+1,k'} \\ \oplus \\ H_\gamma^{k,k'} \end{array}$ it is easy

to see that our problem is equivalent to the following system:
For $t_0 \in I, k \in \mathbb{N}_0, k' \in \mathbb{Z}, v_0 \in \mathcal{K}_\gamma^{k+1,k'}$ there is a unique $v \in C(I, \mathcal{K}_\gamma^{k+1,k'}) \cap C^1(I, \mathcal{K}_\gamma^{k,k'})$ with

$$\frac{dv}{dt}(t) = \underbrace{\begin{pmatrix} 0 & \mathrm{Id} \\ \Delta + q(t, X, D_x) & p(t, X, D_x) \end{pmatrix}}_{=:\mathcal{A}(t)} v(t) + \begin{pmatrix} 0 \\ f(t) \end{pmatrix}, \quad v(t_0) = v_0.$$

We have to show that the Cauchy problem for $\mathcal{A}(t) \in \mathscr{L}(\mathcal{K}_\gamma^{k+1,k'}, \mathcal{K}_\gamma^{k,k'})$ in the scale $(\mathcal{K}_\gamma^{k,k'})_k$ is well-posed with exponential growth for any $k' \in \mathbb{Z}$. Let $\mathcal{K} := \mathcal{K}_\gamma^{0,k'}$ and

$$Z := i \begin{pmatrix} \gamma^{-k'} \langle D_x \rangle \gamma^{k'} & 0 \\ 0 & \gamma^{-k'} \langle D_x \rangle \gamma^{k'} \end{pmatrix}$$

Then Z is skew-selfadjoint in $\mathcal{K}$ with $\mathcal{K}_Z^k = \mathcal{K}_\gamma^{k,k'}$ (cf. 4.2.15).
Moreover, $I \ni t \mapsto \mathcal{A}(t) \in \mathscr{L}(\mathcal{K}_\gamma^{k+1,k'}, \mathcal{K}_\gamma^{k,k'})$ is strongly continuous for any $k \in \mathbb{N}_0$, and

$$[Z, \mathcal{A}(t)] = i \begin{pmatrix} 0 & 0 \\ [\gamma^{-k'} \langle D_x \rangle \gamma^{k'}, \Delta + q(t, X, D_x)] & [\gamma^{-k'} \langle D_x \rangle \gamma^{k'}, p(t, X, D_x)] \end{pmatrix}.$$

Now, lemma 4.2.11 implies that $[\gamma^{-k'} \langle D_x \rangle \gamma^{k'}, \Delta] \in \mathscr{L}(H_\gamma^{k+2,k'}, H_\gamma^{k,k'})$ and $[\gamma^{-k'} \langle D_x \rangle \gamma^{k'}, r(t, X, D_x)] \in \mathscr{L}(H_\gamma^{k+1,k'}, H_\gamma^{k,k'})$ boundedly for $t \in I$ and $r - p, q$.

Hence $[Z, \mathcal{A}(t)] \in \mathcal{L}(K_\gamma^{k+1,k'}, K_\gamma^{k,k'})$ boundedly in $t \in I$ for any $k \in \mathbf{N}_0$. Finally, for $v_1, v_2 \in \mathscr{S}(\mathbf{R}^n)$

$$\mathrm{Re} \left\langle \mathcal{A}(t) \begin{pmatrix} v_1 \\ v_2 \end{pmatrix}, \begin{pmatrix} v_1 \\ v_2 \end{pmatrix} \right\rangle$$

$$= \mathrm{Re} \left(\left\langle \langle D_x \rangle (\gamma^{k'} v_2), \langle D_x \rangle (\gamma^{k'} v_1) \right\rangle + \left\langle \gamma^{k'} \Delta v_1, \gamma^{k'} v_2 \right\rangle \right.$$
$$\left. + \left\langle \gamma^{k'} p(t, X, D_x) v_2, \gamma^{k'} v_2 \right\rangle + \left\langle \gamma^{k'} q(t, X, D_x) v_1, \gamma^{k'} v_2 \right\rangle \right)$$

$$= \mathrm{Re} \left(\left\langle \gamma^{k'} v_2, \gamma^{k'} v_1 \right\rangle - \left\langle \Delta(\gamma^{k'} v_2), \gamma^{k'} v_1 \right\rangle + \left\langle \Delta(\gamma^{k'} v_2), \gamma^{k'} v_1 \right\rangle \right.$$
$$\left. + \left\langle [\gamma^{k'}, \Delta] v_1, \gamma^{k'} v_2 \right\rangle + \left\langle \gamma^{k'} p(t, X, D_x) v_2, \gamma^{k'} v_2 \right\rangle + \left\langle \gamma^{k'} q(t, X, D_x) v_1, \gamma^{k'} v_2 \right\rangle \right)$$

Hence, with suitable $c, d, \beta, \beta' \geq 0$

$$\left| \mathrm{Re} \left\langle \mathcal{A}(t) \begin{pmatrix} v_1 \\ v_2 \end{pmatrix}, \begin{pmatrix} v_1 \\ v_2 \end{pmatrix} \right\rangle \right|$$

$$\leq \|v_2\|_{\gamma,0,k'} \|v_1\|_{\gamma,0,k'} + \left\| [\gamma^{k'}, \Delta] v_1 \right\|_{\gamma,0,0} \|v_2\|_{\gamma,0,k'} + \left| \mathrm{Re} \langle p(t, X, D_x) v_2, v_2 \rangle_{\gamma,0,k'} \right|$$
$$+ \|q(t, X, D_x) v_1\|_{\gamma,0,k'} \|v_2\|_{\gamma,0,k'}$$

$$\leq \|v_1\|_{\gamma,0,k'} \|v_2\|_{\gamma,0,k'} + c \|v_1\|_{\gamma,1,k'} \|v_2\|_{\gamma,0,k'} + \beta' \|v_2\|_{\gamma,0,k'}^2 + d \|v_1\|_{\gamma,1,k'} \|v_2\|_{\gamma,0,k'}$$

$$\leq \beta \left(\|v_1\|_{\gamma,1,k'}^2 + \|v_2\|_{\gamma,0,k'}^2 \right) = \left\| \begin{pmatrix} v_1 \\ v_2 \end{pmatrix} \right\|^2$$

where we have used 4.2.11 and 4.2.16. Hence we can apply 2.3.1 and 2.4.6. □

4.2.20 Bibliographical remarks. Proposition 4.2.4 has been taken from Petersen [123, 5§4], cf. also Hörmander [67, 22.4.1] or Jacob [73]. Corollary 4.2.9 is contained implicitly in [67, 23.1.1] or [97, 7.3.2]. Well-posedness of the linear hyperbolic pseudodifferential Cauchy problem (i.e, $p \in S^1$ with $\mathrm{Re}\, p \in S^0$) is well-known, cf. Hörmander [67, 23.1.2], Taylor [149, 7.7], or Kumano-go [97, 7.4] and the references given therein. Well-posedness of the Cauchy problem for (weakly) degenerate parabolic pseudodifferential equations of arbitrary order (cf. 4.2.10(c)) is due to Iwaski [72]. Theorem 4.2.17 combines and generalizes Hörmander's result on well-posedness of the pseudodifferential, first order hyperbolic Cauchy problem [67, 23.1.2] and the second-order case of Iwasaki's result on the degenerate parabolic Cauchy problem [72]. Whereas Hörmander's result is contained completely in theorem 4.2.17 and corresponds to the special case of first-order symbols, Iwasaki's result is stronger under much stronger assumptions (i.e., assumptions implying hypoellipticity).

4.3 Essential selfadjointness of pseudodifferential operators

In this section we will show how the results in section 2.4 can be used to prove essential selfadjointness of various pseudodifferential operators on $\mathbb{R}^n$ and well-posedness of related time-dependent, linear Schrödinger equations.

4.3.1 Remark. The question whether a symmetric differential operator with smooth coefficients on $\mathbb{R}^n$ is essentially selfadjoint is non-trivial:
There are operators on $\mathbb{R}^n$ having no selfadjoint extensions like $P = 2ie^x\frac{d}{dx}+ie^x$ (cf. [133, I 3.4]) as well as operators with infinitely many selfadjoint extensions like $P = -\frac{d^2}{dx^2} - x^4 + x^2$ (cf. [127, X.1, X.5]. However, symmetric differential operators with real coefficients on $\mathbb{R}^n$ have always selfadjoint extensions (cf. [155, 8.9]).
For differential operators the following result due to Chernoff/Kato [80] is often useful: Let $q \in C^\infty(\mathbb{R}^n, \mathbb{R})$ with $q(x) \geq -a - b|x|^2$ for $x \in \mathbb{R}^n$ and suitable constants $a, b \geq 0$.
Then all powers of $P := -\Delta + q$ are essentially selfadjoint on $C_c^\infty(\mathbb{R}^n)$ in $L^2(\mathbb{R}^n)$.

For pseudodifferential operators the question of essential selfadjointness is interesting in particular for Weyl-quantized operators $p^w(X, D_x)$ because $p^w(X, D_x)$ is symmetric for real-valued $p \in S^m_{\rho,\delta}$, $m > 0, 0 \leq \delta < \rho \leq 1$.

4.3.2 Theorem. Let $0 \leq \delta < \rho \leq 1, 0 \leq m' \leq m$ and let either $p \in S^{2(\rho-\delta)}_{\rho,\delta}$ with $p(x, \xi) \geq -c$ for $x, \xi \in \mathbb{R}^n$ or $p \in HS^{m,m'}_{\rho,\delta}$ be real-valued.
Then $p^w(X, D_x)$ is essentially selfadjoint (and semibounded) on $\mathscr{S}(\mathbb{R}^n)$.

PROOF: If $p \in S^{2(\rho-\delta)}_{\rho,\delta}$ satisfies $p(x, \xi) \geq -c$, then the Fefferman-Phong inequality 4.1.13 shows that there is a $\beta \geq 0$ with

$$\langle p^w(X, D_x)u, u\rangle_{L^2(\mathbb{R}^n)} \geq -\beta \|u\|^2_{L^2(\mathbb{R}^n)} \quad \text{for } u \in \mathscr{S}(\mathbb{R}^n).$$

Hence 4.2.2 and 2.3.2 show that the Cauchy problem for $-p^w(X, D_x)$ in the scale $(H^k)_k$ is well-posed with exponential growth, and 2.1.12 implies that $(\lambda\mathrm{Id} + p^w(X, D_x))(\mathscr{S}(\mathbb{R}^n)) \subset L^2(\mathbb{R}^n)$ densely for large λ. Thus 1.2.5 implies the assertion. The assertion for real-valued $p \in HS^{m,m'}_{\rho,\delta}$ can be proved in the same manner using 4.2.2 and 4.2.6. $\qquad\square$

Next we will give a result similar to 4.3.2 for symbols with spatially growing coefficients. To this end we will show that the scale $\mathcal{H}_p^{2k}$ (cf. 4.1.28) is generated by the harmonic oscillator.

4.3.3 Proposition. Let $1 < p < \infty$ and $Z_p := \Delta - x^2$, $D(Z_p) := \mathcal{H}_p^2$. Then $Z_p : D(Z_p) \to X$ generates a bounded analytic semigroup in $X = L^p(\mathbf{R}^n)$ with angle $\frac{\pi}{2}$. Moreover, we have $X_{Z_p}^k = \mathcal{H}_p^{2k}$ topologically.

PROOF: $Z_2 : D(Z_2) \to L^2(\mathbf{R}^n)$ is a negative, selfadjoint operator in $L^2(\mathbf{R}^n)$ and generates a symmetric Markov semigroup $(e^{tZ_2})_{t \geq 0}$ (cf. e.g. 4.3.1, [37], [141]). This semigroup can be extended to a bounded analytic semigroup in $L^p(\mathbf{R}^n)$ with angle $\frac{\pi}{2}$ (cf. [119] or [62]) and generator $A_p : D(A_p) \to L^p(\mathbf{R}^n)$. Therefore, we only have to show that $Z_p = A_p$ and we have to calculate the generated scale. Since $\mathscr{S}(\mathbf{R}^n)$ is known to be an invariant subspace of $(e^{tZ_2})_{t \geq 0}$, 1.1.14 implies that the closure of $Z_p|_{\mathscr{S}(\mathbf{R}^n)}$ in $L^p(\mathbf{R}^n)$ is equal to A_p, i.e., $u \in D(A_p)$ if and only if there is a sequence $(u_l)_l \subset \mathscr{S}(\mathbf{R}^n)$ with $u_l \xrightarrow[l \to \infty]{L^p} u$ and $(\Delta - x^2)u_l \xrightarrow[l \to \infty]{L^p} A_p u$. Therefore $A_p = Z_p$ and $D(A_p) = \mathcal{H}_p^2$ topologically due to 4.1.24 and 4.1.34. Now assume inductively that $X_{Z_p}^k = \mathcal{H}_p^{2k}$ for a $k \in \mathbf{N}$. Then

$$u \in X_{Z_p}^{k+1} \iff u, (\Delta - x^2)u \in X_{Z_p}^k = \mathcal{H}_p^{2k} \iff u \in \mathcal{H}_p^{2k+2}$$

due to 4.1.34, which proves also the equivalence of topologies. $\qquad\square$

4.3.4 Theorem. Let $p \in G^4, q \in \psi c^{(2,2)}$ with $p(x,\xi) \geq -c$ and $q(x,\xi) \geq -c$ for $x, \xi \in \mathbf{R}^n$ and a suitable $c > 0$.
Then $A := p^w(X, D_x) + q^w(X, D_x)$ is essentially selfadjoint on $\mathscr{S}(\mathbf{R}^n)$ in $L^2(\mathbf{R}^n)$ and its closure is semibounded from below.

PROOF: The Fefferman-Phong (4.1.22 and 4.1.27) inequality shows that A is semibounded from below on $\mathscr{S}(\mathbf{R}^n)$. Thus, due to 1.2.5, to complete the proof we have to show $(\lambda \mathrm{Id} + A)(\mathscr{S}(\mathbf{R}^n)) \subset L^2(\mathbf{R}^n)$ densely for large $\lambda > 0$. Let $H := L^2(\mathbf{R}^n)$, $\Lambda := \Delta - x^2 \in \mathrm{Op}G^2 \cap \mathrm{Op}\psi c^{(2,2)}$, $P := p^w(X, D_x) \in \mathrm{Op}G^4$, and let $Q := q^w(X, D_x) \in \mathrm{Op}\psi c^{(2,2)}$. Then we can choose elliptic operators $S \in \mathrm{Op}G^{2k}$ and $T \in \mathrm{Op}G^{-2k}$ with $R = ST - \mathrm{Id} \in \mathrm{Op}G^{-\infty}$. Then, for $u \in \mathscr{S}(\mathbf{R}^n)$

$$|\langle [\Lambda^k, [\Lambda^k, P]]u, u \rangle_{L^2(\mathbf{R}^n)}|$$
$$\leq |\langle ST[\Lambda^k, [\Lambda^k, P]]u, u \rangle_{L^2(\mathbf{R}^n)}| + |\langle R[\Lambda^k, [\Lambda^k, P]]u, u \rangle_{L^2(\mathbf{R}^n)}|$$
$$\leq \|\underbrace{T[\Lambda^k, [\Lambda^k, P]]}_{\in \mathrm{Op}G^{2k}} u\|_{L^2(\mathbf{R}^n)} \|S^* u\|_{L^2(\mathbf{R}^n)} + \|\underbrace{R[\Lambda^k, [\Lambda^k, P]]}_{\in \mathrm{Op}G^0} u\|_{L^2(\mathbf{R}^n)} \|u\|_{L^2(\mathbf{R}^n)}$$
$$\leq c_k \|u\|_{\mathcal{H}^{2k}}^2 .$$

Moreover, $\Lambda^k = \sum_{j=0}^k \Lambda_j$ with $\Lambda_j \in \mathrm{Op}\psi c^{(2j,2k-2j)}$, hence with 4.1.32

$$|\langle [\Lambda^k, [\Lambda^k, Q]]u, u \rangle_{L^2(\mathbf{R}^n)}| \leq \sum_{j,j'=0}^k |\langle [\Lambda_j, [\Lambda_{j'}, Q]]u, u \rangle_{L^2(\mathbf{R}^n)}| \leq d_k \|u\|_{\mathcal{H}^{2k}}^2 .$$

Hence 4.3.3, 2.3.2, 2.4.6, and 2.1.12 imply $(\lambda \mathrm{Id} + A)(\mathscr{S}(\mathbf{R}^n)) \subset L^2(\mathbf{R}^n)$ densely for large $\lambda > 0$.   □

4.3.5 Examples.   4.3.4 shows essential selfadjointness of operators associated with the following symbols:

(a) positive polynomials p of order 4 in x, ξ.

(b) $\langle (x, \xi) \rangle^\alpha + \langle x \rangle^\beta \langle \xi \rangle^\gamma, 0 \leq \alpha \leq 4, 0 \leq \beta, \gamma \leq 2$.

(c) $a(x)b(\xi)x^2\xi^2$ with $a, b \in C_c^\infty(\mathbf{R}^n), a(x), b(x) \geq 0$ for $x \in \mathbf{R}^n$.

In the sequel we are interested not only in essential selfadjointness of an operator itself but also in essential selfadjointness of all its powers because in this case it is possible to express the scale generated by the selfadjoint operator by distributional conditions. This will be shown in the following.

4.3.6 Proposition.   Let $P \in \mathrm{Op}S_{1,0}^1 \cup \mathrm{Op}\psi c^{(1,1)}$ and $A_P := \overline{P|_{\mathscr{S}}}$ be the closure of $P|_{\mathscr{S}}$ in $L^2(\mathbf{R}^n)$.
Then $H_P^k := D(A_P^k) = \{u \in L^2(\mathbf{R}^n) : P^j u \in L^2(\mathbf{R}^n), 0 \leq j \leq k\}$ and we have $(A_P)^k u = P^k u$ for $u \in H_P^k, k \in \mathbf{N}$.

PROOF:
$k = 1$:
If $u \in D(A_P)$ then there is a sequence $(u_l)_l \subset \mathscr{S}(\mathbf{R}^n)$ such that $u_l \xrightarrow{L^2} u$ and $Pu_l \xrightarrow{L^2} A_P u$. Since $L^2(\mathbf{R}^n) \hookrightarrow \mathscr{S}'(\mathbf{R}^n)$ this implies $u_l \xrightarrow{\mathscr{S}'} u$ (and thus $Pu_l \xrightarrow{\mathscr{S}'} Pu$) and $Pu_l \xrightarrow{\mathscr{S}'} A_P u$, hence $Pu = A_P u \in L^2(\mathbf{R}^n)$.
Now suppose conversely that $u \in L^2(\mathbf{R}^n)$ and $Pu \in L^2(\mathbf{R}^n)$. Let $(J_l)_{l \in \mathbf{N}}$ be the family of 4.1.24. Then $J_l u \in \mathscr{S}(\mathbf{R}^n), J_l u \xrightarrow[l \to \infty]{L^2} u$, and

$$PJ_l u = J_l \underbrace{Pu}_{\in L^2(\mathbf{R}^n)} + [P, J_l]u \xrightarrow{L^2} Pu.$$

This shows $u \in D(A_P)$.

$k \to k+1$:

Let $u \in D((A_P)^{k+1})$, then $u \in D((A_P)^k)$ (i.e., $P^j u \in L^2(\mathbf{R}^n)$, $0 \le j \le k$) and $Pu = A_P u \in D((A_P)^k)$ (i.e., $P^j(Pu) \in L^2(\mathbf{R}^n)$, $0 \le j \le k$). Therefore we have $P^j u \in L^2(\mathbf{R}^n)$, $0 \le j \le k+1$, and $(A_P)^{k+1} u = (A_P)^k(A_P u) = P^{k+1} u$. Next take $u \in L^2(\mathbf{R}^n)$ with $P^j u \in L^2(\mathbf{R}^n)$ for $0 \le j \le k+1$. Then $u \in D((A_P)^k)$ and $P^j(Pu) \in L^2(\mathbf{R}^n)$ for $0 \le j \le k$, hence $A_P u = Pu \in D((A_P)^k)$. This proves $u \in D((A_P)^{k+1})$. □

4.3.7 Lemma. For $P \in \mathrm{Op}S_{1,0}^1$ and $Q \in \mathrm{Op}S_{1,0}^0$ we have

$$Q \in \bigcap_{k \in \mathbb{N}_0} \mathscr{L}(H_P^k).$$

Note that H_P^k is a Hilbert space. Here $H_P^0 := L^2(\mathbf{R}^n)$.

PROOF: Suppose inductively that $Q \in \mathscr{L}(H_P^k)$ for a certain $k \in \mathbb{N}_0$ and any $Q \in \mathrm{Op}S_{1,0}^0$. Take $u \in H_P^{k+1}$, i.e., $u \in H_P^k$ and $Pu \in H_P^k$. Thus $Qu \in H_P^k$ and

$$PQu = Q \underbrace{Pu}_{\in H_P^k} + \underbrace{[P,Q]}_{\in \mathrm{Op}S_{1,0}^0} u \in H_P^k.$$

This shows $Q(H_P^{k+1}) \subset H_P^{k+1}$ and $Q \in \mathscr{L}(H_P^{k+1})$ due to the closed graph theorem. □

4.3.8 Lemma. For $P, P' \in \mathrm{Op}S_{1,0}^1$ with $P - P' \in \mathrm{Op}S_{1,0}^0$ we have $H_P^k = H_{P'}^k$ topologically for $k \in \mathbb{N}_0$.

PROOF: Suppose inductively that $H_P^k = H_{P'}^k$ topologically for a $k \in \mathbb{N}_0$. We only have to show $H_P^{k+1} \hookrightarrow H_{P'}^{k+1}$. Take $u \in H_P^{k+1}$, i.e., $u \in H_P^k = H_{P'}^k$ and $Pu \in H_P^k = H_{P'}^k$. Then, due to 4.3.7 we have $(P - P')u \in H_{P'}^k$ and therefore $P'u = Pu - (P - P')u \in H_{P'}^k$, which proves the assertion. □

4.3.9 Proposition. Suppose that $P \in \mathrm{Op}S_{1,0}^m \cup \mathrm{Op}\psi c^{(m,m')}$, $m, m' > 0$, is symmetric and essentially selfadjoint on $\mathscr{S}(\mathbf{R}^n)$, i.e., the closure $A_P := \overline{P|_{\mathscr{S}}}$ of $P|_{\mathscr{S}}$ is selfadjoint. Then

$$D(A_P) = \{u \in L^2(\mathbf{R}^n) : Pu \in L^2(\mathbf{R}^n)\} \quad \text{and} \quad A_P u = Pu \quad \text{for } u \in D(A_P).$$

PROOF: If $u \in D(A_P)$, then there is a sequence $(u_l)_l \subset \mathcal{S}(\mathbf{R}^n)$ such that $u_l \xrightarrow{L^2} u$ and $Pu_l \xrightarrow{L^2} A_P u$. Since $L^2(\mathbf{R}^n) \hookrightarrow \mathcal{S}'(\mathbf{R}^n)$ this implies $u_l \xrightarrow{\mathcal{S}'} u$ (and thus $Pu_l \xrightarrow{\mathcal{S}'} Pu$) and $Pu_l \xrightarrow{\mathcal{S}'} A_P u$, hence $Pu = A_P u \in L^2(\mathbf{R}^n)$.
Suppose conversely that $u \in L^2(\mathbf{R}^n)$ and $Pu \in L^2(\mathbf{R}^n)$. Then for $v \in \mathcal{S}(\mathbf{R}^n)$

$$\langle Pu, v \rangle_{L^2(\mathbf{R}^n)} = (\underbrace{Pu}_{\in L^2(\mathbf{R}^n)}, \bar{v})_{\mathcal{S}',\mathcal{S}} = (\underbrace{u}_{\in L^2(\mathbf{R}^n)}, \overline{P\bar{v}})_{\mathcal{S}',\mathcal{S}} = \langle u, Pv \rangle_{L^2(\mathbf{R}^n)}.$$

For $v \in D(A_P)$ there is a sequence $(v_l)_{l \in \mathbf{N}} \subset \mathcal{S}(\mathbf{R}^n)$ with $v_l \xrightarrow[l \to \infty]{L^2} v$ and $Pv_l \xrightarrow[l \to \infty]{L^2} A_P v$. Thus

$$\langle Pu, v \rangle_{L^2(\mathbf{R}^n)} = \langle u, A_P v \rangle_{L^2(\mathbf{R}^n)} \qquad \text{for } v \in D(A_P).$$

This shows $u \in D((A_P)^*) = D(A_P)$ and $A_P u = (A_P)^* u = Pu$. □

Finally, we can show that the scale generated by a selfadjoint operator can be characterized by distributional conditions provided that all powers of the operator are essentially selfadjoint on $\mathcal{S}(\mathbf{R}^n)$.

4.3.10 Proposition. Suppose that $P \in \mathrm{OpS}^m_{1,0} \cup \mathrm{Op}\psi c^{(m,m')}$, $m, m' > 0$, is symmetric and every power P^k of P is essentially selfadjoint on $\mathcal{S}(\mathbf{R}^n)$, i.e., for any $k \in \mathbf{N}$ the closure $A_{P^k} := \overline{P^k|_{\mathcal{S}}}$ of $P^k|_{\mathcal{S}}$ is selfadjoint in $L^2(\mathbf{R}^n)$. Then, for any $k \in \mathbf{N}$, the following statements hold:

(a) $B^k_P := A_{P^k} = (A_P)^k$ and $D(A_{P^k}) = D((A_P)^k)$ topologically.

(b) $H^k_P := D(B^k_P) = \{u \in L^2(\mathbf{R}^n) : P^k u \in L^2(\mathbf{R}^n)\}$
  $= \{u \in L^2(\mathbf{R}^n) : P^j u \in L^2(\mathbf{R}^n), 0 \le j \le k\}$.

(c) $B^k_P u = P^k u$ for $u \in H^k_P$.

(d) $\mathcal{S}(\mathbf{R}^n) \subset H^k_P$ densely.

PROOF:
Step 1: $D((A_P)^k) = \{u \in L^2(\mathbf{R}^n) : P^j u \in L^2(\mathbf{R}^n), 0 \le j \le k\}$ and we have $(A_P)^k u = P^k u$ for $u \in D((A_P)^k)$:
The assertion for $k = 1$ follows from 4.3.9. Now suppose step 1 inductively for a $k \in \mathbf{N}$. Let $u \in D((A_P)^{k+1})$, then we have $u \in D((A_P)^k)$ (i.e., $P^j u \in L^2(\mathbf{R}^n)$, $0 \le j \le k$) and $Pu = A_P u \in D((A_P)^k)$ (i.e., $P^j(Pu) \in L^2(\mathbf{R}^n)$, $0 \le j \le k$). Thus $P^j u \in L^2(\mathbf{R}^n)$, $0 \le j \le k+1$ and $(A_P)^{k+1} u = (A_P)^k (A_P u) = P^{k+1} u$. Now take a $u \in L^2(\mathbf{R}^n)$ with $P^j u \in L^2(\mathbf{R}^n)$ for $0 \le j \le k+1$. Then $u \in D((A_P)^k)$

and $P^j(Pu) \in L^2(\mathbf{R}^n)$ for $0 \le j \le k$, hence $A_P u = Pu \in D((A_P)^k)$. This proves $u \in D((A_P)^{k+1})$ and completes the proof of step 1.

Step 2: Proof of the assertion:

Due to 4.3.9 we have $D(A_{P^k}) = \{u \in L^2(\mathbf{R}^n) : P^k u \in L^2(\mathbf{R}^n)\}$ and $A_{P^k} u = P^k u$ for $u \in D(A_{P^k})$. This shows $(A_P)^k \subset A_{P^k}$. Moreover, since A_{P^k} and $(A_P)^k$ are selfadjoint (due to the functional calculus for selfadjoint operators), we have $(A_P)^k = A_{P^k}$ because $A_{P^k} = (A_{P^k})^* \subset ((A_P)^k)^* = (A_P)^k$. In particular, this shows $D((A_P)^k) = D(A_{P^k})$. Obviously we have $D((A_P)^k) \hookrightarrow D(A_{P^k})$, hence the open mapping theorem shows the equivalence of the norms of $D((A_P)^k)$ and $D(A_{P^k})$ because both spaces are Banach spaces. Finally, $\mathscr{S}(\mathbf{R}^n) \subset H_P^k$ densely, because trivially $\mathscr{S}(\mathbf{R}^n) \subset D(A_{P^k})$ densely. $\qquad\square$

4.3.11 Theorem. Let $P \in \mathrm{OpS}_{1,0}^1$ and $Q, R \in \mathrm{OpS}_{1,0}^0$ be symmetric on $\mathscr{S}(\mathbf{R}^n)$, and let $A := P$, $A := QP^2Q + R$, or $A := PQ^2P + R$. Then the following statements hold:

(a) Every power A^k for $k \in \mathbf{N}$ is essentially selfadjoint on $\mathscr{S}(\mathbf{R}^n)$.

(b) $B_A^k := \overline{A^k|_{\mathscr{S}}} = \overline{A|_{\mathscr{S}}}^k$ and $D(B_A^k) := D(\overline{A^k|_{\mathscr{S}}}) = D(\overline{A|_{\mathscr{S}}}^k)$ topologically for $k \in \mathbf{N}$.

(c) $D(B_A^k) = \{u \in L^2(\mathbf{R}^n) : A^k u \in L^2(\mathbf{R}^n)\}$
$= \{u \in L^2(\mathbf{R}^n) : A^j u \in L^2(\mathbf{R}^n), 0 \le j \le k\}, k \in \mathbf{N}_0$.

(d) $B_A^k u = A^k u$ for $u \in D(B_A^k), k \in \mathbf{N}$.

PROOF: Due to 4.3.10 we only have to show that every power of A is essentially selfadjoint of $\mathscr{S}(\mathbf{R}^n)$.

Case 1: $A = P$:

Take $Z = \sqrt{\mathrm{Id} - \Delta}$, then $H_Z^k = H^k$. Moreover, we have $A \in \mathscr{L}(H_Z^{k+1}, H_Z^k)$ and $[Z, A] \in \mathrm{OpS}_{1,0}^1 \hookrightarrow \mathscr{L}(H_Z^{k+1}, H_Z^k)$ for every $k \in \mathbf{N}_0$, and A is symmetric on $\mathscr{S}(\mathbf{R}^n)$. Thus, 2.3.1, 2.4.6, and 2.1.13 can be applied.

Case 2: $A = QP^2Q + R = QPQP + QP[P, Q] + R = (QP)^2 + QP[P, Q] + R$:

Take $Z := \frac{1}{2}(QP + PQ) = QP + \frac{1}{2}[P, Q] \in \mathrm{OpS}_{1,0}^1$. Then Z is symmetric on $\mathscr{S}(\mathbf{R}^n)$, i.e., due to case 1 every power of Z is essentially selfadjoint on $\mathscr{S}(\mathbf{R}^n)$, and we have

$$H_Z^k = D(Z^k) \underset{4.3.8}{=} H_{QP}^k .$$

Since $\mathrm{OpS}_{1,0}^0 \hookrightarrow \mathscr{L}(H_{QP}^k) = \mathscr{L}(H_Z^k)$ for every $k \in \mathbf{N}_0$ due to 4.3.7 we have

$$A = (QP)^2 + QP \underbrace{[P, Q]}_{\in \mathrm{OpS}_{1,0}^0} + R \in \bigcap_{k \in \mathbf{N}_0} \mathscr{L}(H_{QP}^{k+2}, H_{QP}^k)$$

and

$$
\begin{aligned}
[Z, A] \;=\;& [QP, (QP)^2 + QP[P, Q] + R] + \frac{1}{2}[[P, Q], (QP)^2 + QP[P, Q] + R] \\[2mm]
=\;& QP \underbrace{[QP, [P, Q]]}_{\in \mathrm{Op}S^0_{1,0}} + \underbrace{[QP, R]}_{\in \mathrm{Op}S^0_{1,0}} + \frac{1}{2}QP \underbrace{[[P, Q], QP]}_{\in \mathrm{Op}S^0_{1,0}} + \frac{1}{2}\underbrace{[[P, Q], QP]}_{\in \mathrm{Op}S^0_{1,0}} QP \\[2mm]
& + \frac{1}{2}\underbrace{[[P, Q], QP[P, Q] + R]}_{\in \mathrm{Op}S^0_{1,0}} \qquad \in \bigcap_{k \in \mathbb{N}_0} \mathscr{L}(H^{k+1}_{QP}, H^k_{QP})
\end{aligned}
$$

Finally, A is symmetric on $\mathscr{S}(\mathbf{R}^n)$ and thus on H^2_{QP}, and we can apply 2.3.1, 2.4.6, and 2.1.13.

Case 3: $A = PQ^2P + R = PQPQ + PQ[Q, P] + R = (PQ)^2 + PQ[Q, P] + R$
This can be done similarly to case 2. $\qquad\square$

4.3.12 Example. Suppose that $a \in \mathcal{B}^\infty(\mathbf{R}^n, \mathbb{R})$ (i.e., $a \in C^\infty(\mathbf{R}^n, \mathbb{R})$ with bounded derivatives), and $p \in S^1_{1,0}$ is a real-valued symbol depending only on ξ.
Then theorem 4.3.11 can be applied to the following operators:

- $A = a(X)\Delta a(X) = -a(X)\left(\sqrt{\mathrm{Id} - \Delta}\right)^2 a(X) + a^2(X)$

- $A = \partial_j a^2(X)\partial_j$

- $A = a(X)p^2(D_x)a(X)$.

Note that we obtained essential selfadjointness of any power, whereas in 4.3.2 we only obtained essential selfadjointness for the operator itself.
In the last part of this section we study Schrödinger operators. To this end we will use the scale of Hilbert spaces generated by $-\Delta + \gamma^2$, where γ is a weight function. We will study this scale in the following propositions. Define $N = -\Delta + \gamma^2$ for a weight function γ and think of $Nu \in \mathscr{S}', u \in \mathscr{S}'$, in the sense of distributions.

4.3.13 Lemma. There is a $c > 0$ with
$$
N \geq c \text{ on } \mathscr{S}(\mathbf{R}^n) \text{ and } c\|u\|_{L^2(\mathbf{R}^n)} \leq \|Nu\|_{L^2(\mathbf{R}^n)} \text{ for } u \in \mathscr{S}(\mathbf{R}^n).
$$

PROOF: We can calculate
$$
\langle Nu, u\rangle_{L^2} = \sum_{j=1}^n \langle \partial_j u, \partial_j u\rangle_{L^2} + \langle \gamma^2 u, u\rangle_{L^2} \geq 0 + c\|u\|^2_{L^2}
$$

for $u \in \mathscr{S}(\mathbf{R}^n)$, hence $c\|u\|^2_{L^2} \leq \langle Nu, u\rangle_{L^2} \leq \|Nu\|_{L^2}\|u\|_{L^2}$, which proves the assertion. $\qquad\square$

4.3.14 Proposition. Define $N_0 := N|_{C_c^\infty(\mathbf{R}^n)}$. Then

(a) $\overline{N_0^k}$ is selfadjoint in $L^2(\mathbf{R}^n)$ for $k \in \mathbf{N}$. We clearly have $\overline{N_0^k}u = N^k u$ for $u \in D(\overline{N_0^k})$.

(b) $D(\overline{N_0^k}) = \{u \in L^2(\mathbf{R}^n) : N^k u \in L^2(\mathbf{R}^n)\}$, in particular $\mathscr{S}(\mathbf{R}^n) \subset D(\overline{N_0^k})$ densely.

(c) $\overline{N_0^k} = \overline{N_0}^k$ and $D(\overline{N_0^k}) = D(\overline{N_0}^k)$ topologically for $k \in \mathbf{N}$.

PROOF: (a) is a special case of a result of Chernoff [26, 4], cf. 4.3.1. (b) and (c) can be shown as in 4.3.10. $\qquad\qquad\qquad\qquad\qquad\qquad\qquad\qquad\qquad\qquad$ $\square$

4.3.15 Lemma. Assume that γ is a weight function, then there is a $d > 0$ with

$$\|\Delta u\|^2_{L^2(\mathbf{R}^n)} + \|\gamma^2 u\|^2_{L^2(\mathbf{R}^n)} \leq \|(-\Delta + \gamma^2)u\|^2_{L^2(\mathbf{R}^n)} + d\,\|u\|^2_{L^2(\mathbf{R}^n)} \text{ for } u \in \mathscr{S}(\mathbf{R}^n).$$

PROOF: We calculate

$$\begin{aligned}
N^2 &= \left(-\Delta + \gamma^2\right)^2 = \Delta^2 + \gamma^4 - \Delta\gamma^2 - \gamma^2\Delta \\
&= \Delta^2 + \gamma^4 - \gamma\Delta\gamma - [\Delta,\gamma]\gamma - \gamma\Delta\gamma - \gamma[\gamma,\Delta] \\
&= \Delta^2 + \gamma^4 - 2\gamma\Delta\gamma - [\gamma,[\gamma,\Delta]]\,.
\end{aligned}$$

Note that $[\gamma, \Delta] \in \mathrm{Op}S^1_{1,0}$ and $[\gamma,[\gamma,\Delta]] \in \mathrm{Op}S^n_{1,0}$ due to 4.1.7. Hence there is a $d > 0$ with $\|[\gamma,[\gamma,\Delta]]u\|_{L^2} \leq d\,\|u\|_{L^2}$ for $u \in \mathscr{S}(\mathbf{R}^n)$. Thus

$$\begin{aligned}
&\|(-\Delta + \gamma^2)u\|^2_{L^2} = \langle(-\Delta+\gamma^2)u, (-\Delta+\gamma^2)u\rangle_{L^2} = \langle(-\Delta+\gamma^2)^2 u, u\rangle_{L^2} \\
&= \langle\Delta^2 u, u\rangle_{L^2} + \langle\gamma^4 u, u\rangle_{L^2} - 2\langle\gamma\Delta\gamma u, u\rangle_{L^2} - \langle[\gamma,[\gamma,\Delta]]u, u\rangle_{L^2} \\
&\geq \langle\Delta u, \Delta u\rangle_{L^2} + \langle\gamma^2 u, \gamma^2 u\rangle_{L^2} + 2\sum_{j=1}^n \langle\partial_j(\gamma u), \partial_j(\gamma u)\rangle_{L^2} - |\langle[\gamma,[\gamma,\Delta]]u, u\rangle_{L^2}| \\
&\geq \|\Delta^2 u\|^2_{L^2} + \|\gamma^2 u\|^2_{L^2} + 0 - d\,\|u\|^2_{L^2}\,.
\end{aligned}$$

$\qquad\qquad\qquad\qquad\qquad\qquad\qquad\qquad\qquad\qquad\qquad\qquad\qquad\qquad\qquad\qquad$ $\square$

Now we can characterize the scale generated by the anisotropic harmonic oscillator with distributional conditions.

4.3.16 Lemma. Assume that γ is a weight function.

(a) For $s, t \in \mathbf{R}, k \in \mathbb{N}$ there is a $c_{s,t,k} \geq 0$ with

$$\left\|N^k u\right\|_{\gamma,s,t} \leq c_{s,t,k} \sum_{j=0}^{k} \|u\|_{\gamma,s+2j,t+2k-2j} \qquad \text{for } u \in \mathscr{S}(\mathbf{R}^n).$$

(b) For $s, t \in \mathbf{R}, k \in \mathbb{N}$ there is a $d_{s,t,k} \geq 0$ with

$$\left\|[N^k, i\Delta]u\right\|_{\gamma,s,t} \leq d_{s,t,k} \sum_{j=1}^{k} \|u\|_{\gamma,s+2j-1,t+2k-2j+1} \qquad \text{for } u \in \mathscr{S}(\mathbf{R}^n).$$

(c) For $s, t \in \mathbf{R}, k \in \mathbb{N}$, and $P \in \mathrm{Op}S^1_{1,0}$ there is an $e_{s,t,k} \geq 0$ with

$$\left\|[N^k, P]u\right\|_{\gamma,s,t} \leq e_{s,t,k} \sum_{j=0}^{k} \|u\|_{\gamma,s+2j,t+2k-2j} \qquad \text{for } u \in \mathscr{S}(\mathbf{R}^n).$$

PROOF: Follows by simple inductions.   □

4.3.17 Theorem.   If γ is a weight function, then $\Lambda := -\Delta + \gamma^2$ is a
selfadjoint operator in $L^2(\mathbf{R}^n)$ with domain $D(\Lambda) = H^{2,0}_\gamma \cap H^{0,2}_\gamma$.
Moreover, Λ is positive and $D(\Lambda^k) = H^{2k,0}_\gamma \cap H^{0,2k}_\gamma$ topologically for $k \in \mathbb{N}$.

PROOF: Due to 4.3.13, 4.3.14 we only have to show that $D(\Lambda^k) = H^{2k,0}_\gamma \cap H^{0,2k}_\gamma$
topologically for $k \in \mathbb{N}$. Since $\mathscr{S}(\mathbf{R}^n)$ is dense in $D(\Lambda^k)$ (due to 4.3.14) and in
$H^{2k,0}_\gamma \cap H^{0,2k}_\gamma$ (due to 4.1.24) we only have to show (again due to 4.3.14) that
$\|\cdot\|_{\Lambda,k} := \|\cdot\|_{L^2} + \|\Lambda^k\cdot\|_{L^2}$ and $\|\cdot\|_{H,2k} := \|\cdot\|_{\gamma,2k,0} + \|\cdot\|_{\gamma,0,2k}$ define equivalent
norms on $\mathscr{S}(\mathbf{R}^n)$.
First we note that

$$\|u\|_{\Lambda,k} \leq c_k \left\|(-\Delta + \gamma^2)^k u\right\|_{L^2} \leq c'_k \sum_{j=0}^{k} \|u\|_{\gamma,2j,2k-2j} \leq c''_k \|u\|_{H,2k}$$

for $u \in \mathscr{S}(\mathbf{R}^n)$ and constants c_k, c'_k, c''_k due to 4.3.13, 4.3.16, and 4.1.30.
Next we will show that

$$\left\|\Delta^k u\right\|_{L^2} + \left\|\gamma^{2k} u\right\|_{L^2} \leq d_k \left\|\Lambda^k u\right\|_{r,2} \qquad (4.3.1)$$

for $u \in \mathscr{S}(\mathbf{R}^n)$ and suitable constants d_k. For $k = 1$ (4.3.1) is proved in 4.3.13, 4.3.15. Now assume (4.3.1) inductively up to a certain $k \in \mathbf{N}$. Using $[\Lambda, \Delta^k] = [\gamma^2, \Delta^k]$ and 4.2.11 we have

$$\left\|[\Lambda, \Delta^k]u\right\|_{L^2} \leq b_k \left\|u\right\|_{\gamma, 2k-1, 1} \leq b'_k \left(\left\|u\right\|_{\gamma, 2k, 0} + \left\|u\right\|_{\gamma, 0, 2k}\right)$$

$$\leq b''_k \sum_{j=0}^{k} \left(\left\|\Delta^j u\right\|_{L^2} + \left\|\gamma^{2j} u\right\|_{L^2}\right) \leq b'''_k \sum_{j=0}^{k} \left\|\Lambda^j u\right\|_{L^2} \leq b_k \left\|\Lambda^k u\right\|_{L^2}$$

for $u \in \mathscr{S}(\mathbf{R}^n)$ and suitable constants due to 4.1.30, the induction hypothesis and 4.3.13. Moreover

$$\left\|[\Lambda, \gamma^{2k}]u\right\|_{L^2} = \left\|[\Delta, \gamma^{2k}]u\right\|_{L^2} \leq c_k \left\|u\right\|_{\gamma, 1, 2k-1} \leq c'_k \left(\left\|u\right\|_{\gamma, 2k, 0} + \left\|u\right\|_{\gamma, 0, 2k}\right)$$

$$\leq \tilde{b}''_k \sum_{j=0}^{k} \left(\left\|\Delta^j u\right\|_{L^2} + \left\|\gamma^{2j} u\right\|_{L^2}\right) \leq \tilde{b}'''_k \sum_{j=0}^{k} \left\|\Lambda^j u\right\|_{L^2} \leq \tilde{b}_k \left\|\Lambda^k u\right\|_{L^2}$$

due to 4.2.11, 4.1.30, and the induction hypothesis. Therefore we can calculate

$$\left\|\Delta^{k+1} u\right\|_{L^2} + \left\|\gamma^{2k+2} u\right\|_{L^2} \leq d_1 \left(\left\|\Lambda\Delta^k u\right\|_{L^2} + \left\|\Lambda\gamma^{2k} u\right\|_{L^2}\right)$$

$$\leq d_1 \left(\left\|\Delta^k \Lambda u\right\|_{L^2} + \left\|\gamma^{2k} \Lambda u\right\|_{L^2} + \left\|[\Lambda, \Delta^k]u\right\|_{L^2} + \left\|[\Lambda, \gamma^{2k}]u\right\|_{L^2}\right)$$

$$\leq d'_{k+1} \left(\left\|\Lambda^k \Lambda u\right\|_{L^2} + \left\|\Lambda^k u\right\|_{L^2}\right) \leq d_{k+1} \left\|\Lambda^{k+1} u\right\|_{L^2} .$$

This proves (4.3.1) and the assertion because

$$\left\|u\right\|_{H, 2k} \leq D_k \sum_{j=0}^{k} \left(\left\|\Delta^j u\right\|_{L^2} + \left\|\gamma^{2j} u\right\|_{L^2}\right) \leq D'_k \left\|\Lambda^k u\right\|_{L^2} \leq D'_k \left\|u\right\|_{\Lambda, k}$$

for $u \in \mathscr{S}(\mathbf{R}^n)$ and constants D_k, D'_k. $\qquad\square$

Next, we consider Schrodinger equations with first-order pseudodifferential terms in anisotropically weighted Sobolev spaces.

4.3.18 Theorem. Let γ be a weight function, $p \in \mathcal{C}(I \times \mathbf{R}^n_x \times \mathbf{R}^n_\xi)$ such that $\{p(t, \cdot, \cdot) : t \in I\} \subset S^1_{1,0}$ is bounded. Assume that there is a constant $\beta \geq 0$ with

$$\left|\operatorname{Re} \langle p(t, X, D_x)u, u\rangle_{L^2(\mathbf{R}^n)}\right| \leq \beta\langle u, u\rangle_{L^2(\mathbf{R}^n)} \text{ for } u \in \mathscr{S}(\mathbf{R}^n),$$

(e.g. $\{p(t, \cdot, \cdot) + p^*(t, \cdot, \cdot) : t \in I\} \subset S^0_{1,0}$ boundedly, i.e., p is symmetric hyperbolic.) Finally assume that $a \in \mathcal{C}(I)$ is real-valued. Then the Cauchy problem for $\pm(ia(t)\Delta + p(t, X, D_x)) \in \mathscr{L}(H^{2k+2,0}_\gamma \cap H^{0,2k+2}_\gamma, H^{2k,0}_\gamma \cap H^{0,2k}_\gamma)$ in the scale $(H^{2k,0}_\gamma \cap H^{0,2k}_\gamma)_k$ is well-posed with exponential growth.

PROOF: We apply 2.4.6 in the scale generated by Λ of 4.3.17.
4.1.11 shows that $I \ni t \mapsto p(t, X, D_x) \in \mathcal{L}(H_\gamma^{k+1,k'}, H_\gamma^{k,k'})$ is strongly continuous for $k, k' \in \mathbb{R}$. Hence

$$I \ni t \mapsto A(t) := ia(t)\Delta + p(t, X, D_x) \in \mathcal{L}(D(\Lambda^{k+1}), D(\Lambda^k)), k \in \mathbb{N}_0,$$

is strongly continuous due to 4.1.30.
Moreover, 4.3.16 shows that $\{[\Lambda^k, A(t)] : t \in I\} \subset \mathcal{L}(D(\Lambda^k), L^2(\mathbb{R}^n))$ boundedly. Finally,

$$|\mathrm{Re}\, \langle A(t)u, u \rangle_{L^2}| \leq |a(t)|\mathrm{Im}\, \langle \Delta u, u \rangle_{L^2}| + |\mathrm{Re}\, \langle p(t, X, D_x)u, u \rangle_{L^2}| \leq 0 + \beta\langle u, u \rangle_{L^2}$$

for $u \in D(\Lambda)$. Therefore we can apply 2.3.1 and 2.4.6. □

Note in particular that 2.4.6 shows that the Schrödinger propagator $U(t, s)$ exists in the situation of 4.3.18 and that $\mathscr{S}(\mathbb{R}^n)$ and $H_\gamma^{2k,0} \cap H_\gamma^{0,2k}$ for $k \in \mathbb{N}_0$ are invariant subspaces.
Finally, we want to treat operators with spatially growing coefficients. To this end we use Cordes' pseudodifferential calculus for the symbol classes $\psi c^{(m,m')}$ (cf. 4.1.19).

4.3.19 Theorem. Let either $(m, m', \varepsilon, N) = (2, 2, 1, 1)$ be a tuple of numbers or $(m, m', \varepsilon, N) = (3, \frac{3}{2}, \frac{1}{2}, 2)$. Moreover suppose that

(a) $p \in C(I \times \mathbb{R}_\xi^n, \mathbb{R})$ such that $\{p(t, \cdot) : t \in I\} \subset \psi c^{(m,0)}$ is a bounded family of real-valued symbols depending only on ξ.

(b) $q \in C(I \times \mathbb{R}_x^n \times \mathbb{R}_\xi^n)$ such that $\{q(t, \cdot, \cdot) : t \in I\} \subset \psi c^{(1,1)}$ boundedly and $\{q(t, \cdot, \cdot) + q^*(t, \cdot, \cdot) : t \in I\} \subset \psi c^{(0,0)}$ boundedly.

(c) $r \in C(I \times \mathbb{R}_x^n, \mathbb{R})$ such that $\{r(t, \cdot) : t \in I\} \subset \psi c^{(0,m')}$ is a bounded family of real-valued symbols depending only on x.

(d) $A(t) := ip(t, D_x) + q(t, X, D_x) + ir(t, X)$

Then the Cauchy problem for $\pm A(t) \in \mathcal{L}(H^{2(k+N),0} \cap H^{0,\varepsilon 2(k+N)}, H^{2k,0} \cap H^{0,\varepsilon 2k})$ in the scale $(H^{2k,0} \cap H^{0,\varepsilon 2k})_k$ is well-posed with expontial growth.

PROOF: Let $Z_\varepsilon := i(-\Delta + \langle x \rangle^{2\varepsilon})$, then Z_ε generates a unitary C_0-group, and due to 4.3.17 and 4.1.30

$$H^k_{Z_\varepsilon} = H^{2k,0}_{\langle \cdot \rangle^\varepsilon} \cap H^{0,2k}_{\langle \cdot \rangle^\varepsilon} = \bigcap_{j=0}^{2k} H^{j,2k-j}_{\langle \cdot \rangle^\varepsilon} = H^{2k,0} \cap H^{0,\varepsilon 2k} = \bigcap_{j=0}^{2k} H^{j,\varepsilon(2k-j)}.$$

Moreover, there are constants $c_{k,l} \geq 0$ with $\|u\|_{k,\varepsilon l} \leq c_{k,l}(\|u\|_{k+l,0} + \|u\|_{0,\varepsilon(k+l)})$ for $k, l \in \mathbb{N}_0$ and $u \in \mathscr{S}(\mathbb{R}^n)$. Note that $\varepsilon N = 1, m' = \varepsilon m$ and $1 \leq N, m \leq 2N$, hence

$$\begin{aligned}
&\|A(t)u\|_{2k,0} \leq \|p(t, D_x)u\|_{2k,0} + \|q(t, X, D_x)u\|_{2k,0} + \|r(t, X)u\|_{2k,0} \\
&\leq c_k(\|u\|_{2k+m,0} + \|u\|_{2k+1,1} + \|u\|_{2k,m'}) = c_k(\|u\|_{2k+m,0} + \|u\|_{2k+1,\varepsilon N} + \|u\|_{2k,\varepsilon m}) \\
&\leq c'_k(\|u\|_{2k+m,0} + \|u\|_{2k+N+1,0} + \|u\|_{0,\varepsilon(2k+N+1)} + \|u\|_{2k+m,0} + \|u\|_{0,\varepsilon(2k+m)}) \\
&\leq c''_k(\|u\|_{2(k+N),0} + \|u\|_{0,\varepsilon(2(k+N))})
\end{aligned}$$

and

$$\begin{aligned}
&\|A(t)u\|_{0,\varepsilon 2k} \leq \|p(t, D_x)u\|_{0,\varepsilon 2k} + \|q(t, X, D_x)u\|_{0,\varepsilon 2k} + \|r(t, X)u\|_{0,\varepsilon 2k} \\
&\leq c_k(\|u\|_{m,\varepsilon 2k} + \|u\|_{1,\varepsilon 2k+1} + \|u\|_{0,\varepsilon 2k+m'}) \\
&= c_k(\|u\|_{m,\varepsilon 2k} + \|u\|_{1,\varepsilon(2k+N)} + \|u\|_{0,\varepsilon(2k+m)}) \\
&\leq c'_k(\|u\|_{2k+m,0} + \|u\|_{0,\varepsilon(2k+m)} + \|u\|_{2k+N+1,0} + \|u\|_{0,\varepsilon(2k+N+1)} + \|u\|_{0,\varepsilon(2k+m)}) \\
&\leq c''_k(\|u\|_{2(k+N),0} + \|u\|_{0,\varepsilon(2(k+N))})
\end{aligned}$$

for suitable c_k, c'_k, c''_k and $u \in \mathscr{S}(\mathbb{R}^n)$. Therefore $I \ni t \mapsto A(t) \in \mathscr{L}(H^{k+N}_{Z_\varepsilon}, H^k_{Z_\varepsilon})$ is strongly continuous for any $k \in \mathbb{N}_0$. Moreover

$$\begin{aligned}
[p(t, D_x), -\Delta + \langle x \rangle^{2\varepsilon}] &= [p(t, D_x), \langle x \rangle^{2\varepsilon}] \in \mathrm{Op}\psi c^{(m-1,2\varepsilon-1)}, \\
[q(t, X, D_x), -\Delta + \langle x \rangle^{2\varepsilon}] &\in \mathrm{Op}\psi c^{(2,0)} + \mathrm{Op}\psi c^{(0,2\varepsilon)}, \\
[r(t, X), -\Delta + \langle x \rangle^{2\varepsilon}] &= [r(t, X), -\Delta] \in \mathrm{Op}\psi c^{(1,m'-1)}.
\end{aligned}$$

Thus

$$\begin{aligned}
&\left\|[A(t), -\Delta + \langle x \rangle^{2\varepsilon}]u\right\|_{2k,0} \\
&\leq c_k(\|u\|_{2k+(m-1),2\varepsilon-1} + \|u\|_{2k+2,0} + \|u\|_{2k,2\varepsilon} + \|u\|_{2k+1,m'-1}) \\
&\leq \left\{ \begin{array}{ll} c'_k(\|u\|_{2k+1,\varepsilon} + \|u\|_{2k+2,0} + \|u\|_{2k,2\varepsilon} + \|u\|_{2k+1,\varepsilon}) &, m = 2 \\ c'_k(\|u\|_{2k+2,0} + \|u\|_{2k+2,0} + \|u\|_{2k,2\varepsilon} + \|u\|_{2k+1,\varepsilon}) &, m = 3 \end{array} \right\} \\
&\leq c''_k(\|u\|_{2(k+1),0} + \|u\|_{0,\varepsilon(2(k+1))})
\end{aligned}$$

and

$$\left\|[A(t), -\Delta + \langle x \rangle^{2\varepsilon}]u\right\|_{0,\varepsilon 2k}$$

$$\leq \ c_k(\|u\|_{m-1,\varepsilon 2k+2\varepsilon-1} + \|u\|_{2,\varepsilon 2k} + \|u\|_{0,\varepsilon 2k+2\varepsilon} + \|u\|_{1,\varepsilon 2k+m'-1})$$

$$\leq \ \begin{cases} c_k'(\|u\|_{1,\varepsilon(2k+1)} + \|u\|_{2,\varepsilon 2k} + \|u\|_{0,\varepsilon(2k+2)} + \|u\|_{1,\varepsilon(2k+1)}) & , \ m = 2 \\ c_k'(\|u\|_{2,\varepsilon 2k} + \|u\|_{2,\varepsilon 2k} + \|u\|_{0,\varepsilon(2k+2)} + \|u\|_{1,\varepsilon(2k+1)}) & , \ m = 3 \end{cases}$$

$$\leq \ c_k''(\|u\|_{2(k+1),0} + \|u\|_{0,\varepsilon(2(k+1))})$$

for suitable constants c_k, c_k', c_k'' and $u \in \mathscr{S}(\mathbf{R}^n)$. This shows

$$\{[A(t), Z_\varepsilon] : t \in I\} \subset \mathscr{L}(H_{Z_\varepsilon}^{k+1}, H_{Z_\varepsilon}^k)$$

boundedly. Finally, using an argument similar to 4.2.9 we obtain

$$|\mathrm{Re}\ \langle A(t)u, u\rangle_{L^2(\mathbf{R}^n)}| \leq d \, \|u\|_{L^2(\mathbf{R}^n)}^2$$

for a constant $d \geq 0$. Thus 2.3.1 and 2.4.6 imply the assertion.   □

As the last result in this section, we prove essential selfadjointness of all powers of the operators $A(t)$ in 4.3.19 (if $q(t, X, D_x)$ is symmetric).

4.3.20 Theorem. Let $(m, m') = (2, 2)$ or $(m, m') = (3, \frac{3}{2})$ and assume that

(a) $p \in \psi c^{(m,0)}$ is a real-valued symbol depending only on ξ.

(b) $q \in \psi c^{(1,1)}$ is real-valued.

(c) $r \in \psi c^{(0,m')}$ is a real-valued symbol depending only on x.

(d) $A := p(D_x) + q^w(X, D_x) + r(X)$, where $q^w(X, D_x)$ is the Weyl-quantized pseudodifferential operator with symbol q.

Then every power of A is essentially selfadjoint on $\mathscr{S}(\mathbf{R}^n)$, and for $k \in \mathbf{N}$ we have:

(a) $B^k := \overline{A^k|_{\mathscr{S}}} = \overline{A|_{\mathscr{S}}}^k$ and $D(B^k) := D(\overline{A^k|_{\mathscr{S}}}) = D(\overline{A|_{\mathscr{S}}}^k)$ topologically.

(b) $D(B^k) = \{u \in L^2(\mathbf{R}^n) : A^k u \in L^2(\mathbf{R}^n)\}$
$\quad\quad = \{u \in L^2(\mathbf{R}^n) : A^j u \in L^2(\mathbf{R}^n), 0 \leq j \leq k\}$.

(c) $\overline{A^k|_{\mathscr{S}}} = \overline{A|_{\mathscr{S}}}^k u = A^k u$ for $u \in D(\overline{A^k|_{\mathscr{S}}}) = D(\overline{A|_{\mathscr{S}}}^k)$.

PROOF: Let $(\varepsilon, N) = (1, 1)$ resp., $(\varepsilon, N) = (\frac{1}{2}, 2)$ if $m = 2$ resp., $m = 3$, and let $Z_\varepsilon := i(-\triangle + \langle x\rangle^{2\varepsilon})$. In 4.3.19 we have shown that $A \in \mathscr{L}(H_{Z_\varepsilon}^{k+N}, H_{Z_\varepsilon}^k)$ and $[Z_\varepsilon, A] \in \mathscr{L}(H_{Z_\varepsilon}^{k+1}, H_{Z_\varepsilon}^k)$. Moreover, A is symmetric. Thus we can apply 2.3.1, 2.4.6, and 2.1.13 to obtain the essential selfadjointness of all powers of A on $\mathscr{S}(\mathbf{R}^n)$, and we obtain the assertion as in 4.3.10.   □

4.3.21 Bibliographical remarks. Essential selfadjointness of $p^w(X, D_x)$ on $\mathscr{S}(\mathbf{R}^n)$ is a classical result for real-valued symbols $p \in S_{\rho,\delta}^{\rho-\delta}$ or $p \in HS_{\rho,\delta}^{m,m'}$, $m \geq m' \geq 0$, cf. e.g. [97, 3.5.8]. 4.3.5(a) has been obtained by Robinson [130] with a different proof. The statement in 4.3.11 for $A = P \in OpS_{1,0}^1$ goes back essentially to Chernoff [26]. The statements in 4.3.19 and 4.3.20 for the operators themselves (i.e., not their powers) have been obtained by Yamazaki [159] in a more general situation. Moreover, Yamazaki gives examples that show the sharpness of the choice of (m, m').

4.4 Evolution equations in $C_0(\mathbf{R}^n)$ and Feller semigroups

Evolution equations in spaces of continuous or differentiable functions play an important role in applications - in particular in probability theory. But equations in these spaces usually are more difficult to attack than equations in scales of L^p-spaces. Difficulties are caused by the non-reflexivity of such spaces and the absence of natural isomorphisms between spaces of differentiable functions and continuous functions (in dimensions greater than 2, cf. e.g. [87]). In this section we will describe some consequences of the results obtained so far to evolution equations in the scale $C_0^k(\mathbf{R}^n, \mathbf{R})$ of real-valued, k-times continuously differentiable functions vanishing at infinity with all their derivatives. Clearly, $C_0^k(\mathbf{R}^n, \mathbf{R})$ is a real Banach space with norm $\|u\|_{C_0^k} := \sum_{|\alpha| \leq k} \|\partial^\alpha u\|_{L^\infty}$ for $u \in C_0^k(\mathbf{R}^n, \mathbf{R})$. In particular, this will have consequences for Feller semigroups. But first, we have to collect some definitions and results for Banach lattices.

4.4.1 Definition. A real Banach space $(X, \|\cdot\|)$ is called a real Banach lattice, if there is an ordering $\leq$ on X such that

- $x \leq y$ implies $x + z \leq y + z$ and $\lambda x \leq \lambda y$ for $x, y, z \in X, \lambda \in [0, \infty)$.

- for $x, y \in X$ there is a $z \in X$ with $x, y \leq z$ (resp., $z \leq x, y$) and for any $w \in X$ with $x, y \leq w$ (resp., $w \leq x, y$) we have $z \leq w$ (resp., $w \leq z$). $x \vee y := \sup(x, y) := z$ (resp., $x \wedge y := \inf(x, y) := z$) is called the supremum (resp., infimum) of x, y.

- $|x| \leq |y|$ for $x, y \in X$ implies $\|x\| \leq \|y\|$, where $x^+ := x \vee 0$, $x^- := (-x) \vee 0$, $|x| := x^+ + x^-$.

Moreover, a linear operator $A \in \mathscr{L}(X)$ in a Banach lattice X is called positivity preserving, if $Ax \geq 0$ for $x \in X$ with $x \geq 0$.

4.4.2 Lemma. Let $(X, \|\cdot\|)$ be a real Banach lattice and $A \in \mathcal{L}(X)$. Then $\|(Ax)^+\| \le \|x^+\|$ for $x \in X$ if and only if A is a positivity preserving contraction.

PROOF: If A is a positivity preserving contraction, then $0 \le (Ax)^+ \le A(x^+)$ and so $\|(Ax)^+\| \le \|A(x^+)\| \le \|x^+\|$ for $x \in X$.
Conversely, if $\|(Ax)^+\| \le \|A(x^+)\| \le \|x^+\|$ for any $x \in X$, then for every $x \in X$ with $x \ge 0$ we have $\|(Ax)^-\| = \|(A(-x))^+\| \le \|(-x)^+\| = 0$, which shows $(Ax)^- = 0$, i.e. $Ax \ge 0$. This shows $|Ax| \le A|x|$ for $x \in X$. Hence $\|Ax\| = \||Ax|\| \le \|A|x|\| \le \|(A|x|)^+\| \le \||x|^+\| = \|x\|$ for all $x \in X$. $\qquad\square$

4.4.3 Proposition. Let $(X, \|\cdot\|)$ be a real Banach lattice and $A : D(A) \to X$ be a densely defined, linear operator in X. Then the following statements are equivalent:

(a) $\|(\lambda x - Ax)^+\| \ge \lambda \|x^+\|$ for $\lambda > 0, x \in D(A)$.

(b) For $x \in D(A)$ there is a functional $\Phi \in X^*$ with $\langle \Phi, x \rangle_{X^*, X} = \|x^+\|$, $\langle \Phi, y \rangle_{X^*, X} \le \|y^+\|$ for $y \in X$, and $\langle \Phi, Ax \rangle_{X^*, X} \le 0$.

(c) For $x \in D(A)$ and $\Phi \in X^*$ with $\langle \Phi, x \rangle_{X^*, X} = \|x^+\|$ and $\langle \Phi, y \rangle_{X^*, X} \le \|y^+\|$ for $y \in X$, we have $\langle \Phi, Ax \rangle_{X^*, X} \le 0$.

In this case A is called dispersive.

PROOF: We clearly have (c) $\Rightarrow$ (b). (b) $\Rightarrow$ (a) is a consequence of

$$\|x^+\| = \langle \Phi, x \rangle = \langle \Phi, x - \lambda Ax + \lambda Ax \rangle \le \langle \Phi, x - \lambda Ax \rangle \le \|(x - \lambda Ax)^+\|$$

for $x \in X$ and $\lambda > 0$. To prove (a) $\Rightarrow$ (c) let $\Phi \in X^*$ with $\langle \Phi, x \rangle \|x^+\|^2$ and choose $y_k \in D(A)$ with $y_k \xrightarrow[k \to \infty]{} Ax$, then for $t > 0$ and $k \in \mathbb{N}$

$$\langle \Phi, Ax \rangle = \frac{1}{t} (\langle \Phi, x + tAx \rangle - \langle \Phi, x \rangle)$$

$$\le \frac{1}{t} (\|(x + tAx)^+\| - \|x^+\|) = \frac{1}{t} (\|(x + ty_k + tAx - ty_k)^+\| - \|x^+\|)$$

$$\le \frac{1}{t} (\|(x + ty_k)^+\| + t \|(Ax - y_k)^+\| - \|x^+\|)$$

$$\le \frac{1}{t} \left(t \left\| \left(\left(\frac{1}{t} \mathrm{Id} - A \right) (x + ty_k) \right)^+ \right\| + t \|(Ax - y_k)^+\| - \|x^+\| \right)$$

$$\le \frac{1}{t} (\|x^+\| + t \|(y_k - Ax)^+\| + t^2 \|(-Ay_k)^+\| + t \|(Ax - y_k)^+\| - \|x^+\|)$$

$$\le (2 \|y_k - Ax\| + t \|Ay_k\|)$$

Now let first $t \to 0$ and then $k \to \infty$. Then this implies the assertion. $\qquad\square$

For the real Banach lattice $C_0(\mathbb{R}^n, \mathbb{R})$ dispersivity of an operator A in $C_0(\mathbb{R}^n, \mathbb{R})$ is equivalent to the validity of the so-called positive maximum principle. Moreover, operators are dissipative in this case.

4.4.4 Proposition. Let $A : D(A) \to C_0(\mathbb{R}^n, \mathbb{R})$ be a densely defined, linear operator in $C_0(\mathbb{R}^n, \mathbb{R})$. Then the following statements are equivalent:

(a) A is $\|\cdot\|_{L^\infty}$-dispersive.

(b) A satisfies the positive maximum principle, that is for $f \in D(A)$ and $x_0 \in \mathbb{R}^n$ with $f(x_0) = \sup\limits_{x \in \mathbb{R}^n} f(x) \geq 0$ we have $Af(x_0) \leq 0$.

In this case A is also $\|\cdot\|_{L^\infty}$-dissipative.

PROOF: For $f \in D(A)$ let $x_0 \in \mathbb{R}^n$ with $f(x_0) = \sup_{x \in \mathbb{R}^n} f(x) = \|f^+\|_{L^\infty} \geq 0$ and $\Phi(g) := g(x_0)$ for $g \in C_0(\mathbb{R}^n, \mathbb{R})$. Then clearly $\Phi \in C_0(\mathbb{R}^n, \mathbb{R})^*$ is a functional with $\Phi(f) = \|f^+\|_{L^\infty}$ and $\Phi(g) = g(x_0) \leq \|g^+\|_{L^\infty}$ and 4.4.3 implies the equivalence of (a) and (b). Now assume that A is $\|\cdot\|_{L^\infty}$-dispersive, and let $f \in D(A)$. If $\|f\|_{L^\infty} = \|f^+\|_{L^\infty}$, then

$$\lambda \|f\|_{L^\infty} = \lambda \|f^+\|_{L^\infty} \leq \|(\lambda f - Af)^+\|_{L^\infty} \leq \|\lambda f - Af\|_{L^\infty} .$$

Otherwise $\|f\|_{L^\infty} = \|f^-\|_{L^\infty}$, thus $\|f\|_{L^\infty} = \|(-f)^+\|_{L^\infty}$ and

$$\lambda \|f\|_{L^\infty} = \lambda \|(-f)^+\|_{L^\infty} \leq \|(\lambda(-f) - A(-f))^+\|_{L^\infty} \leq \|\lambda f - Af\|_{L^\infty} .$$

This implies the assertion. $\square$

4.4.5 Remark. One importance of dispersive operators lies in the following: Let $A : D(A) \to X$ be a linear operator in a real Banach lattice X. Then A generates a C_0-semigroup of positivity preserving contractions if and only if A is densely defined, closed, and dispersive, and if there is a $\lambda > 0$ with $\mathcal{R}(\lambda \mathrm{Id} - A) = X$. For a proof we refer to [110, C-II, 1.2].

4.4.6 Proposition. Let $(X, \|\cdot\|)$ be a real Banach lattice, $I = [t_0, t_1]$ be a compact interval, and $A(t) : D(A(t)) \to X, t \in I$, be a family of dispersive operators in X. Moreover, let $u \in C^1(I, X)$ be a function with $u(t) \in D(A(t))$ and $u'(t) = A(t)u(t)$ for $t \in I$. Then $\|u(t)^+\| \leq \|u(t_0)^+\|$ for $t \in I$.

PROOF: This follows from 1.1.9 with $\Phi(x) = \|x^+\|$. $\square$

In the sequel we want to work with the Banach lattice $C_0(\mathbb{R}^n, \mathbb{R})$. In order to formulate a more precise result on dispersive operators in $C_0(\mathbb{R}^n, \mathbb{R})$ we will need the notion of negative definite functions.

4.4.7 Definition. A function $a : \mathbb{R}^n \to \mathbb{C}$ is called negative definite, if for $\xi_1, \ldots, \xi_k \in \mathbb{R}^n$ we have

$$\sum_{j,l=1}^{k} (a(\xi_j) + \overline{a(\xi_l)} - a(\xi_j - \xi_l)) \lambda_j \overline{\lambda_l} \geq 0 \quad \text{for } \lambda_j \in \mathbb{C}, j = 1, \ldots, k, k \in \mathbb{N}.$$

4.4.8 Lemma. If $a : \mathbb{R}^n \to \mathbb{C}$ is a continuous, negative definite function, then

(a) $a(\xi) = \overline{a(-\xi)}$ and $\operatorname{Re} a(\xi) \geq a(0) \geq 0$ for $\xi \in \mathbb{R}^n$.

(b) $\sqrt{|a(\xi + \eta)|} \leq \sqrt{|a(\xi)|} + \sqrt{|a(\eta)|}$ for $\xi, \eta \in \mathbb{R}^n$.

(c) There is a $c > 0$ with $|a(\xi)| \leq c_a \langle \xi \rangle^2$ for $\xi \in \mathbb{R}^n$.

PROOF: Clearly, $a(0) \geq 0$. For $\xi \in \mathbb{R}^n$ the matrix

$$\begin{pmatrix} a(\xi) + \overline{a(\xi)} - a(0) & a(\xi) + \overline{a(0)} - a(\xi) \\ a(0) + \overline{a(\xi)} - a(-\xi) & a(0) + \overline{a(0)} - a(0) \end{pmatrix}$$

is non-negative. Therefore $a(\xi) + \overline{a(0)} - a(\xi) = \overline{a(0)} + a(\xi) - \overline{a(-\xi)}$, which implies $a(\xi) = \overline{a(-\xi)}$. Since the matrix has positive determinant, this shows $2(\operatorname{Re} a(\xi))a(0) - 2a(0)^2 \geq 0$, which implies (a). Moreover, since the determinant of

$$\begin{pmatrix} a(\xi) + \overline{a(\xi)} - a(0) & a(\xi) + \overline{a(\eta)} - a(\xi - \eta) \\ a(\eta) + \overline{a(\xi)} - a(\eta - \xi) & a(\eta) + \overline{a(\eta)} - a(0) \end{pmatrix}$$

is non-negative, (a) shows

$$\begin{aligned} |a(\xi) + \overline{a(\eta)} - a(\xi - \eta)|^2 &\leq (2\operatorname{Re} a(\xi) - a(0))(2\operatorname{Re} a(\eta) - a(0)) \\ &\leq (2\operatorname{Re} a(\xi))(2\operatorname{Re} a(\eta)) . \end{aligned}$$

With $-\eta$ at the place of η, this implies $|a(\xi) + a(\eta) - a(\xi + \eta)|^2 \leq 4|a(\xi)||a(\eta)|$ and

$$|a(\xi + \eta)| \leq |a(\xi + \eta) - a(\xi) - a(\eta)| + |a(\xi)| + |a(\eta)| \leq (\sqrt{|a(\xi)|} + \sqrt{|a(\eta)|})^2$$

and we have proved (b). This implies $\sqrt{|a(k\xi)|} \leq k\sqrt{|a(\xi)|}$ or $|a(\xi)| \leq k^2|a(\frac{\xi}{k})|$ for $k \in \mathbb{N}, \xi \in \mathbb{R}^n$. Put $c := \sup_{|\eta| \leq 2} |a(\eta)|$ and let $\xi \in \mathbb{R}^n$ with $|\xi| \geq 1$. There is an $n_0 \in \mathbb{N}$ with $n_0 \leq |\xi| < n_0 + 1$. Hence $|a(\xi)| \leq n_0^2|a(\frac{\xi}{n_0})| \leq |\xi|^2 c$ for $|\xi| \geq 1$. This completes the proof. $\qquad \square$

4.4.9 Examples. The following functions are negative definite:

(a) $a(\xi) = \sum_{j,l=1}^{n} a_{jl}\xi_j\xi_l + \sum_{j=1}^{n} ib_j\xi_j + c$ for $\xi = (\xi_1,\ldots,\xi_n) \in \mathbb{R}^n$ with $a_{jl} = a_{lj}$,

$b_j, c \in \mathbb{R}, j,l = 1,\ldots,n$, and $(a_{jl})_{j,l} \geq 0$ in $M_n(\mathbb{R})$, $c \geq 0$.

(b) $a(\xi) = \langle\xi\rangle^s, 0 \leq s \leq 2$.

(c) $a(\xi) = |\xi|^s, 0 \leq s \leq 2$.

PROOF: Let $l(x) := \sum_{j=1}^{n} b_j x_j$ for $x = (x_1,\ldots,x_n) \in \mathbb{R}^n$ Then, for $k \in \mathbb{N}$ and $\xi_1,\ldots,\xi_k \in \mathbb{R}^n$

$$(il)(\xi_j) + \overline{(il)(\xi_l)} - (il)(\xi_j - \xi_l) = i\left(l(\xi_j) - l(\xi_l) - l(\xi_j) + l(\xi_l)\right) = 0,$$

hence $il : \mathbb{R}^n \to \mathbb{C}$ is negative definite. Moreover, let $q(x) := \sum_{j,l=1}^{n} a_{jl}x_j x_l$ for $x = (x_1,\ldots,x_n) \in \mathbb{R}^n$ and $r(x,y) := \sum_{j,l=1}^{n} a_{jl}x_j\overline{y_l}$ for $x = (x_1,\ldots,x_n)$, $y = (y_1,\ldots,y_n) \in \mathbb{C}^n$. Then $r : \mathbb{C}^n \times \mathbb{C}^n \to \mathbb{C}^n$ is a positive, symmetric sesquilinear form, and for $\xi, \eta \in \mathbb{R}^n$ we have

$$q(\xi) + q(\eta) - q(\xi - \eta) = r(\xi,\xi) + r(\eta,\eta) - r(\xi - \eta, \xi - \eta) = 2r(\xi,\eta).$$

Hence, for $\lambda_1,\ldots,\lambda_k \in \mathbb{C}$ and $\xi_1,\ldots,\xi_k \in \mathbb{R}^n$ we have

$$0 \leq r\left(\sum_{j=1}^{k} \lambda_j\xi_j, \sum_{l=1}^{k} \overline{\lambda_l}\xi_l\right) = \frac{1}{2}\sum_{j,l=1}^{k}\left(q(\xi_j) + \overline{q(\xi_l)} - q(\xi_j - \xi_l)\right)\lambda_j\overline{\lambda_l}$$

This shows that $q : \mathbb{R}^n \to \mathbb{R}$ is negative definite. Hence $a = q + il + c$ is negative definite. Therefore we have proved (a) and also (b),(c) because $a^\alpha : \mathbb{R}^n \to \mathbb{R}$ is negative definite for any $\alpha \in (0,1)$ and continuous, negative definite $a : \mathbb{R}^n \to \mathbb{R}$. We leave this last statement as an exercise. □

Now we can formulate the following result due to Courrège that characterizes many dispersive operators as second-order pseudodifferential operators with non-regular and negative definite symbols.

4.4.10 Theorem.

(a) Let A be a linear, dispersive operator in $C_0(\mathbb{R}^n, \mathbb{R})$ with domain $C_c^\infty(\mathbb{R}^n, \mathbb{R})$.
Then there is an $a : \mathbb{R}^n \times \mathbb{R}^n \to \mathbb{C}$ with

$$Au(x) = \int_{\mathbb{R}^n} e^{ix\cdot\xi}a(x,\xi)\widehat{u}(\xi)d\xi \quad \text{for } u \in C_c^\infty(\mathbb{R}^n, \mathbb{R}) \qquad (4.4.1)$$

such that $\mathbb{R}^n \ni \xi \mapsto -a(x,\xi)$ is continuous and negative definite for any $x \in \mathbb{R}^n$ and there is a locally bounded function $h : \mathbb{R}^n \to [0,\infty)$ with $|a(x,\xi)| \leq h(x)|\xi|^2$ for $x,\xi \in \mathbb{R}^n$.

(b) Let $a : \mathbf{R}^n \times \mathbf{R}^n \to \mathbb{C}$ be continuous and $\mathbf{R}^n \ni \xi \mapsto -a(x, \xi)$ be a negative definite function for any $x \in \mathbf{R}^n$. Then the operator A defined by (4.4.1) with domain $C_c^\infty(\mathbf{R}^n, \mathbf{R})$ is dispersive in $C_0(\mathbf{R}^n, \mathbf{R})$.

PROOF: We will only give the proof of (b), since (a) will not be used in the sequel. We refer to [33] for a proof of (a). Assume that $a \in C(\mathbf{R}^n \times \mathbf{R}^n)$ and that $\mathbf{R}^n \ni \xi \mapsto -a(x, \xi)$ is negative definite for any $x \in \mathbf{R}^n$. Then 4.4.8 shows that there is a locally bounded $c : \mathbf{R} \to [0, \infty)$ with $|a(x, \xi)| \leq c(x)\langle\xi\rangle^2$ for $x, \xi \in \mathbf{R}^n$ and $A : C_c^\infty(\mathbf{R}^n, \mathbf{R}) \to C_0(\mathbf{R}^n, \mathbf{R})$ is well-defined. Let $u \in C_c^\infty(\mathbf{R}^n, \mathbf{R})$ and $x_0 \in \mathbf{R}^n$ with $u(x_0) = \sup_{x \in \mathbf{R}^n} u(x) \geq 0$. Let $b(\xi) := -a(x, \xi)$. Then b is negative definite and [15, II] shows that there is a positivity preserving contractive C_0-semigroup $(T(t))_{t \geq 0} \subset C_0(\mathbf{R}^n, \mathbf{R})$ satisfying $\frac{d}{dt}T(t)f = b(D_x)T(t)f$ for $f \in C_c^\infty(\mathbf{R}^n, \mathbf{R})$. Hence 4.4.4 and 4.4.5 show $(b(D_x)u)(x_0) \leq 0$. But $(a(X, D_x)u)(x_0) = (b(D_x)u)(x_0) \leq 0$ and the assertion is implied by 4.4.4. $\square$

4.4.11 Corollary. Let $a_{jl}, b_j, c \in C(\mathbf{R}^n, \mathbf{R})$ for $j, l = 1, \dots, n$ with $(a_{jl}(x))_{j,l} \geq 0$ and $c(x) \leq 0$ for $x \in \mathbf{R}^n$. Then

$$A := \sum_{j,l=1}^n a_{jl}(\cdot)\partial_j\partial_l + \sum_{j=1}^n b_j(\cdot)\partial_j + c(\cdot)$$

with domain $C_c^\infty(\mathbf{R}^n, \mathbf{R})$ is dispersive and dissipative in $C_0(\mathbf{R}^n, \mathbf{R})$. Moreover, A is also dissipative on $C_c^\infty(\mathbf{R}^n, \mathbb{C})$ in $C_0(\mathbf{R}^n, \mathbb{C})$.

PROOF: The first assertion is a direct consequence of 4.4.9, 4.4.10, and 4.4.4. For the second assertion take $f \in C_c^\infty(\mathbf{R}^n, \mathbb{C})$ with $f \not\equiv 0$ and $x_0 \in \mathbf{R}^n, \theta \in [0, 2\pi]$ with $\|f\|_{L^\infty} = e^{i\theta}f(x_0)$. Then

$$\lambda\|f\|_{L^\infty} = \lambda\mathrm{Re}\,(e^{i\theta}f)(x_0) \leq \lambda\|\mathrm{Re}\,(e^{i\theta}f)\|_{L^\infty} \leq \|(\lambda\mathrm{Id} - A)(\mathrm{Re}\,(e^{i\theta}f))\|_{L^\infty}$$
$$\leq \|\mathrm{Re}\,(e^{i\theta}(\lambda f - Af))\|_{L^\infty} \leq \|\lambda f - Af\|_{L^\infty}.$$

$\square$

4.4.12 Definition. A Feller semigroup on $\mathbf{R}^n$ is a contractive, positivity preserving C_0-semigroup on $C_0(\mathbf{R}^n, \mathbf{R})$. A Feller propagator on an interval I is a propagator $\triangle \ni (t, s) \mapsto U(t, s) \in \mathcal{L}(C_0(\mathbf{R}^n, \mathbf{R}))$ of positivity preserving operators.

4.4.13 Remark. 4.4.5 and 4.4.10 show that generators of Feller semigroups are pseudodifferential operators with non-regular and negative definite symbols if $C_c^\infty(\mathbf{R}^n, \mathbf{R})$ is contained in its domain. Since Feller semigroups play an important role in probability theory (cf. e.g. [143] or [74]), the question whether a given pseudodifferential operator the negative symbol of which is negative definite generates a Feller semigroup is of importance.

4.4.14 Theorem. Let $p \in C(I \times \mathbf{R}^n \times \mathbf{R}^n)$ be with $\{p(t, \cdot, \cdot) : t \in I\} \subset S_{1,0}^2$ and $\{\operatorname{Im} p(t, \cdot, \cdot) : t \in I\} \subset S_{1,0}^1$ boundedly.
If $\mathbf{R}^n \ni \xi \mapsto -p(t, x, \xi)$ is negative definite for any $t \in I, x \in \mathbf{R}^n$, then there is a Feller propagator $U(t, s)$ with $U(t, s)(\mathscr{S}(\mathbf{R}^n, \mathbf{R})) \subset \mathscr{S}(\mathbf{R}^n, \mathbf{R})$ and

$$\frac{\partial}{\partial t} U(t, s)f = p(t, X, D_x)U(t, s)f \quad \text{and} \quad \frac{\partial}{\partial s} U(t, s)f = -U(t, s)p(s, X, D_x)f$$

for $f \in \mathscr{S}(\mathbf{R}^n, \mathbf{R}), t, s \in I, t \geq s$.

PROOF: Due to 4.4.8 and 4.2.17 the Cauchy problem for $p(t, X, D_x)$ in the scale $(H^{k,k'})_k$ is well-posed with exponential growth for any $k' \in \mathbf{Z}$ and the corresponding evolution operator satisfies

$$U(t, s) \in \bigcap_{k, k' \in \mathbb{N}_0} \mathscr{L}(H^{k,k'}[\mathbf{R}]) .$$

Hence $U(t, s)(\mathscr{S}(\mathbf{R}^n, \mathbf{R})) \subset \mathscr{S}(\mathbf{R}^n, \mathbf{R})$ and $[s, t_1] \ni t \mapsto U(t, s)f \in \mathscr{S}(\mathbf{R}^n, \mathbf{R})$ is differentiable for $f \in \mathscr{S}(\mathbf{R}^n, \mathbf{R})$ with derivative $p(t, X, D_x)U(t, s)f$. Therefore 4.4.6, 4.4.4, 4.4.10, and 1.1.10 imply

$$\|U(t, s)f\|_{L^\infty} \leq \|f\|_{L^\infty} \quad \text{and} \quad \left\|(U(t, s)f)^+\right\|_{L^\infty} \leq \left\|f^+\right\|_{L^\infty}$$

for $f \in \mathscr{S}(\mathbf{R}^n, \mathbf{R}), t, s \in I, t \geq s$. Thus 4.4.2 implies that there is a positivity preserving, contractive extension of $U(t, s)$ to $C_0(\mathbf{R}^n, \mathbf{R})$. This extension clearly is a Feller propagator and satisfies the assertions. $\qquad \square$

4.4.15 Corollary. Let $p \in S_{1,0}^2$ with $\operatorname{Im} p \in S_{1,0}^1$ such that $\mathbf{R}^n \ni \xi \mapsto -p(x, \xi)$ is negative definite for any $x \in \mathbf{R}^n$. Then the closure of $p^w(X, D_x)|_{C_c^\infty(\mathbf{R}^n, \mathbf{R})}$ in $C_0(\mathbf{R}^n, \mathbf{R})$ generates a Feller semigroup.

PROOF: This is a consequence of 4.4.14. $\qquad \square$

Finally, we will give a result on scalar, linear hyperbolic first-order evolution equations in the scale $C_0^k(\mathbf{R}^n)$. This will be needed for results on quasilinear, hyperbolic evolution equations in the scale $C_0^k(\mathbf{R}^n)$ that will be given in section 5.6.

4.4.16 Theorem.   Let $K \in \mathbb{N}_0$, $a_j, b \in C(I, \mathcal{B}^{(K+3)}(\mathbf{R}^n))$ with $a_j(t)$ real-valued for $j = 1, \ldots, n$, $t \in I$, and let $A(t) := \sum_{j=1}^n a_j(t)(\cdot)\partial_j + b(t)(\cdot)$.
Then there is a propagator $\triangle \ni (t, s) \mapsto U(t, s) \in \bigcap_{k=0}^{K+1} \mathscr{L}(C_0^k(\mathbf{R}^n))$ satisfying the following properties:

(a) $\triangle \ni (t, s) \mapsto U(t, s) \in \mathscr{L}(C_0^{k+1}(\mathbf{R}^n), C_0^k(\mathbf{R}^n))$ is strongly continuously differentiable for $0 \le k \le K$ with

$$\frac{\partial}{\partial t}U(t, s) = A(t)U(t, s), \qquad \frac{\partial}{\partial s}U(t, s) = -U(t, s)A(s) \quad \text{for } (t, s) \in \triangle.$$

(b) For $0 \le k \le K + 1$ there are $\beta_k \ge 0$ with $\|U(t, s)\|_{\mathscr{L}(C_0^k(\mathbf{R}^n))} \le e^{\beta_k(t-s)}$ for $(t, s) \in \triangle$.

PROOF: Let $X = C_0(\mathbf{R}^n)$ and $D(Z_j) = \{u \in X : \partial_j u \in X \text{ exists}\}$. Then $Z_j = \frac{\partial}{\partial x_j} : D(Z_j) \to X$ define a commuting family $\{Z_1, \ldots, Z_M\}$ of generators of isometric C_0-semigroups in X, and we clearly have $X^k = C_0^k(\mathbf{R}^n)$ for $k \in \mathbb{N}_0$. Let $\rho \in C_c^\infty(\mathbf{R}^n)$ with $\rho(0) = 1$, $0 \le \rho \le 1$, $\int_{\mathbf{R}^n} \rho(x)dx = 1$. Moreover, let $\rho_\varepsilon(x) := \varepsilon^{-n}\rho(\frac{x}{\varepsilon})$ for $\varepsilon > 0$, $x \in \mathbf{R}^n$, and define $a_j^{(\varepsilon)}, b^{(\varepsilon)} \in C(I, \mathcal{B}^\infty(\mathbf{R}^n))$ by $a_j^{(\varepsilon)}(t) := \rho_\varepsilon * a(t)$, $b^{(\varepsilon)}(t) := \rho_\varepsilon * b(t)$. Then

$$I \ni t \mapsto A^{(\varepsilon)}(t) := \sum_{j=1}^n a_j^{(\varepsilon)}(t)(\cdot)\partial_j + b^{(\varepsilon)}(t)(\cdot) \in \bigcap_{k \in \mathbb{N}_0} \mathscr{L}(C_0^{k+1}(\mathbf{R}^n), C_0^k(\mathbf{R}^n))$$

is strongly continuous with $A^{(\varepsilon)}(t) \xrightarrow[\varepsilon \to 0]{} A(t)$ in $\mathscr{L}(C_0^{k+1}(\mathbf{R}^n), C_0^k(\mathbf{R}^n))$ uniformly for $t \in I$ for $0 \le k \le K + 2$, cf. 4.1.23.
For $0 \le k \le K + 2$ there are $\beta_k \ge 0$ (due to 4.4.11 and 2.3.1) with

$$\left\|\lambda f - A^{(\varepsilon)}(t)f\right\|_{C_0^k(\mathbf{R}^n)} \ge (\lambda - \beta_k)\|f\|_{C_0^k(\mathbf{R}^n)} \quad \text{and} \qquad (4.4.2)$$
$$\|\lambda f - A(t)f\|_{C_0^k(\mathbf{R}^n)} \ge (\lambda - \beta_k)\|f\|_{C_0^k(\mathbf{R}^n)}$$

for $f \in \mathscr{S}(\mathbf{R}^n)$, $t \in I$, $\lambda > \beta_k$, $0 \le k \le K + 2$, $\varepsilon > 0$.

Moreover, due to 4.2.17 there are $U^{(\varepsilon)}(t,s) \in \mathscr{L}(\mathscr{S}(\mathbf{R}^n)), (t,s) \in \Delta, \varepsilon > 0$, with

- $U^{(\varepsilon)}(s,s) = \mathrm{Id}$ and $U^{(\varepsilon)}(t,s) = U^{(\varepsilon)}(t,q)U^{(\varepsilon)}(q,s)$ for $t_0 \leq s \leq q \leq t \leq t_1$.

- $\Delta \ni (t,s) \mapsto U^{(\varepsilon)}(t,s) \in \mathscr{L}(\mathscr{S}(\mathbf{R}^n))$ is strongly continuously differentiable with

$$\frac{\partial}{\partial t}U^{(\varepsilon)}(t,s) = A^{(\varepsilon)}(t)U^{(\varepsilon)}(t,s) \qquad \frac{\partial}{\partial s}U^{(\varepsilon)}(t,s) = -U^{(\varepsilon)}(t,s)A^{(\varepsilon)}(s)$$
(4.4.3)

  for $(t,s) \in \Delta$.

Hence, (4.4.2) and 1.1.10 show

$$\left\| U^{(\varepsilon)}(t,s)f \right\|_{C_0^k(\mathbf{R}^n)} \leq e^{\beta_k(t-s)} \|f\|_{C_0^k(\mathbf{R}^n)}$$
(4.4.4)

for $f \in \mathscr{S}(\mathbf{R}^n), \varepsilon > 0, (t,s) \in \Delta, 0 \leq k \leq K+2$. In particular, there are extensions $U^{(\varepsilon)}(t,s) \in \bigcap_{k=0}^{K+2} \mathscr{L}(C_0^k(\mathbf{R}^n))$ to propagators satisfying (4.4.4). Clearly, $\Delta \ni (t,s) \mapsto U^{(\varepsilon)}(t,s) \in \mathscr{L}(C_0^{k+1}(\mathbf{R}^n), C_0^k(\mathbf{R}^n))$ is strongly continuously differentiable for $0 \leq k \leq K+1$ with strong derivatives

$$\frac{\partial}{\partial t}U^{(\varepsilon)}(t,s) = A^{(\varepsilon)}(t)U^{(\varepsilon)}(t,s), \qquad \frac{\partial}{\partial s}U^{(\varepsilon)}(t,s) = -U^{(\varepsilon)}(t,s)A^{(\varepsilon)}(s)$$

for $(t,s) \in \Delta$. Hence 2.5.1 implies the assertion with $m = m' = 1$.   □

4.4.17 Remark. For $n \geq 2$ there seems to exist no natural isomorphism $\Lambda : C_0^{k+1}(\mathbf{R}^n) \to C_0^k(\mathbf{R}^n)$ (cf. [87]). Moreover, the norms

$$\|u\|_{L^\infty} + \|\Delta u\|_{L^\infty} \qquad \text{and} \qquad \sum_{|\alpha| \leq 2} \|\partial^\alpha u\|_{L^\infty}$$

are not equivalent (cf. [133, I.5.7]). This shows the advantage of considering scales that are generated by a family of operators (not only by one operator) in chapter 1.

4.4.18 Bibliographical remarks. For a detailed introduction to Banach lattices we refer to the monographs of Nagel et al. [110] and Clément/Heijmans et al. [28]. Proposition 4.4.3 is due to Arendt/Chernoff/Kato [10], cf. also [110, A-II, 2.3, 2.7]. For more information on dispersive operators we refer to Phillips [125], where also 4.4.4 can be found. A general reference on negative definite functions containing much more than 4.4.8 and 4.4.9 is the book of Berg/Forst [15]. Theorem 4.4.10 is due to Courrège [33]. In the recent years several results concerning the question whether a given pseudodifferential operator generates a Feller semigroup on $\mathbf{R}^n$ for various types of regular and irregular symbols, that additionally were supposed to satisfy certain ellipticity assumptions, have been given e.g. by Jacob [74], Hoh [64], or Baldus [12] and the references given therein. Taira [143] studied similar questions in a different context. Our result on Feller semigroups 4.4.15 contains a result in this direction assuming regularity but no ellipticity.

4.5 Evolution equations in scales of L^q-Sobolev spaces

In this section we will give several applications to differential and pseudodifferential evolution equations in scales of L^q-Sobolev spaces. We will use the following embedding and interpolation results.

4.5.1 Theorem. Let $2 < q < \infty$ and $r > 0$. Then

(a) $H_2^{r'} \hookrightarrow L^q(\mathbf{R}^n)$ densely and continuously for a suitable $r' > 0$.

(b) $[L^2(\mathbf{R}^n), \overline{\mathscr{S}(\mathbf{R}^n)}^{L^\infty(\mathbf{R}^n)}]_{1-\frac{2}{q}} = L^q(\mathbf{R}^n)$ with the complex interpolation functor $[\cdot, \cdot]_\theta$.

(c) $[H_2^{r'}, L^{q'}(\mathbf{R}^n)]_\theta = H_q^r$ with equivalent norms for $r' := [\frac{rq}{2}] + 1 \in \mathbb{N}$ and $\theta := 1 - \frac{r}{r'} \in (0,1)$, where $q' := \frac{2q\theta}{2-q(1-\theta)} \geq 2$ and $[t] := \inf\{t \in \mathbf{Z} : t \geq r\}$ for $t \in \mathbf{R}$.

PROOF: (a) and (b) can be found e.g. in [16, 6.5.1, 6.4.5] or in [127, IX.17]. [16, 6.4.5] shows that $[H_{p_0}^{r_0}, H_{p_1}^{r_1}]_\theta = H_{p_2}^{r_2}$ for $r_0, r_1, r_2 \geq 0$, $p_0, p_1, p_2 \geq 2$, $0 < \theta < 1$ with $r_2 = (1-\theta)r_0 + \theta r_1$ and $\frac{1}{p_2} = \frac{1-\theta}{p_0} + \frac{\theta}{p_1}$. It is simple to check that $r' \in \mathbf{N}, \theta \in (0,1)$, and $q' \in [2,\infty)$ satisfy $r = (1-\theta)r' + \theta 0$ and $\frac{1}{q} = \frac{1-\theta}{2} + \frac{\theta}{q'}$. $\square$

Degenerate-elliptic second-order differential operators are always quasi-dissipative in L^q for $q \geq 2$.

4.5.2 Proposition. For $q \geq 2$ and $R > 0$ there is a $\beta_{q,R} \geq 0$ with

$$\left\| \lambda u - \left(\sum_{j,l=1}^{n} a_{jl}(\cdot)\partial_j\partial_l u + \sum_{j=1}^{n} b_j(\cdot)\partial_j u + c(\cdot)u \right) \right\|_{L^q(\mathbf{R}^n)} \geq (\lambda - \beta_{q,R}) \|u\|_{L^q(\mathbf{R}^n)}$$

for $u \in \mathscr{S}(\mathbf{R}^n)$, $\lambda > \beta_{q,R}$, and $a_{jl} \in C^2(\mathbf{R}^n, \mathbf{R})$, $b_j \in C^1(\mathbf{R}^n, \mathbf{R})$, $c \in C(\mathbf{R}^n, \mathbf{R})$ with $(a_{jl}(x))_{j,l} \geq 0$ in $M_n(\mathbf{R})$ and $\|\partial^\alpha a_{jl}\|_{L^\infty(\mathbf{R}^n)} \leq R$, $\|\partial^\beta b_j\|_{L^\infty(\mathbf{R}^n)} \leq R$, $\|c\|_{L^\infty(\mathbf{R}^n)} \leq R$ for $|\alpha| \leq 2$, $|\beta| \leq 1$, $x \in \mathbf{R}^n$.

PROOF:
Step 1: For $R > 0$ there are $\beta_{2,R} \geq 0$ with

$$\mathrm{Re}\left\langle \sum_{j,l=1}^{n} a_{jl}(\cdot)\partial_j\partial_l u + \sum_{j=1}^{n} b_j(\cdot)\partial_j u, u \right\rangle_{L^2} \leq \beta_{2,R} \|u\|_{L^2}^2$$

for $u \in \mathscr{S}(\mathbf{R}^n)$ and $a_{jl}, b_j \in C^\infty(\mathbf{R}^n, \mathbf{R})$ with $(a_{jl}(x))_{j,l} \geq 0$ in $M_n(\mathbf{R})$ and $\|\partial^\alpha a_{jl}\|_{L^\infty} \leq R$, $\|\partial^\beta b_j\|_{L^\infty} \leq R$ for $|\alpha| \leq 2, |\beta| \leq 1$.
Proof:

$$\mathrm{Re}\left\langle \sum_{j,l=1}^{n} a_{jl}(\cdot)\partial_j\partial_l u + \sum_{j=1}^{n} b_j(\cdot)\partial_j u, u \right\rangle_{L^2}$$

$$= -\sum_{j,l=1}^{n} \langle a_{jl}(\cdot)\partial_l u, \partial_j u\rangle_{L^2} + \sum_{j=1}^{n} \mathrm{Re}\left\langle \left(b_j(\cdot) - \sum_{l=1}^{n}(\partial_l a_{lj})(\cdot) \right) \partial_j u, u \right\rangle_{L^2}$$

$$\leq 0 + \frac{1}{2}\sum_{j=1}^{n}\left(\left\langle \left(b_j(\cdot) - \sum_{l=1}^{n}(\partial_l a_{lj})(\cdot) \right) \partial_j u, u \right\rangle_{L^2} \right.$$

$$\left. -\left\langle \partial_j \left(b_j(\cdot) - \sum_{l=1}^{n}(\partial_l a_{lj})(\cdot) \right) u, u \right\rangle_{L^2} \right)$$

$$= -\frac{1}{2}\sum_{j=1}^{n}\left\langle \left(\partial_j b_j(\cdot) - \sum_{l=1}^{n}(\partial_j\partial_l a_{lj})(\cdot) \right) u, u \right\rangle_{L^2} \leq cR \|u\|_{L^2}^2$$

for a suitable $c > 0$ and $\|\partial^\alpha a_{jl}\|_{L^\infty} \leq R$, $\|\partial^\beta b_j\|_{L^\infty} \leq R$ for $|\alpha| \leq 2, |\beta| \leq 1$.

Step 2: Proof of the assertion for $c = 0$: Let

$$P := \sum_{j,l=1}^{n} a_{jl}(\cdot)\partial_j\partial_l + \sum_{j=1}^{n} b_j(\cdot)\partial_j$$

and let $\rho \in C_c^\infty(\mathbf{R}^n, \mathbb{R})$ with $0 \le \rho \le 1, \rho(0) = 1, \int_{\mathbf{R}^n} \rho(x)dx = 1$.
Moreover, let $a_{jl}^{(\varepsilon)} := \rho_\varepsilon * a_{jl}, b_j^{(\varepsilon)} := \rho_\varepsilon * b_j$, and

$$P^{(\varepsilon)} := \sum_{j,l=1}^{n} a_{jl}^{(\varepsilon)}(\cdot)\partial_j\partial_l + \sum_{j=1}^{n} b_j^{(\varepsilon)}(\cdot)\partial_j.$$

Since $(a_{jl}^{(\varepsilon)}(x))_{j,l} \ge 0$, 4.1.23 and 4.2.17 show that there exist operators
$T^{(\varepsilon)}(t) \in \mathscr{L}(\mathscr{S}(\mathbf{R}^n)), t \ge 0$, with

- $T^{(\varepsilon)}(0) = \mathrm{Id}, T^{(\varepsilon)}(t)T^{(\varepsilon)}(s) = T^{(\varepsilon)}(s+t)$ for $t, s \ge 0$.

- $[0, \infty) \ni t \mapsto T^{(\varepsilon)}(t)f \in \mathscr{S}(\mathbf{R}^n)$ is strongly continuously differentiable for $f \in \mathscr{S}(\mathbf{R}^n)$ with $\frac{d}{dt}T^{(\varepsilon)}(t)f = P^{(\varepsilon)}T^{(\varepsilon)}(t)f = T^{(\varepsilon)}(t)P^{(\varepsilon)}f$.

Now, due to 4.1.23 $\left\|\partial^\alpha a_{jl}^{(\varepsilon)}\right\|_{L^\infty} \le R$ and $\left\|\partial^\beta b_j^{(\varepsilon)}\right\|_{L^\infty} \le R$ for $0 < \varepsilon < 1$,
$|\alpha| \le 2, |\beta| \le 1$, and step 1 shows $\mathrm{Re}\left\langle P^{(\varepsilon)}u, u\right\rangle_{L^2} \le \beta_{2,R}\|u\|_{L^2}^2$ for $0 < \varepsilon < 1$
and $u \in \mathscr{S}(\mathbf{R}^n)$. Moreover, $P^{(\varepsilon)}$ is L^∞- dissipative due to 4.4.11. Hence 1.1.10
shows $\left\|T^{(\varepsilon)}(t)u\right\|_{L^2} \le e^{\beta_{2,R}t}\|u\|_{L^2}$ and $\left\|T^{(\varepsilon)}(t)u\right\|_{L^\infty} \le \|u\|_{L^\infty}$ for $0 < \varepsilon < 1$ and
$u \in \mathscr{S}(\mathbf{R}^n)$. Therefore, 4.5.1 implies $\left\|T^{(\varepsilon)}(t)u\right\|_{L^q} \le e^{\frac{2\beta_{2,R}}{q}t}\|u\|_{L^q}$ for $0 < \varepsilon < 1$
and $u \in \mathscr{S}(\mathbf{R}^n)$ and thus 1.1.11 shows

$$\left\|\lambda u - P^{(\varepsilon)}u\right\|_{L^q} \ge \left(\lambda - \frac{2\beta_{2,R}}{q}\right)\|u\|_{L^q}$$

for $\lambda > \frac{2\beta_{2,R}}{q}$ and we obtain the assertion with $\varepsilon \to 0$ due to 4.1.23.
Step 3: Proof of the assertion: Due to 1.1.7 and step 2 we only have to show

$$\|\lambda u - c(\cdot)u\|_{L^q} \ge (\lambda - \gamma_{q,R})\|u\|_{L^q}$$

for $R > 0, \lambda > \gamma_{q,R}, u \in \mathscr{S}(\mathbf{R}^n), c \in C(\mathbf{R}^n, \mathbb{R})$ with $\|c\|_{L^\infty(\mathbf{R}^n)} \le R$ for suitable
constants $\gamma_{q,R}$.
Proof: Let $u \in \mathscr{S}(\mathbf{R}^n), u^* := \bar{u}|u|^{q-2}\|u\|_{L^q}^{2-q}$, and $1 < r < \infty$ with $\frac{1}{q} + \frac{1}{r} = 1$.
Then

$$\|u^*\|_{L^r} = \left(\int_{\mathbf{R}^n}\left|\overline{u(x)}|u(x)|^{q-2}\|u\|_{L^q}^{2-q}\right|^r dx\right)^{\frac{1}{r}} = \|u\|_{L^q}^{2-q}\left(\int_{\mathbf{R}^n}|u(x)|^{(q-1)r}dx\right)^{\frac{1}{r}}$$

$$= \|u\|_{L^q}^{2-q}\|u\|_{L^q}^{\frac{q}{r}} = \|u\|_{L^q}^{2-q}\|u\|_{L^q}^{q-1} = \|u\|_{L^q}$$

and $\int_{\mathbf{R}^n} u^*(x)u(x)dx = \int_{\mathbf{R}^n} \overline{u(x)}|u(x)|^{q-2} \|u\|_{L^q}^{2-q} u(x)dx = \|u\|_{L^q}^2$. Thus the linear functional $\Phi(v) := \int_{\mathbf{R}^n} u^*(x)v(x)dx$, $v \in L^q(\mathbf{R}^n)$, satisfies $\Phi \in L^q(\mathbf{R}^n)^*$ with $\Phi(u) = \|\Phi\|_{L^q(\mathbf{R}^n)^*}^2 = \|u\|_{L^q(\mathbf{R}^n)}^2$. Since $\Phi((c(\cdot) - R)u) \leq 0$ step 3 is implied by 1.1.6.

□

4.5.3 Remark. Langer and Maz'ya [100] have shown that a L^q-quasi-dissipative differential operator (or even system) with $C_c^\infty(\mathbf{R}^n)$ in its domain for $q \neq 2$ necessarily has order $0, 1,$ or 2.

The next result gives a condition sufficient for L^q-quasi-dissipativity of pseudodifferential operators. Note that this result includes 4.5.2, if the coefficients of P are in $B^\infty(\mathbf{R}^n)$.

4.5.4 Proposition. Let $\mathcal{P} \subset S_{1,0}^2$ boundedly with $\{\operatorname{Im} p : p \in \mathcal{P}\} \subset S_{1,0}^1$ boundedly. Moreover, assume that $\mathbf{R}^n \ni \xi \mapsto -p(x,\xi) \in \mathbf{R}^n$ is negative definite for any $x \in \mathbf{R}^n$ and $p \in \mathcal{P}$. Then there are $\beta_q \geq 0, 2 \leq q < \infty$, with $\|\lambda u - Pu\|_{L^q(\mathbf{R}^n)} \geq (\lambda - \beta_q)\|u\|_{L^q(\mathbf{R}^n)}$ for $u \in \mathscr{S}(\mathbf{R}^n)$.

PROOF: The Cauchy problem for $p(X, D_x) \in \mathscr{L}(H^{k+2}, H^k)$ in the scale $(H^k)_k$ is well-posed with exponential growth (due to 4.4.8 and 4.2.17) and there is an associate semigroup $T(t) \in \mathscr{L}(L^2(\mathbf{R}^n))$ with $\|T(t)\|_{\mathscr{L}(L^2(\mathbf{R}^n))} \leq e^{\beta t}$ and $T(t)(\mathscr{S}(\mathbf{R}^n)) \subset \mathscr{S}(\mathbf{R}^n)$ for $t \geq 0$. Hence 4.4.4, 4.4.10, and 1.1.10 imply the estimates $\|T(t)f\|_{L^\infty(\mathbf{R}^n)} \leq \|f\|_{L^\infty}$ for every $f \in \mathscr{S}(\mathbf{R}^n)$ and $t \geq 0$. Since $\mathscr{S}(\mathbf{R}^n) \subset L^2(\mathbf{R}^n) \cap L^\infty(\mathbf{R}^n)$ densely, 4.5.1 implies $\|T(t)f\|_{L^q(\mathbf{R}^n)} \leq e^{\beta_q t}\|f\|_{L^q(\mathbf{R}^n)}$ for $t \geq 0$ and suitable constants $\beta_q \geq 0, 2 \leq q < \infty$. Hence the assertion is a consequence of 1.1.11.

□

Now, we obtain a result for L^q-well-posedness for a class of pseudodifferential operators.

4.5.5 Theorem.   Let $p \in C(I \times \mathbf{R}^n \times \mathbf{R}^n)$ with $\{p(t, \cdot, \cdot) : t \in I\} \subset S_{1,0}^2$ boundedly and $\{\operatorname{Im} p(t, \cdot, \cdot) : t \in I\} \subset S_{1,0}^1$ boundedly.
If $\mathbf{R}^n \ni \xi \mapsto -p(t, x, \xi) \in \mathbf{C}$ is negative definite for any $t \in I$ and $x \in \mathbf{R}^n$, then there is a propagator $\triangle \ni (t, s) \mapsto U(t, s) \in \bigcap_{r \geq 0, q \geq 2} \mathscr{L}(H_q^r)$ satisfying the following conditions:

(a) $\triangle \ni (t, s) \mapsto U(t, s) \in \mathscr{L}(H_q^{r+2}, H_q^r)$ is strongly continuously differentiable for $r \geq 0, q \geq 2$ with strong derivatives

$$\frac{\partial}{\partial t}U(t, s) = p(t, X, D_x)U(t, s), \quad \frac{\partial}{\partial s}U(t, s) = -U(t, s)p(s, X, D_x)$$

for $(t, s) \in \triangle$.

(b) For every $r \geq 0, q \geq 2$ there are constants $\beta_{r,q}, M_{r,q} \geq 0$ with
$\|U(t,s)\|_{\mathscr{L}(H_q^r)} \leq M_{r,q} e^{\beta_{r,q}(t-s)}$ for $(t,s) \in \Delta$.

PROOF: Let $A(t) := p(t, X, D_x)$, then, due to 4.5.4, for $2 \leq q < \infty$ there are $\beta_{0,q} \geq 0$ with

$$\|(\lambda \mathrm{Id} - A(t))u\|_{L^q} \geq (\lambda - \beta_{0,q}) \|u\|_{L^q} \qquad \text{for } u \in H_q^2, t \in I, \lambda > \beta_{0,q}. \qquad (4.5.1)$$

Moreover, 4.2.17 implies that the Cauchy problem for $A(t) \in \mathscr{L}(H_2^{k+2}, H_2^k)$ in the scale $(H_2^k)_k$ is well-posed with exponential growth and the associated evolution operator satisfies (by interpolation)

$$U(t,s) \in \bigcap_{r \geq 0} \mathscr{L}(H_2^r) \quad, (t,s) \in \Delta$$

with $\|U(t,s)\|_{\mathscr{L}(H_2^r)} \leq e^{\beta_{r,2}(t-s)}$ for $(t,s) \in \Delta$ and constants $\beta_{r,2} \geq 0$ for $r \geq 0$. Let $u_0 \in H_2^\infty$ and $u(t) := U(t,s)u_0$ for $t \in [s,t_1]$. Then $u \in C^1([s,t_1], H_2^\infty)$ with $u'(t) = A(t)u(t)$. Due to 4.5.1 this implies $u \in C^1([s,t_1], L^q(\mathbf{R}^n))$ with derivative $u'(t) = A(t)u(t)$ for any $2 \leq q < \infty$. Hence 1.1.10 and (4.5.1) imply $\|U(t,s)u_0\|_{L^q} = \|u(t)\|_{L^q} \leq e^{\beta_{0,q}(t-s)} \|u_0\|_{L^q}$ and thus by density we find unique extensions of $U(t,s)$ with

$$U(t,s) \in \bigcap_{2 \leq q < \infty} \mathscr{L}(L^q(\mathbf{R}^n))$$

with $\|U(t,s)\|_{\mathscr{L}(L^q(\mathbf{R}^n))} \leq e^{\beta_{0,q}(t-s)}$ for $(t,s) \in \Delta$. Therefore 4.5.1 shows

$$U(t,s) \in \bigcap_{\substack{r \geq 0 \\ 2 \leq q < \infty}} \mathscr{L}(H_q^r)$$

with $\|U(t,s)\|_{\mathscr{L}(H_q^r)} \leq M_{r,q} e^{\beta_{r,q}(t-s)}$ for $(t,s) \in \Delta$ and suitable constants $M_{r,q}, \beta_{r,q}, r \geq 0, 2 \leq q < \infty$ and $U(t,s)$ clearly is the evolution operator for $A(t) \in \mathscr{L}(H_q^{r+2}, H_q^r)$. This implies the assertion.   □

In particular, this result contains degenerate diffusion equations in L^q-spaces.

4.5.6 Corollary. Let $2 \leq q < \infty$ and let $a_{jl}, b_j, c \in C(I, \mathcal{B}^\infty(\mathbf{R}^n))$ be with $a_{jl}(t) = a_{lj}(t)$, $b_j(t)$ real-valued for $j, l = 1, \ldots, n$, and

$$\sum_{j,l=1}^n a_{jl}(t)(x)\xi_j \overline{\xi_l} \geq 0 \quad \text{for } t \in I, x \in \mathbf{R}^n, (\xi_1, \ldots, \xi_n) \in \mathbf{C}^n.$$

Then, for $u_0 \in H_q^{r+2}, r \geq 0$, and $f \in C(I, H_q^r) \cap L^1(I, H_q^{r+2})$ there is a unique $u \in C(I, H_q^{r+2}) \cap C^1(I, H_q^r)$ with

$$\frac{du}{dt}(t) = \sum_{j,l=1}^{n} a_{jl}(t)\partial_j\partial_l u(t) + \sum_{j=1}^{n} b_j(t)\partial_j u(t) + c(t)u(t) + f(t), t \in I, \quad u(t_0) = u_0.$$

PROOF: This is a consequence of 4.4.9 and 4.5.5.   □

Finally, we also want to consider operators with spatially growing coefficients. Here we have the following.

4.5.7 Theorem. Let $1 < q < \infty$ and suppose that

(a) $\{a(t, \cdot) : t \in I\} \subset \psi c^{(2,0)}$ is a bounded family of symbols depending only on ξ such that $I \ni t \mapsto a(t, \xi) \in \mathbb{C}$ is continuous for $\xi \in \mathbf{R}^n$.

(b) $\{b(t, \cdot, \cdot) : t \in I\} \subset \psi c^{(1,1)}$ boundedly such that $I \ni t \mapsto b(t, x, \xi) \in \mathbb{C}$ is continuous for $x, \xi \in \mathbf{R}^n$.

(c) $\{c(t, \cdot) : t \in I\} \subset \psi c^{(0,2)}$ is a bounded family of symbols depending only on x such that $I \ni t \mapsto c(t, x) \in \mathbb{C}$ is continuous for $x \in \mathbf{R}^n$.

(d) $A(t) := a(t, D_x) + b(t, X, D_x) + c(t, X)$.

Then the Cauchy problem for $A(t) \in \mathscr{L}(\mathcal{H}_q^{2(k+1)}, \mathcal{H}_q^{2k})$ in the scale $(\mathcal{H}_q^{2k})_k$ is well-posed with exponential growth if and only if there is a constant $\beta \geq 0$ with

$$\|\lambda u - A(t)u\|_{L^q(\mathbf{R}^n)} \geq (\lambda - \beta)\|u\|_{L^q(\mathbf{R}^n)} \quad \text{for } \lambda > \beta, u \in \mathscr{S}(\mathbf{R}^n), t \in I.$$

Moreover, in this case the closures of $A(t)|_{\mathscr{S}(\mathbf{R}^n)}$ in $\mathcal{H}_q^{2k}$ generate quasi-contractive C_0-semigroups in $\mathcal{H}_q^{2k}$ for any fixed $t \in I$ and $k \in \mathbf{N}_0$.

PROOF: For $0 \leq j \leq 2k$ we have due to 4.1.21 with suitable constants $c, c' \geq 0$

$$\|A(t)u\|_{q,j,2k-j}$$
$$\leq \|a(t, D_x)u\|_{q,j,2k-j} + \|b(t, X, D_x)u\|_{q,j,2k-j} + \|c(t, X)u\|_{q,j,2k-j}$$
$$\leq c\|u\|_{q,j+2,2k-j} + c\|u\|_{q,j+1,2k-j+1} + c\|u\|_{q,j,2k-j+2} \leq c' \sum_{j=0}^{2k+2} \|u\|_{q,j,2k+2-j} .$$

This shows that $A(t) \in \mathscr{L}(\mathcal{H}_q^{2(k+1)}, \mathcal{H}_q^{2k})$ for $k \in \mathbf{N}_0$. Moreover

$$\begin{aligned}
[A(t), \Delta - x^2] &= [a(t, D_x), \Delta - x^2] + [b(t, X, D_x), \Delta - x^2] + [c(t, X), \Delta - x^2] \\
&= -[a(t, D_x), x^2] + [b(t, X, D_x), \Delta - x^2] + [c(t, X), \Delta] \\
&\in \mathrm{Op}\psi c^{(2,0)} + \mathrm{Op}\psi c^{(1,1)} + \mathrm{Op}\psi c^{(0,2)} ,
\end{aligned}$$

hence we obtain $[A(t), \Delta - x^2] \in \mathcal{L}(\mathcal{H}_q^{2(k+1)}, \mathcal{H}_q^{2k})$ for $k \in \mathbb{N}_0$ as before. Therefore 2.4.3 can be applied due to 4.3.3. $\qquad\square$

4.5.8 Bibliographical remarks. For a proof that degenerate- elliptic second-order differential operators are dissipative and dispersive we refer e.g. to Pazy [122, 7.3.6] or Wong-Dzung [158]. Our proof avoids many calculations and shows the statement rather quickly using the results obtained so far and interpolation theory. In [158] a statement similar to 4.5.6 for $r = -1$ is stated. The remaining cases, in particular the existence of L^q-solutions are left as a conjecture.

4.6 An application to a degenerate-elliptic boundary value problem

In this section we will apply the results of chapter 1 to a degenerate-elliptic boundary value problem that is non-characteristic near the boundary (cf. 4.6.7). We always assume that $\Omega \subset \mathbb{R}^n$ is a bounded open set with smooth boundary $\partial\Omega$. With $C^k(\overline{\Omega})$ for $0 \le k \le \infty$ we denote the functions of $C^k(\Omega)$ such that all derivatives have continuous extensions to $\overline{\Omega}$. Let

$$\|u\|_{W^k(\Omega)} := \left(\sum_{|\alpha| \le k} \int_\Omega |\partial^\alpha u(x)|^2 dx \right)^{1/2} \qquad \text{for } u \in C^k(\overline{\Omega}).$$

Then $W^k(\Omega)$ (resp., $W_0^k(\Omega)$) is defined as the closure of $C^\infty(\overline{\Omega})$ (resp., $C_c^\infty(\Omega)$) with respect to $\|\cdot\|_{W^k(\Omega)}$. $W^k(\Omega)$ consists of all $u \in L^2(\Omega)$ such that all weak derivatives $\partial^\alpha u$ for $|\alpha| \le k$ are contained in $L^2(\Omega)$ (cf. [4, theorem 2.2]). $W^k(\Omega)$ is called the Sobolev space of order k on Ω. Moreover, we denote by $\mathcal{D}(\Omega)$ the test functions, i.e., the smooth functions with compact support in Ω, by $\mathcal{D}'(\Omega)$ the Schwartz distributions on Ω, and by $(\cdot, \cdot)_{\mathcal{D}'(\Omega),\mathcal{D}(\Omega)}$ the natural pairing on $\mathcal{D}'(\Omega) \times \mathcal{D}(\Omega)$ (cf. [134, chapter 6]).

4.6.1 Definition. A matrix of C^∞-functions $(b_{jl})_{j,l}$ with $b_{jl} \in C^\infty(\overline{\Omega})$ is called positive definite, if $b_{jl} \in C^\infty(\overline{\Omega})$ with $b_{jl}(x) = b_{lj}(x)$ for $j, l = 1, \ldots, n$ and if there is a $c > 0$ with $\sum_{j,l=1}^n b_{jl}(x)\xi_j\overline{\xi_l} \ge c|\xi|^2$ for $x \in \Omega$, $\xi = (\xi_1, \ldots, \xi_n) \in \mathbb{C}^n$. The matrix is called positive semi-definite, if the same statement holds with $c = 0$.

We will use the following elliptic regularity result.

4.6.2 Theorem.   Let $b_{jl} \in C^\infty(\overline{\Omega})$ for $j, l = 1, \ldots, n$ such that $(b_{jl})_{j,l}$ is positive definite. Moreover, let $b_j, b \in C^\infty(\overline{\Omega})$ for $j = 1, \ldots, n$, $u \in W_0^1(\Omega)$, and $f \in W^k(\Omega)$ for a $k \in \mathbb{N}_0$.

(a) If $\displaystyle\sum_{j,l=1}^{n} (b_{jl}\partial_l u, \partial_j v)_{\mathcal{D}'(\Omega), \mathcal{D}(\Omega)} + \sum_{j=1}^{n} (b_j \partial_j u + bu, v)_{\mathcal{D}'(\Omega), \mathcal{D}(\Omega)} = (f, v)_{\mathcal{D}'(\Omega), \mathcal{D}(\Omega)}$

    for $v \in \mathcal{D}(\Omega)$, then $u \in W^{k+2}(\Omega)$ and there is a constant γ independent of u, f with $\|u\|_{W^{k+2}(\Omega)} \leq \gamma(\|f\|_{W^k(\Omega)} + \|u\|_{W^0(\Omega)})$.

(b) If $\displaystyle Bu := \sum_{j,l=1}^{n} \partial_j(b_{jl}\partial_l u) + \sum_{j=1}^{n} b_j \partial_j u + bu = f$ in the sense of Schwartz dis-

    tributions, then $u \in W^{k+2}(\Omega)$ and there is a constant γ independent of u, f with $\|u\|_{W^{k+2}(\Omega)} \leq \gamma(\|f\|_{W^k(\Omega)} + \|u\|_{W^0(\Omega)})$.

PROOF: (a) is a consequence of [4, 9.8], (b) is a corollary of (a).   □

Next, we can use elliptic operators B to generate a scale of Hilbert spaces.

4.6.3 Theorem.   Let $b_{jl} \in C^\infty(\overline{\Omega}, \mathbb{R})$ for $j, l = 1, \ldots, n$ such that $(b_{jl})_{j,l}$ is positive definite, and let $\displaystyle Bu := \sum_{j,l=1}^{n} \partial_j(b_{jl}\partial_l u)$ for $u \in D(B) := W^2(\Omega) \cap W_0^1(\Omega)$. Then the following statements hold:

(a) $c\mathrm{Id} - B \cdot D(B) \to L^2(\Omega)$ is strictly positive and self-adjoint with domain $D((c\mathrm{Id} - B)^{1/2}) = W_0^1(\Omega)$.
    Let $Z := (c\mathrm{Id} - B)^{1/2}$ with domain $D(Z) := D((c\mathrm{Id} - B)^{1/2})$.

(b) $D(Z^{2k-1}) = \{u \in W^{2k-1}(\Omega) : u, Bu, \ldots, B^{k-1}u \in W_0^1(\Omega)\}$ and
    $D(Z^{2k}) = \{u \in W^{2k}(\Omega) : u, Bu, \ldots, B^{k-1}u \in W_0^1(\Omega)\}$ for $k \in \mathbb{N}$.

(c) There are $c_k, C_k \geq 0$ with $c_k \|u\|_{W^k(\Omega)} \leq \|u\|_{D(Z^k)} \leq C_k \|u\|_{W^k(\Omega)}$ for $u \in D(Z^k), k \in \mathbb{N}_0$.

(d) $B \in \mathscr{L}(D(Z^{k+2}), D(Z^k))$ for $k \in \mathbb{N}_0$.

PROOF: Let $V := W_0^1(\Omega)$ and let $a(u, v) := \sum_{j,l=1}^{n} \langle b_{jl} \partial_l u, \partial_j v \rangle_{L^2(\Omega)} + c \langle u, v \rangle_{L^2(\Omega)}$

for $u, v \in V$. Then $a : V \times V \to \mathbb{C}$ is a sesquilinear form with

- $\overline{a(v, u)} = a(u, v)$ for $u, v \in V$ because

$$\sum_{j,l=1}^{n} \overline{\langle b_{jl} \partial_l v, \partial_j u \rangle_{L^2(\Omega)}} = \sum_{j,l=1}^{n} \langle \partial_l u, b_{lj} \partial_j v \rangle_{L^2(\Omega)} = \sum_{j,l=1}^{n} \langle b_{jl} \partial_l u, \partial_j v \rangle_{L^2(\Omega)} .$$

- $|a(u, v)| \leq M \|u\|_{W^1(\Omega)} \|v\|_{W^1(\Omega)}$ for $u, v \in V$.

- $a(u, u) \geq c \|u\|_{W^1(\Omega)}^2$ for $u \in V$ because

$$\int_{\Omega} \left(\sum_{j,l=1}^{n} b_{jl}(x) \, (\partial_l u(x)) \, \overline{\partial_j u(x)} \right) dx + c \|u\|_{L^2(\Omega)}^2$$

$$\geq \int_{\Omega} c \left(\sum_{j=1}^{n} |\partial_j u(x)|^2 \right) dx + c \|u\|_{L^2(\Omega)}^2 .$$

Hence, due to 1.2.10 and 1.4.10, the operator $A : D(A) \to L^2(\Omega)$ defined by

$$D(A) := \{ u \in W_0^1(\Omega) : \exists f_u \in L^2(\Omega) \; \forall v \in W_0^1(\Omega) : \langle f_u, v \rangle_{L^2(\Omega)} = a(u, v) \}$$

and $Au := f_u$ for $u \in D(A)$ is strictly positive and selfadjoint with domain $D(A^{1/2}) = W_0^1(\Omega)$. Let $u \in D(A) \subset W_0^1(\Omega)$. For $v \in \mathcal{D}(\Omega) \subset W_0^1(\Omega)$ we have

$$(Au, v)_{\mathcal{D}'(\Omega), \mathcal{D}(\Omega)} = \langle Au, \overline{v} \rangle_{L^2(\Omega)} = \sum_{j,l=1}^{n} \langle b_{jl} \partial_l u, \partial_j \overline{v} \rangle_{L^2(\Omega)} + c \langle u, \overline{v} \rangle_{L^2(\Omega)}$$

$$= \sum_{j,l=1}^{n} (b_{jl} \partial_l u, \partial_j v)_{\mathcal{D}'(\Omega), \mathcal{D}(\Omega)} + c(u, v)_{\mathcal{D}'(\Omega), \mathcal{D}(\Omega)} .$$

Therefore 4.6.2 implies $u \in W^2(\Omega)$ and $D(A) \subset W^2(\Omega) \cap W_0^1(\Omega)$. Conversely, for $u \in W^2(\Omega) \cap W_0^1(\Omega)$ and $v \in \mathcal{D}(\Omega)$ we have

$$\langle (c\mathrm{Id} - B)u, v \rangle_{L^2(\Omega)} = c \langle u, v \rangle_{L^2(\Omega)} - (Bu, \overline{v})_{\mathcal{D}'(\Omega), \mathcal{D}(\Omega)}$$

$$= c \langle u, v \rangle_{L^2(\Omega)} + \sum_{j,l=1}^{n} (b_{jl} \partial_l u, \partial_j \overline{v})_{\mathcal{D}'(\Omega), \mathcal{D}(\Omega)}$$

$$= c \langle u, v \rangle_{L^2(\Omega)} + \sum_{j,l=1}^{n} \langle b_{jl} \partial_l u, \partial_j v \rangle_{L^2(\Omega)} = a(u, v) .$$

Since $\mathcal{D}(\Omega) \subset W_0^1(\Omega)$ densely we obtain $\langle (c\mathrm{Id} - B)u, v \rangle_{L^2(\Omega)} = a(u, v)$ for $v \in W_0^1(\Omega)$. Hence $u \in D(A)$ with $Au = (c\mathrm{Id} - B)u$. Thus we have proved $A = c\mathrm{Id} - B$. This implies (a).

(b) and (c) for $k = 1$ are a consequence of 4.6.2. Now assume (b) and (c) inductively for a $k \in \mathbb{N}$. Then, with 4.6.2

$$
\begin{aligned}
u \in D(Z^{2k+1}) &\iff u \in D(Z^{2k}), Zu \in D(Z^{2k}) \\
&\iff u \in D(Z^{2k}), Zu \in D(Z^{2k-1}), Z^2u \in D(Z^{2k-1}) \\
&\iff u \in D(Z^{2k}), (c\mathrm{Id} - B)u \in D(Z^{2k-1}) \iff u \in D(Z^{2k}), Bu \in D(Z^{2k-1}) \\
&\iff u \in W^{2k}(\Omega), Bu \in W^{2k-1}(\Omega), \\
&\qquad B^j u \in W_0^1(\Omega), 0 \le j \le k-1, B^j Bu \in W_0^1(\Omega), 0 \le j \le k-1 \\
&\iff u \in W^{2k+1}(\Omega), B^j u \in W_0^1(\Omega), 0 \le j \le k
\end{aligned}
$$

and

$$
\begin{aligned}
u \in D(Z^{2k+2}) &\iff u, (c\mathrm{Id} - B)u \in D(Z^{2k}) \iff u, Bu \in D(Z^{2k}) \\
&\iff u, Bu \in W^{2k}(\Omega), B^j u, B^j Bu \in W_0^1(\Omega), 0 \le j \le k-1 \\
&\iff u \in W^{2k+2}(\Omega), B^j u \in W_0^1(\Omega), 0 \le j \le k .
\end{aligned}
$$

Moreover, for $u \in D(Z^{2k+1})$ we have due to 4.6.2

$$
\begin{aligned}
\|u\|_{W^{2k+1}(\Omega)} &\le c_1 (\|Bu\|_{W^{2k-1}(\Omega)} + \|u\|_{W^0(\Omega)}) \\
&\le c_2 \left(\|Z^2 u\|_{D(Z^{2k-1})} + \|u\|_{D(Z^{2k-1})} \right) \\
&\le c_3 \|u\|_{D(Z^{2k+1})} \le d \left(\|Z^2 u\|_{D(Z^{2k-1})} + \|u\|_{D(Z^{2k-1})} \right) \\
&\le d_1 (\|Bu\|_{W^{2k-1}(\Omega)} + \|u\|_{W^{2k-1}(\Omega)}) \le d_2 \|u\|_{W^{2k+1}(\Omega)}
\end{aligned}
$$

with suitable constants $c_l, d_l \ge 0$, and tor $u \in D(Z^{2k+2})$ we have with constants $c_l \ge 0$ similarly

$$
\begin{aligned}
\|u\|_{W^{2k+2}(\Omega)} &\le c_1 (\|Bu\|_{W^{2k}(\Omega)} + \|u\|_{W^0(\Omega)}) \\
&\le c_2 (\|(c\mathrm{Id} - B)u\|_{D(Z^{2k})} + \|u\|_{D(Z^{2k})}) \le c_3 \|u\|_{W^{2k+2}(\Omega)} .
\end{aligned}
$$

This proves (b) and (c) for any $k \in \mathbb{N}$. Finally, this clearly implies (d). $\qquad \square$

4.6.4 Proposition. Let $C := \displaystyle\sum_{|\alpha| \le m} c_\alpha \partial^\alpha$ with $c_\alpha \in C_c^\infty(\Omega), m \in \mathbb{N}$.
Then $C \in \mathscr{L}(D(Z^{k+m}), D(Z^k))$ for any $k \in \mathbb{N}_0$.

PROOF: We clearly have $\|Cu\|_{W^k(\Omega)} \leq c\|u\|_{W^{k+m}(\Omega)}$ for $u \in W^{k+m}(\Omega)$. Hence we only have to show $C(D(Z^{k+m})) \subset D(Z^k)$. Let $u \in D(Z^{k+m}) \subset W^{k+m}(\Omega)$. Then there is a sequence $(u_l)_l \subset C^\infty(\overline{\Omega})$ with $u_l \underset{l\to\infty}{\longrightarrow} u$ in $W^{k+m}(\Omega)$. Therefore, for $0 \leq 2j \leq k-1$ we have $B^j C u_l \in C_c^\infty(\Omega)$ and $B^j C u_l \underset{l\to\infty}{\longrightarrow} B^j C u$ in $W^{k-2j}(\Omega)$. This shows $B^j C u \in W_0^1(\Omega)$ for $0 \leq j \leq \frac{k}{2} - \frac{1}{2}$.   □

4.6.5 Proposition. Let $a_{jl} \in C^\infty(\overline{\Omega}, \mathbb{R})$ for $j,l = 1,\ldots,n$ such that $(a_{jl})_{j,l}$ is positive semi-definite and let $A = \sum_{j,l=1}^{n} \partial_j(a_{jl}\partial_l)$ Assume that there is a constant $\rho > 0$ with $a_{jl}(x) = b_{jl}(x)$ for every $x \in \overline{\Omega}$ with $\text{dist}(x,\partial\Omega) < \rho$. Here $\text{dist}(x,\partial\Omega) := \inf\{|x - \omega| : \omega \in \partial\Omega\}$.
Then $A \in \mathscr{L}(D(Z^{k+2}), D(Z^k))$ for any $k \in \mathbb{N}_0$, and for $k \in \mathbb{N}_0$ there are $\beta_k \geq 0$ with $\text{Re}\,\langle Z^k A u, Z^k u\rangle_{L^2(\Omega)} \leq \beta_k \|Z^k u\|_{L^2(\Omega)}^2$ for $u \in D(Z^{k+2})$. Moreover,

$$D(Z^{2k-1}) = \{u \in W^{2k-1}(\Omega) : A^j u \in W_0^1(\Omega), 0 \leq j \leq k-1\}$$
$$\text{and}\quad D(Z^{2k}) = \{u \in W^{2k}(\Omega) : A^j u \in W_0^1(\Omega), 0 \leq j \leq k-1\}\quad \text{for } k \in \mathbb{N}.$$

PROOF: We have $A = B + (A - B) =: B + C$. Since $a_{jl} - b_{jl} \in C_c^\infty(\Omega)$ the first assertion is a consequence of 4.6.3 and 4.6.4. Moreover, for $u \in W^2(\Omega) \cap W_0^1(\Omega)$ and $v \in D(\Omega)$ we have

$$\langle Au, v\rangle_{L^2(\Omega)} = -\sum_{j,l=1}^{n} (a_{jl}\partial_l u, \partial_j \overline{v})_{\mathcal{D}'(\Omega),\mathcal{D}(\Omega)} = -\sum_{j,l=1}^{n} \langle a_{jl}\partial_l u, \partial_j v\rangle_{L^2(\Omega)}.$$

Hence, since $\mathcal{D}(\Omega) \subset W_0^1(\Omega)$ densely, we obtain

$$\langle Au, v\rangle_{L^2(\Omega)} = -\sum_{j,l=1}^{n} \langle a_{jl}\partial_l u, \partial_j v\rangle_{L^2(\Omega)}\quad \text{for } u \in W^2(\Omega) \cap W_0^1(\Omega), v \in W_0^1(\Omega).$$

Thus $\langle Au, u\rangle_{L^2(\Omega)} \leq 0$ for $u \in W^2(\Omega) \cap W_0^1(\Omega)$. With $Z := (c\text{Id} - B)^{1/2}$ we have for $j \in \mathbb{N}$

$$Z^{2k}\text{ad}^j(Z^2)(A) = (c\text{Id} - B)^k\text{ad}^j(B)(A) = \sum_{|\alpha|\leq 2k+2+j} c_\alpha \partial^\alpha$$

with $c_\alpha \in C_c^\infty(\Omega)$. Hence $\|Z^{2k}\text{ad}^j(Z^2)(A)u\|_{L^2(\Omega)} \leq d_{k,j}\|u\|_{D(Z^{2k+2+j})}$ with constants $d_{k,j} \geq 0$ for $u \in D(Z^{2k+2+2j})$, and 2.3.10 shows that for $k \in \mathbb{N}_0$ there are $\beta_k \geq 0$ with

$$\text{Re}\,\langle Z^k A u, Z^k u\rangle_{L^2(\Omega)} \leq \beta_k \|Z^k u\|_{L^2(\Omega)}^2\quad \text{for } u \in D(Z^{k+2}).$$

Now, let $j \in \mathbb{N}$ and $C := A^j - B^j = \sum_{|\alpha| \leq 2j} c_\alpha$ with $c_\alpha \in C_c^\infty(\Omega)$. Then one can show as in 4.6.4 that $Cu \in W_0^1(\Omega)$ for $u \in W^l(\Omega)$ with $l \geq 2j + 1$. Due to $A^j = B^j + C$ this proves the assertion. $\qquad\square$

4.6.6 Proposition. Let $a_j \in C_c^\infty(\Omega)$, $j = 1, \ldots, n$ be real-valued functions, $a \in C_c^\infty(\Omega)$, and $A := \sum_{j=1}^n a_j \partial_j + a$.
Then we have $A \in \mathscr{L}(D(Z^{k+1}), D(Z^k))$ for any $k \in \mathbb{N}_0$, and for $k \in \mathbb{N}_0$ there are $\beta_k \geq 0$ with $\operatorname{Re} \langle Z^k Au, Z^k u \rangle_{L^2(\Omega)} \leq \beta_k \|u\|_{D(Z^k)}^2$ for $u \in D(Z^{k+1})$.

PROOF: The first assertion is a consequence of 4.6.4. For $u \in C_c^\infty(\Omega)$ we have with constants $\gamma, \gamma' \geq 0$

$$
2\operatorname{Re} \langle Au, u \rangle_{L^2(\Omega)} = \sum_{j=1}^n \left(\langle a_j \partial_j u, u \rangle_{L^2(\Omega)} + \langle u, a_j \partial_j u \rangle_{L^2(\Omega)} \right) + 2\operatorname{Re} \langle au, u \rangle_{L^2(\Omega)}
$$

$$
\leq \sum_{j=1}^n \left(\langle a_j \partial_j u, u \rangle_{L^2(\Omega)} - \langle \partial_j(a_j u), u \rangle_{L^2(\Omega)} \right) + \gamma' \|u\|_{L^2(\Omega)}^2
$$

$$
= - \sum_{j=1}^n \left(\langle (\partial_j a_j)u, u \rangle_{L^2(\Omega)} \right) + \gamma' \|u\|_{L^2(\Omega)}^2 \leq \gamma \|u\|_{L^2(\Omega)}^2 .
$$

Since $C_c^\infty(\Omega) \subset D(Z)$ densely, this implies $2\operatorname{Re} \langle Au, u \rangle_{L^2(\Omega)} \leq \gamma \|u\|_{L^2(\Omega)}^2$ for $u \in D(Z)$. Moreover

$$
Z^{2k} \operatorname{ad}^j(Z^2)(A) = (c\operatorname{Id} - B)^k \operatorname{ad}^j(B)(A) = \sum_{|\alpha| \leq 2k+j+1} c_\alpha \partial^\alpha
$$

with $c_\alpha \in C_c^\infty(\Omega)$. Hence 4.6.4 and 2.3.9 imply the assertion. $\qquad\square$

4.6.7 Definition. A second-order differential operator

$$
Au := \sum_{j,l=1}^n \partial_j(a_{jl} \partial_l u) + \sum_{j=1}^n a_j \partial_j u + au
$$

on a bounded, open set $\Omega \subset \mathbb{R}^n$ with smooth boundary $\partial\Omega$ and with $a_{jl}, a_j, a \in C^\infty(\overline{\Omega}, \mathbb{R})$ is called non-characteristic at the boundary, if there are $\rho > 0$, $c > 0$ with

$$
\sum_{j,l=1}^n a_{jl}(x) \xi_j \overline{\xi_l} \geq c|\xi|^2
$$

for $x \in \Omega$ with $\operatorname{dist}(x, \partial\Omega) \leq \rho$ and $(\xi_1, \ldots, \xi_n) \in \mathbb{C}^n$.

Finally, we can prove the main result of this section.

4.6.8 Theorem. Assume that

(a) $\Omega \subset \mathbb{R}^n$ is a bounded, open set with smooth boundary $\partial \Omega$.

(b) $a_{jl} \in \mathcal{C}^\infty(\overline{\Omega}, \mathbb{R})$ with $\sum\limits_{j,l=1}^n a_{jl}(x)\xi_j \overline{\xi_l} \geq 0$ and $a_{lj}(x) = a_{jl}(x)$ for $x \in \Omega$, $j, l = 1, \ldots, n$, $(\xi_1, \ldots, \xi_n) \in \mathbb{C}^n$.

(c) $a_j \in \mathcal{C}_c^\infty(\Omega)$ are real-valued, $j = 1, \ldots, n$, and $a \in \mathcal{C}_c^\infty(\overline{\Omega})$.

(d) $Au := \sum\limits_{j,l=1}^n \partial_j(a_{jl}\partial_l u) + \sum\limits_{j=1}^n a_j \partial_j u + au$ for $u \in D := W^2(\Omega) \cap W_0^1(\Omega)$.

(e) A is non-characteristic at the boundary.

Then, for $u_0 \in D$, there is a unique $u \in \mathcal{C}^1([0, \infty), L^2(\Omega)) \cap \mathcal{C}([0, \infty), D)$ with

$$\frac{du}{dt}(t) = Au(t) \quad \text{for } t \in [0, \infty), \qquad u(0) = u_0 . \tag{4.6.1}$$

More precisely, the closure of $A|_D$ in $L^2(\Omega)$ generates a quasi-contractive $\mathcal{C}_0$-semigroup $(T(t))_{t \geq 0}$ in $L^2(\Omega)$ and D is an invariant subspace. In particular, this implies $(\lambda \mathrm{Id} - A)(D) \subset L^2(\Omega)$ densely for large $\lambda > 0$. Moreover, we have the following regularity results: Let

$$D^{2k-1} := \{u \in W^{2k-1}(\Omega) : A^j u \in W_0^1(\Omega), 0 \leq j \leq k - 1\}$$
$$D^{2k} := \{u \in W^{2k}(\Omega) : A^j u \in W_0^1(\Omega), 0 \leq j \leq k - 1\}$$

for $k \in \mathbb{N}$. Then D^k for $k \in \mathbb{N}_0$ is an invariant subspace of $(T(t))_{t \geq 0}$, and for $k \in \mathbb{N}_0, u_0 \in D^{k+2}$ there is a unique $u \in \mathcal{C}^1([0, \infty), D^k) \cap \mathcal{C}([0, \infty), \overline{D}^{k+2})$ with (4.6.1), where D^k carries the topology of $W^k(\Omega)$.

PROOF: Let $\rho' := \frac{\rho}{2}$ with ρ as in 4.6.7 and choose a $\chi \in \mathcal{C}_c^\infty(\Omega)$ with $0 \leq \chi \leq 1$ and $\chi(x) = 1$ for $x \in \Omega$ with $\mathrm{dist}(x, \partial\Omega) \geq 2\rho'$, $\chi(x) = 0$ for $x \in \Omega$ with $\mathrm{dist}(x, \partial\Omega) \leq \rho'$. Let $b_{jj} := a_{jj} + \chi$ and $b_{jl} := a_{jl}$ for $j, l = 1, \ldots, n, j \neq l$. Then, with c as if 4.6.7

$$\sum_{j,l=1}^n b_{jl}(x)\xi_j \xi_l = \sum_{j,l=1}^n a_{jl}(x)\xi_j \xi_l + \chi(x) \sum_{j-1}^n \xi_j^2 \geq \min(c, 1)|\xi|^2$$

for $x \in \Omega, (\xi_1, \ldots, \xi_n) \in \mathbf{C}^n$. Moreover $a_{jl}(x) = b_{jl}(x)$ for $j, l = 1, \ldots, n, x \in \Omega$ with $\text{dist}(x, \partial\Omega) \leq \rho'$. Let B and Z be as in 4.6.3. Then 4.6.5 and 4.6.6 imply that $A \in \mathscr{L}(D(Z^{k+2}), D(Z^k))$ for any $k \in \mathbf{N}_0$ and that for $k \in \mathbf{N}_0$ there exist $\beta_k \geq 0$ with $\text{Re}\, \langle Z^k Au, Z^k u \rangle_{L^2(\Omega)} \leq \beta_k \|u\|^2_{D(Z^k)}$ for $u \in D(Z^{k+2})$. Hence 2.4.8 implies the assertion. $\square$

4.6.9 Corollary. Suppose with the assumptions of 4.6.8 that A is symmetric and semibounded from above, then A is essentially selfadjoint on D in $L^2(\Omega)$.

PROOF: This is implied by 1.2.5. $\square$

4.6.10 Bibliographical remarks. For an introduction to distribution theory and Sobolev spaces we refer to the books of Rudin [134], Adams [3], Agmon [4]. Elliptic boundary value problems are discussed e.g. by Agmon [4] or Friedman [49] For results related to 4.6.8 in sub-elliptic situations we refer to Parmeggiani/Xu [121] and the references given therein. Note that our result includes operators that are completely non-elliptic on some part of Ω. Further results on degenerate elliptic boundary value problems can be found in Oleinik/Radkevic [117], Oleinik [116].

4.7 Evolution equations on networks

The results of chapter 2 can also be applied to evolution equations on complicated geometric objects. To this end we adopt some ideas of F. Ali Mehmeti who studied wave equations on networks and introduced generalized Laplacians on networks, so called interaction operators. We will use the generated scale and 2.4.6 to prove well-posedness of the Cauchy problem for special time-dependent evolution equations of Schrödinger and (degenerate) diffusion type on networks.

4.7.1 Definition.

- Let $n \in \mathbf{N}$ be fixed and consider disjoint sets $N_1, \ldots, N_n$ such that there exist bijective mappings $\sigma_j : [0, 1] \to N_j$ for $j = 1, \ldots, n$. Moreover, let $\partial N_j := \{\sigma_j(s) : s \in \{0, 1\}\}$ and $\Gamma_0 := \bigcup_{j=1}^n \partial N_j$. Consider an equivalence relation $\sim$ on Γ_0 and let $\Gamma := \Gamma_0/\sim$ be the set of equivalence classes. Assume in addition that $x_1 \not\sim x_2$ for $x_1, x_2 \in \partial N_j, j = 1, \ldots, n$. The elements $\gamma \in \Gamma$ are called *nodes*. If $|\gamma| > 1$, γ is called *interior node*, otherwise *exterior node*; here $|\gamma|$ denotes the number of elements of γ. We do not distinguish between endpoints of N_j and nodes γ. In this sense $\Gamma = \bigcup_{j=1}^n \partial N_j$. $(N := \bigcup_{j=1}^n N_j, \Gamma)$ is called a *one-dimensional network*.

- Consider functions $u_j : N_j \to \mathbb{C}$ and write shortly $(u_j) : N \to \mathbb{C}$ and $u_j(x) = u_j(\sigma_j(x))$ for $x \in [0,1], j = 1, \ldots, n$. We say that (u_j) satisfies (T) if

$$u_{j_1}(\gamma) = u_{j_2}(\gamma) \text{ for } \gamma \in \Gamma \text{ and } j_1, j_2 \text{ with } \gamma \in N_{j_1} \cap N_{j_2}.$$

  The system $(N, \Gamma, (T))$ is called a *network with transmission condition* (T).

- We write $u_j \in C^k(N_j)$ for $k \in \mathbb{N}_0 \cup \{\infty\}$, if $u_j \circ \sigma_j \in C^k[0,1]$ and set $\partial_x^k u_j(x) = \frac{d^k}{ds^k}(u_j \circ \sigma_j)(s)$ for $x = \sigma_j(s)$. $C^k(N_j)$ is a Banach space with the norm $\|u_j\|_{C^k(N_j)} := \|u_j \circ \sigma_j\|_{C^k[0,1]}$. Moreover, let

$$H^k(N_j) := \{u : N_j \to \mathbb{C} : u \circ \sigma_j \in W^k(0,1)\}$$

  and $\langle u, v \rangle_{H^k(N_j)} := \langle u \circ \sigma_j, v \circ \sigma_j \rangle_{W^k(0,1)}$ and $\|u\|_{H^k(N_j)} := \sqrt{\langle u, u \rangle_{H^k(N_j)}}$ for $u, v \in H^k(N_j)$.
  Let $H^k(N) := \prod_{j=1}^n H^k(N_j)$ and $\langle (u_j), (v_j) \rangle_{H^k(N)} := \sum_{j=1}^n \langle u_j, v_j \rangle_{H^k(N_j)}$ and $\|(u_j)\|_{H^k(N)} := \sqrt{\langle (u_j), (u_j) \rangle_{H^k(N)}}$ for $(u_j), (v_j) \in H^k(N)$.
  Clearly, $(H^k(N), \langle \cdot, \cdot \rangle_{H^k(N)}, \|\cdot\|_{H^k(N)})$ and $(H^k(N_j), \langle \cdot, \cdot \rangle_{H^k(N_j)}, \|\cdot\|_{H^k(N_j)})$ are Hilbert spaces.

First we introduce interaction operators on networks with transmission conditions.

4.7.2 Proposition. Let (N, Γ) be a one-dimensional network with transmission condition (T). Let $V_T := \{(u_j) \in H^1(N) : (u_j) \text{ satisfies } (T)\}$ and

$$a_T((u_j), (v_j)) := \sum_{j=1}^n \left(\langle \partial_x u_j, \partial_x v_j \rangle_{H^0(N_j)} + \langle u_j, v_j \rangle_{H^0(N_j)} \right) \text{ for } (u_j), (v_j) \in V_T.$$

Then a_T is a closed, positive quadratic form with domain V_T, and its associated, positive selfadjoint operator $\Lambda_T : D(\Lambda_T) \to H^0(N)$ exists due to 1.2.9. Let $Z_T := \sqrt{\Lambda_T}$ and $H_T^k(N) := D((Z_T)^k)$ for $k \in \mathbb{N}_0$. Then Z_T is called interaction operator and $H_T^k(N)$ is called the Sobolev space of order k on the network N with transmission condition (T).
With $\left(\frac{\partial}{\partial \nu} \right) u_j(\gamma) := (-1)^{s_0} \frac{d}{ds}(u_j \circ \sigma_j)(s)|_{s=s_0}$ for $\gamma = \sigma_j(s_0)$ we have

$$H_T^k(N) = \Big\{ (u_j) \in H^k(N) | \forall 0 \leq l \leq k-1, \gamma \in \Gamma :$$

$$u_{j_1}^{(l)}(\gamma) = u_{j_2}^{(l)}(\gamma) \text{ for } \gamma \in N_{j_1} \cap N_{j_2}, \text{ if } l \text{ is even, and}$$

$$\sum_{j \text{ with } \gamma \in N_j} \left(\frac{\partial}{\partial \nu} \right)^l u_j(\gamma) = 0, \text{ if } l \text{ is odd} \Big\}. \tag{4.7.1}$$

Moreover, the graph topology of $H_T^k(N)$ is equivalent to the induced topology of $H^k(N)$ and $H_T^k(N)$ is a closed subspace of $H^k(N)$.
Finally, for $(u_j) \in H_T^2(N)$ we have $Z_T^2(u_j) = (u_j - \partial_x^2 u_j)$ in the sense of distributions.

PROOF: Due to the Sobolev embedding theorem $H^1(N_j) \hookrightarrow C(N_j)$ (because $1 > 1/2$). Hence V_T is a closed subspace of $H^1(N)$, i.e., it is a Hilbert space. Moreover, a_T is sesquilinear and symmetric on V_T, and there exists a constant $M > 0$ with

$$|a_T(u,v)| \leq M \, \|u\|_{H^1(N)} \, \|v\|_{H^1(N)}$$

$$a_T(u,u) = \sum_{j=1}^n \left(\|\partial_x u_j\|_{H^0(N_j)}^2 + \|u_j\|_{H^0(N_j)}^2 \right) = \|u\|_{H^1(N)}^2$$

for $u = (u_j), v = (v_j) \in V_T$. Therefore, the operator Λ_T associated to a_T exists. Now, for $(u_j) \in D(\Lambda_T)$ there is an $(f_j) \in H^0(N)$ with

$$a_T((u_j),(v_j)) = \langle (f_j),(v_j) \rangle_{H^0(N)} \qquad \text{for } (v_j) \in V_T . \tag{4.7.2}$$

In particular, for $(\varphi_j) \in C_c^\infty(0,1)^n$ we have

$$\sum_{j=1}^n \langle f_j, \varphi_j \rangle_{H^0(N_j)} = = \sum_{j=1}^n \left(\langle \partial_x u_j, \partial_x \varphi_j \rangle_{H^0(N_j)} + \langle u_j, \varphi_j \rangle_{H^0(N_j)} \right)$$

$$= \sum_{j=1}^n \langle u_j, -\partial_x^2 \varphi_j + \varphi_j \rangle_{H^0(N_j)}$$

by partial integration, and therefore $f_j = -\partial_x^2 u_j + u_j$ in the sense of distributions. Due to $f_j, u_j \in H^0(N_j)$ we obtain $(u_j) \in H^2(N)$, and we can calculate for $(v_j) \in V_T$:

$$\langle (f_j),(v_j) \rangle_{H^0(N)} = \sum_{j=1}^n \left[\langle -\partial_x^2 u_j, v_j \rangle_{H^0(N_j)} + \langle u_j, v_j \rangle_{H^0(N_j)} \right]$$

$$= \sum_{j=1}^n \left[\langle \partial_x u_j, \partial_x v_j \rangle_{H^0(N_j)} - (\partial_x u_j)\overline{v_j}\big|_{\sigma_j(0)}^{\sigma_j(1)} + \langle u_j, v_j \rangle_{H^0(N_j)} \right]$$

$$= a_T((u_j),(v_j)) + \sum_{\gamma \in \Gamma} \sum_{\substack{j=1,\dots,n \\ \gamma \in N_j}} \underbrace{(\partial_x u_j)(\gamma)(-1)^{\sigma_j^{-1}(\gamma)}}_{=\frac{\partial}{\partial \nu} u_j(\gamma)} \overline{v_j(\gamma)} .$$

Hence, by (4.7.2) we have

$$\sum_{\substack{\gamma \in \Gamma \\ }} \sum_{\substack{j=1,\dots,n \\ \gamma \in N_j}} \left(\frac{\partial}{\partial \nu} u_j(\gamma) \right) \overline{v_j(\gamma)} = 0 \quad \text{for } (v_j) \in V_T.$$

Now we choose a $(v_j) \in V_T$ which is non zero only near a fixed node $\gamma \in \Gamma$ to obtain

$$\sum_{\substack{j=1,\dots,n \\ \gamma \in N_j}} \left(\frac{\partial}{\partial \nu} u_j(\gamma) \right) = 0 \quad \text{for } \gamma \in \Gamma,$$

thus

$$D(A_T) \subset \left\{ (u_j) \in H^2(N) \cap V_T : \sum_{\substack{j=1,\dots,n \\ \gamma \in N_j}} \left(\frac{\partial}{\partial \nu} u_j(\gamma) \right) = 0 \quad \text{for } \gamma \in \Gamma \right\} =: W_T.$$

Now take conversely $(u_j) \in W_T$ and let $f_j := u_j - \partial_x^2 u_j$, then

$$a_T((u_j),(v_j)) = \sum_{j=1}^{n} \left(\langle \partial_x u_j, \partial_x v_j \rangle_{H^0(N_j)} + \langle u_j, v_j \rangle_{H^0(N_j)} \right)$$

$$= \sum_{j=1}^{n} \left(-\langle \partial_x^2 u_j, v_j \rangle_{H^0(N_j)} + (\partial_x u_j) \overline{v_j} |_{\sigma_j(0)}^{\sigma_j(1)} + \langle u_j, v_j \rangle_{H^0(N_j)} \right)$$

$$= \sum_{j=1}^{n} \langle f_j, v_j \rangle_{H^0(N_j)} - \sum_{\substack{\gamma \in \Gamma}} \sum_{\substack{j=1,\dots,n \\ \gamma \in N_j}} \underbrace{(\partial_x u_j)(\gamma)(-1)^{\sigma_j^{-1}(\gamma)}}_{= \frac{\partial}{\partial \nu} u_j(\gamma)} \overline{v_j(\gamma)}$$

$$= \langle (f_j),(v_j) \rangle_{H^0(N)}$$

for $(v_j) \in V_T$, which implies $(u_j) \in D(\Lambda_T)$. This proves (4.7.1) in the case $k = 2$, (4.7.1) in the case $k = 1$ is a consequence of 1.4.10. Moreover, for $(u_j) \in H_T^2(N)$ we have $Z_T^2(u_j) = (u_j - \partial_x^2 u_j)$ in the distribution sense. Thus, the assertion for general k follows inductively from

$$H_T^{2k+2}(N) = \{ u \in H_T^{2k}(N) : Z_T^2 u \in H_T^{2k}(N) \}$$

and

$$u \in H_T^{2k+1}(N) \iff u \in H_T^{2k}(N), Z_T^2 u \in H_T^{2k-1}(N).$$

For $u \in H_T^{2k+l}(N), l \in \mathbb{N}_0$ we have further with a constant $c_l \geq 0$

$$\left\| Z_T^{2k} u \right\|_{H^l(N)}^2 = \sum_{j=1}^{n} \left\| (\mathrm{Id} - \partial_x^2)^k u_j \right\|_{H^l(N_j)}^2 \leq c_l^2 \left\| u \right\|_{H^{2k+l}(N)}^2,$$

and for $u \in V_T$ we know that

$$\|(Z_T)u\|^2_{H^0(N)} = \langle (Z_T)u, (Z_T)u \rangle_{H^0(N)} = a_T(u, u) \leq M \|u\|^2_{H^1(N)} \ .$$

Therefore

$$\left\|Z_T^{2k}u\right\|_{H^0(N)} \leq c_0 \|u\|_{H^{2k}(N)} \ \text{ for } u \in H_T^{2k}(N)$$

$$\text{and } \ \left\|Z_T^{2k+1}u\right\|_{H^0(N)} \leq \sqrt{M} \left\|Z_T^{2k}u\right\|_{H^1(N)} \leq \sqrt{M}c_1 \|u\|_{H^{2k+1}(N)}$$

for $u \in H_T^{2k+1}(N)$. Now, $H_T^k(N) = D(Z_T^k)$ is complete with respect to the graph norm and complete with respect to $\|\cdot\|_{H^k(N)}$ (because $H_T^k(N)$ is closed in $H^k(N)$ due to $H^k(N_j) \hookrightarrow C^{k-1}(N_j)$). Therefore the open mapping theorem implies the assertion.   □

Now we introduce differential operators in the scale $H_T^k(N)$.

4.7.3 Proposition. Let (N, Γ) be a network with transmission condition (T) as in 4.7.1.
For $b_j \in C^\infty(N_j)$ let $b(X)(u_j) := (b_j u_j)$ for $u_j \in H^0(N)$. Suppose that

$$\text{for } \gamma \in \Gamma, k \in \mathbb{N}_0 \begin{cases} b_{j_1}^{(k)}(\gamma) = b_{j_2}^{(k)}(\gamma) & \text{for } j_1, j_2 \text{ with } \gamma \in N_{j_1} \cap N_{j_2} \ ,k \text{ even} \\ b_j^{(k)}(\gamma) = 0 & \text{for } j \text{ with } \gamma \in N_j \ \qquad ,k \text{ odd} \end{cases}$$

Then $b(X) \in \mathscr{L}(H_T^k(N))$ for $k \in \mathbb{N}_0$.

PROOF: Clearly, $b(X) \in \mathscr{L}(H_T^0(N))$. Note that $b(X) \in \mathscr{L}(H_T^k(N))$ for $k \in \mathbb{N}$ if and only if $b(X)(H_T^k(N)) \subset H_T^k(N)$ (due to the closed graph theorem). For $u \in H_T^k(N), 0 \leq l \leq k-1, l$ even, and $\gamma \in \Gamma$ with $\gamma \in N_{j_1} \cap N_{j_2}$ we have

$$\left(b(X)(u_j)\right)_{j_1}^{(l)}(\gamma) = \sum_{m=0}^{l} \binom{l}{m} b_{j_1}^{(m)}(\gamma) u_{j_1}^{(l-m)}(\gamma)$$

$$= \sum_{m=0}^{l/2} \binom{l}{2m} \underbrace{b_{j_1}^{(2m)}(\gamma)}_{=b_{j_2}^{(2m)}(\gamma)} \underbrace{u_{j_1}^{(2(l/2-m))}(\gamma)}_{=u_{j_2}^{(2(l/2-m))}(\gamma)} + \sum_{m=0}^{l/2-1} \binom{l}{2m+1} \underbrace{b_{j_1}^{(2m+1)}(\gamma)}_{=0} u_{j_1}^{(2(l/2-m)-1)}(\gamma)$$

$$= \left(b(X)(u_j)\right)_{j_2}^{(l)}(\gamma) \ .$$

For $0 \le l \le k-1$, l odd, $\gamma \in \Gamma$, j_0 with $\gamma \in N_{j_0}$ we have

$$\sum_{j \text{ with } \gamma \in N_j} \left(\frac{\partial}{\partial \nu}\right)^l (b_j u_j)(\gamma)$$

$$= \sum_{m=0}^{l} \binom{l}{m} \sum_{j \text{ with } \gamma \in N_j} \left[\left(\frac{\partial}{\partial \nu}\right)^m b_j\right](\gamma) \left[\left(\frac{\partial}{\partial \nu}\right)^{l-m} u_j\right](\gamma)$$

$$= \sum_{m=0}^{(l-1)/2} \binom{l}{2m} b_{j_0}^{(2m)}(\gamma) \underbrace{\sum_{j \text{ with } \gamma \in N_j} \left[\left(\frac{\partial}{\partial \nu}\right)^{2((l-1)/2-m)+1} u_j\right](\gamma)}_{=0}$$

$$+ \sum_{m=0}^{(l-1)/2} \binom{l}{2m+1} \sum_{j \text{ with } \gamma \in N_j} \underbrace{\left[\left(\frac{\partial}{\partial \nu}\right)^{2m+1} b_j\right](\gamma)}_{=0} \left[\left(\frac{\partial}{\partial \nu}\right)^{2((l-1)/2-m)} u_j\right](\gamma) = 0.$$

This proves the assertion. □

4.7.4 Proposition. Let (N, Γ) be a network with transmission condition (T) as in 4.7.1.
For $a_j \in C^\infty(N_j)$ let $a(X)\partial_x(u_j) := (a_j(x)\partial_x u_j)$ for $u_j \in H^1(N)$. Suppose that

$$\text{for } \gamma \in \Gamma, k \in \mathbb{N}_0 \begin{cases} a_j^{(k)}(\gamma) = 0 & \text{for } j \text{ with } \gamma \in N_j & , k \text{ even} \\ a_{j_1}^{(k)}(\gamma) = a_{j_2}^{(k)}(\gamma) & \text{for } j_1, j_2 \text{ with } \gamma \in N_{j_1} \cap N_{j_2} & , k \text{ odd} \end{cases}$$

Then $a(X)\partial_x \in \mathscr{L}(H_T^{k+1}(N), H_T^k(N))$ for $k \in \mathbb{N}_0$.

PROOF: Clearly, we have $a(X)\partial_x \in \mathscr{L}(H_T^1(N), H_T^0(N))$ in any case. Moreover, due to the closed graph theorem $a(X)\partial_x \in \mathscr{L}(H_T^{k+1}(N), H_T^k(N))$ for $k \in \mathbb{N}$ if and only if $a(X)\partial_x(H_T^{k+1}(N)) \subset H_T^k(N)$.
For $u \in H_T^{k+1}(N), 0 \le l \le k-1$, l even, and $\gamma \in \Gamma$ with $\gamma \in N_{j_1} \cap N_{j_2}$ we have

$$\left(a(X)\partial_x(u_j)\right)_{j_1}^{(l)}(\gamma) = \sum_{m=0}^{l} \binom{l}{m} a_{j_1}^{(m)}(\gamma) u_{j_1}^{(l-m+1)}(\gamma)$$

$$= \sum_{m=0}^{l/2} \binom{l}{2m} \underbrace{a_{j_1}^{(2m)}(\gamma)}_{=0} u_{j_1}^{(2(l/2-m)+1)}(\gamma) + \sum_{m=0}^{l/2-1} \binom{l}{2m+1} \underbrace{a_{j_1}^{(2m+1)}(\gamma)}_{=a_{j_2}^{(2m+1)}(\gamma)} \underbrace{u_{j_1}^{(2(l/2-m))}(\gamma)}_{=u_{j_2}^{(2(l/2-m))}(\gamma)}$$

$$= \left(a(X)\partial_x(u_j)\right)_{j_2}^{(l)}(\gamma).$$

For $0 \leq l \leq k-1$, l odd, $\gamma \in \Gamma$, j_0 with $\gamma \in N_{j_0}$ we have

$$\sum_{j \text{ with } \gamma \in N_j} \left(\frac{\partial}{\partial \nu}\right)^l (a_j \partial_x u_j)(\gamma)$$

$$= \sum_{m=0}^{l} \binom{l}{m} \sum_{j \text{ with } \gamma \in N_j} \left[\left(\frac{\partial}{\partial \nu}\right)^m a_j\right](\gamma) \left[\left(\frac{\partial}{\partial \nu}\right)^{l-m} \partial_x u_j\right](\gamma)$$

$$= \sum_{m=0}^{(l-1)/2} \binom{l}{2m} \sum_{j \text{ with } \gamma \in N_j} \underbrace{\left[\left(\frac{\partial}{\partial \nu}\right)^{2m} a_j\right](\gamma)}_{=0} \left[\left(\frac{\partial}{\partial \nu}\right)^{2((l-1)/2-m)+1} \partial_x u_j\right](\gamma)$$

$$+ \sum_{m=0}^{(l-1)/2} \binom{l}{2m+1} \sum_{j \text{ with } \gamma \in N_j} \left[\left(\frac{\partial}{\partial \nu}\right)^{2m+1} a_j\right](\gamma) \left[\left(\frac{\partial}{\partial \nu}\right)^{2((l-1)/2-m)} \partial_x u_j\right](\gamma)$$

$$= \sum_{m=0}^{(l-1)/2} \binom{l}{2m+1} a_{j_0}^{(2m+1)}(\gamma) \underbrace{\sum_{j \text{ with } \gamma \in N_j} \left[\left(\frac{\partial}{\partial \nu}\right)^{2((l-1)/2-m)+1} u_j\right](\gamma)}_{=0} = 0.$$

This proves the assertion. $\qquad\qquad\qquad\qquad\qquad\qquad\qquad\qquad\qquad\qquad$ $\square$

Now we can define differential operators on networks with transmission conditions that are operators of order m in the scale $H_T^k(N)$.

4.7.5 Definition. Let (N, Γ) be a network with transmission condition (T) as in 4.7.1 and $m \in \mathbb{N}_0$.
Let $a_{l,j} \in C^\infty(N_j)$, $l = 0, \dots, m$, $j = 1, \dots, n$, such that for $\gamma \in \Gamma$, $k \in \mathbb{N}_0$, $l = 1, \dots, m$ the following properties are satisfied:

- $a_{l,j}^{(k)}(\gamma) = 0$ for j with $\gamma \in N_j$ and k even, l odd.

- $a_{l,j_1}^{(k)}(\gamma) = a_{l,j_2}^{(k)}(\gamma)$ for j_1, j_2 with $\gamma \in N_{j_1} \cap N_{j_2}$ and k odd, l odd.

- $a_{l,j_1}^{(k)}(\gamma) = a_{l,j_2}^{(k)}(\gamma)$ for j_1, j_2 with $\gamma \in N_{j_1} \cap N_{j_2}$ and k even, l even.

- $a_{l,j}^{(k)}(\gamma) = 0$ for j with $\gamma \in N_j$ and k odd, l even.

Then 4.7.3 and 4.7.4 show that for any $k \in \mathbb{N}_0$

$$P := \sum_{l=0}^{m} (a_{l,j}(x))_j \partial_x^l \in \mathcal{L}(H_T^{k+m}(N), H_T^k(N)),$$

where $P(u_j) = \left(\sum_{l=0}^{m} a_{l,j}(x)\partial_x^l u_j\right)$. We write shortly $P \in \text{Diff}_T^m(N)$.

4.7.6 Corollary. For $P \in \mathrm{Diff}_T^m(N), m \in \mathbb{N}_0, j \in \mathbb{N}_0$ let $\mathrm{ad}^0(Z_T^2)(P) := P$ and $\mathrm{ad}^{j+1}(Z_T^2)(P) := [Z_T^2, \mathrm{ad}^j(Z_T^2)(P)]$. Then $\mathrm{ad}^j(Z_T^2)(P) \in \mathrm{Diff}_T^{m+j}(N)$. Moreover, for $P \in \mathrm{Diff}_T^m(N)$ and $k \in \mathbb{Z}, j \in \mathbb{N}_0$ there are $d_{k,j} \geq 0$ with

$$\left\| Z_T^k \mathrm{ad}^j(Z_T)(P) u \right\|_{H^0(N)} \leq d_{k,j} \left\| Z_T^{k+m} u \right\|_{H^0(N)}$$

for $k \in \mathbb{Z}, j \in \mathbb{N}_0, u \in H_T^\infty(N) := \bigcap_{l \in \mathbb{N}_0} H_T^l(N)$.

PROOF: We only have to prove the first statement for $j = 1$. Let

$$P = \sum_{l=0}^{m} (a_{l,j}(x))_j \partial_x^l \in \mathrm{Diff}_T^m(N) \,,$$

then

$$[Z_T^2, P] = -\sum_{l=0}^{m} (\partial_x^2 a_{l,j}(x))_j \partial_x^l - 2\sum_{l=0}^{m} (\partial_x a_{l,j}(x))_j \partial_x^{l+1} \,.$$

This implies the first statement. The other statements are consequences of 2.3.8 $\qquad\square$

Now we can treat special time-dependent Schrödinger type equations in the scale $H_T^k(N)$.

4.7.7 Theorem.   Let $a_j(t,x), b_j(t,x) \in C^\infty(I \times N_j)$, $c_j \in C(I)$, a_j, c_j, $j = 1, \ldots, n$, be real-valued such that

$$A(t) := i(c_j(t))_j \partial_x^2 + (a_j(t,x))_j \partial_x + (b_j(t,x))_j \in \mathrm{Diff}_T^2(N), t \in I.$$

Then the Cauchy problem for $\pm A(t) \in \mathscr{L}(H_T^{k+2}(N), H_T^k(N))$ in the scale $(H_T^k(N))_k$ is well-posed with exponential growth. In particular, for any $k \in \mathbb{N}_0$, $u_0 \in H_T^{k+2}(N)$, and $f \in C(I, H_T^{k+2}(N))$ there exists a unique function $u = (u_j) \in C(I, H_T^{k+2}(N)) \cap C^1(I, H_T^k(N))$ with

$$\frac{du_j}{dt}(t) = ic_j(t)\partial_x^2 u_j(t) + a_j(t,x)\partial_x u_j(t) + b_j(t,x)u_j(t) + f_j(t), j = 1, \ldots, n,$$

$$u_j(t_0) = u_{0,j}, \quad j = 1, \ldots, n \,.$$

PROOF: First, $I \ni t \mapsto A(t) \in \bigcap_{k \in \mathbb{N}_0} \mathscr{L}(H_T^{k+2}(N), H_T^k(N))$ is strongly continuous.

Moreover, for $(u_j) \in H_T^2(N)$

$$\langle \partial_x^2 (u_j), (u_j) \rangle_{H_T^0(N)} = \sum_{j=1}^{n} \left[-\langle \partial_x u_j, \partial_x u_j \rangle_{H^0(N_j)} + (\partial_x u_j) \overline{u_j}|_{\sigma_j(0)}^{\sigma_j(1)} \right]$$

$$= -\langle (\partial_x u_j), (\partial_x u_j) \rangle_{H_T^0(N)} - \sum_{\gamma \in \Gamma} \sum_{\substack{j=1,\dots,n \\ \gamma \in N_j}} \underbrace{(\partial_x u_j)(\gamma)(-1)^{\sigma_j^{-1}(\gamma)}}_{= \frac{\partial}{\partial \nu} u_j(\gamma)} \overline{u_j(\gamma)}$$

$$= -\langle (\partial_x u_j), (\partial_x u_j) \rangle_{H_T^0(N)} .$$

Hence, with suitable constants $\beta, \beta' \geq 0$ for $(u_j) \in H_T^2(N)$

$$|2\mathrm{Re}\, \langle A(t)(u_j), (u_j) \rangle_{H_T^0(N)}|$$

$$\leq 0 + \left| \sum_{j=1}^{n} \left(\langle a_j(t, x) \partial_x u_j, u_j \rangle_{H^0(N_j)} + \langle u_j, a_j(t, x) \partial_x u_j \rangle_{H^0(N_j)} \right) \right|$$

$$+ \left| \langle (b_j(t, x) u_j), (u_j) \rangle_{H_T^0(N)} \right|$$

$$\leq \left| \sum_{j=1}^{n} \langle a_j(t, x) \partial_x u_j - \partial_x (a_j(t, x) u_j), u_j \rangle_{H^0(N_j)} \right.$$

$$\left. - \sum_{\gamma \in \Gamma} \sum_{\substack{j=1,\dots,n \\ \gamma \in N_j}} \underbrace{a_j(t, \gamma)(-1)^{\sigma_j^{-1}(\gamma)} |u_j(\gamma)|^2}_{=0} \right| + \beta' \, \|(u_j)\|_{H_T^0(N)}^2$$

$$\leq 2\beta \, \|(u_j)\|_{H_T^0(N)}^2 .$$

This proves $|\mathrm{Re}\, \langle A(t)u, u \rangle_{H_T^0(N)}| \leq \beta \|u\|_{H_T^0(N)}^2$ for $u \in H_T^2(N), t \in I$. Moreover, with 4.7.6 we clearly have

$$\mathrm{ad}^j(Z_T^2)((a_j(t, \cdot)) \partial_x + (b_j(t, \cdot))) \in \mathrm{Diff}_T^{j+1}(N) \quad \text{for } j \in \mathbb{N}_0$$

and 2.3.9 implies the existence of constants $\beta_k \geq 0$ for $k \in \mathbb{N}_0$ with

$$|\mathrm{Re}\, \langle A(t)u, u \rangle_{H_T^k(N)}| \leq \beta_k \|u\|_{H_T^k(N)}^2 \quad \text{for } u \in H_T^{k+2}(N), t \in I .$$

Hence 2.4.6 implies the assertion. $\square$

Finally, we can treat special evolution equations of (degenerate) diffusion type on networks. Note that in the following theorem c_j my be zero.

4.7.8 Theorem. Let $a_j(t,x), b_j(t,x), c_j(t,x) \in C^\infty(I \times N_j)$ with a_j real-valued and c_j non-negative for $j = 1, \ldots, n$ such that

$$A(t) := (c_j(t,x))_j \partial_x^2 + (a_j(t,x))_j \partial_x + (b_j(t,x))_j \in \mathrm{Diff}_T^2(N), t \in I.$$

Then the Cauchy problem for $A(t) \in \mathscr{L}(H_T^{k+2}(N), H_T^k(N))$ in the scale $(H_T^k(N))_k$ is well-posed with exponential growth. In particular, for any $k \in \mathbb{N}_0, u_0 \in H_T^{k+2}(N)$, and $f \in C(I, H_T^{k+2}(N))$ there exists a unique function $u = (u_j) \in \mathcal{C}(I, H_T^{k+2}(N)) \cap C^1(I, H_T^k(N))$ with

$$\frac{du_j}{dt}(t) = c_j(t,x) \partial_x^2 u_j(t) + a_j(t,x) \partial_x u_j(t) + b_j(t,x) u_j(t) + f_j(t), j = 1, \ldots, n,$$
$$u_j(t_0) = u_{0,j}, \quad j = 1, \ldots, n.$$

PROOF: Let $A_1(t) := \partial_x[(c_j(t,x)) \partial_x] = (c_j(t,x)) \partial_x^2 + (\partial_x c_j(t,x)) \partial_x \in \mathrm{Diff}_T^2(N)$ and
$A_2(t) = (a_j(t,x) - (\partial_x c_j(t,x))) \partial_x + (b_j(t,x)) \in \mathrm{Diff}_T^1(N)$ for $t \in I$.
Then $A(t) = A_1(t) + A_2(t)$, and due to 2.4.6 we have to show that for $k \in \mathbb{N}_0$ there are $\beta_k \geq 0$ with

$$\mathrm{Re} \left\langle A_l(t)(u_j), (u_j) \right\rangle_{H_T^k(N)} \leq \beta_k \left\| (u_j) \right\|_{H_T^k(N)}^2 \text{ for } (u_j) \in H_T^{k+2}(N), t \in I, l = 1, 2.$$
$$(4.7.3)$$

For $l = 2$ this can be shown as in 4.7.7. Moreover

$$\left\langle A_1(t)(u_j), (v_j) \right\rangle_{H_T^0(N)} = \sum_{j=1}^n \left\langle \partial_x(c_j(t,x) \partial_x u_j), v_j \right\rangle_{H^0(N_j)}$$

$$= -\sum_{j=1}^n \left\langle c_j(t,x) \partial_x u_j, \partial_x v_j \right\rangle_{H^0(N_j)} - \sum_{\gamma \in \Gamma} \sum_{\substack{j=1,\ldots,n \\ \gamma \in N_j}} c_j(t,\gamma) \left(\frac{\partial}{\partial \nu} u_j(\gamma) \right) \overline{v_j(\gamma)}$$

$$= -\sum_{j=1}^n \left\langle c_j(t,x) \partial_x u_j, \partial_x v_j \right\rangle_{H^0(N_j)} - \underbrace{\sum_{\gamma \in \Gamma} c_{j_\gamma}(t,\gamma) \overline{v_{j_\gamma}(\gamma)} \sum_{\substack{j=1,\ldots,n \\ \gamma \in N_j}} \left(\frac{\partial}{\partial \nu} u_j(\gamma) \right)}_{=0}$$

$$= -\sum_{j=1}^n \left\langle c_j(t,x) \partial_x u_j, \partial_x v_j \right\rangle_{H^0(N_j)}$$

for $(u_j), (v_j) \in H_T^2(N)$, where j_γ is chosen with $\gamma \in N_{j_\gamma}$. This shows

$$\langle A_1(t)u, v \rangle_{H_T^0(N)} = \langle u, A_1(t)v \rangle_{H_T^0(N)} \quad \text{and} \quad \langle A_1(t)u, u \rangle_{H_T^0(N)} \leq 0$$

for $u, v \in H_T^2(N)$. Since $\mathrm{ad}^j(Z_T^2)(A_1(t)) \in \mathrm{Diff}_T^{j+2}(N)$ (due to 4.7.6), 2.3.10 implies (4.7.3), and we have finished the proof. $\square$

4.7.9 Remark.

(a) In this section we have used a different method for applications of 2.4.6 than in section 4.5. Whereas in section 4.5 we started with a (complicated) operator A and constructed a scale of Hilbert spaces in which A operates, we started in this section with a scale of Hilbert spaces and determined operators A in this scale afterwards. Clearly, the first method could also be applied to evolution equations on networks.

(b) Similar results are obtained, if we substitute the transmission condition (T) by

$$(u_j) \text{ satisfies } (T') \quad :\Longleftrightarrow \quad \sum_{\substack{j=1 \\ \gamma \in N_j}}^{n} u_j(\gamma) = 0 \quad \text{for } \gamma \in \Gamma.$$

4.7.10 Bibliographical remarks.   Definition 4.7.1 is taken from Ali Mehmeti [6] (cf. also [5]). Proposition 4.7.2 is a slight generalizations of [6, 4.4.1]. One can also introduce interaction operators on more complicated ramified spaces, cf. Ali Mehemti [5, ch. 3] and the references given in [6]. Thus, the methods presented in this section could be a starting point for a study of evolution equations on more complicated ramified spaces, also on non-compact ramified spaces.

Chapter 5

Applications to quasilinear evolution equations

The fifth and last chapter is devoted to applications to quasilinear differential and pseudodifferential evolution equations. In section 5.1 we prove several inequalities that are based on Gagliardo-Moser-Nirenberg estimates, i.e., estimates like

$$\|uv\|_{H_q^k} \leq c\left(\|u\|_{H_q^k}\|v\|_{L^\infty(\mathbf{R}^n)} + \|u\|_{L^\infty(\mathbf{R}^n)}\|v\|_{H_q^k}\right).$$

These inequalities are fundamental for applications of the results in chapter 3, in particular for the application of regularity results. Then, in section 5.2 we prove well-posedness of the Cauchy problem for

$$u_t = Pu + Qu + \sum_{j=1}^{n} a_j(u)u_{x_j} + f(u), \qquad (5.0.1)$$

where P is a quasi-dissipative, second-order pseudodifferential operator in the spatial variables as in section 4.2 and Q is a quasi-dissipative Fourier multiplier in the spatial variables. Many equations of fluid dynamics are contained in (5.0.1), e.g. the Burger equation, the Korteweg-de Vries equation, or the Benjamin-Ono equation. Moreover, as in section 4.2 equations are described by this result that are parabolic in some parts of the space and hyperbolic in others. An important further example with this property is described in section 5.3. Here a unified approach to the Navier-Stokes equation and the Euler equation is given. More precisely, we prove well-posedness of the Cauchy problem for

$$u_t = \nu\Delta_x - (u \cdot \operatorname{grad}_x)u - \operatorname{grad}_x \pi, \qquad \operatorname{div}_x u = 0$$

with a non-negative and smooth function $\nu : \mathbf{R}^n \to [0, \infty)$. The usual (parabolic) Navier-Stokes equation corresponds to the case $\nu = \mathrm{const.} > 0$ and the usual (hyperbolic) Euler equation to the case $\nu = 0$. In section 5.3 a unified approach to both types of Kadomtsev-Petiviashvili equations

$$(u_t + u_{xxx} + uu_x)_x \mp u_{yy} = 0$$

with periodic boundary conditions is described. The chapter concludes with simple proofs of results on well-posedness of the Cauchy problem for first-order, quasilinear, hyperbolic evolution equations in scales of L^q-Sobolev spaces in section 5.5 and in the scale $C_0^k(\mathbf{R}^n)$ of k-times continuously differentiable functions vanishing at infinity with all their derivatives in section 5.6.

5.1   Estimates of Nash-Moser type for differential operators

To give applications to quasilinear evolution equations we will have to use a number of inequalities. Therefore we will collect several inequalities for this purpose in this section. Our main goal will be to prove the following estimates. First, for multiplication operators we have the next theorem.

5.1.1 Theorem.   Let $K \in \mathbf{N}, 1 < p < \infty$ with $K > \frac{n}{p}$ and $\mathcal{F} \subset C^\infty(\mathbf{R}^N, \mathbb{C})$ boundedly on compact subsets with $f(0) = 0$ for $f \in \mathcal{F}$.
Then, for $k \geq K$ there are functions $G_k : [0, \infty) \longrightarrow [0, \infty)$ with the following properties:

- $f(u) \in H_p^k$ for $f \in \mathcal{F}, u \in H_p^k[\mathbf{R}^N], k \geq K$.

- $\|f(u)\|_{H_p^k} \leq G_k(R) \|u\|_{H_p^k[\mathbf{R}^N]}$ for $f \in \mathcal{F}$, $u \in H_p^k[\mathbf{R}^N]$, $\|u\|_{H_p^K[\mathbf{R}^N]} \leq R$, $k \geq K$, $R > 0$.

- $\|f(u) - f(v)\|_{H_p^k} \leq G_k(R) \|u - v\|_{H_p^k[\mathbf{R}^N]}$ for $f \in \mathcal{F}, u, v \in H_p^k[\mathbf{R}^N]$, $\|u\|_{H_p^k[\mathbf{R}^N]} \leq R, \|v\|_{H_p^k[\mathbf{R}^N]} \leq R, k \geq K, R > 0$.

Then, for first-order differential operators, the following theorem will be crucial for applications.

5.1.2 Theorem.   Let $K \in \mathbf{N}, 1 < p < \infty$ with $K > \frac{n}{p}$ and $\mathcal{F} \subset C^\infty(\mathbf{R}^N, \mathbb{C})$ boundedly on compact subsets. Moreover, let $[P(u)]v(x) := \sum_{j=1}^{n} a_j(u(x)) \partial_j v(x)$

with $a_j \in \mathcal{F}$ and let $m \in \mathbf{N}$ be fixed.

Then, for $k \geq K$ there are functions $G_k : [0, \infty) \longrightarrow [0, \infty)$ with the following properties:

(a) $[P(u)]w \in H_p^k$ with $\|P(u)w\|_{H_p^k} \leq G_k(R)(1 + \|u\|_{H_p^k[\mathbf{R}^N]})\|w\|_{H_p^{k+1}}$
for $u \in H_p^k[\mathbf{R}^N], \|u\|_{H_p^K[\mathbf{R}^N]} \leq R, w \in H_p^{k+1}[\mathbf{R}], k \geq K, R > 0$.

(b) $\|P(u)w - P(v)w\|_{H_p^k} \leq G_k(R)\|u - v\|_{H_p^k[\mathbf{R}^N]}\|w\|_{H_p^{k+1}}$
for $u, v \in H_p^k[\mathbf{R}^N], \|u\|_{H_p^k[\mathbf{R}^N]} \leq R, \|v\|_{H_p^k[\mathbf{R}^N]} \leq R, w \in H_p^{k+1}[\mathbf{R}], k \geq K$,
$R > 0$.

(c) $\|\partial^\alpha(P(u)w) - P(u)\partial^\alpha w\|_{L^p(\mathbf{R}^n)} \leq G_k(R)\left(\|w\|_{H_p^{k+m}} + \|u\|_{H_p^{k+m}[\mathbf{R}^N]}\|w\|_{H_p^k}\right)$
for $u \in H_p^{k+m}[\mathbf{R}^N], \|u\|_{H_p^k[\mathbf{R}^N]} \leq R, w \in H_p^\infty[\mathbf{R}], |\alpha| \leq k+m, k \geq K+1$,
$R > 0$.

(d) $\left\|\langle D_x\rangle^l(P(u)w) - P(u)(\langle D_x\rangle^l w)\right\|_{L^2(\mathbf{R}^n)} \leq G_k(R)\|w\|_{H^k}$ for $p = 2, l \leq k$,
$u \in H^k[\mathbf{R}^N]$ with $\|u\|_{H^k[\mathbf{R}^N]} \leq R, w \in H^\infty[\mathbf{R}^n], k \geq K+1, R > 0$, if
$a_j(0) = 0$ for $j = 1, \ldots, n$.

(e) $\left\|\langle D_x\rangle^l(P(u)w) - P(u)(\langle D_x\rangle^l w)\right\|_{L^2} \leq G_k(R)(\|w\|_{H^{k+m}} + \|u\|_{H^{k+m}[\mathbf{R}^N]}\|w\|_{H^k})$
for $p = 2, l \leq k+m, u \in H^{k+m}[\mathbf{R}^N], \|u\|_{H^k[\mathbf{R}^N]} \leq R, w \in H^\infty[\mathbf{R}]$,
$k \geq K+1, R > 0$, if $a_j(0) = 0$ for $j = 1, \ldots, n$.

Now we start the lengthy proof of these estimates. First, we have a generalization of Hölder's inequality.

5.1.3 Lemma. Let $1 \leq p < \infty, 1 < q_j < \infty, j = 1, \ldots, m$, with $\sum_{j=1}^m \frac{1}{q_j} = 1$.
Then, for $u_j \in L^{pq_j}(\mathbf{R}^n), j = 1, \ldots, m$, we have $u_1 \cdots u_m \in L^p(\mathbf{R}^n)$ with

$$\|u_1 \cdots u_m\|_{L^p(\mathbf{R}^n)} \leq \|u_1\|_{L^{pq_1}(\mathbf{R}^n)} \cdots \|u_m\|_{L^{pq_m}(\mathbf{R}^n)} . \tag{5.1.1}$$

PROOF: We proceed inductively for $m \geq 2$. First, for $m = 2$ we have with Hölder's inequality

$$\|u_1 u_2\|_{L^p} = \||u_1|^p|u_2|^p\|_{L^1}^{1/p} \leq (\||u_1|^p\|_{L^{q_1}}\||u_2|^p\|_{L^{q_2}})^{1/p} = \|u_1\|_{L^{pq_1}}\|u_2\|_{L^{pq_2}} .$$

Now assume (5.1.1) for an $m \geq 2$, then $\frac{1}{q_m} + \frac{1}{q_{m+1}} = \frac{q_{m+1}+q_m}{q_m q_{m+1}} = \frac{1}{q_m'}$ with
$q_m' := \frac{q_m q_{m+1}}{q_{m+1}+q_m}$.
Hence, by the induction hypothesis

$$\|u_1 \cdots u_{m+1}\|_{L^p} \leq \|u_1\|_{L^{pq_1}} \cdots \|u_{m-1}\|_{L^{pq_{m-1}}}\|u_m u_{m+1}\|_{L^{p\frac{q_m q_{m+1}}{q_{m+1}+q_m}}} . \tag{5.1.2}$$

Let $q_1' := \frac{q_m + q_{m+1}}{q_m q_{m+1}} q_m$ and $q_2' := \frac{q_m + q_{m+1}}{q_m q_{m+1}} q_{m+1}$, then

$$\frac{1}{q_1'} + \frac{1}{q_2'} = \frac{q_m q_{m+1}}{q_m + q_{m+1}} \left(\frac{1}{q_m} + \frac{1}{q_{m+1}} \right) = 1$$

and (5.1.1) for $m = 2$ implies

$$\left\| u_m u_{m+1} \right\|_{L^{p \frac{q_m q_{m+1}}{q_{m+1} + q_m}}} = \left\| |u_m|^{\frac{q_m q_{m+1}}{q_{m+1} + q_m}} |u_{m+1}|^{\frac{q_m q_{m+1}}{q_{m+1} + q_m}} \right\|_{L^p}^{\frac{q_m + q_{m+1}}{q_m q_{m+1}}}$$

$$\leq \left\| |u_m|^{\frac{q_m q_{m+1}}{q_{m+1} + q_m}} \right\|_{L^{p \frac{q_m + q_{m+1}}{q_m q_{m+1}} q_m}}^{\frac{q_m + q_{m+1}}{q_m q_{m+1}}} \left\| |u_{m+1}|^{\frac{q_m q_{m+1}}{q_{m+1} + q_m}} \right\|_{L^{p \frac{q_m + q_{m+1}}{q_m q_{m+1}} q_{m+1}}}^{\frac{q_m + q_{m+1}}{q_m q_{m+1}}}$$

$$= \left\| u_m \right\|_{L^{p q_m}} \left\| u_{m+1} \right\|_{L^{p q_{m+1}}}$$

Together with (5.1.2) this implies the assertion.   □

5.1.4 Lemma. Define $g(x) := x|x|^s$ for $x \in \mathbf{R}, s > 0$. Then $g \in C^1(\mathbf{R})$ with $g'(x) = (s+1)|x|^s$. In particular, if $f \in C^1(\mathbf{R}^n, \mathbf{R})$, then $f|f|^s \in C^1(\mathbf{R}^n)$ and $\partial_j(f|f|^s) = (s+1)(\partial_j f)|f|^s$.

PROOF: For $x > 0$ we have $g(x) = x^{s+1}$ and $g'(x) = (s+1)x^s = (s+1)|x|^s$. Since $g(-x) = -g(x)$ for $x \in \mathbf{R}$ we have $g'(x) = g'(-x) = (s+1)|x|^s$ for $x < 0$. For $x = 0$ we have

$$\frac{g(x+h) - g(x)}{h} = \frac{h|h|^s - 0}{h} = |h|^s \xrightarrow[h \to 0]{} 0 \qquad \text{because } s \geq 0.$$

This proves $g \in C^1(\mathbf{R})$ with $g'(x) = (s+1)|x|^s$. The second assertion is implied by the chain rule.   □

5.1.5 Proposition. For $1 < p < \infty$ and $u \in \mathscr{S}(\mathbf{R}^n, \mathbf{R})$ we have

$$\left\| \partial_j u \right\|_{L^{2p}(\mathbf{R}^n)} \leq \sqrt{2p - 1} \left\| \partial_j^2 u \right\|_{L^p(\mathbf{R}^n)}^{1/2} \left\| u \right\|_{L^\infty(\mathbf{R}^n)}^{1/2}.$$

PROOF:
5.1.4 implies that $(\partial_j u)|\partial_j u|^{2p-2} \in C^1(\mathbf{R}^n)$ and for its derivative we have $\partial_j[(\partial_j u)|\partial_j u|^{2p-2}] = (2p-1)(\partial_j^2 u)|\partial_j u|^{2p-2}$, hence

$$|\partial_j u|^{2p} = (\partial_j u)(\partial_j u)|\partial_j u|^{2p-2} = \partial_j[u(\partial_j u)|\partial_j u|^{2p-2}] - u\partial_j[(\partial_j u)|\partial_j u|^{2p-2}]$$

$$= \partial_j[u(\partial_j u)|\partial_j u|^{2p-2}] - (2p-1)u(\partial_j^2 u)|\partial_j u|^{2p-2}.$$

Therefore, due to $\frac{1}{p} + \frac{p-1}{p} = 1$

$$\|\partial_j u\|_{L^{2p}}^{2p} = \int_{\mathbf{R}^n} |\partial_j u(x)|^{2p} dx$$

$$\leq \underbrace{\int_{\mathbf{R}^n} \partial_j [u(x)(\partial_j u(x))|\partial_j u(x)|^{2p-2}] dx}_{=0} + (2p-1) \int_{\mathbf{R}^n} |u(x)||\partial_j^2 u(x)||\partial_j u(x)|^{2p-2} dx$$

$$\leq (2p-1) \|u\|_{L^\infty} \int_{\mathbf{R}^n} |\partial_j^2 u(x)||\partial_j u(x)|^{2p-2} dx$$

$$\leq (2p-1) \|u\|_{L^\infty} \|\partial_j^2 u\|_{L^p} \||\partial_j u|^{2p-2}\|_{L^{\frac{p}{p-1}}} = (2p-1) \|u\|_{L^\infty} \|\partial_j^2 u\|_{L^p} \|\partial_j u\|_{L^{2p}}^{2p-2}.$$

This proves $\|\partial_j u\|_{L^{2p}}^2 \leq (2p-1) \|u\|_{L^\infty} \|\partial_j^2 u\|_{L^p}$ and hence the assertion. $\square$

5.1.6 Lemma. $(a+b)^s \leq 2^{s-1}(a^s + b^s)$ for $a, b \geq 0, s \geq 1$.

PROOF: $f(x) := x^s$ is convex, hence $f(a + t(b-a)) \leq f(a) + t(f(b) - f(a))$
for $0 \leq t \leq 1$.
Let $t := 1/2$, then

$$\left(\frac{1}{2}(a+b)\right)^s = f\left(a + \frac{1}{2}(b-a)\right) \leq f(a) + \frac{1}{2}(f(b) - f(a)) = \frac{1}{2}(a^s + b^s)$$

This implies the assertion. $\square$

5.1.7 Lemma. Let $f \in C^2([a, b])$, $a, b \in \mathbf{R}, a < b$, and $1 \leq p, q, r < \infty$ with $\frac{2}{q} = \frac{1}{r} + \frac{1}{p}$. Then

$$\|f'\|_{L^q(a,b)}^q \leq 18^q \left[(b-a)^{1+q-q/p} \|f''\|_{L^p(a,b)}^q + (b-a)^{-(1+q-q/p)} \|f\|_{L^r(a,b)}^q\right]$$

PROOF: The mean value theorem shows that for $\xi, \eta \in \left(0, \frac{b-a}{3}\right)$ there is a $\lambda \in (a+\xi, b-\eta)$ with

$$|f'(\lambda)| = \left|\frac{f(b-\eta) - f(a+\xi)}{(b-\eta) - (a+\xi)}\right| \leq \frac{3(|f(b-\eta)| + |f(a+\xi)|)}{b-a}$$

because $|b - \eta - (a + \xi)| \geq \frac{b-a}{3}$. Hence

$$|f'(x)| = \left|f'(\lambda) + \int_\lambda^x f''(t) dt\right| \leq \frac{3}{b-a}(|f(b-\eta)| + |f(a+\xi)|) + \int_a^b |f''(t)| dt$$

for $x \in [a, b]$, which implies

$$\frac{(b-a)^2}{9}|f'(x)| = \int_0^{\frac{b-a}{3}} \int_0^{\frac{b-a}{3}} |f'(x)| d\eta d\xi$$

$$\leq \frac{3}{b-a} \int_0^{\frac{b-a}{3}} \int_0^{\frac{b-a}{3}} |f(b-\eta)| d\eta d\xi + \frac{3}{b-a} \int_0^{\frac{b-a}{3}} \int_0^{\frac{b-a}{3}} |f(a+\xi)| d\eta d\xi$$

$$+ \int_0^{\frac{b-a}{3}} \int_0^{\frac{b-a}{3}} \int_a^b |f''(t)| dt d\eta d\xi$$

$$\leq \int_a^b |f(t)| dt + \frac{(b-a)^2}{9} \int_a^b |f''(t)| dt$$

$$\leq \|f\|_{L^r(a,b)} \left(\int_a^b 1 dt \right)^{1-1/r} + \frac{(b-a)^2}{9} \|f''\|_{L^p(a,b)} \left(\int_a^b 1 dt \right)^{1-1/p}$$

Thus

$$|f'(x)|^q \leq \left(9(b-a)^{1-1/r-2} \|f\|_{L^r(a,b)} + (b-a)^{1-1/p} \|f''\|_{L^p(a,b)} \right)^q$$

$$\underset{5.1.6}{\leq} 2^{q-1} \left(9^q (b-a)^{q(-1-\frac{1}{r})} \|f\|_{L^r(a,b)}^q + (b-a)^{q-\frac{q}{p}} \|f''\|_{L^p(a,b)}^q \right).$$

Noting that $q\left(-1-\frac{1}{r}\right) = -q + q/p - 2$ this shows

$$\|f'\|_{L^q(a,b)}^q \leq 18^q \left((b-a)^{-q+q/p-2+1} \|f\|_{L^r(a,b)}^q + (b-a)^{q-q/p+1} \|f''\|_{L^p(a,b)}^q \right),$$

which implies the assertion. $\square$

5.1.8 Lemma. $\sum_{j=1}^n a_j^s b_j^t \leq \left(\sum_{j=1}^n a_j \right)^s \left(\sum_{j=1}^n b_j \right)^t$ for $0 < s, t < 1$, $s+t=1$,

$a_j, b_j \geq 0$, $j = 1, \ldots, n$,

PROOF: Due to $\frac{1}{1/s} + \frac{1}{1/t} = 1$ and $1/s, 1/t > 1$ Hölder's inequality implies

$$\sum_{j=1}^n a_j^s b_j^t \leq \left[\sum_{j=1}^n (a_j^s)^{1/s} \right]^{\frac{1}{1/s}} \left[\sum_{j=1}^n (b_j^t)^{1/t} \right]^{\frac{1}{1/t}} = \left(\sum_{j=1}^n a_j \right)^s \left(\sum_{j=1}^n b_j \right)^t.$$

$\square$

5.1.9 Lemma. Let $1 < p, q, r < \infty$ with $\frac{2}{q} = \frac{1}{r} + \frac{1}{p}$. Then

$$\|f'\|_{L^q(\mathbf{R}^n)} \leq 2 \cdot 18^q \|f''\|_{L^p(\mathbf{R}^n)}^{1/2} \|f\|_{L^r(\mathbf{R}^n)}^{1/2} \qquad \text{for } f \in \mathscr{S}(\mathbf{R}, \mathbf{R}).$$

PROOF: For $a, b \in \mathbb{R}$ with $a < b$ let $T_1(a, b) := (b-a)^{1+q-q/p} \|f''\|^q_{L^p(a,b)}$ and $T_2(a, b) := (b-a)^{-(1+q-q/p)} \|f\|^q_{L^r(a,b)}$. Then 5.1.7 shows

$$\|f'\|^q_{L^q(a,b)} \le 18^q (T_1(a, b) + T_2(a, b)) . \tag{5.1.3}$$

Choose arbitrary $R > 0, k \in \mathbb{N}$, and let $a_0 := -R, a_1' := -R + \frac{2R}{k}$.
If $T_1(a_0, a_1') > T_2(a_0, a_1')$ then with $a_1 := a_1'$ estimate (5.1.3) implies

$$\|f'\|^q_{L^q(a_0,a_1)} \le 18^q \cdot 2 T_1(a_0, a_1) \le 2 \cdot 18^q \left(\frac{2R}{k}\right)^{1+q-q/p} \|f''\|^q_{L^p(a_0,a_1)} . \tag{5.1.4}$$

If $T_1(a_0, a_1') \le T_2(a_0, a_1')$, choose a large $a_1 \ge a_1'$ with $T_1(a_0, a_1) = T_2(a_0, a_1)$. This is possible because $1 + q - \frac{q}{p} = 1 + q + \frac{q}{r} - 2 = q - 1 + \frac{q}{r} > 0$. Then

$$\|f'\|^q_{L^q(a_0,a_1)} \le 18^q (2 T_1(a_0, a_1)) = 2 \cdot 18^q \sqrt{T_1(a_0, a_1)} \sqrt{T_2(a_0, a_1)}$$

$$\le 2 \cdot 18^q \|f''\|^{q/2}_{L^p(a_0,a_1)} \|f\|^{q/2}_{L^r(a_0,a_1)} . \tag{5.1.5}$$

Proceeding in the same way for $i = 2, \dots, k', k' \le k$, with a_{i-1} replacing a_0 and $a_i' := a_{i-1} + \frac{2R}{k}$ replacing a_1', as long as $a_{i-1} < R$ holds, we obtain

$$\|f'\|^q_{L^q(-R,R)} \le \sum_{i=1}^{k'} \|f'\|^q_{L^q(a_{i-1},a_i)} \le 2 \cdot 18^q \left(\sum_{i=1}^{k'} \|f''\|^{q/2}_{L^p(a_{i-1},a_i)} \|f\|^{q/2}_{L^r(a_{i-1},a_i)}\right.$$

$$\left. + \sum_{i=1}^{k'} \left(\frac{2R}{k}\right)^{1+q-q/p} \|f''\|^q_{L^p(a_{i-1},a_i)}\right) . \tag{5.1.6}$$

Now we have $\frac{q}{2p} + \frac{q}{2r} = 1$, hence

$$\sum_{i=1}^{k'} \|f''\|^{q/2}_{L^p(a_{i-1},a_i)} \|f\|^{q/2}_{L^r(a_{i-1},a_i)} = \sum_{i=1}^{k'} \left(\int_{a_{i-1}}^{a_i} |f''(x)|^p dx\right)^{\frac{q}{2p}} \left(\int_{a_{i-1}}^{a_i} |f(x)|^r dx\right)^{\frac{q}{2r}}$$

$$\underset{5.1.8}{\le} \left(\sum_{i=1}^{k'} \int_{a_{i-1}}^{a_i} |f''(x)|^p dx\right)^{\frac{q}{2p}} \left(\sum_{i=1}^{k'} \int_{a_{i-1}}^{a_i} |f(x)|^r dx\right)^{\frac{q}{2r}} \le \|f''\|^{q/2}_{L^p} \|f\|^{q/2}_{L^r} . \tag{5.1.7}$$

Moreover

$$\sum_{i=1}^{k'} \left(\frac{2R}{k}\right)^{1+q-q/p} \|f''\|^q_{L^p(a_{i-1},a_i)} = \left(\frac{2R}{k}\right)^{1+q-q/p} \sum_{i=1}^{k'} \left(\int_{a_{i-1}}^{a_i} |f''(x)|^p dx\right)^{q/p}$$

$$\le \left(\frac{2R}{k}\right)^{1+q-q/p} \sum_{i=1}^{k'} \|f''\|^q_{L^p} \le \left(\frac{2R}{k}\right)^{1+q-q/p} k \|f''\|^q_{L^p}$$

$$= (2R)^{1+q-q/p} \left(\frac{1}{k}\right)^{q-q/p} \|f''\|^q_{L^p} \xrightarrow[k\to\infty]{} 0 \tag{5.1.8}$$

because $q > q/p$ (since $p > 1$).

Hence (with $k \to \infty$) (5.1.6), (5.1.7), and (5.1.8) imply $\|f'\|_{L^q(-R,R)}^q \leq 2 \cdot$ $18^q \|f''\|_{L^p}^{q/2} \|f\|_{L^r}^{q/2}$ for any $R > 0$ and with $R \longrightarrow \infty$ we obtain the assertion. $\square$

5.1.10 Lemma. For $1 < p, q, r < \infty$ with $\frac{2}{q} = \frac{1}{p} + \frac{1}{r}$ and $u \in \mathscr{S}(\mathbf{R}^n, \mathbf{R})$ we have

$$\|\partial_j u\|_{L^q(\mathbf{R}^n)} \leq 2 \cdot 18^q \|\partial_j^2 u\|_{L^p(\mathbf{R}^n)}^{1/2} \|u\|_{L^r(\mathbf{R}^n)}^{1/2} .$$

PROOF: Let $\hat{x}_j := (x_1, \ldots, x_{j-1}, x_{j+1}, \ldots, x_n)$, then

$$\|\partial_j u\|_{L^q}^q = \int_{\mathbf{R}^{n-1}} \int_{\mathbf{R}} |\partial_j u(x)|^q dx_j d\hat{x}_j$$

$$\leq (2 \cdot 18^q)^q \int_{\mathbf{R}^{n-1}} \left(\int_{\mathbf{R}} |\partial_j^2 u(x)|^p dx_j \right)^{q/(2p)} \left(\int_{\mathbf{R}} |u(x)|^r dx_j \right)^{q/(2r)} d\hat{x}_j$$

$$\leq (2 \cdot 18^q)^q \left(\int_{\mathbf{R}^{n-1}} \int_{\mathbf{R}} |\partial_j^2 u(x)|^p dx_j d\hat{x}_j \right)^{q/(2p)} \left(\int_{\mathbf{R}^{n-1}} \int_{\mathbf{R}} |u(x)|^r dx_j d\hat{x}_j \right)^{q/(2r)}$$

$$\leq (2 \cdot 18^q)^q \|\partial_j^2 u\|_{L^p}^{q/2} \|u\|_{L^r}^{q/2}$$

$\square$

5.1.11 Remark. If we assume in 5.1.10 in addition $q \geq 2$, then the proof of 5.1.10 and 5.1.9 can be simplified and one obtains the sharper estimate

$$\|\partial_j u\|_{L^q(\mathbf{R}^n)} \leq \sqrt{q-1} \|\partial_j^2 u\|_{L^p(\mathbf{R}^n)}^{1/2} \|u\|_{L^r(\mathbf{R}^n)}^{1/2} \quad \text{for } u \in \mathscr{S}(\mathbf{R}^n, \mathbf{R}), j = 1, \ldots, m .$$

The proof is an adaption of the proof of 5.1.5.

PROOF: If $q = 2$ we clearly obtain the assertion by partial integration and Hölder's inequality.
Now assume $q > 2$, then $\frac{1}{p} + \frac{1}{r} + \frac{1}{\frac{q}{q-2}} = \frac{1}{p} + \frac{1}{r} + 1 - \frac{2}{q} = 1$. Moreover, 5.1.4 implies that $(\partial_j u)|\partial_j u|^{q-2} \in C^1(\mathbf{R}^n)$ with $\partial_j[(\partial_j u)|\partial_j u|^{q-2}] = (q-1)(\partial_j^2 u)|\partial_j u|^{q-2}$, hence

$$|\partial_j u|^q = (\partial_j u)(\partial_j u)|\partial_j u|^{q-2} = \partial_j[u(\partial_j u)|\partial_j u|^{q-2}] - u\partial_j[(\partial_j u)|\partial_j u|^{q-2}]$$
$$= \partial_j[u(\partial_j u)|\partial_j u|^{q-2}] - (q-1)u(\partial_j^2 u)|\partial_j u|^{q-2} .$$

Therefore

$$\|\partial_j u\|_{L^q}^q = \int_{\mathbf{R}^n} |\partial_j u(x)|^q dx$$

$$\leq \underbrace{\int_{\mathbf{R}^n} \partial_j[u(x)(\partial_j u(x))|\partial_j u(x)|^{q-2}] dx}_{=0} + (q-1) \int_{\mathbf{R}^n} |u(x)||\partial_j^2 u(x)||\partial_j u(x)|^{q-2} dx$$

$$\underset{5.1.3}{\leq} (q-1) \|u\|_{L^r} \|\partial_j^2 u\|_{L^p} \||\partial_j u|^{q-2}\|_{L^{\frac{q}{q-2}}} = (q-1) \|u\|_{L^r} \|\partial_j^2 u\|_{L^p} \|\partial_j u\|_{L^q}^{q-2} .$$

This proves $\left\|\partial_j u\right\|_{L^q}^2 \leq (q-1)\left\|u\right\|_{L^r}\left\|\partial_j^2 u\right\|_{L^p}$ for $u \in \mathscr{S}(\mathbf{R}^n, \mathbf{R})$ and all $q \geq 2$. $\square$

5.1.12 Definition. For $u \in H_p^k, 1 < p < \infty, k \in \mathbf{N}_0$ let

$$\left\|D^k u\right\|_{L^p(\mathbf{R}^n)} := \left(\sum_{|\gamma|=k} \left\|\partial_x^\gamma u\right\|_{L^p(\mathbf{R}^n)}^p \right)^{1/p}.$$

Finally, we obtain the so-called Gagliardo-Moser-Nirenberg estimate.

5.1.13 Proposition. For $1 < p < \infty$ and $m \in \mathbf{N}$ there are constants $c = c(p, m)$ with

$$\left\|D^j u\right\|_{L^{\frac{m}{j}p}(\mathbf{R}^n)} \leq c(p, m)\left\|D^m u\right\|_{L^p(\mathbf{R}^n)}^{j/m}\left\|u\right\|_{L^\infty(\mathbf{R}^n)}^{1-j/m} \tag{5.1.9}$$

for $1 \leq j \leq m, u \in \mathscr{S}(\mathbf{R}^n, \mathbf{R})$.

PROOF: Note that the assertion for $j = m$ is trivial. To prove the assertion for $j \leq m-1$ we proceed inductively for $m \geq 2$. (5.1.9) for $m = 2$ is a consequence of 5.1.5.

Now assume (5.1.9) for an $m \geq 2$. Let $q := \frac{m+1}{m}p$ and $r := \frac{m+1}{m-1}p$, then

$$\frac{1}{p} + \frac{1}{r} = \frac{1}{p} + \frac{m-1}{(m+1)p} = 2\frac{m}{(m+1)p} = \frac{2}{q}.$$

Now, by the induction hypothesis we have for $\alpha \in \mathbf{N}_0^n, |\alpha| = m-1, e_j \in \mathbf{N}_0^n, |e_j| = 1$

$$\left\|\partial^{e_j}\partial^\alpha u\right\|_{L^{\frac{m+1}{m}p}} = \left\|\partial^{e_j}\partial^\alpha u\right\|_{L^q} \underset{5.1.10}{\leq} 2 \cdot 18^q \left\|\partial^{2e_j}\partial^\alpha u\right\|_{L^p}^{1/2}\left\|\partial^\alpha u\right\|_{L^r}^{1/2}$$

$$\leq 2 \cdot 18^q \left\|D^{m+1}u\right\|_{L^p}^{1/2}\left\|D^{m-1}u\right\|_{L^{\frac{m}{m-1}\frac{m+1}{m}p}}^{1/2}$$

$$\leq 2 \cdot 18^q \left\|D^{m+1}u\right\|_{L^p}^{1/2} c\left(\frac{m+1}{m}p, m\right)\left\|D^m u\right\|_{L^{\frac{m+1}{m}p}}^{\frac{m-1}{2m}}\left\|u\right\|_{L^\infty}^{\frac{1}{2}\left(1-\frac{m-1}{m}\right)}.$$

Hence

$$\left\|D^m u\right\|_{L^{\frac{m+1}{m}p}}^{\frac{m+1}{2m}} = \left\|D^m u\right\|_{L^{\frac{m+1}{m}p}}^{1-\frac{m-1}{2m}} \leq c'\left\|D^{m+1}u\right\|_{L^p}^{1/2}\left\|u\right\|_{L^\infty}^{\frac{1}{2m}}$$

and

$$\left\|D^m u\right\|_{L^{\frac{m+1}{m}p}} \leq (c')^{\frac{2m}{m+1}}\left\|D^{m+1}u\right\|_{L^p}^{\frac{m}{m+1}}\left\|u\right\|_{L^\infty}^{\frac{1}{m+1}}.$$

Thus, using the induction hypothesis again we obtain for $j \leq m$

$$\|D^j u\|_{L^{\frac{m+1}{j}p}} = \|D^j u\|_{L^{\frac{m}{j}\frac{m+1}{m}p}} \leq c\left(\frac{m+1}{m}p, m\right) \|D^m u\|_{L^{\frac{m+1}{m}p}}^{j/m} \|u\|_{L^\infty}^{1-j/m}$$

$$\leq c\left(\frac{m+1}{m}p, m\right) (c')^{\frac{2mj}{m(m+1)}} \|D^{m+1}u\|_{L^p}^{\frac{j}{m+1}} \|u\|_{L^\infty}^{\frac{j}{m(m+1)}+1-\frac{j}{m}}$$

$$\leq c(p, m+1) \|D^{m+1}u\|_{L^p}^{\frac{j}{m+1}} \|u\|_{L^\infty}^{1-\frac{j}{m+1}} .$$

This implies the assertion. $\square$

5.1.14 Remark. In the same manner one can show for $1 < p < \infty$ that there are constants with

$$\|\partial_k^j u\|_{L^{\frac{m}{j}p}(\mathbf{R}^n)} \leq c(p, m) \|\partial_k^m u\|_{L^p(\mathbf{R}^n)}^{j/m} \|u\|_{L^\infty(\mathbf{R}^n)}^{1-j/m}$$

for $k = 1, \ldots, n, 1 \leq j \leq m, u \in \mathscr{S}(\mathbf{R}^n, \mathbf{R})$.

The geometric mean value can always be estimated by the arithmetic mean value.

5.1.15 Lemma. $a^s b^{1-s} \leq sa + (1-s)b \leq a + b$ for $a, b \geq 0, 0 \leq s \leq 1$.

5.1.16 Proposition. For $1 < p < \infty, \alpha, \beta \in \mathbf{N}_0^n, k, m \in \mathbf{N}_0$ with $|\alpha|+|\beta| = m$ there is a constant $c = c_{k,m,p} > 0$ such that for $u, v \in H_p^{k+m} \cap C_0(\mathbf{R}^n, \mathbf{R})$ we have $(\partial^\alpha u)(\partial^\beta v) \in H_p^k$ with

$$\|(\partial^\alpha u)(\partial^\beta v)\|_{H_p^k} \leq c\left(\|u\|_{L^\infty(\mathbf{R}^n)} \|v\|_{H_p^{k+m}} + \|u\|_{H_p^{k+m}} \|v\|_{L^\infty(\mathbf{R}^n)}\right) \qquad (5.1.10)$$

for $u, v \in H_p^{k+m} \cap C_0(\mathbf{R}^n, \mathbf{R})$.

PROOF: Step 1: Proof of (5.1.10) for $u, v \in \mathscr{S}(\mathbf{R}^n, \mathbf{R})$:
We start with the case $k = 0$. This is trivial for $|\alpha| = 0$ or $|\beta| = 0$ and we can assume $m, |\alpha|, |\beta| > 0$. Then 5.1.3 implies due to $\frac{1}{m/|\alpha|} + \frac{1}{m/|\beta|} = \frac{|\alpha|+|\beta|}{m} = 1$

$$\|(\partial^\alpha u)(\partial^\beta v)\|_{L^p} \leq \|\partial^\alpha u\|_{L^p \frac{m}{|\alpha|}} \|\partial^\beta v\|_{L^p \frac{m}{|\beta|}} \leq \left\|D^{|\alpha|}u\right\|_{L^p \frac{m}{|\alpha|}} \left\|D^{|\beta|}v\right\|_{L^p \frac{m}{|\beta|}}$$

$$\underset{5.\overline{1}.13}{\leq} c\|D^m u\|_{L^p}^{\frac{|\alpha|}{m}} \|u\|_{L^\infty}^{1-\frac{|\alpha|}{m}} \|D^m v\|_{L^p}^{\frac{|\beta|}{m}} \|v\|_{L^\infty}^{1-\frac{|\beta|}{m}}$$

$$\leq c\left(\|u\|_{H_p^m} \|v\|_{L^\infty}\right)^{\frac{|\alpha|}{m}} \left(\|u\|_{L^\infty} \|v\|_{H_p^m}\right)^{1-\frac{|\alpha|}{m}}$$

$$\underset{5.\overline{1}.15}{\leq} c\left(\|u\|_{L^\infty} \|v\|_{H_p^m} + \|u\|_{H_p^m} \|v\|_{L^\infty}\right) .$$

This proves step 1 for $k = 0$. For $k > 0$ we have with suitable constants $c, c', c'', c''' \geq 0$

$$\left\| (\partial^\alpha u)(\partial^\beta v) \right\|_{H_p^k} \leq c \sum_{|\gamma| \leq k} \left\| \partial^\gamma [(\partial^\alpha u)(\partial^\beta v)] \right\|_{L^p}$$

$$\leq c' \sum_{l=0}^{k} \sum_{|\gamma_1| + |\gamma_2| = l} \left\| (\partial^{\alpha + \gamma_1} u)(\partial^{\beta + \gamma_2} v) \right\|_{L^p}$$

$$\leq c'' \sum_{l=0}^{k} \sum_{|\gamma_1| + |\gamma_2| = l} \left(\|u\|_{L^\infty} \|v\|_{H_p^{k+m}} + \|u\|_{H_p^{k+m}} \|v\|_{L^\infty} \right)$$

$$\leq c''' \left(\|u\|_{L^\infty} \|v\|_{H_p^{k+m}} + \|u\|_{H_p^{k+m}} \|v\|_{L^\infty} \right).$$

Step 2: Proof of the assertion:
Let $u, v \in H_p^{k+m} \cap C_0(\mathbf{R}^n, \mathbf{R})$. There are sequences $(u_j)_j, (v_j)_j \subset \mathscr{S}(\mathbf{R}^n, \mathbf{R})$ (due to 4.1.23) such that $u_j \xrightarrow[j \to \infty]{} u$ and $v_j \xrightarrow[j \to \infty]{} v$ in H_p^{k+m} and $C_0(\mathbf{R}^n)$ with $\|u_j\|_{H_p^{k+m}} \leq d \|u\|_{H_p^{k+m}}$, $\|v_j\|_{H_p^{k+m}} \leq d \|v\|_{H_p^{k+m}}$, $\|u_j\|_{L^\infty} \leq d \|u\|_{L^\infty}$, and $\|v_j\|_{L^\infty} \leq d \|v\|_{L^\infty}$ for a suitable $d \geq 0$. Then

$$\left\| (\partial^\alpha u_j)(\partial^\beta v_j) - (\partial^\alpha u_l)(\partial^\beta v_l) \right\|_{H_p^k}$$

$$\leq \left\| (\partial^\alpha u_j)(\partial^\beta (v_j - v_l)) \right\|_{H_p^k} + \left\| \partial^\alpha (u_j - u_l)(\partial^\beta v_l) \right\|_{H_p^k}$$

$$\underset{\text{Step 1}}{\leq} cd \left(\|u\|_{L^\infty} \|v_j - v_l\|_{H_p^{k+m}} + \|u\|_{H_p^{k+m}} \|v_j - v_l\|_{L^\infty} \right.$$

$$\left. + \|u_j - u_l\|_{L^\infty} \|v\|_{H_p^{k+m}} + \|u_j - u_l\|_{H_p^{k+m}} \|v\|_{L^\infty} \right).$$

Hence $((\partial^\alpha u_j)(\partial^\beta v_j))_j \subset H_p^k$ is a Cauchy sequence and there is a $w \in H_p^k$ with $(\partial^\alpha u_j)(\partial^\beta v_j) \xrightarrow[j \to \infty]{} w$ in H_p^k and $\|w\|_{H_p^k} \leq c \left(\|u\|_{L^\infty} \|v\|_{H_p^{k+m}} + \|u\|_{H_p^{k+m}} \|v\|_{L^\infty} \right)$. Moreover, since $\partial^\alpha u_j \xrightarrow[j \to \infty]{} \partial^\alpha u$ and $\partial^\beta v_j \xrightarrow[j \to \infty]{} \partial^\beta v$ in H_p^k and $H_p^k \hookrightarrow L^p(\mathbf{R}^n)$ we can choose subsequently a sequence $(j_l)_l$ with $\partial^\alpha u_{j_l} \xrightarrow[l \to \infty]{} \partial^\alpha u$, $\partial^\beta v_{j_l} \xrightarrow[l \to \infty]{} \partial^\beta v$, and $(\partial^\alpha u_{j_l})(\partial^\beta v_{j_l}) \xrightarrow[l \to \infty]{} w$ point-wise almost everywhere. But this implies $(\partial^\alpha u_{j_l})(\partial^\beta v_{j_l}) \xrightarrow[l \to \infty]{} (\partial^\alpha u)(\partial^\beta v)$ point-wise almost everywhere and thus we have $w = (\partial^\alpha u)(\partial^\beta v)$ point-wise almost everywhere. This proves the assertion. $\square$

Finally, in 5.1.17 and 5.1.18 we prove similar estimates for fractional L^2-Sobolev spaces.

5.1.17 Lemma. Fix $K > \frac{n}{2}$. Then, for $s \geq 0$, there is a $c_s \geq 0$ with

$$\|uv\|_{H^s} \leq c_s \left(\|u\|_{H^K} \|v\|_{H^s} + \|u\|_{H^s} \|v\|_{H^K} \right)$$

for $u, v \in H^\infty$.

PROOF: We only have to prove the assertion for $u, v \in \mathscr{S}(\mathbf{R}^n)$. First note that there is a $d_s \geq 0$ with

$$
\begin{aligned}
\langle \xi \rangle^s &= (1 + |\xi|^2)^{s/2} \leq (1 + 2|\eta|^2 + 2|\xi - \eta|^2)^{s/2} \\
&\leq d_s \left((1 + |\eta|^2)^{s/2} + (1 + |\xi - \eta|^2)^{s/2} \right) = d_s (\langle \eta \rangle^s + \langle \xi - \eta \rangle^s).
\end{aligned}
$$

Hence, if $u * v$ denotes the convolution of u, v we have with suitable constants d, d'_s, c_s

$$
\begin{aligned}
\langle \xi \rangle^s \widehat{uv}(\xi) &= d\langle \xi \rangle^s \hat{u} * \hat{v}(\xi) \leq dd_s \int_{\mathbf{R}^n} (\langle \eta \rangle^s + \langle \xi - \eta \rangle^s) \hat{u}(\xi - \eta) \hat{v}(\eta) d\eta \\
&= d'_s \left((((\langle \cdot \rangle^s \hat{u}) * \hat{v})(\xi) + (\hat{u} * (\langle \cdot \rangle^s \hat{v}))(\xi)) \right)
\end{aligned}
$$

and thus

$$
\begin{aligned}
\|uv\|_{H^s} &= \left(\int_{\mathbf{R}^n} \langle \xi \rangle^{2s} |\widehat{uv}(\xi)|^2 d\xi \right)^{1/2} \\
&\leq d'_s \left(\|(\langle \cdot \rangle^s \hat{u}) * \hat{v}\|_{L^2(\mathbf{R}^n)} + \|\hat{u} * (\langle \cdot \rangle^s \hat{v})\|_{L^2(\mathbf{R}^n)} \right) \\
&= c_s \left(\|(\langle D_x \rangle^s u)v\|_{L^2(\mathbf{R}^n)} + \|u(\langle D_x \rangle^s v)\|_{L^2(\mathbf{R}^n)} \right) \\
&\leq c_s \left(\|u\|_{L^\infty(\mathbf{R}^n)} \|v\|_{H^s} + \|u\|_{H^s} \|v\|_{L^\infty(\mathbf{R}^n)} \right).
\end{aligned}
$$

This implies the assertion.   □

5.1.18 Lemma. Let $K > \frac{n}{2}, t \geq 1$, then there is a $c_t \geq 0$ with

$$\left\| \langle D_x \rangle^t (uv) - u(\langle D_x \rangle^t v) \right\|_{L^2(\mathbf{R}^n)} \leq c_t \left(\|u\|_{H^{K+1}} \|v\|_{H^{t-1}} + \|u\|_{H^t} \|v\|_{H^K} \right)$$

for $u \in H^{K+1} \cap H^t, v \in \mathscr{S}(\mathbf{R}^n)$.

PROOF: We only have to prove the assertion for $u \in \mathscr{S}(\mathbf{R}^n)$.
Let $w := \langle D_x \rangle^t (uv) - u(\langle D_x \rangle^t v) \in \mathscr{S}(\mathbf{R}^n)$ and note that with suitable $c, c' \geq 0$

$$|\langle \xi \rangle^t - \langle \eta \rangle^t| = \left| \int_0^1 \frac{d}{d\tau} \langle \eta + \tau(\xi - \eta) \rangle^t d\tau \right| \leq c \int_0^1 \langle \eta + \tau(\xi - \eta) \rangle^{t-1} d\tau |\xi - \eta|$$
$$\leq c' (\langle \xi \rangle^{t-1} + \langle \eta \rangle^{t-1}) |\xi - \eta| .$$

Hence

$$|\widehat{w}(\xi)| = \left| \langle \xi \rangle^t \widehat{uv}(\xi) - (u\langle D_x \rangle^t v)\widehat{\;}(\xi) \right| = \left| d \int_{\mathbf{R}^n} (\langle \xi \rangle^t - \langle \eta \rangle^t) \widehat{u}(\xi - \eta) \widehat{v}(\eta) d\eta \right|$$
$$\leq c'' \int_{\mathbf{R}^n} (\langle \xi \rangle^{t-1} + \langle \eta \rangle^{t-1}) |\xi - \eta| |\widehat{u}(\xi - \eta)| |\widehat{v}(\eta)| d\eta$$

for suitable $d, c'' \geq 0$. Now $|\xi| |\widehat{u}(\xi)|, |\widehat{v}(\xi)| \in L^2(\mathbf{R}^n)$, hence there are functions $f, g \in L^2(\mathbf{R}^n)$ with $\widehat{f}(\xi) = |\xi| |\widehat{u}(\xi)|$ and $\widehat{g}(\eta) = |\widehat{v}(\eta)|$. Thus, for $s \geq 0$

$$\|f\|_{H^s}^2 = \int_{\mathbf{R}^n} \langle \xi \rangle^{2s} |\xi|^2 |\widehat{u}(\xi)|^2 d\xi \leq \int_{\mathbf{R}^n} \langle \xi \rangle^{2(s+1)} |\widehat{u}(\xi)|^2 d\xi = \|u\|_{H^{s+1}}^2$$

and

$$\|g\|_{H^s}^2 = \int_{\mathbf{R}^n} \langle \xi \rangle^{2s} |\widehat{g}(\xi)|^2 d\xi = \int_{\mathbf{R}^n} \langle \xi \rangle^{2s} |\widehat{v}(\xi)|^2 d\xi = \|v\|_{H^s}^2 .$$

Hence $f, g \in H^\infty$, and we have

$$|\widehat{w}(\xi)| \leq c'' \left(\langle \xi \rangle^{t-1} (\widehat{f} * \widehat{g})(\xi) + (\widehat{f} * ((\cdot)^{t-1} \widehat{g}))(\xi) \right)$$
$$= d' \left(\langle \xi \rangle^{t-1} \widehat{fg}(\xi) + (f(\langle D_x \rangle^{t-1} \widehat{g}))\widehat{\;}(\xi) \right)$$

for a $d' \geq 0$, and hence for suitable $d'', c_t \geq 0$ using 5.1.17

$$\|w\|_{L^2(\mathbf{R}^n)} = \left(\int_{\mathbf{R}^n} |\widehat{w}(\xi)|^2 d\xi \right)^{1/2}$$
$$\leq d' \left(\left(\int_{\mathbf{R}^n} \langle \xi \rangle^{2(t-1)} |\widehat{fg}(\xi)|^2 d\xi \right)^{1/2} + \left(\int_{\mathbf{R}^n} |(f\langle D_x \rangle^{t-1} g)\widehat{\;}(\xi)|^2 d\xi \right)^{1/2} \right)$$
$$\leq d'' \left(\|fg\|_{H^{t-1}} + \|f\|_{L^\infty(\mathbf{R}^n)} \|g\|_{H^{t-1}} \right) \leq c_t \left(\|f\|_{H^K} \|g\|_{H^{t-1}} + \|f\|_{H^{t-1}} \|g\|_{H^K} \right)$$
$$\leq c_t \left(\|u\|_{H^{K+1}} \|v\|_{H^{t-1}} + \|u\|_{H^t} \|v\|_{H^K} \right) .$$

This proves the assertion. □

5.1.19 Definition. For a Banach space $(X, \|\cdot\|)$ and $K \in \mathbb{N}_0 \cup \{\infty\}$ we call $\mathcal{F} \subset C^K(\mathbb{R}^n, X)$ bounded on compact subsets if for $k \in \mathbb{N}_0$ with $k \leq K$ and $R > 0$ there are $F_k(R) \geq 0$ with $\|\partial^\alpha f(x)\| \leq F_k(R)$ for $\alpha \in \mathbb{N}_0^n, |\alpha| \leq k, f \in \mathcal{F}$, and $x \in \mathbb{R}^n$ with $|x| \leq R$.

5.1.20 Lemma. Let $\alpha \in \mathbb{N}_0^n$ with $\alpha \neq 0$, let $f \in C^{|\alpha|}(\mathbb{R}_y^N, \mathbb{C})$, and let $u = (u_1, \ldots, u_N) \in C^{|\alpha|}(\mathbb{R}_x^n, \mathbb{R}_y^N)$, then

$$\partial_x^\alpha(f(u)) = \sum_{\substack{1 \leq |\beta| \leq |\alpha| \\ \beta \in \mathbb{N}_0^N}} (\partial_y^\beta f)(u) \sum_{\substack{\gamma_1 + \ldots + \gamma_{|\beta|} = \alpha, \gamma_l \in \mathbb{N}_0^n \setminus \{0\} \\ l_1, \ldots, l_{|\beta|} \in \{1, \ldots, N\}}} c_{\gamma, l} \prod_{j=1}^{|\beta|} \partial_x^{\gamma_j} u_{l_j} \qquad (5.1.11)$$

with suitable $c_{\gamma, l} = c_{\gamma_1, \ldots, \gamma_{|\beta|}, l_1, \ldots, l_{|\beta|}}$ for $\gamma = (\gamma_1, \ldots, \gamma_{|\beta|}), l = (l_1, \ldots, l_{|\beta|})$.

PROOF: Assume (5.1.11) for an $\alpha \in \mathbb{N}_0^n, \alpha \neq 0$. Then, for $i \in \{1, \ldots, n\}$

$$\partial_x^{\alpha + e_i}(f(u)) = \partial_x^{e_i}\left[\sum_{\substack{1 \leq |\beta| \leq |\alpha| \\ \beta \in \mathbb{N}_0^N}} (\partial_y^\beta f)(u) \sum_{\substack{\gamma_1 + \ldots + \gamma_{|\beta|} = \alpha, \gamma_l \in \mathbb{N}_0^n \setminus \{0\} \\ l_1, \ldots, l_{|\beta|} \in \{1, \ldots, N\}}} c_{\gamma, l} \prod_{j=1}^{|\beta|} \partial_x^{\gamma_j} u_{l_j} \right]$$

$$= \sum_{\substack{1 \leq |\beta| \leq |\alpha| \\ \beta \in \mathbb{N}_0^N}} \sum_{k=1}^N (\partial_y^{\beta + e_k} f)(u)(\partial_x^{e_i} u_k) \sum_{\substack{\gamma_1 + \ldots + \gamma_{|\beta|} = \alpha, \gamma_l \in \mathbb{N}_0^n \setminus \{0\} \\ l_1, \ldots, l_{|\beta|} \in \{1, \ldots, N\}}} c_{\gamma, l} \prod_{j=1}^{|\beta|} \partial_x^{\gamma_j} u_{l_j}$$

$$+ \sum_{\substack{1 \leq |\beta| \leq |\alpha| \\ \beta \in \mathbb{N}_0^N}} (\partial_y^\beta f)(u) \sum_{\substack{\gamma_1 + \ldots + \gamma_{|\beta|} = \alpha, \gamma_l \in \mathbb{N}_0^n \setminus \{0\} \\ l_1, \ldots, l_{|\beta|} \in \{1, \ldots, N\}}} c_{\gamma, l} \sum_{l_0 = 1}^{|\beta|} \partial_x^{\gamma_k + e_i} u_{l_k} \prod_{\substack{j=1 \\ j \neq k}}^{|\beta|} \partial_x^{\gamma_j} u_{l_j}$$

This implies the assertion. $\square$

5.1.21 Lemma. Let $1 < p < \infty, k \in \mathbb{N}$, and $\mathcal{F} \subset C^k(\mathbb{R}_y^N, \mathbb{C})$ boundedly on compact subsets. Then there is a function $G_k : [0, \infty) \longrightarrow [0, \infty)$ such that for $\alpha \in \mathbb{N}_0^n$ with $0 < |\alpha| \leq k, R > 0$, and $u = (u_1, \ldots, u_N) \in \mathscr{S}(\mathbb{R}_x^n, \mathbb{R}_y^N)$ with $\|u\|_{L^\infty(\mathbb{R}^n, \mathbb{R}^N)} \leq R$ we have $\partial_x^\alpha f(u) \in L^p(\mathbb{R}_x^n)$ with

$$\|\partial_x^\alpha f(u)\|_{L^p(\mathbb{R}^n)} \leq G_k(R) \|u\|_{H_p^k[\mathbb{R}^N]} .$$

PROOF: We have

$$\left\|\partial_x^\alpha f(u)\right\|_{L^p(\mathbf{R}^n)}$$

$$\leq \sum_{\substack{1\leq|\beta|\leq|\alpha|\\ \beta\in\mathbf{N}_0^N}} \left\|(\partial_y^\beta f)(u)\right\|_{L^\infty} \sum_{\substack{\gamma_1+\ldots+\gamma_{|\beta|}=\alpha,\gamma_l\in\mathbf{N}_0^n\setminus\{0\}\\ l_1,\ldots,l_{|\beta|}\in\{1,\ldots,N\}}} |c_{\gamma,l}| \left\|\prod_{j=1}^{|\beta|}\partial_x^{\gamma_j} u_{l_j}\right\|_{L^p}$$

$$\underset{5.1.3}{\leq} \sum_{\substack{1\leq|\beta|\leq|\alpha|\\ \beta\in\mathbf{N}_0^N}} F_k(R) \sum_{\substack{\gamma_1+\ldots+\gamma_{|\beta|}=\alpha,\gamma_l\in\mathbf{N}_0^n\setminus\{0\}\\ l_1,\ldots,l_{|\beta|}\in\{1,\ldots,N\}}} |c_{\gamma,l}| \prod_{j=1}^{|\beta|}\left\|\partial_x^{\gamma_j} u_{l_j}\right\|_{L^p\frac{|\alpha|}{|\gamma_j|}}$$

$$\underset{5.1.13}{\leq} \sum_{\substack{1\leq|\beta|\leq|\alpha|\\ \beta\in\mathbf{N}_0^N}} F_k'(R) \sum_{\substack{\gamma_1+\ldots+\gamma_{|\beta|}=\alpha,\gamma_l\in\mathbf{N}_0^n\setminus\{0\}\\ l_1,\ldots,l_{|\beta|}\in\{1,\ldots,N\}}} \prod_{j=1}^{|\beta|}\left\|u_{l_j}\right\|_{H_p^k}^{\frac{|\gamma_j|}{|\alpha|}} \left\|u_{l_j}\right\|_{L^\infty}^{1-\frac{|\gamma_j|}{|\alpha|}}$$

$$\leq \sum_{\substack{1\leq|\beta|\leq|\alpha|\\ \beta\in\mathbf{N}_0^N}} F_k'(R) \sum_{\substack{\gamma_1+\ldots+\gamma_{|\beta|}=\alpha,\gamma_l\in\mathbf{N}_0^n\setminus\{0\}\\ l_1,\ldots,l_{|\beta|}\in\{1,\ldots,N\}}} \left\|u\right\|_{H_p^k[\mathbf{R}^N]} R^{|\beta|-1} \leq G_k(R)\left\|u\right\|_{H_p^k[\mathbf{R}^N]}$$

with suitable $F_k(R), F_k'(R), G_k(R) \geq 0$. $\qquad\square$

The following lemma is called Minkowski's inequality for integrals.

5.1.22 Lemma. Let $\Omega_j \subset \mathbf{R}^{n_j}_{x_j}$ be measurable sets, $j = 1, 2, 1 \leq p < \infty$, $f(\cdot, x_2) \in L^p(\Omega_1)$ for a.e. $x_2 \in \Omega_2$, and $[\Omega_2 \ni x_2 \mapsto \|f(\cdot, x_2)\|_{L^p(\Omega_1)}] \in L^1(\Omega_2)$. Then $f(x_1, \cdot) \in L^1(\Omega_2)$ for a.e. $x_1 \in \Omega_1, [\Omega_1 \ni x_1 \mapsto \int_{\Omega_2} f(x_1, x_2)dx_2] \in L^p(\Omega_1)$, and

$$\left\|\int_{\Omega_2} f(\cdot, x_2)dx_2\right\|_{L^p(\Omega_1)} \leq \int_{\Omega_2} \|f(\cdot, x_2)\|_{L^p(\Omega_1)} dx_2$$

PROOF: [47, 6.19]. $\qquad\square$

5.1.23 Proposition. Let $k \in \mathbf{N}, 1 < p < \infty$ with $k > \frac{n}{p}$ and $\mathcal{F} \subset C^k(\mathbf{R}^N, \mathbf{C})$ boundedly on compact subsets with $f(0) = 0$ for $f \in \mathcal{F}$. Then $f(u) \in H_p^k$ for $u \in H_p^k[\mathbf{R}^N]$ and $f \in \mathcal{F}$, and there is a function $G_k : [0, \infty) \longrightarrow [0, \infty)$ with

$$\|f(u)\|_{H_p^k} \leq G_k(R)\|u\|_{H_p^k[\mathbf{R}^N]}$$

for $f \in \mathcal{F}, u = (u_1, \ldots, u_N) \in H_p^k[\mathbf{R}^N], \|u\|_{L^\infty(\mathbf{R}^n, \mathbf{R}^N)} \leq R, R > 0$.

PROOF:

Step 1: There is a function $G_k : [0, \infty) \longrightarrow [0, \infty)$ with

$$\|f(u)\|_{H_p^k} \leq G_k(R)\, \|u\|_{H_p^k[\mathbf{R}^N]}$$

for $f \in \mathcal{F}, u = (u_1, \dots, u_N) \in \mathscr{S}(\mathbf{R}^n, \mathbf{R}^N), \|u\|_{L^\infty(\mathbf{R}^n, \mathbf{R}^N)} \leq R$.

Proof: For $u \in \mathscr{S}(\mathbf{R}^n, \mathbf{R}^N)$ with $\|u\|_{L^\infty(\mathbf{R}^n, \mathbf{R}^N)} \leq R$ we have

$$f(u(x)) = f(0) + \int_0^1 \frac{d}{dt} f(tu(x)) dt = \sum_{j=1}^N \int_0^1 \frac{\partial f}{\partial y_j}(tu(x)) u_j(x) dt \in L^p(\mathbf{R}_x^n).$$

Hence with suitable $F_k(R), F_k'(R) \geq 0$

$$\|f(u(x))\|_{L^p(\mathbf{R}_x^n)} \quad \leq \quad \sum_{j=1}^N \left\| \int_0^1 \frac{\partial f}{\partial y_j}(tu(x)) u_j(x) dt \right\|_{L^p(\mathbf{R}_x^n)}$$

$$\underset{5.1.22}{\leq} \quad \sum_{j=1}^N \int_0^1 \left\| \frac{\partial f}{\partial y_j}(tu(x)) u_j(x) \right\|_{L^p(\mathbf{R}_x^n)} dt$$

$$\leq \quad \sum_{j=1}^N \int_0^1 F_k(R) \|u_j\|_{L^p(\mathbf{R}_x^n)} dt \leq F_k'(R) \|u\|_{L^p(\mathbf{R}_x^n, \mathbf{R}^N)}$$

Hence, step 1 follows with 5.1.21.

Step 2: Proof of the assertion.

Let $u \in H_p^k[\mathbf{R}^N]$. Due to 4.1.23 there is a sequence $(u_{(j)})_j \subset \mathscr{S}(\mathbf{R}^n, \mathbf{R}^N)$ with $u_{(j)} \underset{j \to \infty}{\longrightarrow} u$ in $H_p^k[\mathbf{R}^N]$ and $\|u_{(j)}\|_{L^\infty(\mathbf{R}^n, \mathbf{R}^N)} \leq c \|u\|_{L^\infty(\mathbf{R}^n, \mathbf{R}^N)}$. Since $H_p^k \hookrightarrow C_0(\mathbf{R}^n)$ we also have $u_{(j)} \underset{j \to \infty}{\longrightarrow} u$ uniformly, thus $f(u_{(j)}) \underset{j \to \infty}{\longrightarrow} f(u)$ in $C_0(\mathbf{R}^n)$. Moreover, $\|f(u_{(j)})\|_{H_p^k} \leq G_k(R) \|u\|_{H_p^k[\mathbf{R}^N]}$. Hence, since H_p^k is reflexive and $H_p^k \hookrightarrow C_0(\mathbf{R}^n)$ this shows due to 3.3.2 that $f(u) \in H_p^k$ and $f(u_{(j)}) \underset{j \to \infty}{\longrightarrow} f(u)$ weakly in H_p^k. In particular, this implies

$$\|f(u)\|_{H_p^k} \leq G_k(R) \|u\|_{H_p^k[\mathbf{R}^N]}$$

and proves 5.1.23.   □

5.1.24 Proposition. Let $k \in \mathbb{N}, 1 < p < \infty$ with $k > \frac{n}{p}$ and $\mathcal{F} \subset C^k(\mathbf{R}^N, \mathbb{C})$ boundedly on compact subsets. Then $f(u)w \in H_p^k$ for $f \in \mathcal{F}, u \in H_p^k[\mathbf{R}^N]$, $w \in H_p^k$, and there is a function $G_k : [0, \infty) \to [0, \infty)$ with

$$\|f(u)w\|_{H_p^k} \leq G_k(R)(1 + \|u\|_{H_p^k[\mathbf{R}^N]}) \|w\|_{H_p^k}$$

for $u \in H_p^k[\mathbf{R}^N], \|u\|_{L^\infty(\mathbf{R}^n, \mathbf{R}^N)} \leq R, w \in H_p^k[\mathbf{R}], R > 0$.

PROOF: We only have to prove the assertion for $u \in \mathscr{S}(\mathbf{R}^n_x, \mathbf{R}^N_y), w \in \mathscr{S}(\mathbf{R}^n_x)$. First, for $|\alpha_1| \le |\alpha| - 1, 1 \le |\alpha| \le k$ we have

$$\left\| \partial_x^{\alpha_1} \partial_x^{e_j} f(u) \right\|_{L^{\frac{|\alpha|}{|\alpha_1|+1}p}}$$

$$\underset{5.1.20}{\le} \sum_{\substack{1 \le |\beta| \le |\alpha_1|+1 \\ \beta \in \mathbf{N}_0^N}} \left\| (\partial_y^\beta f)(u) \right\|_{L^\infty} \sum_{\substack{\gamma_1+\dots+\gamma_{|\beta|}=\alpha_1+e_j \\ l_1,\dots,l_{|\beta|} \in \{1,\dots,N\}}} |c_{\gamma,l}| \left\| \prod_{j=1}^{|\beta|} \partial_x^{\gamma_j} u_{l_j} \right\|_{L^{\frac{|\alpha|}{|\alpha_1|+1}p}}$$

$$\underset{5.1.3}{\le} \sum_{\substack{1 \le |\beta| \le |\alpha_1|+1 \\ \beta \in \mathbf{N}_0^N}} F_k(R) \sum_{\substack{\gamma_1+\dots+\gamma_{|\beta|}=\alpha_1+e_j \\ l_1,\dots,l_{|\beta|} \in \{1,\dots,N\}}} |c_{\gamma,l}| \prod_{j=1}^{|\beta|} \left\| \partial_x^{\gamma_j} u_{l_j} \right\|_{L^{\frac{|\alpha_1|+1}{|\gamma_j|}\frac{|\alpha|}{|\alpha_1|+1}p}}$$

$$\underset{5.1.13}{\le} \sum_{\substack{1 \le |\beta| \le |\alpha_1|+1 \\ \beta \in \mathbf{N}_0^N}} F_k'(R) \sum_{\substack{\gamma_1+\dots+\gamma_{|\beta|}=\alpha_1+e_j \\ l_1,\dots,l_{|\beta|} \in \{1,\dots,N\}}} \prod_{j=1}^{|\beta|} \|u_{l_j}\|_{H_p^{|\alpha|}}^{\frac{|\gamma_j|}{|\alpha|}} \|u_{l_j}\|_{L^\infty}^{1-\frac{|\gamma_j|}{|\alpha|}}$$

$$\le F_k''(R) \|u\|_{H_p^{|\alpha|}[\mathbf{R}^N]}^{\frac{|\alpha_1|+1}{|\alpha|}}$$

with constants $F_k(R), F_k'(R), F_k''(R) \ge 0$. Hence for $0 < |\alpha| \le k$ we have with suitable constants $c' \ge 0, G_k(R) \ge 0$

$$\left\| \partial_x^\alpha [f(u)w] \right\|_{L^p}$$

$$\le \|f(u)\partial_x^\alpha w\|_{L^p} + c \sum_{j=1}^{N} \sum_{|\alpha_1|+|\alpha_2|=|\alpha|-1} \left\| (\partial_x^{\alpha_1} \partial_x^{e_j} f(u))(\partial_x^{\alpha_2} w) \right\|_{L^p}$$

$$\underset{5.1.3}{\le} \|f(u)\|_{L^\infty} \|w\|_{H_p^k} + c \sum_{j=1}^{N} \sum_{|\alpha_1|+|\alpha_2|=|\alpha|-1} \left\| \partial_x^{\alpha_1} \partial_x^{e_j} f(u) \right\|_{L^{\frac{|\alpha|}{|\alpha_1|+1}p}} \left\| \partial_x^{\alpha_2} w \right\|_{L^{\frac{|\alpha|}{|\alpha_2|}p}}$$

$$\underset{5.1.13}{\le} F_k(R) \|w\|_{H_p^k} + c' \sum_{j=1}^{N} \sum_{|\alpha_1|+|\alpha_2|=|\alpha|-1} F_k''(R) \|u\|_{H_p^{|\alpha|}[\mathbf{R}^N]}^{\frac{|\alpha_1|+1}{|\alpha|}} \|w\|_{H_p^k}^{\frac{|\alpha_2|}{|\alpha|}} \|w\|_{L^\infty}^{1-\frac{|\alpha_2|}{|\alpha|}}$$

$$\le G_k(R) \left(1 + \|u\|_{H_p^k[\mathbf{R}^N]} \right) \|w\|_{H_p^k}.$$

$\square$

5.1.25 Proposition. Let $k \in \mathbf{N}, 1 < p < \infty$ with $k > \frac{n}{p}$ and $\mathcal{F} \subset C^{k+1}(\mathbf{R}^N, \mathbf{C})$ boundedly on compact subsets. Then $f(u) - f(v) \in H_p^k$ for $u, v \in H_p^k[\mathbf{R}^N]$ and $f \in \mathcal{F}$, and there is a function $G_k : [0, \infty) \longrightarrow [0, \infty)$ with

$$\|f(u) - f(v)\|_{H_p^k} \le G_k(R) \|u - v\|_{H_p^k[\mathbf{R}^N]}$$

for $f \in \mathcal{F}, u, v \in H_p^k[\mathbf{R}^N], \|u\|_{H_p^k[\mathbf{R}^N]} \le R, \|v\|_{H_p^k[\mathbf{R}^N]} \le R, R > 0$.

PROOF: As in 5.1.23 we only have to prove the assertion for $u, v \in \mathscr{S}(\mathbf{R}^n, \mathbf{R}^N)$. Now

$$f(u(x)) - f(v(x)) = \sum_{j=1}^{N} \int_0^1 \frac{\partial f}{\partial y_j}(v(x) + t(u(x) - v(x)))dt(u_j(x) - v_j(x)).$$

Hence $f(u) - f(v) \in C_0(\mathbf{R}^n)$, and for $\alpha \in \mathbf{N}_0^n, |\alpha| \leq k$ we have with suitable $G_k'(R) \geq 0$

$$\|f(u) - f(v)\|_{H_p^k} \leq \sum_{j=1}^{N} \left\| \int_0^1 \left[\frac{\partial f}{\partial y_j}(v + t(u-v))(u_j - v_j) \right] dt \right\|_{H_p^k}$$

$$\underset{5.\overline{1.22}}{\leq} \sum_{j=1}^{N} \int_0^1 \left\| \frac{\partial f}{\partial y_j}(v + t(u-v))(u_j - v_j) \right\|_{H_p^k} dt$$

$$\underset{5.\overline{1.24}}{\leq} \sum_{j=1}^{N} \int_0^1 G_k(3R)(1 + \|v + t(u-v)\|_{H_p^k[\mathbf{R}^N]}) \|u_j - v_j\|_{H_p^k} dt$$

$$\leq G_k'(R) \|u - v\|_{H_p^k[\mathbf{R}^N]}.$$

$\square$

5.1.26 Proposition. Let $k \in \mathbf{N}, 1 < p < \infty$ with $k > \frac{n}{p}$ and $\mathcal{F} \subset C^{k+1}(\mathbf{R}^N, \mathbf{C})$ boundedly on compact subsets. Then $f(u)w - f(v)w \in H_p^k$ for $u, v \in H_p^k[\mathbf{R}^N]$, $w \in H_p^k[\mathbf{R}]$, and $f \in \mathcal{F}$, and there is a function $G_k : [0, \infty) \longrightarrow [0, \infty)$ with

$$\|f(u)w - f(v)w\|_{H_p^k} \leq G_k(R) \|u - v\|_{H_p^k[\mathbf{R}^N]} \|w\|_{H_p^k}$$

for $f \in \mathcal{F}, u, v \in H_p^k[\mathbf{R}^N], \|u\|_{H_p^k[\mathbf{R}^N]} \leq R, \|v\|_{H_p^k[\mathbf{R}^N]} \leq R, w \in H_p^k[\mathbf{R}], R > 0$.

PROOF: The assertion follows with suitable $c, F_k'(R), G_k(R) \geq 0$ from

$$\|f(u)w - f(v)w\|_{H_p^k} = \|(f(u) - f(v))w\|_{H_p^k}$$

$$\underset{5.\overline{1.25}}{\overset{5.1.16}{\leq}} c \left(\|f(u) - f(v)\|_{L^\infty} \|w\|_{H_p^k} + \|f(u) - f(v)\|_{H_p^k} \|w\|_{L^\infty} \right)$$

$$\underset{5.\overline{1.25}}{\leq} c \left(F_k'(R) \|u - v\|_{L^\infty(\mathbf{R}^n, \mathbf{R}^N)} \|w\|_{H_p^k} + G_k(R) \|u - v\|_{H_p^k[\mathbf{R}^N]} \|w\|_{H_p^k} \right).$$

$\square$

5.1.27 Proposition. Let $K \in \mathbb{N}, p \geq 2$ with $K > \frac{n}{p} + 1$ and $\mathcal{F} \subset C^1(\mathbb{R}^N, \mathbb{R})$ boundedly on compact subsets. Then there is a function $\beta_p : [0, \infty) \to [0, \infty)$ with

$$\left\| \lambda w - \sum_{j=1}^n a_j(u) \partial_j w \right\|_{L^p(\mathbb{R}^n)} \geq (\lambda - \beta_p(R)) \|w\|_{L^p(\mathbb{R}^n)}$$

for $R > 0, \lambda > \beta_p(R), w \in H_p^1, a_j \in \mathcal{F}, u \in H_p^K$ with $\|u\|_{H_p^K} \leq R$.

PROOF: $\|a_j(u)\|_{L^\infty} \leq G(R)$ and $\|\partial_k a_j(u)\|_{L^\infty} \leq G(R)$ for $a_j \in \mathcal{F}$, $u \in H_p^K$, $j, k = 1, \ldots, n$ with $\|u\|_{H_p^K} \leq R$ and a suitable function $G : [0, \infty) \to [0, \infty)$. Hence 4.5.2 implies the assertion. □

5.1.28 Proposition. Let $K \in \mathbb{N}$ with $K > \frac{n}{2} + 1$, and $\mathcal{F} \subset C^1(\mathbb{R}^N, \mathscr{L}(\mathbb{C}^M))$ boundedly on compact subsets with $a(x)^* = a(x)$ for $x \in \mathbb{R}^N$ and $a \in \mathcal{F}$. Then there is a function $\beta : [0, \infty) \to [0, \infty)$ with

$$\left| \operatorname{Re} \left\langle \sum_{j=1}^n a_j(u) \partial_j w, w \right\rangle_{L^2(\mathbb{R}^n, \mathbb{R}^M)} \right| \leq \beta(R) \|w\|_{L^2(\mathbb{R}^n, \mathbb{R}^M)}^2$$

for $R > 0, w \in H^1[\mathbb{R}^M], a_j \in \mathcal{F}, j = 1, \ldots, n, u \in H^K[\mathbb{R}^N]$ with $\|u\|_{H^K[\mathbb{R}^N]} \leq R$.

PROOF:

$$2 \left| \operatorname{Re} \left\langle \sum_{j=1}^n a_j(u) \partial_j w, w \right\rangle_{L^2(\mathbb{R}^n, \mathbb{R}^M)} \right|$$

$$= \left| \sum_{j=1}^n \left(\langle a_j(u) \partial_j w, w \rangle_{L^2(\mathbb{R}^n, \mathbb{R}^M)} - \langle \partial_j(a_j(u)w), w \rangle_{L^2(\mathbb{R}^n, \mathbb{R}^M)} \right) \right|$$

$$\leq \left| \sum_{j=1}^n \|(\partial_j a_j(u))w\|_{L^2(\mathbb{R}^n, \mathbb{R}^M)} \|w\|_{L^2(\mathbb{R}^n, \mathbb{R}^M)} \right|$$

$$\leq \sum_{j=1}^n \|\partial_j a_j(u)\|_{L^\infty(\mathbb{R}^n, \mathscr{L}(\mathbb{C}^M))} \|w\|_{L^2(\mathbb{R}^n, \mathbb{R}^M)}^2 \leq \beta(R) \|w\|_{L^2(\mathbb{R}^n, \mathbb{R}^M)}^2$$

for a suitable $\beta(R) \geq 0$. □

5.1.29 Proposition. Let $k, K \in \mathbb{N}, 1 < p < \infty$ with $k \geq K > \frac{n}{p} + 1$ and $\mathcal{F} \subset C^k(\mathbb{R}^N, \mathbb{C})$ boundedly on compact subsets. Then there is a function $G_k : [0, \infty) \longrightarrow [0, \infty)$ with

$$\|\partial^\alpha(f(u)w) - f(u)\partial^\alpha w\|_{L^p} \leq G_k(R) \left(\|w\|_{H_p^{k-1}} + \|u\|_{H_p^k[\mathbb{R}^N]} \|w\|_{L^\infty(\mathbb{R}^n)} \right)$$

for $f \in \mathcal{F}, u \in H_p^k[\mathbb{R}^N], \|u\|_{H_p^K[\mathbb{R}^N]} \leq R, w \in H_p^k, |\alpha| \leq k, R > 0$.

PROOF: We only have to prove the assertion for $u \in \mathscr{S}(\mathbb{R}^n, \mathbb{R}^N), w \in \mathscr{S}(\mathbb{R}^n)$. Now

$$\|\partial_x^\alpha(f(u)w) - f(u)\partial_x^\alpha w\|_{L^p}$$

$$\leq c \sum_{l=1}^{n} \sum_{|\alpha_1|+|\alpha_2|=|\alpha|-1} \|(\partial_x^{\alpha_1}\partial_x^{e_l}f(u))(\partial_x^{\alpha_2}w)\|_{L^p}$$

$$\overset{5.1.21}{\underset{5.1.16}{\leq}} c' \sum_{l=1}^{n} \sum_{|\alpha_1|+|\alpha_2|=|\alpha|-1} \left(\|\partial_x^{e_l}f(u)\|_{L^\infty} \|w\|_{H_p^{k-1}} + \|\partial_x^{e_l}f(u)\|_{H_p^{k-1}} \|w\|_{L^\infty} \right)$$

$$\overset{}{\underset{5.1.21}{\leq}} c'' \left(G'(R) \|w\|_{H_p^{k-1}} + G''(R) \|u\|_{H_p^k[\mathbb{R}^N]} \|w\|_{L^\infty} \right)$$

with constants $c, c', c'', G_k'(R), G_k''(R) \geq 0$. This implies the assertion. $\square$

Finally, we can summarize the results obtained so far. 5.1.23 and 5.1.25 imply 5.1.1. We still have to finish the proof of 5.1.2.

PROOF OF 5.1.2: (a) is a consequence of 5.1.24, (b) of 5.1.26, and (c) of 5.1.29. Moreover, (d) follows from

$$\left\| \langle D_x \rangle^l a_j(u) \partial_j w - a_j(u) \partial_j \langle D_x \rangle^l w \right\|_{L^2}$$

$$\overset{}{\underset{5.1.18}{\leq}} c_k \left(\|a_j(u)\|_{H^{K+1}} \|\partial_j w\|_{H^{l-1}} + \|a_j(u)\|_{H^k} \|\partial_j w\|_{H^K} \right)$$

$$\overset{}{\underset{5.1.23}{\leq}} c_k \|a_j(u)\|_{H^k} \|w\|_{H^k} \leq G_k(R) \|w\|_{H^k}$$

for $l \leq k, u \in H^k[\mathbb{R}^N], \|u\|_{H^k[\mathbb{R}^N]} \leq R, w \in H^\infty$, and (e) follows from

$$\left\| \langle D_x \rangle^l a_j(u) \partial_j w - a_j(u) \partial_j \langle D_x \rangle^l w \right\|_{L^2}$$

$$\overset{}{\underset{5.1.18}{\leq}} c_k \left(\|a_j(u)\|_{H^{K+1}} \|\partial_j w\|_{H^{l-1}} + \|a_j(u)\|_{H^{k+m}} \|\partial_j w\|_{H^K} \right)$$

$$\overset{}{\underset{5.1.23}{\leq}} G_k(R) \left(\|u\|_{H^{k+m}[\mathbb{R}^N]} \|w\|_{H^k} + \|w\|_{H^{k+m}} \right)$$

for $l \leq k+m, u \in H^{k+m}[\mathbb{R}^N], \|u\|_{H^k[\mathbb{R}^N]} \leq R, w \in H^\infty, k \geq K+1$. $\square$

5.1.30 Bibliographical remarks. For the proof of the Gagliardo-Moser-Nirenberg estimate 5.1.13 including the lemmata we followed in part the books of Taylor [148] and Racke [126] and we refer to these books for more references of estimates of that type. 5.1.17 and 5.1.18 are due to Kato [84, appendix].

5.2 Quasilinear evolution equations in Sobolev spaces

In this section we describe applications to quasilinear systems of evolution equations in the scale of L^2-Sobolev spaces. We consider equations that are of the same type as in 4.2.17, but now we admit nonlinear terms up to order 1. Recall that $S^m_{1,0}[E]$ denotes $\mathscr{L}(E)$-valued symbols and $H^s[E]$ denotes the E-valued Sobolev space of order s, if E is a Hilbert space (cf. section 4.1). In particular, $S^m_{1,0}[\mathbb{C}^N]$ and $H^s[\mathbb{R}^N]$ are defined in that way.

5.2.1 Theorem. Assume that

(a) $N \in \mathbb{N}, m \geq 2$ are fixed.

(b) $p = (p_{jl})_{j,l} \in C(I \times \mathbb{R}^n \times \mathbb{R}^n, \mathscr{L}(\mathbb{C}^N))$ with $\{p(t,\cdot,\cdot) : t \in I\} \subset S^2_{1,0}[\mathbb{C}^N]$ boundedly and $\{\mathrm{Im}\, p(t,\cdot,\cdot) : t \in I\} \subset S^1_{1,0}[\mathbb{C}^N]$ boundedly such that $\overline{p_{jl}(t,x,\xi)} = p_{jl}(t,x,-\xi)$ for $t \in I$, $x,\xi \in \mathbb{R}^n$ and such that there is a $\beta \geq 0$ with $\mathrm{Re}\, \langle p(t,X,D_x)u, u \rangle_{L^2(\mathbb{R}^n,\mathbb{R}^N)} \leq \beta \|u\|^2_{L^2(\mathbb{R}^n,\mathbb{R}^N)}$ for $t \in I, u \in \mathscr{S}(\mathbb{R}^n,\mathbb{R}^N)$.

(c) $q = (q_{jl})_{j,l} : I \times \mathbb{R}^n \to \mathscr{L}(\mathbb{C}^N)$ such that $q(t,\cdot) : \mathbb{R}^n \to \mathscr{L}(\mathbb{C}^N)$ is measurable for $t \in I$ and $q(\cdot,\xi) : I \to \mathscr{L}(\mathbb{C}^N)$ is continuous with $\|q(t,\xi)\|_{\mathscr{L}(\mathbb{C}^N)} \leq c\langle\xi\rangle^m$, $\mathrm{Re}\, q(t,\xi) \leq c\mathrm{Id}$, and with $\overline{q_{jl}(t,\xi)} = q_{jl}(t,-\xi)$ for $t \in I, \xi \in \mathbb{R}^n, j,l = 1,\ldots,n$, and suitable $c \geq 0$.

(d) $a_j \in C(I \times \mathbb{R}^N, \mathscr{L}(\mathbb{R}^N))$ with $\{a_j(t,\cdot) : t \in I\} \subset C^\infty(\mathbb{R}^N, \mathscr{L}(\mathbb{R}^N))$ boundedly on compact sets and with $a_j(t,x)^* = a_j(t,x)$ and $a_j(t,0) = 0$ for $t \in I$, $x \in \mathbb{R}^N$, $j = 1\ldots,N$.

(e) $f \in C(I \times \mathbb{R}^N, \mathbb{R}^N)$ with $\{f(t,\cdot) : t \in I\} \subset C^\infty(\mathbb{R}^N, \mathbb{R}^N)$ boundedly on compact subsets with $f(t,0) = 0$ for $t \in I$.

Then, for $u_0 \in H^{k+2m}[\mathbf{R}^N]$, $k > n/2+1$, there is an interval $I_\varepsilon = [t_0, t_0+\varepsilon] \subset I$ and a unique $u \in C^1(I_\varepsilon, H^k[\mathbf{R}^N]) \cap C(I_\varepsilon, H^{k+m}[\mathbf{R}^N]) \cap C_w(I_\varepsilon, H^{k+2m}[\mathbf{R}^N])$ with

$$\frac{du}{dt}(t) = p(t, X, D_x)u(t) + q(t, D_x)u(t) + \sum_{j=1}^{n} a_j(t, u(t))\partial_j u(t) + f(t, u(t))$$

$$u(t_0) = u_0 .$$

Moreover, if $u_0 \in H^\infty[\mathbf{R}^N]$, there is an interval $I_\varepsilon = [t_0, t_0+\varepsilon] \subset I$ and a unique solution $u \in C^1(I_\varepsilon, H^\infty[\mathbf{R}^N])$ of this equation.

PROOF: Let $H := L^2(\mathbf{R}^n, \mathbf{R}^N)$ and

$$\Lambda := \begin{pmatrix} -\langle D_x \rangle & & \\ & \ddots & \\ & & -\langle D_x \rangle \end{pmatrix} .$$

Then $(\Lambda, H^1[\mathbf{R}^N])$ is strictly negative and selfadjoint in H with $H_\Lambda^k = H^k[\mathbf{R}^N]$ topologically for $k \in \mathbf{N}_0$. Moreover, let $S_l := (\mathrm{Id} - \frac{1}{l}\Lambda)^{-3m}$ and $K \in \mathbf{N}$ with $K > \frac{n}{2}+1$.

Step 1: Let $P(t) := p(t, X, D_x)$, then

(a) $I \ni t \mapsto P(t) \in \mathscr{L}(H_\Lambda^{k+m}, H_\Lambda^k)$ is strongly continuous for any $k \in \mathbf{N}_0$.

(b) For $k \in \mathbf{N}_0$ there are $\beta_k \geq 0$ with

$$\mathrm{Re} \langle P(t)u, u \rangle_{H_\Lambda^k} \leq \beta_k \|u\|_{H_\Lambda^k}^2 \quad \text{and} \quad \mathrm{Re} \langle S_l P(t)u, S_l u \rangle_{H_\Lambda^k} \leq \beta_k \|u\|_{H_\Lambda^k}^2 \tag{5.2.1}$$

for $u \in H_\Lambda^{k+m}, t \in I, l \in \mathbf{N}$.

Proof: We clearly have (a). Moreover, 4.2.7 shows $\mathrm{Re} \langle P(t)u, u \rangle_{H_\Lambda^k} \leq \beta_k \|u\|_{H_\Lambda^k}^2$ for $u \in H_\Lambda^{k+2}, t \in I$, and suitable constants $\beta_k \geq 0$ for any $k \in \mathbf{N}_0$. Now let $P_1(t) := \frac{P(t)+P(t)^*}{2}$ and $P_2(t) := \frac{P(t)-P(t)^*}{2}$ for $t \in I$. Then the proof of 4.2.7 shows that $P_1(t) = P_1(t)^* \in \mathrm{Op}S_{1,0}^2[\mathbf{C}^N], P_2(t) \in \mathrm{Op}S_{1,0}^1[\mathbf{C}^N]$ and

$$\mathrm{Re} \langle P_j(t)u, u \rangle_H \leq \beta' \|u\|_H^2 \quad \text{for } u \in \mathscr{S}(\mathbf{R}^n, \mathbf{R}^N), j = 1, 2 ,$$

and a suitable $\beta' \geq 0$. Moreover, $\Lambda^k[\Lambda, [\Lambda, P_1(t)]] \in \mathrm{Op}S_{1,0}^{k+2}[\mathbf{C}^N]$ and $[\Lambda^k, P_2(t)] \in \mathrm{Op}S_{1,0}^k[\mathbf{C}^N]$ for any $k \in \mathbf{N}_0$. Hence 3.2.3 and 3.2.4 imply the existence of $\gamma_k \geq 0$ for any $k \in \mathbf{N}_0$ with

$$\mathrm{Re} \langle S_l P_j(t)u, S_l u \rangle_{H_\Lambda^k} \leq \gamma_k \|u\|_{H_\Lambda^k}^2$$

for $t \in I, l \in \mathbf{N}, u \in H_\Lambda^{k+2}, j = 1, 2$. This implies the assertion of step 1.

Step 2: Let $Q(t) := q(t, D_x)$, then

(a) $I \ni t \mapsto Q(t) \in \mathscr{L}(H_\Lambda^{k+m}, H_\Lambda^k)$ is strongly continuous for $k \in \mathbb{N}_0$.

(b) For any $k \in \mathbb{N}_0$ there are $\gamma_k \geq 0$ with

$$\mathrm{Re}\, \langle Q(t)u, u \rangle_{H_\Lambda^k} \leq \gamma_k \|u\|_{H_\Lambda^k}^2 \quad \text{and} \quad \mathrm{Re}\, \langle S_l Q(t)u, S_l u \rangle_{H_\Lambda^k} \leq \gamma_k \|u\|_{H_\Lambda^k}^2$$

for $u \in H_\Lambda^{k+m}, t \in I, l \in \mathbb{N}$.

Proof: We clearly have (a). There is a $\gamma \geq 0$ with $\mathrm{Re}\, \langle Q(t)u, u \rangle_H \leq \gamma \|u\|_H^2$ for $u \in H_\Lambda^m, t \in I$. Since $\Lambda^k Q(t)u = Q(t)\Lambda^k u$ for $u \in H_\Lambda^{k+m}$ and $S_l Q(t)u = Q(t)S_l u$ for $u \in H_\Lambda^m$, $t \in I$, this immediately implies (b).

Step 3: Let $A(t, v)u := \sum_{j=1}^{n} a_j(t, v)\partial_j u$, then

(a) $I \ni t \mapsto A(t, v) \in \mathscr{L}(H_\Lambda^{k+m}, H_\Lambda^k)$ strongly continuously for $v \in H_\Lambda^k, k \geq K$.

(b) For $k \geq K, R > 0$ there are $\tilde{a}_{k,R} \geq 0$ with

$$\|A(t, v) - A(t, w)\|_{\mathscr{L}(H_\Lambda^{k+m}, H_\Lambda^k)} \leq \tilde{a}_{k,R} \|v - w\|_{H_\Lambda^k}$$

for $t \in I, v, w \in H_\Lambda^k, \|v\|_{H_\Lambda^k} \leq R, \|w\|_{H_\Lambda^k} \leq R$.

(c) For $k \geq K, R > 0$ there are $\beta_{k,R} \geq 0$ with

$$\mathrm{Re}\, \langle A(t, v)w, w \rangle_{H_\Lambda^k} \leq \beta_{k,R} \|w\|_{H_\Lambda^k}^2 \tag{5.2.2}$$

for $t \in I, w \in H_\Lambda^{k+m}, v \in H_\Lambda^k, \|v\|_{H_\Lambda^k} \leq R$.

(d) For $k \geq K + 1, R > 0$ there are $\gamma_{k,R} \geq 0$ with

$$\mathrm{Re}\, \langle S_l A(t, u)u, S_l u \rangle_{H_\Lambda^{k+m}} \leq \gamma_{k,R} \|u\|_{H_\Lambda^{k+m}}^2 \tag{5.2.3}$$

for $u \in H_\Lambda^{k+m}, \|u\|_{H_\Lambda^k} \leq R$.

Proof: (a) and (b) are consequences of 5.1.2. Moreover, due to 5.1.28 for $R > 0$ there are $\beta_R \geq 0$ with $\mathrm{Re}\, \langle A(t, u)w, w \rangle_H \leq \beta_R \|w\|_H^2$ for $t \in I, u \in H_\Lambda^K$, $\|u\|_{H_\Lambda^K} \leq R$, and $w \in H_\Lambda^{K+1}$. Due to 5.1.2 for $R > 0; k \geq K$ there are $G_k(R) \geq 0$ with $\|\Lambda^j(A(t, u)w) - A(t, u)(\Lambda^j w)\|_H \leq G_k(R) \|w\|_{H_\Lambda^k}$ for $j \leq k$, $u \in H_\Lambda^k, \|u\|_{H_\Lambda^k} \leq R, w \in \mathscr{S}(\mathbb{R}^n, \mathbb{R}^N), k \geq K$, and

$$\|\Lambda^j(A(t, u)w) - A(t, u)(\Lambda^j w)\|_H \leq G_k(R) \left(\|w\|_{H_\Lambda^{k+m}} + \|u\|_{H_\Lambda^{k+m}} \|w\|_{H_\Lambda^k} \right)$$

for $j \leq k + m, u \in H_\Lambda^{k+m}$ with $\|u\|_{H_\Lambda^k} \leq R, w \in \mathscr{S}(\mathbb{R}^n, \mathbb{R}^N), k \geq K$. Hence 3.2.1 and 3.2.2 imply (c) and (d).

Step 4: Let $f(t, v)(x) := f(t, v(x))$, then

(a) $I \ni t \mapsto f(t, v) \in H_\Lambda^k$ is continuous for $v \in H_\Lambda^k, k \geq K$.

(b) For $k \geq K, R > 0$ there are $\widetilde{F}_k(R) \geq 0$ with

$$\|f(t, v) - f(t, w)\|_{H_\Lambda^k} \leq \widetilde{F}_k(R) \|v - w\|_{H_\Lambda^k}$$

for $v, w \in H_\Lambda^k$ with $\|v\|_{H_\Lambda^k} \leq R, \|w\|_{H_\Lambda^k} \leq R$.

(c) For $k \geq K, R > 0$ there are $G_k(R) \geq 0$ with

$$\|f(t, u)\|_{H_\Lambda^{k+m}} \leq G_k(R) \|u\|_{H_\Lambda^{k+m}}$$

for $u \in H_\Lambda^{k+m}, \|u\|_{H_\Lambda^k} \leq R$.

Proof: This is a consequence of 5.1.1.
Step 5: Proof of the assertion:
The assertion is implied by 3.3.9, 3.4.4, and the previous steps.   □

5.2.2 Corollary. Assume that

(a) $t_0 \in \mathbf{R}, m \geq 2$, and $c > 0$.

(b) $p \in S_{1,0}^2$ with Re $p(x, \xi) \leq c$, Im $p \in S_{1,0}^1$, and $\overline{p(x, \xi)} = p(x, -\xi)$ for $x, \xi \in \mathbb{R}^n$.

(c) $q : \mathbf{R}^n \to \mathbf{C}$ is measurable with $|q(\xi)| \leq c\langle \xi \rangle^m$, Re $q(x, \xi) \leq c$, and $\overline{q(\xi)} = q(-\xi)$ for $\xi \in \mathbb{R}^n$.

(d) $a_j, f \in C^\infty(\mathbf{R}, \mathbf{R})$ with $a_j(0) = 0$ for $j = 1, \ldots, n$ and $f(0) = 0$.

Then, for $u_0 \in H^{k+2m}[\mathbf{R}], k > \frac{n}{2} + 1$, there is an interval $I_\varepsilon = [\ell_0, \ell_0 + \varepsilon]$ and a unique $u \in C^1(I_\varepsilon, H^k[\mathbf{R}]) \cap C(I_\varepsilon, H^{k+m}[\mathbf{R}]) \cap C_w(I_\varepsilon, H^{k+2m}[\mathbf{R}])$ with

$$\frac{du}{dt}(t) = p(X, D_x)u(t) + q(D_x)u(t) + \sum_{j=1}^n a_j(u(t))\partial_j u(t) + f(u(t)), \quad t \in I_\varepsilon,$$

$$u(t_0) = u_0 .$$

Moreover, if $u_0 \in H^\infty[\mathbf{R}]$ there is an interval $I_\varepsilon = [t_0, t_0 + \varepsilon]$ and a unique solution $u \in C^1(I_\varepsilon, H^\infty[\mathbf{R}])$ of this equation.

PROOF: This is implied by 5.2.1 and 4.2.8.   □

5.2.3 Examples. In fluid dynamics many local and nonlocal evolution equations occur which in part have also applications in other fields of physics. We do not want to go into physical details here and merely list some equations of importance. Let $a \in C^\infty(\mathbb{R}, \mathbb{R})$ with $a(0) = 0$.

(a) Strongly degenerate Burger equation $u_t = \nu(x)u_{xx} + a(u)u_x$, where $\nu \in B^\infty(\mathbb{R}^n), \nu(x) \geq 0$ for $x \in \mathbb{R}$.

Often, only the case $\nu(x) = \nu = \text{const.} \in [0, \infty)$ is treated in the literature. Then, in the case $\nu > 0$ the equation is treated with parabolic methods and for $\nu = 0$ with hyperbolic methods.

(b) Korteweg-de Vries equation $u_t + u_{xxx} + a(u)u_x = 0$.

(c) Benjamin-Ono-equation $u_t + (u^2 + 2iHu_x)_x = 0$, where $H = h(D_x)$ with $h(\xi) = \text{signum}(\xi)$ is the Hilbert transform.

(d) Smith equation $u_t + \partial_x(\langle D_x \rangle - \text{Id})u + a(u)u_x = 0$.

If we do not consider second-order degenerate operators on the right-hand side of the evolution equations the regularity statement can be refined. Moreover, the condition $a_j(t, 0) = 0$ is no longer needed in the proofs. Note that this result includes symmetric hyperbolic quasilinear evolution equations.

5.2.4 Theorem. Assume that

(a) $I = [t_0, t_1]$ is a compact interval, $N \in \mathbb{N}, c > 0$, and $m \geq 1$ are fixed.

(b) $q = (q_{jl})_{j,l} : I \times \mathbb{R}^n \to \mathscr{L}(\mathbb{C}^N)$ such that $q(t, \cdot) : \mathbb{R}^n \to \mathscr{L}(\mathbb{C}^N)$ is measurable and $q(\cdot, \xi) : I \to \mathscr{L}(\mathbb{C}^N)$ is continuous with $\|q(t, \xi)\|_{\mathscr{L}(\mathbb{C}^N)} \leq c\langle\xi\rangle^m$, $\|\text{Re}\, q(t, \xi)\|_{\mathscr{L}(\mathbb{C}^N)} \leq c$, and with $\overline{q_{jl}(t, \xi)} = q_{jl}(t, -\xi)$ for $t \in I, \xi \in \mathbb{R}^n$, $j, l = 1, \ldots, n$, and suitable $c \geq 0$.

(c) $a_j \in C(I \times \mathbb{R}^N, \mathscr{L}(\mathbb{R}^N))$ with $\{a_j(t, \cdot) : t \in I\} \subset C^\infty(\mathbb{R}^N, \mathscr{L}(\mathbb{R}^N))$ boundedly on compact sets with $a_j(t, x)^* = a_j(t, x)$ for $t \in I, x \in \mathbb{R}^N$, $j = 1 \ldots, N$.

(d) $f \in C(I \times \mathbb{R}^N, \mathbb{R}^N)$ with $\{f(t, \cdot) : t \in I\} \subset C^\infty(\mathbb{R}^N, \mathbb{R}^N)$ boundedly on compact subsets with $f(t, 0) = 0$ for $t \in I$.

Then, for any $u_0 \in H^{k+m}[\mathbb{R}^N], k > n/2 + m + 1, \tilde{t}_0 \in I$ there is an interval $I_\varepsilon = [\tilde{t}_0 - \varepsilon, \tilde{t}_0 + \varepsilon] \cap I \subset I$ and a unique $u \in C^1(I_\varepsilon, H^k[\mathbb{R}^N]) \cap C(I_\varepsilon, H^{k+m}[\mathbb{R}^N])$ with

$$\frac{du}{dt}(t) = q(t, D_x)u(t) + \sum_{j=1}^{n} a_j(t, u(t))\partial_j u(t) + f(t, u(t)), \quad t \in I_\varepsilon,$$

$$u(\tilde{t}_0) = u_0.$$

PROOF: Let $H := L^2(\mathbf{R}^n, \mathbf{R}^N)$, $K > \frac{n}{2} + 1$, and $Z = \{Z_1, \ldots, Z_n\}$ with

$$Z_j := \begin{pmatrix} \frac{\partial}{\partial x_j} & & \\ & \ddots & \\ & & \frac{\partial}{\partial x_j} \end{pmatrix}, \qquad D(Z_j) = \{u \in H : Z_j u \in H\}, j = 1, \ldots, n.$$

Then Z is a commuting family of infinitesimal generators of isometric C_0-groups in H with $H_Z^k = H^k[\mathbf{R}^N]$ topologically for $k \in \mathbf{N}_0$.

Step 1: Let $Q(t) := q(t, D_x)$, then

(a) $I \ni t \mapsto Q(t) \in \mathcal{L}(H_Z^{k+m}, H_Z^k)$ is strongly continuous for $k \in \mathbf{N}_0$.

(b) $Z^\alpha Q(t)u - Q(t)Z^\alpha u = 0$ for $u \in H_Z^\infty, t \in I, \alpha \in \mathbf{N}_0^n$.

(c) For $k \in \mathbf{N}_0$ there are $\gamma_k \geq 0$ with $|\text{Re } \langle Q(t)u, u \rangle_{H_Z^k}| \leq \gamma_k \|u\|_{H_Z^k}^2$ for $u \in H_\Lambda^{k+m}, t \in I$.

Proof: This can be shown as step 3 of 5.2.1.

Step 2: Let $A(t, v)u := \sum_{j=1}^n a_j(t, v)\partial_j u$, then

(a) $I \ni t \mapsto A(t, v) \in \mathcal{L}(H_Z^{k+m}, H_Z^k)$ is strongly continuous for $v \in H_Z^k, k \geq K$.

(b) For $k \geq K, R > 0$ there are $\tilde{a}_{k,R} \geq 0$ with

$$\|A(t, v) - A(t, w)\|_{\mathcal{L}(H_Z^{k+m}, H_Z^k)} \leq \tilde{a}_{k,R} \|v - w\|_{H_Z^k}$$

for $t \in I, v, w \in H_Z^k, \|v\|_{H_Z^k} \leq R, \|w\|_{H_Z^k} \leq R$.

(c) For $k \geq K, R > 0$ there are $d_{k,R} \geq 0$ with

$$\|Z^\alpha A(t, v)u - A(t, v)Z^\alpha u\|_H \leq d_{k,R} \|u\|_{H_Z^k}$$

for $t \in I, u \in H_Z^\infty, v \in H_Z^k, \|v\|_{H_Z^k} \leq R, |\alpha| \leq k.$

(d) For $R > 0$ there are $\beta_R \geq 0$ with

$$|\text{Re } \langle A(t, v)w, w \rangle_H| \leq \beta_R \|w\|_H^2$$

for $t \in I, w \in H_Z^\infty, v \in H_Z^K, \|v\|_{H_Z^K} \leq R.$

Proof: (a), (b) are consequences of 5.1.2, (c) of 5.1.29, and (d) of 5.1.28.

Step 3: Let $f(t,v)(x) := f(t,v(x))$, then

(a) $I \ni t \mapsto f(t,v) \in H_Z^k$ is continuous for $v \in H_Z^k, k \geq K$.

(b) For $k \geq K, R > 0$ there are $\widetilde{F}_k(R) \geq 0$ with

$$\|f(t,v) - f(t,w)\|_{H_Z^k} \leq \widetilde{F}_k(R)\,\|v - w\|_{H_Z^k}$$

for $v, w \in H_Z^k, \|v\|_{H_Z^k} \leq R, \|w\|_{H_Z^k} \leq R$.

Proof: This is a consequence of 5.1.1.
Step 4: Now the assertion is a consequence of 3.3.12. □

5.2.5 Bibliographical remarks. We will give only a few references to the huge amount of literature available on equations of fluid dynamics. General references for the equations of fluid dynamics are the monographs of Whitham [156] and Dodd/Eilbeck/Gibbon/Morris [39]. In particular, important results and properties of the Burger equation can be found in Whitham [156], Hörmander [68], of the Korteweg-de Vries equation in Whitham [156], Kato [84], Bourgain [18], of the Benjamin-Ono-equation in Benjamin [14], Ono [118], Kenig/Ponce/Vega [90], Iorio [69], of the Smith equation in Iorio [69], Abdelouhab [1], and the references given therein. 5.2.4 could also be proved with Kato's results on abstract quasilinear evolution equations [82].

5.3   Degenerate Navier-Stokes equations

This section deals with a unified approach to the Euler equation

$$u_t = -(u \cdot \mathrm{grad}_x\,)u - \mathrm{grad}_x\,\pi, \qquad \mathrm{div}_x\,u = 0$$

and the Navier-Stokes equation

$$u_t = \nu\triangle_x u - (u \cdot \mathrm{grad}_x\,)u - \mathrm{grad}_x\,\pi, \qquad \mathrm{div}_x\,u = 0\,.$$

Here $\triangle_x = \sum_{j=1}^n \frac{\partial^2}{\partial x_j^2}$ denotes the Laplacian in the spatial variables and grad_x, div_x denote the gradient and the divergence in the spatial variables, respectively. These equations describe the motion of an incompressible fluid without (i.e., $\nu = 0$) or with (i.e., $\nu = \mathrm{const} > 0$) viscosity. We will show that local existence and regularity results for the Euler equation and the Navier-Stokes equation still hold, if we admit a space-varying viscosity $\nu(x) \geq 0$ that may vanish on some parts of the space. In particular, this includes local existence and regularity results for the Euler equation and the Navier-Stokes equation.

But first let us prove some properties of solenoidal vector fields. Recall again that $H^s[\mathbf{R}^n]$ denotes the Sobolev space of order s with values in $\mathbf{R}^n$.

5.3.1 Definition. For $s \in [0, \infty]$ the Sobolev space of solenoidal vector fields is defined as the closure

$$H^s_\sigma := \overline{\left\{ u = (u_1, \ldots, u_n) \in H^\infty[\mathbf{R}^n] : \operatorname{div}_x u = \sum_{j=1}^n \frac{\partial u_j}{\partial x_j} = 0 \right\}}^{H^s[\mathbf{R}^n]}$$

in $H^s[\mathbf{R}^n]$.

5.3.2 Lemma. Let $\varphi \in \mathscr{S}(\mathbf{R}^n)$ with $\varphi(0) = 1$ and let $J_\varepsilon u := \varphi(\varepsilon D_x)u$ for $u \in H^0[\mathbf{R}^n]$. Then

(a) $J_\varepsilon u \in H^\infty[\mathbf{R}^n]$ for $u \in H^s[\mathbf{R}^n], s \geq 0$.

(b) $J_\varepsilon u \xrightarrow[\varepsilon \to 0]{} u$ in $H^s[\mathbf{R}^n]$ for $u \in H^s[\mathbf{R}^n], s \geq 0$.

(c) $\operatorname{div}_x J_\varepsilon u = \varphi(\varepsilon D_x)(\operatorname{div}_x u)$ for $u \in H^s[\mathbf{R}^n], s \geq 0$.

PROOF: We clearly have (a) and (b) due to the proof of 4.1.24. Moreover, for $u \in H^s[\mathbf{R}^n]$

$$\operatorname{div}_x J_\varepsilon u = \sum_{j=1}^n \frac{\partial}{\partial x_j} \varphi(\varepsilon D_x)u_j = \varphi(\varepsilon D_x) \sum_{j=1}^n \frac{\partial}{\partial x_j} u_j = \varphi(\varepsilon D_x)(\operatorname{div}_x u).$$

$\square$

5.3.3 Corollary. $H^s_\sigma = \{u \in H^s[\mathbf{R}^n] : \operatorname{div}_x u = 0\} = H^s[\mathbf{R}^n] \cap H^0_\sigma$ for $s \geq 0$.

PROOF: The assertion is a consequence of 5.3.2. $\square$

5.3.4 Proposition. Let $p_{jk}(\xi) := \delta_{jk} - \frac{\xi_j \xi_k}{|\xi|^2}$ for $j, k = 1, \ldots, n$, $\xi \in \mathbf{R}^n$, and $P_\sigma := (p_{jk}(D_x))_{j,k}$. Then $P_\sigma \in \bigcap_{s \geq 0} \mathscr{L}(H^s[\mathbf{R}^n])$ is the orthogonal projection from $H^s[\mathbf{R}^n]$ onto H^s_σ for $s \geq 0$.

PROOF: Since $\left|\delta_{jk} - \frac{\xi_j \xi_k}{|\xi|^2}\right| \leq 1 + \frac{|\xi||\xi|}{|\xi|^2} \leq 2$ we clearly have $P_\sigma \in \mathscr{L}(H^s[\mathbf{R}^n])$ for any $s \geq 0$. Moreover, for $u \in H^\infty[\mathbf{R}^n]$ we have

$$\operatorname{div}_x (P_\sigma u) = \sum_{j=1}^n \partial_j (P_\sigma u)_j = \sum_{j=1}^n \sum_{k=1}^n \partial_j p_{jk}(D_x) u_k$$

$$= \sum_{j=1}^n \partial_j u_j - \sum_{j=1}^n \sum_{k=1}^n i\mathcal{F}^{-1}\left(\xi_j \frac{\xi_j \xi_k}{|\xi|^2} \widehat{u}_k\right) = \operatorname{div}_x u - \sum_{k=1}^n i\mathcal{F}^{-1}\left(\xi_k \frac{\sum_{j=1}^n \xi_j^2}{|\xi|^2} \widehat{u}_k\right)$$

$$= \operatorname{div}_x u - \operatorname{div}_x u = 0.$$

Now, for $u \in H^s[\mathbf{R}^n]$ choose a sequence $(u^{(l)})_l \subset H^\infty[\mathbf{R}^n]$ with $u^{(l)} \xrightarrow[l \to \infty]{} u$ in $H^s[\mathbf{R}^n]$. Then we have $P_\sigma u^{(l)} \longrightarrow P_\sigma u$ for $l \to \infty$ in $H^s[\mathbf{R}^n]$ which shows that $0 = \operatorname{div}_x (P_\sigma u_l) \longrightarrow \operatorname{div}_x P_\sigma u$ for $l \to \infty$ in $H^{s-1}[\mathbf{R}]$, i.e., $P_\sigma(H^s[\mathbf{R}^n]) \subset H^s_\sigma$. For $u \in H^\infty[\mathbf{R}^n]$ with $\operatorname{div}_x u = 0$ we have further

$$(P_\sigma u)_j = \sum_{k=1}^n p_{jk}(D_x) u_k = u_j - \sum_{k=1}^n \mathcal{F}^{-1}\left(\frac{\xi_j \xi_k}{|\xi|^2} \widehat{u}_k\right)$$

$$= u_j - \mathcal{F}^{-1}\left(\frac{1}{i} \frac{\xi_j}{|\xi|^2} \mathcal{F}\left(\sum_{k=1}^n \partial_k u_k\right)\right) = u_j - \mathcal{F}^{-1}\left(\frac{1}{i} \frac{\xi_j}{|\xi|^2} \mathcal{F}(\operatorname{div}_x u)\right) = u_j.$$

Now, let $u \in H^s_\sigma$, then there is a sequence $(u^{(l)})_l \subset H^\infty[\mathbf{R}^n]$ with $\operatorname{div}_x u^{(l)} = 0$ and $u^{(l)} \xrightarrow[l \to \infty]{} u$ for $l \to \infty$ in $H^s[\mathbf{R}^n]$, hence

$$P_\sigma u \xleftarrow[\infty \leftarrow l]{H^s[\mathbf{R}^n]} P_\sigma u^{(l)} = u^{(l)} \xrightarrow[l \to \infty]{H^s[\mathbf{R}^n]} u.$$

Thus $P_\sigma u = u$ for $u \in H^s_\sigma$ and we have proved the assertion.   □

Closed differential forms on $\mathbf{R}^n$ are exact, in other words:

5.3.5 Lemma. Let $f_1, \ldots, f_n : \mathbf{R}^n \to \mathbb{C}$ be continuously differentiable with $\frac{\partial f_j}{\partial x_l} = \frac{\partial f_l}{\partial x_j}$ for $j, l = 1, \ldots, n$.

Then $F(x) := \sum_{j=1}^n \int_0^1 f_j(\tau x) d\tau x_j$ satisfies $\operatorname{grad} F = \begin{pmatrix} f_1 \\ \vdots \\ f_n \end{pmatrix}$.

PROOF: Clearly, F is differentiable with

$$\frac{\partial F}{\partial x_l}(x) = \sum_{j=1}^{n} \int_0^1 \tau(\partial_l f_j)(\tau x)d\tau x_j + \int_0^1 f_l(\tau x)d\tau$$

$$= \sum_{j=1}^{n} \int_0^1 \tau(\partial_j f_l)(\tau x)d\tau x_j + \int_0^1 f_l(\tau x)d\tau = \int_0^1 \frac{d}{d\tau}(\tau f_l(\tau x))d\tau = f_l(x) .$$

$\square$

5.3.6 Lemma. Let $u \in \mathcal{C}(I, H^\infty)$ with a compact interval $I \subset \mathbb{R}$ and let

$$w(t)(x) := \int_0^1 u(t)(\tau x)d\tau \quad \text{for } t \in I, x \in \mathbb{R}^n .$$

Then $w \in \mathcal{C}(I, \mathcal{B}^\infty(\mathbb{R}^n))$.

PROOF: Since $u(t)(\cdot) \in H^\infty \subset \mathcal{B}^\infty(\mathbb{R}^n)$ we clearly have $w(t)(\cdot) \in \mathcal{B}^\infty(\mathbb{R}^n)$ for any $t \in I$ with

$$\partial_x^\alpha w(t)(x) = \int_0^1 \tau^{|\alpha|}(\partial_x^\alpha u(t))(\tau x)d\tau .$$

Hence, for $x \in \mathbb{R}^n$ we have

$$\left|\partial_x^\alpha w(t')(x) - \partial_x^\alpha w(t)(x)\right| \leq \int_0^1 \tau^{|\alpha|}\left|(\partial_x^\alpha u(t'))(\tau x) - (\partial_x^\alpha u(t))(\tau x)\right|d\tau$$

$$\leq \int_0^1 \tau^{|\alpha|}\left\|\partial_x^\alpha u(t') - \partial_x^\alpha u(t)\right\|_{L^\infty}d\tau < \varepsilon$$

for $|t' \quad t| < \delta$ with a suitable $\delta > 0$. This implies the assertion. $\square$

5.3.7 Definition. Let $\mathcal{B}^{\infty,1}(\mathbb{R}^n) := \{u \in \mathcal{C}^\infty(\mathbb{R}^n) : \langle\cdot\rangle^{-1}u \in \mathcal{B}^\infty(\mathbb{R}^n)\}$ topologized with its natural Fréchet topology.

5.3.8 Proposition. For an interval $I \subset \mathbb{R}$ and $v \in \mathcal{C}(I, H^\infty[\mathbb{R}^n])$ let

$$\pi(t)(x) := \sum_{j=1}^{n} \int_0^1 ((\text{Id} - P_\sigma)v(t))_j(\tau x)d\tau x_j$$

with. Then $\pi \in \mathcal{C}(I, \mathcal{B}^{\infty,1}(\mathbb{R}^n))$ with $\text{grad}_x \pi(t) = (\text{Id} - P_\sigma)v(t)$.

PROOF: Due to 5.3.6 we have $\pi \in C(I, \mathcal{B}^{\infty,1}(\mathbf{R}^n))$. Moreover,

$$\partial_l((\mathrm{Id} - P_\sigma)v(t))_j = \sum_{k=1}^{n} \mathcal{F}^{-1}\left(i\frac{\xi_l\xi_j\xi_k}{|\xi|^2}\widehat{v_k(t)}\right) = \partial_j((\mathrm{Id} - P_\sigma)v(t))_l$$

and 5.3.5 shows $\mathrm{grad}_x \, \pi(t) = (\mathrm{Id} - P_\sigma)v(t)$. This proves the assertion. $\quad\square$

5.3.9 Corollary.

(a) $P_\sigma(H^\infty[\mathbf{R}^n]) = \{u \in H^\infty[\mathbf{R}^n] : \mathrm{div}_x \, u = 0\}$.

(b) $P_\sigma(\, \mathrm{grad}_x \, \pi) = 0$ for $\pi \in C^\infty(\mathbf{R}^n)$ with $\mathrm{grad}_x \, \pi \in L^2(\mathbf{R}^n, \mathbf{R}^n)$.

(c) $(\mathrm{Id} - P_\sigma)(H^\infty[\mathbf{R}^n]) = \{\, \mathrm{grad}_x \, \pi : \pi \in \mathcal{B}^{\infty,1}(\mathbf{R}^n) \text{ with } \mathrm{grad}_x \, \pi \in H^\infty[\mathbf{R}^n]\}$.

PROOF: (a) is a consequence of 5.3.4 and (b), (c) follow from 5.3.8 with $v(t) = v$ and

$$\begin{aligned}
(P_\sigma \, \mathrm{grad}_x \, \pi)_j &= \partial_j\pi - \sum_{k=1}^{n} \mathcal{F}^{-1}\left(\frac{\xi_j\xi_k}{|\xi|^2}\widehat{\partial_k\pi}\right) = \partial_j\pi - \sum_{k=1}^{n} \mathcal{F}^{-1}\left(\frac{\xi_j\xi_k}{|\xi|^2}i\xi_k\widehat{\pi}\right) \\
&= \partial_j\pi - \mathcal{F}^{-1}(i\xi_j\widehat{\pi}) = 0
\end{aligned}$$

for $\pi \in C^\infty(\mathbf{R}^n)$ with $\mathrm{grad}_x \, \pi \in L^2(\mathbf{R}^n, \mathbf{R}^n)$. $\quad\square$

Finally, we can state and prove is the main result of this section.

5.3.10 Theorem. Assume that

(a) $p = (p_{jl})_{j,l} \in C([0, t_1] \times \mathbf{R}^n \times \mathbf{R}^n, \mathscr{L}(\mathbf{C}^n))$ is a matrix of symbols such that $\{p(t, \cdot, \cdot) : t \in [0, t_1]\} \subset S_{1,0}^2[\mathbf{C}^n]$ and $\{\mathrm{Im} \, p(t, \cdot, \cdot) : t \in [0, t_1]\} \subset S_{1,0}^1[\mathbf{C}^n]$ boundedly, and $\overline{p_{jl}(t, x, \xi)} = p_{jl}(t, x, -\xi)$ for $t \in [0, t_1]$, $j, l = 1, \ldots, n$, $x, \xi \in \mathbf{R}^n$, where $t_1 > 0$ is fixed.

(b) There is a $\beta \geq 0$ with

$$\mathrm{Re} \, \langle p(t, X, D_x)u, u\rangle_{L^2(\mathbf{R}^n, \mathbf{R}^n)} \leq \beta \|u\|_{L^2(\mathbf{R}^n, \mathbf{R}^n)}^2 \text{ for } u \in H^2[\mathbf{R}^n], t \in I \, .$$

(c) $a_j \in C^\infty(\mathbf{R}^n, \mathscr{L}(\mathbf{R}^n))$ with $a_j(x)^* = a_j(x)$ and $a_j(0) = 0$ for $x \in \mathbf{R}^n$ and $j = 1, \ldots, n$.

(d) $f \in C^\infty(\mathbf{R}^n, \mathbf{R}^n)$ with $f(0) = 0$.

(e) $u_0 \in H^\infty[\mathbf{R}^n]$ with $\mathrm{div}_x \, u_0 = 0$.

Then there are $0 < \varepsilon < t_1$ and $u \in C^1([0,\varepsilon], H^\infty[\mathbf{R}^n])$, $\pi \in C([0,\varepsilon], \mathcal{B}^{\infty,1}(\mathbf{R}^n))$ with $\mathrm{grad}_x \, \pi \in C([0,\varepsilon], H^\infty[\mathbf{R}^n])$ and

$$\frac{du}{dt}(t) = p(t, X, D_x)u(t) + \sum_{j=1}^{n} a_j(u(t))\partial_j u(t) - \mathrm{grad}_x \, \pi(t) + f(u(t))$$

$$\mathrm{div}_x \, u(t) = 0, \quad t \in [0,\varepsilon]$$

$$u(0) = u_0$$

u and $\mathrm{grad}_x \, \pi$ are unique with these properties.

PROOF:

Step 1:

- $\Lambda := \begin{pmatrix} -\langle D_x \rangle & & \\ & \ddots & \\ & & -\langle D_x \rangle \end{pmatrix}$ is selfadjoint and negative in H_σ^0 with

 domain H_σ^1, and we have $(H_\sigma^0)_\Lambda^k = H_\sigma^k$ topologically for $k \in \mathbb{N}_0$:
 Proof: Clearly, Λ is symmetric in H_σ^0 on H_σ^1. Moreover, for $u \in H_\sigma^0$ and $\lambda \in \mathbb{C}$ with $\mathrm{Im}\,\lambda \neq 0$ there is a $v \in H^1[\mathbf{R}^n]$ with $(\lambda \mathrm{Id} - \Lambda)v = u$. Then $(\lambda \mathrm{Id} - \Lambda)P_\sigma v = P_\sigma(\lambda \mathrm{Id} - \Lambda)v = P_\sigma u = u$, which proves that (Λ, H_σ^1) is selfadjoint in H_σ^0. Clearly, we have $(H_\sigma^0)_\Lambda^k = H_\sigma^k$.

- Let $S_l := (\mathrm{Id} - \frac{1}{l}\Lambda)^{-4}$ for $l \in \mathbb{N}$, $I = [0, t_1]$, and $K > \frac{n}{2} + 1$.

- $I \ni t \mapsto P_\sigma p(t, X, D_x) \in \mathscr{L}(H_\sigma^{k+2}, H_\sigma^k)$ is strongly continuous for any $k \in \mathbb{N}_0$.

- For $k \in \mathbb{N}_0$ there are $\beta_k \geq 0$ with $\mathrm{Re}\,\langle P_\sigma p(t, X, D_x)u, u \rangle_{H_\sigma^k} \leq \beta_k \|u\|_{H_\sigma^k}^2$ for $u \in H_\sigma^{k+2}, t \in I$.
 Proof: Using the assumptions and (5.2.1) we obtain

$$\mathrm{Re}\,\langle P_\sigma p(t, X, D_x)u, u \rangle_{H_\sigma^k} = \mathrm{Re}\,\langle p(t, X, D_x)u, u \rangle_{H^k[\mathbf{R}^n]} \leq \beta_k \|u\|_{H^k[\mathbf{R}^n]}^2$$

 for $u \in H_\sigma^{k+2}, t \in I, k \in \mathbb{N}_0$.

- For $k \in \mathbb{N}_0$ there are $\beta_k' \geq 0$ with $\mathrm{Re}\,\langle S_l P_\sigma p(t, X, D_x)u, S_l u \rangle_{H_\sigma^k} \leq \beta_k' \|u\|_{H_\sigma^k}^2$ for $l \in \mathbb{N}, u \in H_\sigma^\infty, t \in I$.
 Proof: $\langle S_l P_\sigma p(t, X, D_x)u, S_l u \rangle_{H_\sigma^k} = \langle S_l p(t, X, D_x)u, S_l u \rangle_{H^k[\mathbf{R}^n]}$ for $u \in H_\sigma^\infty$ and (5.2.1) implies the assertion.

- $A_j(v)u := P_\sigma(a_j(v(\cdot))\partial_j u(\cdot))$ satisfies $A_j(v) \in \mathscr{L}(H_\sigma^{k+2}, H_\sigma^k)$ for $v \in H_\sigma^k$, $k \geq K > \frac{n}{2} + 1$ and $\|A_j(v) - A_j(w)\|_{\mathscr{L}(H_\sigma^{k+2}, H_\sigma^k)} \leq \tilde{a}_{k,R} \|v - w\|_{H_\sigma^k}$ for $v, w \in H_\sigma^k, \|v\|_{H_\sigma^k} \leq R, \|w\|_{H_\sigma^k} \leq R, k \geq K$, and suitable $\tilde{a}_{k,R} \geq 0$. This is a consequence of 5.1.2.

- For $k \geq K + 1, R > 0$ there are $\beta_{k,R} \geq 0$ and $\gamma_{k,R} \geq 0$ with

$$\text{Re} \left\langle A_j(v)w, w \right\rangle_{H_\sigma^k} \leq \beta_{k,R} \|w\|_{H_\sigma^k}^2, \; \text{Re} \left\langle S_l A_j(u)u, S_l u \right\rangle_{H_\sigma^{k+2}} \leq \gamma_{k,R} \|u\|_{H_\sigma^{k+2}}^2$$

for $w \in H_\sigma^{k+2}, v \in H_\sigma^k, \|v\|_{H_\sigma^k} \leq R, u \in H_\sigma^{k+2}, \|u\|_{H_\sigma^k} \leq R, k \geq K + 1$.
Proof: $\left\langle A_j(v)w, w \right\rangle_{H_\sigma^k} = \left\langle a_j(v)\partial_j w, P_\sigma w \right\rangle_{H^k[\mathbf{R}^n]} = \left\langle a_j(v)\partial_j w, w \right\rangle_{H^k[\mathbf{R}^n]}$
and $\left\langle S_l A_j(u)u, S_l u \right\rangle_{H_\sigma^k} = \left\langle S_l a_j(u)\partial_j u, S_l u \right\rangle_{H^k[\mathbf{R}^n]}$, hence the assertion follows from (5.2.2) and (5.2.3).

- $P_\sigma f(v) \in H_\sigma^k$ for $v \in H_\sigma^k, k \geq K$, and

$$\|P_\sigma f(v) - P_\sigma f(w)\|_{H_\sigma^k} \leq \tilde{F}_k(R) \|v - w\|_{H_\sigma^k}$$

for $v, w \in H_\sigma^k, \|v\|_{H_\sigma^k} \leq R, \|w\|_{H_\sigma^k} \leq R, R > 0$, and suitable $\tilde{F}_k(R) \geq 0$. This is a consequence of 5.1.1.

- For every $k \geq K$, $R > 0$ there are constants $G_k(R) \geq 0$ with $\|P_\sigma f(u)\|_{H_\sigma^{k+2}} \leq G_k(R) \|u\|_{H_\sigma^{k+2}}$ for $u \in H_\sigma^{k+2}$, $\|u\|_{H_\sigma^k} \leq R$, which is a consequence of 5.1.1 as well.

Step 2: For $u_0 \in H_\sigma^\infty$ there is an $\varepsilon > 0$ and a unique $u \in C^1([0, \varepsilon], H_\sigma^\infty)$ with

$$\frac{du}{dt}(t) = P_\sigma \left(p(t, X, D_x)u(t) + \sum_{j=1}^n a_j(u(t))\partial_j u(t) + f(u(t)) \right) \quad (5.3.1)$$

$$u(0) = u_0.$$

Proof: This is implied by 3.3.9, 3.4.4, and step 1.
Step 3: Proof of the assertion:
Let u be the function from step 2. Then due to 5.3.8 there is a function $\pi \in C([0, \varepsilon], B^{\infty,1}(\mathbf{R}^n))$ with

$$\text{grad}_x \, \pi(t) = (\text{Id} - P_\sigma) \left(p(t, X, D_x)u(t) + \sum_{j=1}^n a_j(u(t))\partial_j u(t) + f(u(t)) \right).$$

Then u and π satisfy the assertion, because $u(t) \in H_\sigma^\infty$. This implies the existence statement of the assertion.
Conversely, if u and π satisfy the assertion, then $u(t) = P_\sigma u(t) \in C^1([0, \varepsilon], H_\sigma^\infty)$ and application of P_σ shows (5.3.1). This implies the uniqueness statement of the assertion. $\qquad \square$

5.3.11 Remark. If $p \in C([0, t_1] \times \mathbf{R}^n \times \mathbf{R}^n, \mathbf{C})$ with $\{p(t, \cdot, \cdot) : t \in [0, t_1]\} \subset S^2$ boundedly, $\{\text{Im } p(t, \cdot, \cdot) : t \in [0, t_1]\} \subset S^1$ boundedly, $\text{Re } p(t, x, \xi) \leq c$, and $p(t, x, \xi) = p(t, x, -\xi)$ for $t \in [0, t_1], x, \xi \in \mathbf{R}^n$, then (b) in 5.3.10 is satisfied due to 4.2.8 using $S^m \hookrightarrow\hookrightarrow S^2[\mathbf{C}^n]$.

5.3.12 Corollary. Let $\nu \in \mathcal{B}^\infty(\mathbf{R}^n)$ be a function with $\nu(x) \geq 0$ for $x \in \mathbf{R}^n$, $f \in C^\infty(\mathbf{R}^n, \mathbf{R}^n)$ with $f(0) = 0$, and $u_0 \in H^\infty[\mathbf{R}^n]$ with $\text{div}_x \, u_0 = 0$. Then there are $\varepsilon > 0$ and $u \in C^1([0, \varepsilon], H^\infty[\mathbf{R}^n])$, $\pi \in C([0, \varepsilon], \mathcal{B}^{\infty,1}(\mathbf{R}^n))$ with $\text{grad}_x \, \pi \in C([0, \varepsilon], H^\infty(\mathbf{R}^n))$ and

$$\frac{du}{dt}(t) = \nu(x)\Delta_x u(t) - (u(t) \cdot \text{grad}_x)u(t) - \text{grad}_x \, \pi(t) + f(u(t))$$
$$\text{div}_x \, u(t) = 0 \quad , t \in [0, \varepsilon]$$
$$u(0) = u_0 \, .$$

u and $\text{grad}_x \, \pi$ are unique with these properties.

5.3.13 Remark. Note that ν may vanish on subsets of $\mathbf{R}^n$. In particular, the statement in 5.3.12 contains – as mentioned at the beginning of the section – the usual Navier-Stokes equation (if $\nu(x) \equiv \nu > 0$) and the Euler equation (if $\nu(x) \equiv 0$). In particular, fluid motions are described by this statement that are viscous in some parts of the space and non-viscous in other parts.

5.3.14 Bibliographical remarks. There are many results on the Navier-Stokes equation and the Euler equation in numerous research articles and monographs. We mention only Fujita/Kato [50], Ladyzhenskaya [99], Temam [150], von Wahl [154], Taylor [148] and refer to the references given therein. There one can also find the results on Sobolev spaces of solenoidal vector fields that we collected at the beginning of this section.

5.4   The generalized Kadomtsev-Petviashvili equation

The Cauchy problem for the Kadomtsev-Petviashvili equation (KP equation) in $\mathbf{R}^2$ is given by

$$(u_t + u_{xxx} + uu_x)_x \mp u_{yy} = 0 \qquad u(0) = u_0 \, . \tag{5.4.1}$$

Kadomtsev/Petviashvili [75] introduced this equation as a two-dimensional generalization of the Korteweg-de Vries equation modelling long waves of small amplitudes that propagate in one direction in a two-dimensional fluid of small and constant depth. Problem (5.4.1) with "−" resp., "+" is called KP-I resp., KP-II equation. Moreover, both equations are considered with periodic boundary conditions and on $\mathbf{R}^n$. In this section we will describe a unified approach to both types of KP-equations admitting also equations of "mixed type" in the periodic case. Denote by

$$C^k(T^2) := \{u \in C^k(\mathbf{R}^2, \mathbf{C}) : u(x + 2\pi l, y + 2\pi j) = u(x, y) \text{ for } x, y \in \mathbf{R}, l, j \in \mathbf{Z}\}$$

for $k \in \mathbf{N}_0 \cup \{\infty\}$ the k-times continuously differentiable functions on the two-dimensional torus and by

$$H^k(T^2) := \overline{C^\infty(T^2)}^{\|\cdot\|_{H^k(T^2)}}, \|u\|_{H^k(T^2)} := \sqrt{\sum_{|\alpha| \le k} \int_0^{2\pi} \int_0^{2\pi} |\partial^\alpha u(x, y)|^2 dx dy}$$

$u \in C^\infty(T^2)$ the Sobolev space of order $k \in \mathbf{N}_0$ on the torus.
Clearly, $H^k(T^2)$ is a Hilbert space with respect to the scalar product $\langle u, v \rangle_{H^k(T^2)} := \sum_{|\alpha| \le k} \int_0^{2\pi} \int_0^{2\pi} (\partial^\alpha u)(x, y)(\partial^\alpha v)(x, y) dx dy$ for $u, v \in H^k(T^2)$. As usual, we write $L^2(T^2) := H^0(T^2)$. Let $e_{\mu,\nu}(x, y) := \frac{1}{2\pi} e^{i\mu x} e^{i\nu y}$ for $\mu, \nu \in \mathbf{Z}, x, y \in \mathbf{R}$. Then, with $\hat{u}(\mu, \nu) := \langle u, e_{\mu,\nu} \rangle_{L^2(T^2)}$ for $u \in L^2(T^2)$ we have $u = \sum_{\mu,\nu \in \mathbf{Z}} \hat{u}(\mu, \nu) e_{\mu,\nu}$ in $L^2(T^2)$.
Throughout this section we will use the following assumptions.

5.4.1 Assumptions. Suppose that

- $P(t) = \sum_{0 < |\alpha| \le m} a_\alpha(t) \partial^\alpha_{(x,y)}$, $t \in I$, where $m \in \mathbf{N}$ and $I \ni t \mapsto a_\alpha(t) \in \mathbf{R}$ continuously.

- there is a $\beta \ge 0$ with $|\text{Re } \langle P(t)u, u \rangle_{L^2(T^2)}| \le \beta \|u\|^2_{L^2(T^2)}$ for $u \in C^\infty(T^2)$, $t \in I$.

- $q(t, x) = \sum_{j=0}^{m'} q_j(t) x^j$, $t \in I$, with $I \ni t \mapsto q_j(t) \in \mathbf{R}$ continuously for $j = 0, \ldots, m'$. \hfill (5.4.2)

- $I \ni t \mapsto c(t) \in \mathbf{R}$ is a continuous mapping.

Now we can state our main result on generalized Kadomtsev-Petviashvili equations. It will be proved in the remaining part of this section.

5.4.2 Theorem. Suppose that $m_0 = \max(m, 2)$, $k \geq 3 + m_0$, and $\alpha \in \mathbb{R}$. Furthermore, suppose that $u_0 \in H^{k+m_0}(T^2)$ is real-valued and satisfies

$$\int_0^{2\pi} u_0(x, y)dx = \alpha \quad \text{for } y \in [0, 2\pi]. \tag{5.4.3}$$

Then, there is an $\varepsilon > 0$ such that there is a unique, real-valued solution $u \in C([t_0, t_0 + \varepsilon], H^{k+m_0}(T^2)) \cap C^1([t_0, t_0 + \varepsilon], H^k(T^2))$ of

$$\partial_x \left(\frac{du}{dt}(t) + q(t, u(t))\partial_x u(t) + \partial_x^3 u(t) + P(t)u(t) \right) + c(t)\partial_y^2 u(t) = 0 \tag{5.4.4}$$

$$\int_0^{2\pi} u(t)(x, y)dx = \alpha \quad \text{for } t \in [t_0, t_0 + \varepsilon], y \in [0, 2\pi] \tag{5.4.5}$$

$$u(t_0) = u_0. \tag{5.4.6}$$

Moreover, $u \in C^1([t_0, t_0 + \varepsilon], C^\infty(T^2))$ provided that $u_0 \in C^\infty(T^2)$.

5.4.3 Remark.

(a) Clearly, both types of KP-equations are contained in 5.4.2 as well as equations of mixed type.

(b) (5.4.3) is a necessary condition for a periodic solution u of (5.4.4) if $c(t_0) \neq 0$. More precisely, if $c(t_0) \neq 0$ and $u \in C([t_0, t_0 + \varepsilon], H^{k+m_0}(T^2)) \cap C^1([t_0, t_0 + \varepsilon], H^k(T^2))$ satisfies (5.4.4) and (5.4.6) for a $u_0 \in H^{k+m_0}(T^2)$ and $k \geq 3 + m_0$, then u_0 satisfies (5.4.3) for a suitable α.
PROOF: We have

$$c(t)\partial_y^2 \int_0^{2\pi} u(t)(x, y)dx = \int_0^{2\pi} c(t)\partial_y^2 u(t)(x, y)dx$$

$$\underset{(5.4.4)}{=} -\int_0^{2\pi} \partial_x \left(\frac{du}{dt}(t)(x, y) + q(t, u(t)(x, y))\partial_x u(t)(x, y) \right.$$

$$\left. + \partial_x^3 u(t)(x, y) + P(t)u(t)(x, y) \right) dx$$

$$= -\left[\frac{du}{dt}(t)(x, y) + q(t, u(t)(x, y))\partial_x u(t)(x, y) \right.$$

$$\left. + \partial_x^3 u(t)(x, y) + P(t)u(t)(x, y) \right]_{x=0}^{x=2\pi} = 0$$

due to the periodicity. Hence, again due to the periodicity

$$c(t) \int_0^{2\pi} u(t)(x,y)dx = \tilde{\alpha} \quad \text{for } y \in [0, 2\pi]$$

for a suitable constant $\tilde{\alpha}$. In particular

$$c(t_0) \int_0^{2\pi} u_0(x,y)dx = c(t_0) \int_0^{2\pi} u(t_0)(x,y)dx = \tilde{\alpha} \quad \text{for } y \in [0, 2\pi]$$

and (5.4.3) is fulfilled with $\alpha = \frac{\tilde{\alpha}}{c(t_0)}$.   □

5.4.4 Definition. We define

- $X^\infty(T^2) := \{u \in C^\infty(T^2) : \int_0^{2\pi} u(x,y)dx = 0 \text{ for } y \in [0, 2\pi]\}$

- $X^k(T^2) := \overline{X^\infty(T^2)}^{H^k(T^2)}$ for $k \geq 0$.

5.4.5 Lemma. Define $Iu(x,y) := \int_0^{2\pi} u(z,y)dz$ for $u \in L^2(T^2)$. Then $I \in \mathcal{L}(H^k(T^2))$ for every $k \in \mathbb{N}_0$ and

$$\widehat{Iu}(\mu, \nu) = \begin{cases} 0 & , \mu \neq 0 \\ 2\pi \widehat{u}(0, \nu) & , \mu = 0 \end{cases}.$$

PROOF: For $u \in L^2(T^2)$ we have

$$\|Iu\|^2_{L^2(T^2)} \leq 2\pi \int_0^{2\pi} \left(\int_0^{2\pi} |1||u(z,y)|dz \right)^2 dy$$

$$= (2\pi)^2 \int_0^{2\pi} \int_0^{2\pi} |u(z,y)|^2 dzdy = (2\pi)^2 \|u\|^2_{L^2(T^2)}.$$

In particular, $Iu \in L^2(T^2)$ and we can calculate $\widehat{Iu}(m,n)$:

$$\widehat{Iu}(\mu, \nu) = \int_0^{2\pi} \int_0^{2\pi} u(z,y) \left(\int_0^{2\pi} \overline{e_{\mu,\nu}(x,y)}dx \right) dzdy = \begin{cases} 0 & , \mu \neq 0 \\ 2\pi \widehat{u}(0, \nu) & , \mu = 0 \end{cases}.$$

This proves the assertion.   □

5.4.6 Lemma. We have $X^k(T^2) \hookrightarrow X^{k'}(T^2)$ densely and continuously for $k \geq k'$, and

$$
\begin{aligned}
X^k(T^2) &= \{u \in H^k(T^2) : \int_0^{2\pi} u(x,y)dx = 0 \text{ for a.e. } y \in [0, 2\pi]\} \\
&= \{u \in H^k(T^2) : Iu = 0\} \\
&= \{u \in H^k(T^2) : \widehat{u}(0, \nu) = 0 \text{ for } \nu \in \mathbb{Z}\} .
\end{aligned}
$$

Moreover, $\partial^\alpha_{(x,y)} \in \mathscr{L}(X^{k+|\alpha|}(T^2), X^k(T^2))$ for $\alpha \in \mathbb{N}_0^2$ and $k \in \mathbb{N}_0$.

PROOF: This is obvious. $\square$

5.4.7 Lemma. Define $(Vu)(x,y) := \int_0^x u(w,y)dw$ for $u \in L^2(T^2)$ and

$$
(A_1(t)u)(x,y) := c(t)\int_0^x (\partial_y^2 u)(w,y)dw - c(t)\frac{1}{2\pi}\int_0^{2\pi}\int_0^z (\partial_y^2 u)(w,y)dwdz
$$

for $t \in I, u \in X^2(T^2)$. Then we have $(\mathrm{Id} - \frac{1}{2\pi}I)V\partial_x u = u$ for $u \in X^1(T^2)$ and $\mathbb{R} \ni t \mapsto A_1(t) \in \mathscr{L}(X^{k+2}(T^2), X^k(T^2))$ strongly continuously for $k \in \mathbb{N}_0$ with

$$
(A_1(t)u)\widehat{\ }(\mu, \nu) = \begin{cases} ic(t)\frac{\nu^2}{\mu}\widehat{u}(\mu, \nu) & , \mu \neq 0 \\ 0 & , \mu = 0 \end{cases} .
$$

Moreover, $\mathrm{Re}\,\langle A_1(t)u, u\rangle_{L^2(T^2)} = 0$ and $\partial^\alpha A_1(t)u - A_1(t)\partial^\alpha u = 0$ for $\alpha \in \mathbb{N}_0^2$, $u \in X^\infty(T^2)$, $t \in I$.

PROOF: Clearly (cf. the proof of 5.4.5) $V \in \mathscr{L}(L^2(T^2))$ with

$$
(Ve_{k,l})(x,y) = \int_0^x e_{k,l}(w,y)dw = \frac{1}{ik}(e_{k,l}(x,y) - e_{0,l}(x,y))
$$

for $k \in \mathbb{Z} \setminus \{0\}, l \subset \mathbb{Z}$. Hence, for $u \in X^0(T^2)$ and $\mu, \nu \in \mathbb{Z}, \mu \neq 0$ we have

$$
\left(\left(\mathrm{Id} - \frac{1}{2\pi}I\right)Vu\right)\widehat{\ }(\mu, \nu)
$$

$$
= \langle Vu, e_{\mu,\nu}\rangle_{L^2(T^2)} - \frac{1}{2\pi}\int_0^{2\pi}\int_0^z\int_0^{2\pi}\underbrace{\int_0^{2\pi}\overline{e_{\mu,\nu}(x,y)}dx}_{=0(\mu\neq 0)}\, u(w,y)dydwdz
$$

$$
= \sum_{\substack{k,l\in\mathbb{Z}\\k\neq 0}} \widehat{u}(k,l)\langle Ve_{k,l}, e_{\mu,\nu}\rangle_{L^2(T^2)}
$$

$$
= \sum_{\substack{k,l\in\mathbb{Z}\\k\neq 0}} \frac{\widehat{u}(k,l)}{ik}\left(\langle e_{k,l}, e_{\mu,\nu}\rangle_{L^2(T^2)} - \langle e_{0,l}, e_{\mu,\nu}\rangle_{L^2(T^2)}\right) = \frac{\widehat{u}(\mu,\nu)}{i\mu}
$$

and for $\mu = 0$ we have

$$\left(\left(\text{Id} - \frac{1}{2\pi}I\right)Vu\right)^{\widehat{}}(\mu, \nu) = \widehat{Vu}(0, \nu) - \widehat{Vu}(0, \nu) = 0.$$

As $A_1(t)u = c(t)(\text{Id} - \frac{1}{2\pi}I)V\partial_y^2$ this implies the assertion. □

On the torus inequalities of Gagliardo-Moser-Nirenberg type also hold.

5.4.8 Proposition.

(a) $H^k(T^2) \hookrightarrow C^l(T^2)$ for $k, l \in \mathbb{N}_0$ with $k \geq l + 2$.

(b) $H^k(T^2)$ is an algebra for $k \geq 2$ with

$$\|uv\|_{H^k(T^2)} \leq c_k \left(\|u\|_{H^k(T^2)}\|v\|_{H^2(T^2)} + \|u\|_{H^2(T^2)}\|v\|_{H^k(T^2)}\right)$$
$$\leq c_k' \|u\|_{H^k(T^2)}\|v\|_{H^k(T^2)}$$

for $u, v \in H^k(T^2)$ and suitable constants $c_k, c_k' \geq 0$.

(c) $\|\partial^\alpha(uv) - u\partial^\alpha v\|_{L^2(T^2)} \leq c_k \left(\|u\|_{H^k(T^2)}\|v\|_{H^2(T^2)} + \|u\|_{H^3(T^2)}\|v\|_{H^{k-1}(T^2)}\right)$
for $u, v \in H^k(T^2), \alpha \in \mathbb{N}_0^2, |\alpha| \leq k, k \geq 3$, and suitable $c_k \geq 0$.

PROOF: This can be proved in the same way as 5.1.17 and 5.1.18. □

5.4.9 Lemma. Define

$$A_2(t, v)u(x, y) := q(t, v(x, y))(\partial_x u)(x, y) - \frac{1}{2\pi}\int_0^{2\pi} q(t, v(z, y))(\partial_x u)(z, y)dz$$
$$= \left(\text{Id} - \frac{1}{2\pi}I\right)(q(t, v(x, y))(\partial_x u)(x, y))$$

for $v \in X^2(T^2), u \in X^3(T^2)$.
Then $I \ni t \mapsto A_2(t, v) \in \mathcal{L}(X^{k+1}(T^2), X^k(T^2))$ strongly continuously for $v \in X^k(T^2)$ and $k \geq 2$. Moreover, for $R > 0, k \geq 2$ there is a constant $c_{k,R} > 0$ with

$$\|A_2(t, v) - A_2(t, w)\|_{\mathcal{L}(X^{k+1}(T^2), X^k(T^2))} \leq c_{k,R}\|v - w\|_{H^k(T^2)}$$

for $t \in I, v, w \in X^k(T^2)$ with $\|v\|_{H^k(T^2)} \leq R, \|w\|_{H^k(T^2)} \leq R$.

PROOF: Due to 5.4.8 have $q(t,v)\partial_x u \in H^k(T^2)$ for every $u \in X^{k+1}(T^2)$, $v \in X^k(T^2), k \geq 2$. Thus

$$A_2(t,v)u = \left(\mathrm{Id} - \frac{1}{2\pi}I\right) q(t,v)\partial_x u \in H^k(T^2)$$

and $\|A_2(t,v)u\|_{H^k(T^2)} \leq c\,\|q(t,v)\partial_x u\|_{H^k(T^2)} \leq c'\left(\sum_{j=0}^{m'}\|v\|_{H^k(T^2)}^j\right)\|u\|_{H^{k+1}(T^2)}$
for constants $c, c' \geq 0$. The mapping $I \ni t \mapsto A_2(t,v) \in \mathcal{L}(H^{k+1}(T^2), H^k(T^2))$
is strongly continuous for $v \in X^k(T^2), k \geq 2$. Moreover $((\mathrm{Id}-\frac{1}{2\pi}I)w)\hat{\ }(0,\nu) = 0$
for $w \in L^2(T^2)$, hence $(A_2(t,v)u)\hat{\ }(0,\nu) = 0$ for $\nu \in \mathbb{Z}$, i.e., $A_2(t,v)u \in X^k(T^2)$.
Therefore $I \ni t \mapsto A_2(t,v) \in \mathcal{L}(X^{k+1}(T^2), X^k(T^2))$ strongly continuously for
$k \geq 2$ and $v \in X^k(T^2)$. Finally, with suitable constants $c, c', c_{k,R} \geq 0$ we have

$$\|A_2(t,v) - A_2(t,w)\|_{\mathcal{L}(X^{k+1}(T^2),X^k(T^2))} \leq c\,\|q(t,v) - q(t,w)\|_{H^k(T^2)}$$

$$\leq c'\sum_{j=1}^{m'}\|(v-w)(v^{j-1}+v^{j-2}w+\ldots+vw^{j-2}+w^{j-1})\|_{H^k(T^2)}$$

$$\leq c_{k,R}\,\|v-w\|_{H^k(T^2)}$$

for $t \in I, v, w \in X^k(T^2)$ with $\|v\|_{H^k(T^2)} \leq R, \|w\|_{H^k(T^2)} \leq R$. $\qquad\square$

5.4.10 Lemma. For $R > 0$ there is $\beta_R > 0$ with

$$|\mathrm{Re}\,\langle A_2(t,v)u, u\rangle_{L^2(T^2)}| \leq \beta_R\,\|u\|_{L^2(T^2)}^2$$

for $u \in X^3(T^2), v \in X^3(T^2)$ with $\|v\|_{H^3(T^2)} \leq R$ and v real-valued.

PROOF: For $v \in X^3(T^2), u \in X^3(T^2)$ and $h(y) := \int_0^{2\pi} q(t,v(z,y))(\partial_x u)(z,y)dz$
we have

$$\mathrm{Re}\,\langle A_2(t,v)u, u\rangle_{L^2(T^2)}$$

$$= \mathrm{Re}\,\langle q(t,v)\partial_x u, u\rangle_{L^2(T^2)} - \frac{1}{2\pi}\mathrm{Re}\,\int_0^{2\pi} h(y)\underbrace{\int_0^{2\pi} u(x,y)dxdy}_{=0 \text{ a.e.}}$$

$$= \mathrm{Re}\,\langle q(t,v)\partial_x u, u\rangle_{L^2(T^2)}.$$

Hence, with suitable constants $c', c'', \beta_R \geq 0$ we have

$$|2\mathrm{Re}\,\langle A_2(t,v)u, u\rangle_{L^2(T^2)}|$$
$$= |\langle q(t,v)\partial_x u, u\rangle_{L^2(T^2)} - \langle \partial_x(q(t,v)u), u\rangle_{L^2(T^2)}|$$

$$= \quad |\langle(\partial_x q(t,v))u, u\rangle_{L^2(T^2)}| \leq \|(\partial_x q(t,v))u\|_{L^2(T^2)} \|u\|_{L^2(T^2)}$$

$$= \quad c' \sum_{j=1}^{m'} \left\||jv^{j-1}(\partial_x v)|\right\|_{H^2(T^2)} \|u\|_{L^2(T^2)}^2 \leq c'' \sum_{j=1}^{m'} \|v\|_{H^2(T^2)}^{j-1} \|v\|_{H^3(T^2)} \|u\|_{L^2(T^2)}^2$$

$$\leq \quad 2\beta_R \|u\|_{L^2(T^2)}$$

for $u, v \in X^3(T^2)$ with $\|v\|_{H^3(T^2)} \leq R$ and v real-valued. $\qquad\qquad \square$

5.4.11 Lemma. For $k \geq 3, R \geq 0$ there is a $d_{k,R} > 0$ with

$$\left\|\partial^\alpha_{(x,y)} A_2(t,v)u - A_2(t,v)\partial^\alpha_{(x,y)}u\right\|_{L^2(T^2)} \leq d_{k,R} \|u\|_{H^k(T^2)}$$

for $|\alpha| \leq k,\ u \in X^\infty(T^2), v \in X^k(T^2)$ with $\|v\|_{H^k(T^2)} \leq R$ and for $m_0 \in \mathbb{N}$ there is a d'_{k,R,m_0} with

$$\left\|\partial^\alpha_{(x,y)} A_2(t,v)u - A_2(t,v)\partial^\alpha_{(x,y)}u\right\|_{L^2(T^2)}$$

$$\leq \quad d'_{k,R,m_0} \left(\|v\|_{H^{k+m_0}(T^2)} \|u\|_{H^3(T^2)} + \|u\|_{H^{k+m_0}(T^2)}\right)$$

for $|\alpha| \leq k + m_0$, and $u \in X^\infty(T^2), v \in X^{k+m_0}(T^2)$ with $\|v\|_{H^k(T^2)} \leq R$.

PROOF: Using $\partial^\alpha_{(x,y)}I - I\partial^\alpha_{(x,y)} = 0$ we have

$$\left\|\partial^\alpha_{(x,y)} A_2(t,v)u - A_2(t,v)\partial^\alpha_{(x,y)}u\right\|_{L^2(T^2)}$$

$$= \quad \left\|\partial^\alpha_{(x,y)}\left(\mathrm{Id} - \frac{1}{2\pi}I\right) q(t,v)\partial_x u - \left(\mathrm{Id} - \frac{1}{2\pi}I\right) q(t,v)\partial_x \partial^\alpha_{(x,y)}u\right\|_{L^2(T^2)}$$

$$= \quad \left\|\left(\mathrm{Id} - \frac{1}{2\pi}I\right)\left(\partial^\alpha_{(x,y)}(q(t,v)\partial_x u) - q(t,v)\partial^\alpha_{(x,y)}\partial_x u\right)\right\|_{L^2(T^2)}$$

$$\leq \quad c \sum_{j=1}^{m'} \left\|\partial^\alpha_{(x,y)}(v^j \partial_x u) - v^j \partial^\alpha_{(x,y)}\partial_x u\right\|_{L^2(T^2)}$$

for a suitable constant $c \geq 0$. Thus, for $|\alpha| \leq k$ and $u \in X^\infty(T^2), v \in X^k(T^2)$ with $\|v\|_{H^k(T^2)} \leq R$ we have using 5.4.8 with constants $c', c'', d_{k,R} \geq 0$

$$\left\|\partial^\alpha_{(x,y)} A_2(t,v)u - A_2(t,v)\partial^\alpha_{(x,y)}u\right\|_{L^2(T^2)}$$

$$\leq \quad c' \sum_{j=1}^{m'} \left(\|v^j\|_{H^k(T^2)} \|\partial_x u\|_{H^2(T^2)} + \|v^j\|_{H^3(T^2)} \|\partial_x u\|_{H^{k-1}(T^2)}\right)$$

$$\leq \quad c'' \sum_{j=1}^{m'} \left(\|v\|_{H^k(T^2)}^j \|u\|_{H^3(T^2)} + \|v\|_{H^3(T^2)}^j \|u\|_{H^k(T^2)}\right) \leq d_{k,R} \|u\|_{H^k(T^2)}$$

and for $|\alpha| \leq k + m_0$, $u \in X^\infty(T^2), v \in X^{k+m_0}(T^2)$ with $\|v\|_{H^k(T^2)} \leq R$ we have with constants $c', c'', d_{k,R,m_0} \geq 0$

$$\left\| \partial^\alpha_{(x,y)} A_2(t,v)u - A_2(t,v)\partial^\alpha_{(x,y)}u \right\|_{L^2(T^2)}$$

$$\leq c' \sum_{j=1}^{m'} \left(\|v^j\|_{H^{k+m_0}(T^2)} \|\partial_x u\|_{H^2(T^2)} + \|v^j\|_{H^3(T^2)} \|\partial_x u\|_{H^{k+m_0-1}(T^2)} \right)$$

$$\leq c'' \sum_{j=1}^{m'} \left(\|v\|^{j-1}_{H^2(T^2)} \|v\|_{H^{k+m_0}(T^2)} \|u\|_{H^3(T^2)} + \|v\|^j_{H^3(T^2)} \|u\|_{H^{k+m_0}(T^2)} \right)$$

$$\leq d'_{k,R,m_0} \left(\|v\|_{H^{k+m_0}(T^2)} \|u\|_{H^3(T^2)} + \|u\|_{H^{k+m_0}(T^2)} \right).$$

$\square$

5.4.12 Lemma. Let $H^k := \{u \in X^k(T^2) : u$ real-valued$\}$ for $k \in \mathbb{N}_0$. Then $(H^k, \langle \cdot, \cdot \rangle_{H^k(T^2)})$ for $k \in \mathbb{N}_0$ is a real Hilbert space and the operators ∂_z with domains $D(\partial_z) = \{w \in H^0 : \partial_z w \in H^0\}$ for $z = x,y$ are generators of commuting isometric C_0-groups in H^0 with $(\partial_z)^* = -\partial_z$ for $z = x,y$. Moreover, $H^k = (H^0)^k_{\{\partial_x, \partial_y\}}$ topologically.

PROOF: Clearly, H^k is a closed subspace of $X^k(T^2)$, i.e., it is a real Hilbert space. Moreover, ∂_x is the infinitesimal generator of the C_0-group $[T_x(t)f](x,y) = f(x+t,y)$ and ∂_y is the infinitesimal generator of the C_0-group $[T_y(t)f](x,y) = f(x,y+t)$ in $L^2(T^2)$, where the domain of ∂_z is given by $\{u \in L^2(T^2) : \partial_z u \in L^2(T^2)\}$ for $z = x,y$. Since $T_z(t)(H^0) \subset H^0$ for $z = x,y$ (due to the periodicity), ∂_z are infinitesimal generators of commuting isometric C_0-groups in H^0 with domain $D(\partial_z) = \{u \in H^0 : \partial_z u \in H^0\}$, $z = x,y$. Clearly, $(\partial_z)^* = -\partial_z$ for $z = x,y$.
Finally, $(H^0)^1_{\{\partial_x, \partial_y\}} = \{u \in H^0 : \partial_x u, \partial_y u \in H^0\} = H^1$ and inductively for $k \geq 1$

$$u \in (H^0)^{k+1}_{\{\partial_x, \partial_y\}} \iff u, \partial_x u, \partial_y u \in (H^0)^k_{\{\partial_x, \partial_y\}} = H^k \iff u \in H^{k+1}.$$

Thus $H^k = (H^0)^k_{\{\partial_x, \partial_y\}}$ for $k \in \mathbb{N}_0$. Clearly, both spaces have the same topologies.
$\square$

5.4.13 Proposition. Suppose that $m_0 = \max(m, 2), k \geq 3 + m_0$. Furthermore, suppose that $u_0 \in X^{k+m_0}(T^2)$ is real-valued. Then, there is an $\varepsilon > 0$ such that there is a unique, real-valued solution $u \in C([t_0, t_0 + \varepsilon], X^{k+m_0}(T^2)) \cap C^1([t_0, t_0 + \varepsilon], X^k(T^2))$ of

$$\frac{du}{dt}(t) + q(t, u(t))\partial_x u(t) + P(t)u(t) + A_1(t)u(t) = 0, \quad t \in [t_0, t_0 + \varepsilon]$$

$$u(t_0) = u_0 .$$

Moreover, $u \in C^1([t_0, t_0 + \varepsilon], X^\infty(T^2))$ provided that $u_0 \in X^\infty(T^2)$.

PROOF: Define $A(t, v)u := -P(t)u - A_2(t, v) - A_1(t), K := 3$, then

- $I \ni t \mapsto A(t, v) \in \mathscr{L}(H^{k+m_0}, H^k)$ is strongly continuous for any $v \in H^k$, $k \geq K$, due to 5.4.1, 5.4.7, 5.4.9.

- For $R > 0, k \geq K$ there is a constant $\tilde{a}_{k,R} \geq 0$ with

$$\|A(t, v) - A(t, w)\|_{\mathscr{L}(H^{k+m_0}, H^k)} \leq \tilde{a}_{k,R} \|v - w\|_{H^k}$$

  for $t \in I, v, w \in H^k$ with $\|v\|_{H^k(T^2)} \leq R, \|w\|_{H^k(T^2)} \leq R$ due to 5.4.9.

- For $R > 0, k \geq K$ there is a $d_{k,R} > 0$ with

$$\|\partial_{(x,y)}^\alpha A(t, v)u - A(t, v)\partial_{(x,y)}^\alpha u\|_{H^0(T^2)} \leq d_{k,R} \|u\|_{H^k(T^2)}$$

  for $t \in I, v \in H^k, \|v\|_{H^k(T^2)} \leq R, u \in H^\infty, \alpha \in \mathbb{N}_0^2, |\alpha| \leq k$ due to 5.4.7, 5.4.11.

- For $R > 0, k \geq K$ there is a $d'_{k,R} > 0$ with

$$\|\partial_{(x,y)}^\alpha A(t, v)u - A(t, v)\partial_{(x,y)}^\alpha u\|_{L^2(T^2)}$$
$$\leq d'_{k,R}(\|v\|_{H^{k+m_0}(T^2)} \|u\|_{H^k(T^2)} + \|u\|_{H^{k+m_0}(T^2)})$$

  for $t \in I, v \in H^{k+m_0}, \|v\|_{H^k(T^2)} \leq R, u \in H^\infty, \alpha \in \mathbb{N}_0^2, |\alpha| \leq k + m_0$ due to 5.4.7, 5.4.11.

- For $R > 0$ there is a $\beta_{0,R} \geq 0$ with

$$|\mathrm{Re}\,\langle A(t, v)u, u\rangle_{L^2(T^2)}| \leq \beta_{0,R} \|u\|_{L^2(T^2)}^2$$

  for $u \in H^\infty, v \in H^K, \|v\|_{H^K(T^2)} \leq R, t \in I$ due to 5.4.1, 5.4.7, 5.4.10.

Thus 3.3.12 shows that there is an $\varepsilon > 0$ such that there is a unique, real-valued solution
$u \in C([t_0, t_0 + \varepsilon], X^{k+m_0}(T^2)) \cap C^1([t_0, t_0 + \varepsilon], X^k(T^2))$ of

$$\frac{du}{dt}(t) = A(t, u(t))u(t) = -P(t)u(t) - A_2(t, u(t))u(t) - A_1(t)u(t)$$

$$= -P(t)u(t) - \left(\mathrm{Id} - \frac{1}{2\pi}I\right)(q(t, u(t))\partial_x u(t)) - A_1(t)u(t)$$

$$u(t_0) = u_0 \,.$$

Moreover, 3.2.2 and 3.4.4 show that $u \in C^1([t_0, t_0+\varepsilon], X^\infty(T^2))$ if $u_0 \in X^\infty(T^2)$. Now, for $v \in H^5(T^2)$ we have $I(v^0\partial_x v) = I(\partial_x v) = 0$ due to 5.4.5, and for $j \in \mathbb{N}$ we have

$$I(v^j \partial_x v) = \int_0^{2\pi} v(x,y)^j \partial_x v(x,y) dx = -\int_0^{2\pi} (\partial_x(v(x,y)^j))v(x,y) dx$$

$$= -j \int_0^{2\pi} v(x,y)^{j-1}(\partial_x v(x,y))v(x,y) dx = -jI(v^j\partial_x v) \,.$$

Thus $I(v^j \partial_x v) = 0$ and $I(q(t,v)\partial_x v) = 0$, hence
$$\frac{du}{dt}(t) = -P(t)u(t) - q(t, u(t))\partial_x u(t) - A_1(t)u(t) \,. \qquad \square$$

Finally, we can prove 5.4.2:

PROOF:

Existence:
There is a polynomial $\tilde{q}_\alpha(t, x)$ in x satisfying assumption (5.4.2) in 5.4.1 with $\tilde{q}_\alpha(t, x) = q(t, x + \frac{\alpha}{2\pi})$ for $x \in \mathbb{R}, t \in I$. Now, due to 5.4.13 there is an $\varepsilon > 0$ and a real-valued $\tilde{u} \in C([t_0, t_0 + \varepsilon], X^{k+m_0}(T^2)) \cap C^1([t_0, t_0 + \varepsilon], X^k(T^2))$ with

$$\frac{d(\tilde{u} + \frac{\alpha}{2\pi})}{dt}(t) + q\left(t, \tilde{u}(t) + \frac{\alpha}{2\pi}\right)\partial_x\left(\tilde{u}(t) + \frac{\alpha}{2\pi}\right) + \partial_x^3\left(\tilde{u}(t) + \frac{\alpha}{2\pi}\right)$$
$$+ P(t)\left(\tilde{u}(t) + \frac{\alpha}{2\pi}\right) + A_1(t)\left(\tilde{u}(t) + \frac{\alpha}{2\pi}\right)$$
$$= \frac{d\tilde{u}}{dt}(t) + \tilde{q}_\alpha(t, \tilde{u}(t))\partial_x \tilde{u}(t) + \partial_x^3 \tilde{u}(t) + P(t)\tilde{u}(t) + A_1(t)\tilde{u}(t) = 0$$
$$\tilde{u}(t_0) = u_0 - \frac{\alpha}{2\pi} \,.$$

Now apply ∂_x, then 5.4.7 yields

$$\partial_x \left(\frac{d(\tilde{u} + \frac{\alpha}{2\pi})}{dt}(t) + q\left(t, \tilde{u}(t) + \frac{\alpha}{2\pi}\right)\partial_x\left(\tilde{u}(t) + \frac{\alpha}{2\pi}\right) + \partial_x^3\left(\tilde{u}(t) + \frac{\alpha}{2\pi}\right) \right.$$
$$\left. + P(t)\left(\tilde{u}(t) + \frac{\alpha}{2\pi}\right)\right) + c(t)\partial_y^2\left(\tilde{u}(t) + \frac{\alpha}{2\pi}\right) = 0$$

Let $u(t) := \tilde{u}(t) + \frac{\alpha}{2\pi}$, then $u \in C([t_0, t_0+\varepsilon], H^{k+m_0}(T^2)) \cap C^1([t_0, t_0+\varepsilon], H^k(T^2))$ satisfies

$$\partial_x \left(\frac{du}{dt}(t) + q(t, u(t))\partial_x u(t) + \partial_x^3 u(t) + P(t)u(t) \right) + c(t)\partial_y^2 u(t) = 0$$

$$u(t_0) = u_0.$$

Clearly, u satisfies (5.4.5).
Uniqueness:
Suppose that there is an $\varepsilon > 0$ and a real-valued $u \in C([t_0, t_0+\varepsilon], H^{k+m_0}(T^2)) \cap C^1([t_0, t_0 + \varepsilon], H^k(T^2))$ with

$$\partial_x \left(\frac{du}{dt}(t) + q(t, u(t))\partial_x u(t) + \partial_x^3 u(t) + P(t)u(t) \right) + c(t)\partial_y^2 u(t) = 0$$

$$\int_0^{2\pi} u(t)(x, y)dx = \alpha \quad \text{for } y \in [0, 2\pi]$$

$$u(t_0) = u_0.$$

Then

$$\partial_x \left(\frac{d(u - \frac{\alpha}{2\pi})}{dt}(t) + \tilde{q}_\alpha \left(t, u(t) - \frac{\alpha}{2\pi} \right) \partial_x \left(u(t) - \frac{\alpha}{2\pi} \right) + \partial_x^3 \left(u(t) - \frac{\alpha}{2\pi} \right) \right.$$
$$\left. + P(t) \left(u(t) - \frac{\alpha}{2\pi} \right) \right) + c(t)\partial_y^2 \left(u(t) - \frac{\alpha}{2\pi} \right)$$
$$= \partial_x \left(\frac{du}{dt}(t) + q(t, u(t))\partial_x u(t) + \partial_x^3 u(t) + P(t)u(t) \right) + c(t)\partial_y^2 u(t) = 0.$$

Let $\tilde{u} := u - \frac{\alpha}{2\pi} \in C([t_0, t_0 + \varepsilon], X^{k+m_0}(T^2)) \cap C^1([t_0, t_0 + \varepsilon], X^k(T^2))$, then application of $(\text{Id} - \frac{1}{2\pi}I)V$ yields (due to 5.4.7)

$$\frac{d\tilde{u}}{dt}(t) + \tilde{q}_\alpha(t, \tilde{u}(t))\partial_x \tilde{u}(t) + \partial_x^3 \tilde{u}(t) + P(t)\tilde{u}(t) + A_1(t)\tilde{u}(t) = 0$$

$$\tilde{u}(t_0) = u_0 - \frac{\alpha}{2\pi}.$$

Hence, 5.4.13 shows the uniqueness of $\tilde{u}$ and u. $\qquad\qquad \square$

5.4.14 Bibliographical remarks. In this section we applied 2.4.6 in a similar way as Isaza/Mejía/Stallbohm [70] applied Kato's theory for evolution equations [78], [79] in the KP-I case. Whereas their application of Kato's theory works only in the KP-I case, our application works in both cases. Definition 5.4.4 and some of the properties of operators in these spaces are due to [70].

In general, most methods for treating (5.4.1) work only for equations of either type I or type II. For results on well-posedness of (5.4.1) for the KP-I equation or the KP-II equation we refer e.g. to Ukai [152], Zhou [161], Bourgain [19], Isaza/Mejía/Stallbohm [70], [71], Schleinkofer [136], Wickerhauser [157].

5.5 Quasilinear evolution equations in scales of L^q-Sobolev spaces

In this section we will give an application of our abstract theory in chapter 3 to the L^q-theory of nonlinear evolution equations of the form

$$u_t = \sum_{j,l=1}^n a_{jl}\partial_j\partial_l u + \sum_{j=1}^n a_j(u)\partial_j u + a(u) .$$

Unlike the L^2-theory in section 5.2 we have to assume for L^q-results constant second-order coefficients a_{jl} for technical reasons.

5.5.1 Theorem.   Let $(a_{jl})_{j,l} \in M_n(\mathbf{R})$ with $(a_{jl})_{j,l} \geq 0$, $a_j \in C^\infty(\mathbf{R},\mathbf{R})$, and $a \in C^\infty(\mathbf{R},\mathbf{R})$ with $a(0) = 0$. Then, for every $k \in \mathbf{N}_0$, $k > \frac{n}{q}+1$, $q \geq 2$, and initial value $u_0 \in H_q^{k+4}[\mathbf{R}]$ there is a $T_k(u_0) \in (0,\infty]$ and a unique function $u \in C^1([0,T_k(u_0)), H_q^k[\mathbf{R}]) \cap C([0,T_k(u_0)), H_q^{k+2}[\mathbf{R}]) \cap C_w([0,T_k(u_0)), H_q^{k+4}[\mathbf{R}])$ with

$$\frac{du}{dt}(t) = \sum_{j,l=1}^n a_{jl}\partial_j\partial_l u(t) + \sum_{j=1}^n a_j(u(t))\partial_j u(t) + a(u(t)), t \in [0,T_k(u_0)), u(0) = u_0.$$

Moreover, we either have $T_k(u_0) = \infty$ or $\limsup\limits_{t\nearrow T_k(u_0)} \|u(t)\|_{H_q^{k+4}} = \infty$.

PROOF: Let $X = L^q(\mathbf{R}^n,\mathbf{R})$, $Z_j u := \partial_j u$ for $u \in D(Z_j) = \{u \in X : \partial_j u \in X\}$, $j = 1,\ldots,n$, and $Z = \{Z_1,\ldots,Z_n\}$. Moreover, let

$$A(v)u(x) := \sum_{j,l=1}^n a_{jl}\partial_j\partial_l u(x) + \sum_{j=1}^n a_j(v(x))\partial_j u(x), \quad f(v)(x) := a(v(x)),$$

and $K \in \mathbf{N}_0$ with $K > \frac{n}{q}+1, m = 2$. Then, the following holds:

- Z is a commuting family of generators of isometric C_0-groups in X with $X_Z^k = H_q^k[\mathbf{R}]$ topologically for $k \in \mathbf{N}_0$. In particular, due to 2.2.9 there is a mollifier in the scale X_Z^k.

- $A(v) \in \mathcal{L}(X_{\tilde{Z}}^{k+2}, X_{\tilde{Z}}^k)$ for $v \in X_{\tilde{Z}}^k, k \geq K$, and for $k \geq K, R > 0$ there are $\tilde{a}_{k,R} \geq 0$ with

$$\|A(v) - A(w)\|_{\mathcal{L}(X_{\tilde{Z}}^{k+2}, X_{\tilde{Z}}^k)} \leq \tilde{a}_{k,R} \|v - w\|_k$$

for $v, w \in X_{\tilde{Z}}^k, \|v\|_{X_{\tilde{Z}}^k} \leq R, \|w\|_{X_{\tilde{Z}}^k} \leq R$. This is implied by 5.1.2.

- For $R > 0, k \geq K$ there are $\beta_{k,R} \geq 0$ with

$$\|\lambda u - A(v)u\|_{H_q^k} \geq (\lambda - \beta_{k,R}) \|u\|_{H_q^k}$$

for $u \in H_q^{k+2}, v \in H_q^k, \|v\|_{H_q^k} \leq R$, and $\lambda > \beta_{k,R}$.

Proof: $\|\lambda u - A(v)u\|_{L^q} \geq (\lambda - \beta_R) \|u\|_{L^q}$ for $\lambda > \beta_R, v \in H_q^K, R > 0$, $\|v\|_{H_q^K} \leq R, u \in \mathscr{S}(\mathbf{R}^n)$ with suitable $\beta_R \geq 0$ due to 4.5.2 and 5.1.27. Moreover, for $R > 0, k \geq K$ there are $d_{k,R} \geq 0$ with

$$\|\partial^\alpha A(v)u - A(v)\partial^\alpha u\|_{L^q} \leq d_{k,R} \|u\|_{H_q^k}$$

for $v \in H_q^k$ with $\|v\|_{H_q^k} \leq R, \alpha \in \mathbf{N}_0^n, |\alpha| \leq k, u \in \mathscr{S}(\mathbf{R}^n)$ due to 5.1.29. Hence 3.2.1 implies the assertion.

- For $K \leq K_1 \leq K_2$, compact intervals $I \subset \mathbf{R}$, and $v \in C(I, X^{K_2+7m+6})$ the Cauchy problem for $A(v(t))$ in the scale $(X_{\tilde{Z}}^k)_k$ is well-posed in the part $(K_1, K_2, 7m + 6)$.

Proof: First, there are $c_1, c_2 \geq 0$ with

$$c_1 \left(\sum_{|\alpha| \leq 2} \|\partial^\alpha u\|_{L^q}^2 \right)^{1/2} \leq \|u\|_{L^q} + \left\| \sum_{j=1}^n \partial_j^2 u \right\|_{L^q} \leq c_2 \left(\sum_{|\alpha| \leq 2} \|\partial^\alpha u\|_{L^q}^2 \right)^{1/2}$$

for $u \in X_{\tilde{Z}}^2$. Moreover, for $K_1 \leq k \leq K_2 + 7m + 5, v \in C(I, X_{\tilde{Z}}^{K_2+7m+6})$, and $u \in \mathscr{S}(\mathbf{R}^n)$ we have due to 5.1.29

$$\|\partial_j A(v(t))u - A(v(t))\partial_j u\|_{H_q^k}$$

$$\leq c \sum_{l=1}^n \sum_{|\alpha| \leq k} [\|\partial^{\alpha+e_j}[a_l(v(t))\partial_l u] - a_l(v(t))\partial^\alpha \partial_j \partial_l u\|_{L^q}$$

$$+ \|a_l(v(t))\partial^\alpha(\partial_j \partial_l u) - \partial^\alpha[a_l(v(t))\partial_j \partial_l u]\|_{L^q}]$$

$$\leq c \|u\|_{H_q^{k+1}}$$

with a suitable constant $c \geq 0$. Hence the assertion is implied by 2.5.14.

- 5.1.1 shows that $f(v) \in H_q^k$ for $v \in H_q^k, k \geq K$, and for $R > 0, k \geq K$ there are $\widetilde{F}_{k,R} \geq 0$ with

$$\|f(v) - f(w)\|_{H_q^k} \leq \widetilde{F}_{k,R} \|v - w\|_{H_q^k}$$

for $v, w \in H_q^k, \|v\|_{H_q^k} \leq R, \|w\|_{H_q^k} \leq R$.

Hence, the assertion of the theorem is implied by 3.3.9.   □

5.5.2 Bibliographical remarks. The result in this section could also be proved with Kato's theory for quasilinear evolution equations [82], [87] and pseudodifferential Nash-Moser estimates like

$$\|\langle D_x \rangle^s (fg) - f(\langle D_x \rangle^s g)\|_{L^q} \leq c \left(\|\operatorname{grad} f\|_{L^\infty} \left\|\langle D_x \rangle^{s-1} g\right\|_{L^q} + \|\langle D_x \rangle^s f\|_{L^q} \|g\|_{L^\infty} \right).$$
$$(5.5.1)$$

which are quite hard to prove, cf. Kato/Ponce [88]. This section shows advantages in considering scales to be generated by a family of operators (like the derivatives) and not to be generated by one operator (like $\langle D_x \rangle$). Since we chose the derivatives to generate the scale H_q^k, the assumptions of the abstract results in chapter 3 could be verified rather quickly with the Leibniz formula. Clearly, time-dependent coefficients could be treated as well in 5.5.1. Moreover, similarly to Kato/Ponce [88] L^q-results for the Navier-Stokes and Euler equations can be obtained. Again one can avoid pseudodifferential estimates like (5.5.1) and use only differential estimates as in the proof of 5.5.1.

5.6 First order hyperbolic evolution equations in the C_0^k-scale

The purpose of this section is to give an application of the abstract results in chapter 3 to hyperbolic quasilinear first-order evolution equations in the scale $(C_0^k(\mathbf{R}^n, \mathbb{R}))_k$ of k-times continuously differentiable functions vanishing at infinity with their derivatives. Questions of this type generally are supposed to be rather difficult due to the non-reflexivity of $C_0^k(\mathbf{R}^n, \mathbb{R})$ and due to the lack of a natural isomorphism $\Lambda : C_0^{k+1}(\mathbf{R}^n, \mathbb{R}) \to C_0^k(\mathbf{R}^n, \mathbb{R})$ in the case $n > 1$ (cf. e.g. the introduction of [87]). As for applications in L^q-Sobolev spaces we again have to use several inequalities.

5.6.1 Lemma. For $a \in C^\infty(\mathbb{R}), k \in \mathbb{N}_0, R > 0$ there are $F_k(R) \geq 0$ with

$$a(v) \in C^k(\mathbb{R}^n) \quad \text{and} \quad \|\partial^\alpha a(v)\|_{L^\infty} \leq F_k(R)$$

for $v \in C^k(\mathbb{R}^n, \mathbb{R})$ with $\|\partial^\alpha v\|_{L^\infty} \leq R, |\alpha| \leq k$. Moreover, $\partial^\alpha(a(v)) \in C_0(\mathbb{R}^n)$ for $v \in C_0^k(\mathbb{R}^n, \mathbb{R})$ and $0 < |\alpha| \leq k$.

PROOF: The assertion for $k = 0$ is trivial and for $\alpha \neq 0$ it is implied by

$$\|\partial^\alpha a(v)\|_{L^\infty} \leq c \sum_{j=1}^{|\alpha|} \sum_{\substack{\beta_1 + \ldots + \beta_j = \alpha \\ \beta_l \neq 0}} \left\| a^{(j)}(v)(\partial^{\beta_1} v) \cdots (\partial^{\beta_j} v) \right\|_{L^\infty} .$$

$\square$

5.6.2 Lemma. For $a \in C^\infty(\mathbb{R}), k \in \mathbb{N}_0, R > 0$, there are $F_k(R) \geq 0$ with

$$a(v) - a(w) \in C_0^k(\mathbb{R}^n) \quad \text{and} \quad \|a(v) - a(w)\|_{C_0^k} \leq F_k(R) \|v - w\|_{C_0^k}$$

for $v, w \in C_0^k(\mathbb{R}^n, \mathbb{R})$ with $\|v\|_{C_0^k} \leq R$ and $\|w\|_{C_0^k} \leq R$.

PROOF: The assertion is implied by

$$a(v(x)) - a(w(x)) = \int_0^1 a'(w(x) + t(v(x) - w(x)))(v(x) - w(x)) dt .$$

$\square$

Now we can formulate the main result of this section.

5.6.3 Theorem. Assume $a_j, b \in C^\infty(\mathbb{R}, \mathbb{R})$ for $j = 1, \ldots, n$ and $b(0) = 0$. Then, for $u_0 \in C_0^{k+2}(\mathbb{R}^n, \mathbb{R}), k \in \mathbb{N}_0$, there is an $\varepsilon > 0$ and a unique function $u \in C^1([0, \varepsilon], C_0^k(\mathbb{R}^n, \mathbb{R})) \cap C([0, \varepsilon], C_0^{k+1}(\mathbb{R}^n, \mathbb{R}))$ with

$$\frac{du}{dt}(t) = \sum_{j=1}^n a_j(u(t))\partial_j u(t) + b(u(t)), t \in [0, \varepsilon], \qquad u(0) = u_0 .$$

PROOF: First $\mathcal{C}_0^k(\mathbb{R}^n, \mathbb{R}) = X_Z^k$, where $X = \mathcal{C}_0(\mathbb{R}^n, \mathbb{R})$ and $Z = \{Z_1, \ldots, Z_n\}$ with $Z_j f = \partial_j f$ for $f \in D(\partial_j) = \{f \in \mathcal{C}_0(\mathbb{R}^n, \mathbb{R}) : \partial_j f \in \mathcal{C}_0(\mathbb{R}^n, \mathbb{R}) \text{ exists}\}$. As $Z_j : D(Z_j) \to X$ generates an isometric $\mathcal{C}_0$-group in X, 2.2.9 shows that there is a mollifier in the scale $(X^k)_k$. Now, let

$$A(v)u(\cdot) := \sum_{j=1}^n a_j(v(\cdot))\partial_j u(\cdot)$$

and $K = 0$. Then, due to 5.6.1 and 5.6.2, for $k \in \mathbb{N}_0, R > 0$ there are $F_k(R) \geq 0$ with

$$A(v) \in \mathscr{L}(X_Z^{k+1}, X_Z^k) \text{ and } \|A(v) - A(w)\|_{\mathscr{L}(X_Z^{k+1}, X_Z^k)} \leq F_k(R)\|v - w\|_{X_Z^k}$$

for $v, w \in \mathcal{C}_0^k(\mathbb{R}^n, \mathbb{R})$ with $\|v\|_{X_Z^k} \leq R, \|w\|_{X_Z^k} \leq R$. Due to 4.4.11 $A(v)$ is dissipative in $\mathcal{C}_0(\mathbb{R}^n, \mathbb{R})$ on $\mathcal{C}_0^1(\mathbb{R}^n, \mathbb{R})$ for any $v \in \mathcal{C}_0(\mathbb{R}^n, \mathbb{R})$. Moreover, for $R > 0, k \in \mathbb{N}_0$ there are $d_{k,R} \geq 0$ with

$$\|\partial^\alpha A(v)u - A(v)\partial^\alpha u\|_X \leq c \sum_{\substack{\alpha_1 + \alpha_2 = \alpha \\ \alpha_2 \neq \alpha}} \sum_{j=1}^n \|(\partial^{\alpha_1} a_j(v))(\partial^{\alpha_2}\partial_j u)\|_{L^\infty} \leq d_{k,R}\|u\|_{X_Z^k}$$

for $\alpha \in \mathbb{N}_0^n, |\alpha| \leq k, u \in \mathcal{C}_0^\infty(\mathbb{R}^n, \mathbb{R}), v \in \mathcal{C}_0^k(\mathbb{R}^n, \mathbb{R}), \|v\|_{X_Z^k} \leq R$. Hence 3.2.1 shows that for $R > 0, k \in \mathbb{N}_0$ there are $\beta_{k,R} \geq 0$ with

$$\|\lambda u - A(v)u\|_{X_Z^k} \geq (\lambda - \beta_{k,R})\|u\|_{X_Z^k}$$

for $\lambda > \beta_{k,R}, u \in X_Z^{k+1}, v \in X_Z^k$ with $\|v\|_{X_Z^k} \leq R$. Moreover, for $0 \leq K_1 \leq K_2$ theorem 4.4.16 shows that the Cauchy problem for $A_v(t) := A(v(t))$ in the scale $(X_Z^k)_k$ is well-posed in the part $(K_1, K_2, 3)$ for any $v \in \mathcal{C}(I, \mathcal{C}_0^{K_2+3}(\mathbb{R}^n))$, where $I \subset [0, \infty)$ is any compact subinterval. Finally, let $f(v)(\cdot) := b(v(\cdot))$, then $f(v) \in \mathcal{C}_0^k(\mathbb{R}^n)$ with $\|f(v) - f(w)\|_{X_Z^k} \leq \tilde{F}_k(R)\|v - w\|_{X_Z^k}$ for $v, w \in \mathcal{C}_0^k(\mathbb{R}^n)$ with $\|v\|_{X_Z^k} \leq R$ and $\|w\|_{X_Z^k} \leq R$ for suitable $\tilde{F}_k(R) \geq 0$ due to 5.6.1 and 5.6.2. Hence the assertion is implied by 3.3.9. $\qquad\square$

5.6.4 Bibliographical remarks. Our result is rather weak because we have to assume the initial values to be more regular than the solutions we can construct. However, we could obtain this $\mathcal{C}^k$-result rather easily. Better results for $\mathcal{C}^1$-theory of such equations have been obtained by Kato [87], Cinquini-Cibrario [27], Douglis [41], Hartman/Wintner [58].

Bibliography

[1] L. Abdelouhab. Nonlocal dispersive equations in weighted Sobolev spaces. *Differential Integral Equations*, 5:307–338, 1992.

[2] N.I. Achieser and I.M. Glasmann. *Theorie der linearen Operatoren im Hilbert-Raum*. Akademie-Verlag, Berlin, 1981.

[3] R.A. Adams. *Sobolev Spaces*. Academic Press, 1975.

[4] S. Agmon. *Lectures on Elliptic Boundary Value Problems*. D. van Nostrand Company, Inc., 1965.

[5] F. Ali Mehmeti. *Lokale und globale Lösungen linearer und nichtlinearer hyperbolischer Evolutionsgleichungen mit Transmission*. PhD thesis, Johannes Gutenberg-Universität Mainz, 1987.

[6] F. Ali Mehmeti. *Nonlinear Waves in Networks*. Mathematical Research, volume 80, Akademie-Verlag, 1994.

[7] J. Alvarez and J. Hounie. Spectral invariance and tameness of pseudo-differential operators on weighted Sobolev spaces. *J. Operator Theory*, 30:41–67, 1993.

[8] H. Amann. *Gewöhnliche Differentialgleichungen*. Walther de Gruyter, Berlin, New York, 1983.

[9] H. Amann. *Linear and Quasilinear Parabolic Problems, I, Abstract Linear Theory*. Monographs in Mathematics, 89. Birkhäuser, 1995.

[10] W. Arendt, P.R. Chernoff, and T. Kato. A generalization of dissipativity and positive semigroups. *J. Operator Theory*, 8:167–180, 1992.

[11] D. Ascoli. Global wellposedness for a class of quasi-linear abstract evolution problems of hyperbolic type. *Boll. Un. Mat. Ital. B(7)*, 4:661–677, 1990.

[12] F. Baldus. $S(M, g)$-pseudo-differential calculus with spectral invariance on $\mathbb{R}^n$ and manifolds for Banach function spaces. PhD thesis, Johannes Gutenberg-Universität Mainz, 2000.

[13] C.J.K. Batty and D.W. Robinson. Commutators and generators. Math. Scand., 62:303–326, 1988.

[14] T.B. Benjamin. Internal waves of permanent form in fluids of great depth. J. Fluid Mech., 29:559–592, 1967.

[15] C. Berg and G. Forst. Potential Theory on Locally Compact Abelian Groups. Springer, 1975.

[16] J. Bergh and J. Löfström. Interpolation Spaces, An Introduction. Springer, 1976.

[17] J. Bokobza and A. Unterberger. Les opérateurs pseudo-différentiels d'ordre variable. C.R. Acad. Sci. Paris Sér. I Math., 261:2271–2273, 1965.

[18] J. Bourgain. Fourier transform restriction phenomena for certain lattice subsets and applications to nonlinear evolution equations II, the KdV equation. Geom. Funct. Anal., 3:209–262, 1993.

[19] J. Bourgain. On the Cauchy problem for the Kadomtsev-Petviashvili equation. Geom. Funct. Anal., 3:315–341, 1993.

[20] R. Brummelhuis. A counterexample to the Fefferman-Phong inequality for systems. C. R. Acad. Sci. Paris Sér. I Math., 310:95–98, 1990.

[21] A.P. Calderón. Intermediate spaces and complex interpolation. Studia Math., 24:113–190, 1964.

[22] O. Caps. Well-posedness of the time-dependent Cauchy problem in scales of Banach spaces. Preprint-Reihe des Fachbereiches Mathematik, Universität Mainz, Nr. 1/99, 1999.

[23] O. Caps. Well-posedness of the time-dependent Cauchy problem in scales of Hilbert spaces. Preprint-Reihe des Fachbereiches Mathematik, Universität Mainz, Nr. 16/98, 1998.

[24] O. Caps. Well-posedness of linear and quasilinear evolution equations in scales of Banach spaces. PhD thesis, Johannes Gutenberg-Universität Mainz, 2000.

[25] T. Cazenave and A. Haraux. *An Introduction to Semilinear Evolution Equations*. Clarendon Press, Oxford, 1998.

[26] P. Chernoff. Essential self-adjointness of powers of generators of hyperbolic equations. *J. Funct. Anal.*, 12:401–414, 1973.

[27] M. Cinquini-Cibrario. Ulteriori resultati per sistemi di equazioni quasi lineari a derivate parziali in più variabili indipendenti. *Istit. Lombardo Accad. Sci. Lett. Rend. A*, 103:373–407, 1969.

[28] Ph. Clément, H.J.A.M. Heijmans et al. *One-Parameter Semigroups*. North-Holland, 1987.

[29] F. Colombini, D. Del Santo, and C. Zuily. The Fefferman-Phong inequality in the locally temperate Weyl calculus. *Osaka J. Math.*, 33:847–861, 1996.

[30] A. Constantin and J. Escher. Well-posedness, global existence, and blowup phenomena for a periodic quasi-linear hyperbolic equation. *Comm. Pure Appl. Math.*, 51:475–504, 1998.

[31] J.B. Conway. *Functions of One Complex Variable I*. Springer, 1978.

[32] H.O. Cordes. *The Technique of Pseudodifferential Operators*. London Math. Soc. Lecture Note Ser. 202, 1995.

[33] P. Courrège. Sur la forme intégro-différentielle des opérateurs de C_k^∞ dans C satisfaisant au principe du maximum. *Séminaire Brelot-Choquet-Deny, Théorie du potentiel*, 10e année:2–01 – 2–38, 1965/1966.

[34] G. Da Prato and E. Sinestrari. Differential operators with nondense domain. *Ann. Scuola Norm. Sup. Pisa Cl. Sci. (4)*, 14:285–344, 1988.

[35] G. Da Prato and E. Sinestrari. Nonautonomous evolution operators of hyperbolic type. *Semigroup Forum*, 45:302–321, 1992.

[36] E.B. Davies. *One-Parameter Semigroups*. Academic Press, 1980.

[37] E.B. Davies. *Heat kernels and spectral theory*. Cambridge University Press, 1989.

[38] R. de Laubenfels. Powers of generators of holomorphic semigroups. *Proc. Amer. Math. Soc.*, 99:105–108, 1987.

[39] R.K. Dodd, J.C. Eilbeck, J.D. Gibbon, and H.C. Morris. *Solitons and Nonlinear Wave Equations*. Academic Press, 1982.

[40] J.R. Dorroh. A simplified proof of a theorem of Kato on linear evolution equations. *J. Math. Soc. Japan*, 27:474–478, 1975.

[41] A. Douglis. Some existence theorems for hyperbolic systems of partial differential equations in two independent variables. *Comm. Pure Appl. Math.*, 5:119–154, 1952.

[42] N. Dunford and J.T. Schwartz. *Linear Operators, Part I: General Theory*. Interscience, New York, 1958.

[43] K.J. Engel and R. Nagel. *One-parameter semigroups for linear evolution equations*. Graduate Texts in Mathematics, 194, Springer-Verlag, 2000.

[44] H.O. Fattorini. *The Cauchy Problem*. Encyclopedia of Mathematics and its Applications, 18, Addison-Wesley, 1983.

[45] C. Fefferman and D.H. Phong. On positivity of pseudo-differential operators. *Proc. Natl. Acad. Sci. USA*, 75:4673–4674, 1978.

[46] C. Fefferman and D.H. Phong. Sharp Gårding inequalities. *Proc. Sympos. Pure Math.*, 35, Part 2:137–141, 1979.

[47] G.B. Folland. *Real Analysis*. John Wiley & Sons, 1984.

[48] G.B. Folland. *Harmonic Analysis in Phase Space*. Annals of Mathematics Studies, 122, Princeton University Press, 1989.

[49] Avner Friedman. *Partial differential equations of parabolic type*. Prentice-Hall Inc., 1964.

[50] H. Fujita and T. Kato. On the Navier-Stokes initial value problem I. *Arch. Rational Mech. Anal.*, 16:269–315, 1964.

[51] J.A. Goldstein. On the absence of necessary conditions for linear evolution operators. *Proc. Amer. Math. Soc.*, 64:77–80, 1977.

[52] J.A. Goldstein. *Semigroups of Linear Operators and Applications*. Oxford Univ. Press, 1985.

[53] R. Goodman. One-parameter groups generated by operators in an enveloping algebra. *J. Funct. Anal.*, 6:218–236, 1970.

[54] B. Gramsch. Relative Inversion in der Störungstheorie von Operatoren und Ψ-algebren. *Math. Ann.*, 269:27–71, 1984.

[55] B. Gramsch and K.G. Kalb. Pseudo-locality and hypoellipticity in operator algebras. *Semesterbericht Funktionalanalysis, Tübingen*, 8:51–61, 1985.

[56] B. Gramsch, J. Ueberberg, and K. Wagner. Spectral invariance and submultiplicativity for Fréchet algebras with applications to pseudo-differential operators and Ψ*-quantization. In *Operator Theory: Advances and Applications, vol. 57*, pages 71–98. Birkhäuser, Basel, 1992.

[57] P. Hartman. *Ordinary Differential Equations*. John Wiley & Sons, 1964.

[58] P. Hartman and A. Wintner. On hyperbolic partial differential equations. *Amer. J. Math.*, 74:834–864, 1952.

[59] B. Helffer. *Théorie spectrale pour des opérateurs globallement elliptiques*. Astérisque 112, Société Mathématique de France, 1984.

[60] J.V. Herod and R.W. McKelvey. A Hille-Yosida theory for evolutions. *Israel J. Math*, 36:13–40, 1980.

[61] G. Herzog and R. Lemmert. Ordinary differential equations in Fréchet spaces. In V.Covachev D. Bainov, editor, *Third International Colloquium on Differential Equations, Plovdiv, Bulgaria*, pages 89–93, 1993.

[62] M. Hieber. Gaussian estimates and holomorphy of semigroups on L^p spaces. *J. London Math. Soc. (2)*, 54:148–160, 1996.

[63] E. Hille and K.S. Phillips. *Functional Analysis and Semi-groups*. Amer. Math. Soc. Coll. Publ. Vol. 31, Providence, R.I., 1957.

[64] W. Hoh. A symbolic calculus for pseudo-differential operators generating Feller semigroups. *Osaka J. Math.*, 35:789–820, 1998.

[65] L. Hörmander. The Cauchy problem for differential equations with double characteristics. *J. Anal. Math.*, 32:118–196, 1977.

[66] L. Hörmander. The Weyl calculus of pseudodifferential operators. *Comm. Pure Appl. Math.*, 32:360–444, 1979.

[67] L. Hörmander. *The Analysis of Linear Partial Differential Operators*, volume III. Springer, 1985.

[68] L. Hörmander. *Lectures on Nonlinear Hyperbolic Differential Equations.* Mathématiques & Applications 26, Springer, 1997.

[69] R.J. Iório. KdV, BO and friends in weighted Sobolev spaces. In *Functional-Analytic Methods for Partial Differential Equations*, Lecture Notes in Math., 1450, pages 104–121, 1989.

[70] P. Isaza, J. Mejía, and V. Stallbohm. Local solution for the Kadomtsev-Petviashvili equation with periodic boundary conditions. *Manuscripta Math.*, 75:383–393, 1992.

[71] P. Isaza, J. Mejía, and V. Stallbohm. A regularity theorem for the Kadomtsev-Petviashvili equation with periodic boundary conditions. *Nonlinear Anal.*, 23:683–687, 1994.

[72] C. Iwasaki. The fundamental solution for pseudo-differential operators of parabolic type. *Osaka J. Math.*, 14:569–592, 1977.

[73] N. Jacob. A converse to Gårding's inequality for a class of formally hypoelliptic differential operators. *Arch. Math. (Basel)*, 49:309–315, 1987.

[74] N. Jacob. *Pseudo-Differential Operators and Markov Processes.* Mathematical Research, Vol. 94, Akademie-Verlag, 1996.

[75] B.B. Kadomtsev and V.I. Petviashvili. On the stability of solitary waves in weakly dispersive media. *Soviet Phys. Dokl.*, 15:539–543, 1970.

[76] T. Kato. Fractional powers of dissipative operators, I-II. *J. Math. Soc. Japan*, 13:246-274, 1961; *J. Math. Soc. Japan*, 14:242-248, 1962.

[77] T. Kato. Integration of the equation of evolution in a Banach space. *J. Math. Soc. Japan*, 5:208–234, 1953.

[78] T. Kato. Linear evolution equations of "hyperbolic" type. *J. Fac. Sci. Univ. Tokyo*, 17:241–258, 1970.

[79] T. Kato. Linear evolution equations of "hyperbolic" type, II. *J. Math. Soc. Japan*, 25:648–666, 1973.

[80] T. Kato. A remark to the preceding paper by Chernoff. *J. Funct. Anal.*, 12:415–417, 1973.

[81] T. Kato. The Cauchy problem for quasi-linear symmetric hyperbolic systems. *Arch. Rational Mech. Anal.*, 58:181–205, 1975.

[82] T. Kato. Quasi-linear equations of evolution, with applications to partial differential equations. In W.H. Everitt, editor, *Spectral Theory and Differential Equations*, Lecture Notes in Math., 448, pages 25–70. Springer, 1975.

[83] T. Kato. Linear and quasi-linear equations of evolution of hyperbolic type. In *C.I.M.E., II ciclo, Cortona*, pages 127–191, 1976.

[84] T. Kato. On the Cauchy problem for the (generalized) Korteweg-de Vries equation. *Studies in Applied Mathematics, Advances in Mathematics, supplementary studies*, 8:93–128, 1983.

[85] T. Kato. *Abstract Differential Equations and Nonlinear Mixed Problems*. Lezioni Fermiane, Academia Nazionale dei Lincei, Scuola Normale Superiore, 1985.

[86] T. Kato. Nonlinear equations of evolution in Banach spaces. *Proc. Sympos. Pure Math.*, 45:9–23, 1986.

[87] T. Kato. Abstract evolution equations, linear and quasilinear, revisited. In H. Komatsu, editor, *Functional Analysis and Related Topics*, Lecture Notes in Math., 1540, pages 103–125. Springer, 1993.

[88] T. Kato and G. Ponce. Commutator estimates and the Euler and Navier-Stokes equations. *Comm. Pure Appl. Math.*, 41:891–907, 1988.

[89] H. Kellermann. Linear evolution equations with time-dependent domain. *Semesterbericht Funktionalanalysis, Tübingen*, 9:15–30, 1985/86.

[90] C.E. Kenig, G. Ponce, and L. Vega. On the generalized Benjamin-Ono equation. *Trans. Amer. Math. Soc.*, 342:155–172, 1994.

[91] J. Kisyński. Sur les opérateurs de Green des problèmes de Cauchy abstraits. *Studia Math.*, 23:285–328, 1964.

[92] K. Kobayasi. On a theorem for linear evolution equations of hyperbolic type. *J. Math. Soc. Japan*, 31:647–654, 1979.

[93] K. Kobayasi and N. Sanekata. A method of iterations for quasi-linear evolution equations in nonreflexive Banach spaces. *Hiroshima Math. J.*, 19:521–540, 1989.

[94] H. Komatsu. Fractional powers of operators, I-VI. *Pacific J. Math.*, 19:285-346, 1966; *Pacific J. Math.*, 21:89-111, 1967; *J. Math. Soc. Japan*, 21:205-220, 1969; *J. Math. Soc. Japan*, 21:221-228, 1969; *J. Fac. Sci. Univ. Tokyo Sect. I*, 17:373-396, 1970; *J. Fac. Sci. Univ. Tokyo Sect. IA Math.*, 19:1-63, 1972.

[95] Y. Komura. On linear evolution operators in reflexive Banach spaces. *J. Fac. Sci. Univ. Tokyo*, 17:529–542, 1970.

[96] S.G. Kreĭn. *Linear Differential Equations in Banach Space.* Translations of Mathematical Monographs 29, American Mathematical Society, 1971.

[97] H. Kumano-go. *Pseudo-Differential Operators.* MIT Press, Cambridge, MA, 1981.

[98] H. Kumano-go and C. Tsutsumi. Complex powers of hypoelliptic pseudo-differential operators with applications. *Osaka J. Math.*, 10:147–174, 1973.

[99] O. Ladyzhenskaya. *The Mathematical Theory of Viscous Incrompressible Flow.* Gordon and Breach, New York, 1969.

[100] M. Langer and V. Maz'ya. On L^p-contractivity of semigroups generated by linear partial differential operators. *J. Funct. Anal.*, 164:73–109, 1999.

[101] R. Lauter. *Holomorphic functional calculus in several variables and Ψ^*-algebras of totally characteristic operators on manifolds with boundary.* PhD thesis, Johannes Gutenberg-Universität Mainz, 1996.

[102] C. Lennard. A Baire category theorem for the domains of iterates of a linear operator. *Rocky Mountain J. Math.*, 24:615–627, 1994.

[103] N. Lerner and J. Nourrigat. Lower bounds for pseudo-differential operators. *Ann. Inst. Fourier (Grenoble)*, 40:657–682, 1990.

[104] J.L. Lions. Une construction d'espaces d'interpolation. *C.R. Acad. Sci. Paris*, 251:1853–1855, 1960.

[105] V.A. Liskevich and Y.A. Semenov. Some problems on Markov semigroups. In M. Demuth et al., editor, *Schrödinger Operators, Markov Semigroups, Wavelet Analysis, Operator Algebras*, pages 163–217. Advances in Partial Differential Equations 11, Akademie Verlag, 1996.

[106] F.J. Massey III. Abstract evolution equations and the mixed problem for symmetric hyperbolic systems. *Trans. Amer. Math. Soc.*, 168:165–188, 1972.

[107] A. Melin. Lower bounds for pseudo-differential operators. *Ark. Mat.*, 9:117–140, 1971.

[108] M. Nagase. A new proof of sharp Gårding inequality. *Funkcial. Ekvac.*, 20:259–271, 1977.

[109] M. Nagase and T. Umeda. Weyl quantized Hamiltonians of relativistic spinless particles in magnetic fields. *J. Funct. Anal.*, 92:136–154, 1990.

[110] R. Nagel, editor. *One-Parameter Semigroups of Positive Operators*. Lecture Notes in Math., 1184, 1986.

[111] R. Nagel. Extrapolation spaces for semigroups. *Semesterbericht Funktionalanalysis, Tübingen*, 1997.

[112] G. Nickel. *On Evolution Semigroups and Wellposedness of Nonautonomous Cauchy problems*. PhD thesis, Tübingen, 1996.

[113] G. Nickel and R. Schnaubelt. An extension of Kato's stability condition for nonautonomous Cauchy problems. *Semesterbericht Funktionalanalysis, Tübingen*, 1995.

[114] N. Okazawa and A. Unai. Abstract quasilinear evolution equations in a Hilbert space, with applications to symmetric hyperbolic systems. *SUT J. Math.*, 29:263–290, 1993.

[115] N. Okazawa and A. Unai. Linear evolution equations of hyperbolic type in a Hilbert space. *SUT J. Math.*, 29:51–70, 1993.

[116] O.A. Oleinik. On the smoothness of solutions of degenerate elliptic and parabolic equations. *Soviet Math. Dokl.*, 6:972–975, 1965.

[117] O.A. Oleinik and E.V. Radkevic. *Second Order Equations with Nonnegative Characteristic Form*. Plenum Press, 1973.

[118] H. Ono. Algebraic solitary waves in stratified fluids. *J. Phys. Soc. Japan*, 39:1082–1091, 1975.

[119] E. Ouhabaz. Gaussian estimates and holomorphy of semigroups. *Proc. Amer. Math. Soc.*, 123:1465–1474, 1995.

[120] C. Parenti and A. Parmeggiani. Some remarks on almost-positivity of ψdo's. *Boll. Unione Mat. Ital. Sez. B Artic. Ric. Mat.*, 8:187–215, 1998.

[121] A. Parmeggiani and C.-J. Xu. The Dirichlet problem for sub-elliptic second order equations. *Ann. Mat. Pura Appl. (4)*, 173:233–243, 1997.

[122] A. Pazy. *Semigroups of Linear Operators and Applications to Partial Differential Equations.* Springer, 1983.

[123] B. Petersen. *Introduction to the Fourier transform & Pseudo-Differential Operators.* Pitman, 1983.

[124] R.S. Phillips. Perturbation theory for semi-groups of linear operators. *Trans. Amer. Math. Soc.*, 74:199–221, 1953.

[125] R.S. Phillips. Semi-groups of positive contraction operators. *Czechoslovak Math. J.*, 12:294–313, 1962.

[126] R. Racke. *Lectures on Nonlinear Evolution Equations.* Aspects of Mathematics, Vol. 19, Vieweg, 1992.

[127] M. Reed and B. Simon. *Methods of Modern Mathematical Physics*, volume II. Academic Press, 1975.

[128] M. Reed and B. Simon. *Methods of Modern Mathematical Physics*, volume I. Academic Press, 1980.

[129] D. Robert. Propriétés spectrales d'opérateurs pseudo-différentiels. *Comm. Partial Differential Equations*, 3:755–826, 1978.

[130] D.W. Robinson. Commutator theory on Hilbert space. *Canad. J. Math.*, 39:1235–1280, 1987.

[131] D.W. Robinson. The differential and integral structure of representations of Lie groups. *J. Operator Theory*, 19:95–128, 1988.

[132] D.W. Robinson. Commutators and generators II. *Math. Scand.*, 64:87–108, 1989.

[133] D.W. Robinson. *Elliptic Operators and Lie Groups.* Clarendon Press, Oxford, 1991.

[134] W. Rudin. *Functional Analysis.* McGraw-Hill, 1991.

[135] X. Saint Raymond. *Elementary Introduction to the Theory of Pseudodifferential Operators.* CRC Press, 1991.

[136] G. Schleinkofer. The generalized Kadomtsev-Petviashvili equation I. *Nonlinear Anal.*, 30:1245–1249, 1997.

[137] H. Schröder. *Funktionalanalyis.* Akademie-Verlag, 1997.

[138] E. Schrohe. Boundedness and spectral invariance for standard pseudodifferential operators on anisotropically weighted L^p-Sobolev spaces. *Integral Equations Operator Theory*, 13:271–284, 1990.

[139] I.E. Segal. Non-linear semi-groups. *Ann. of Math. (2)*, 78:339–364, 1963.

[140] M.A. Shubin. *Pseudodifferential Operators and Spectral Theory.* Springer, 1987.

[141] E. Stein. *Topics in Harmonic Analysis related to the Littlewood-Paley Theory.* Annals of Mathematical Studies 63, Princeton University Press, 1970.

[142] L. Sung. Semiboundedness of systems of differential operators. *J. Differential Equations*, 65:427–434, 1986.

[143] K. Taira. *Diffusion Processes and Partial Differential Equations.* Academic Press, 1988.

[144] H. Tanabe. *Functional Analytic Methods for Partial Differential Equations.* Monographs and Textbooks in Pure and Applied Mathematics, 204, Marcel Dekker, 1997.

[145] N. Tanaka. Quasilinear evolution equations with non-densely defined operators. *Differential Integral Equations*, 9:1067–1106, 1996.

[146] M. Taylor. *Pseudodifferential Operators.* Princeton University Press, 1981.

[147] M. Taylor. Fefferman-Phong inequalities in diffraction theory. *Proc. Sympos. Pure Math.*, 43:261–300, 1985.

[148] M. Taylor. *Partial Differential Equations*, volume III. Springer, 1996.

[149] M. Taylor. *Partial Differential Equations*, volume II. Springer, 1996.

[150] R. Temam. *Navier Stokes Equations.* Studies in Mathematics and its Applications, 2, North-Holland, 1977.

[151] H. Triebel. *Interpolation Theory, Function Spaces, Differential Operators.* North-Holland, 1978.

[152] S. Ukai. Local solutions of the Kadomtsev-Petviashvili equation. *J. Fac. Sci. Univ. Tokyo*, 36:193–209, 1989.

[153] J. Voigt. The sector of holomorphy for symmetric submarkovian semi-groups. *Semesterbericht Funktionalanalysis, Tübingen*, 1995.

[154] W. von Wahl. *The Equations of Navier-Stokes and Abstract Parabolic Equations*. Aspects of Mathematics, 8, Vieweg, 1985.

[155] J. Weidmann. *Linear Operators in Hilbert Spaces*. Springer, 1980.

[156] G.B. Whitham. *Linear and Nonlinear Waves*. John Wiley & Sons, 1974.

[157] M.V. Wickerhauser. Inverse scattering for the heat operator and evolutions in $2 + 1$ variables. *Comm. Math. Phys.*, 108:67–87, 1987.

[158] B. Wong-Dzung. L^p-theory of degenerate-elliptic and parabolic operators of second order. *Proc. Roy. Soc. Edinburgh Sect. A*, 95:95–113, 1983.

[159] M. Yamazaki. The essential self-adjointness of pseudodifferential operators associated with non-elliptic Weyl symbols with large potentials. *Osaka J. Math.*, 29:175–202, 1992.

[160] K. Yosida. *Functional Analysis*. Sixth edition, Springer, 1980.

[161] X. Zhou. Inverse scattering transform for the time-dependent Schrödinger equation with applications to the KPI equation. *Comm. Math. Phys.*, 128:551–564, 1990.

Index

Stiftung Benedictus Gotthelf Teubner
(i. G.)

„Was gemacht werden kann, wird gemacht."
Leitspruch des Leipziger Stadtrats B. G. Teubner
(Teubner-Firmengründung am 21.02.1811 in Leipzig)

Die Gründungsphase der künftigen Stiftung Benedictus Gotthelf Teubner
begann am 01.07.2000 im Haus des Buches in Leipzig.
Am 21.02.2001 wurde die Stiftungs-Homepage eröffnet, und pünktlich am 21.02.2002
ist das Internet-Portal www.stiftung-teubner-leipzig.de deutlich erweitert worden.
Unter anderem finden Sie hier die folgenden Rubriken:
Firma B. G. Teubner / Zeitzeugen / Leipzig / Mathematik / Aktuelles / Überraschungen.

Eine am 21.02.2002 neu eröffnete, assoziierte Homepage,
die ebenfalls vom Haus des Buches aus betreut wird, enthält u. a.:
Teubner-Buchreihen und -Einzeltitel Mathematik / Naturwissenschaften (1845 – 1945),
Teubner-Herausgeber und -Autoren Mathematik / Naturwissenschaften (1845 – 1945),
Teubner-Buchreihen und -Einzeltitel Mathematik / Naturwiss. / Informatik (1948 – 2000),
Teubner-Herausgeber und -Autoren Mathematik / Naturwiss. / Informatik (1948 – 2000),
Hall of Fame: B. G. Teubner 1845 – 2000.

Wichtigste Zielstellung der in Gründung befindlichen Teubner-Stiftung ist die Wissenschaftsförderung.

Haus des Buches. Gerichtsweg 28. D-04103 Leipzig
e-mail: weiss@stiftung-teubner-leipzig.de
Internet: www.stiftung-teubner-leipzig.de